WAREHOUSE DISTRIBUTION AND OPERATIONS HANDBOOK

Other Handbooks of Interest from McGraw-Hill

Avallone, Baumeister • MARKS' STANDARD HANDBOOK FOR MECHANICAL ENGINEERS

Bhushan, Gupta • HANDBOOK OF TRIBOLOGY

Brady, Clauser • MATERIALS HANDBOOK

Bralla • HANDBOOK OF PRODUCT DESIGN FOR MANUFACTURING

Brink • HANDBOOK OF FLUID SEALING

Brunner • HANDBOOK OF INCINERATION SYSTEMS

Corbitt • STANDARD HANDBOOK OF ENVIRONMENTAL ENGINEERING

Ehrich • HANDBOOK OF ROTORDYNAMICS

Elliot • STANDARD HANDBOOK OF POWERPLANT ENGINEERING

Freeman • STANDARD HANDBOOK OF HAZARDOUS WASTE TREATMENT AND DISPOSAL

Ganic, Hicks • THE MCGRAW-HILL HANDBOOK OF ESSENTIAL ENGINEERING INFORMATION AND DATA

Gieck • ENGINEERING FORMULAS

Grimm, Rosaler • HANDBOOK OF HVAC DESIGN

Harris • HANDBOOK OF ACOUSTICAL MEASUREMENTS ON NOISE CONTROL

Harris • SHOCK AND VIBRATION HANDBOOK

Hicks • STANDARD HANDBOOK OF ENGINEERING CALCULATIONS

Hodson • MAYNARD'S INDUSTRIAL ENGINEERING HANDBOOK

Jones • DIESEL PLANT OPERATIONS HANDBOOK

Juran, Gryna • JURAN'S QUALITY CONTROL HANDBOOK

Kurtz • HANDBOOK OF APPLIED MATHEMATICS FOR ENGINEERS AND SCIENTISTS

Karassik et al. • PUMP HANDBOOK

Mason • SWITCH ENGINEERING HANDBOOK

Nayyar • PIPING HANDBOOK

Parmley • STANDARD HANDBOOK OF FASTENING AND JOINING

Rosaler • STANDARD HANDBOOK OF PLANT ENGINEERING

Schwartz • COMPOSITE MATERIALS HANDBOOK

Schwartz • HANDBOOK OF STRUCTURAL CERAMICS

Shigley, Mischke • STANDARD HANDBOOK OF MACHINE DESIGN

Townsend • DUDLEY'S GEAR HANDBOOK

Tuma • HANDBOOK OF NUMERICAL CALCULATIONS IN ENGINEERING

Tuma • ENGINEERING MATHEMATICS HANDBOOK

Wadsworth • HANDBOOK OF STATISTICAL METHODS FOR ENGINEERS AND SCIENTISTS

Walsh • MCGRAW-HILL MACHINING AND METALWORKING HANDBOOK

Wang • HANDBOOK OF AIR CONDITIONING AND REFRIGERATION

Woodruff, Lammers • STEAM-PLANT OPERATION

Young • ROARK'S FORMULAS FOR STRESS AND STRAIN

WAREHOUSE DISTRIBUTION AND OPERATIONS HANDBOOK

David E. Mulcahy

Grand Rapids, Michigan

McGraw-Hill, Inc.

New York   San Francisco   Washington, D.C.   Auckland   Bogotá
Caracas   Lisbon   London   Madrid   Mexico City   Milan
Montreal   New Delhi   San Juan   Singapore
Sydney   Tokyo   Toronto

Library of Congress Cataloging-in-Publication Data

Mulcahy, David E.
    Warehouse and distribution operations handbook / David E. Mulcahy.
        p.   cm.
    Includes index.
    ISBN 0-07-044002-6 (alk. paper)
      1. Warehouses—Management—Handbooks, manuals, etc.
    2. Distribution of goods—Management—Handbooks, manuals, etc.
    I. Title.
    TS189.6.M85   1994
    658.7′85—dc20   93-2578
                                           CIP

1 2 3 4 5 6 7 8 9 0   DOC/DOC   9 9 8 7 6 5 4 3

ISBN 0-07-044002-6

The sponsoring editor for this book was Robert W. Hauserman, the editing supervisor was Paul R. Sobel, and the production supervisor was Pamela A. Pelton. It was set in Times Roman by McGraw-Hill's Professional Book Group composition unit.

Printed and bound by R. R. Donnelley & Sons Company.

This book is printed on acid-free paper.

CONTENTS

Chapter 4. Truck and Rail Receiving and Shipping Dock Areas 4.1

Chapter 5. Routing and Organizing the Work of Order Pickers 5.1

Chapter 6. Organizing Small-Item Facility Handling Activities 6.1

Chapter 7. Carton (Full-Case) Handling in Warehouse and Distribution Operations 7.1

Chapter 8. Arranging Key Pallet (Unit Load) Functions to Control Operating Costs, Improve Inventory Control, and Maximize Storage Space 8.1

Chapter 9. Which Horizontal or Vertical Transportation Method Controls Internal Transportation Costs and Handles the Volume and Product   9.1

Chapter 10. The Best Lift Truck and Mobile Warehouse Equipment for an Operation   10.1

Chapter 15. Facility Site Selection   15.1

PREFACE

The objective for writing this practioner (practical how-to-do-it) book is to provide a book that contains insights and tips for warehouse and distribution professionals to make their operation more efficient and cost effective. The chapters follow a sequence taking the product flow through a warehouse distribution operation and focuses on a particular topic for the reader's quick and easy reference to a key warehouse function. These chapters cover equipment applications, material handling concepts, and practices that are considered for implementation whether your business is large, medium, or small and handles raw materials, goods-in-process, or finished goods. The book contains illustrations and tables that assist in developing logistic strategies to improve profits, enhance product flow, increase employee productivity, improve customer service, reduce operating costs, increase space utilization, and improve asset protection.

It is necessary to understand that the purpose of this book is to help develop the skill and knowledge of its readers. Since the profession of logistics management, material handling concept design, and logistics is constantly changing, the book may not include the latest changes in the state of the art references to new technologies, the various equipment applications, or material handling concepts.

It is also necessary to recognize that this book cannot cover all of the available equipment applications, technologies, and material handling concepts in the field of warehouse distribution and transportation. The book does assist in the training and obtaining of practical experience which has no substitute. To assist in this objective, line art illustrations and sketches are used to visually present a piece of equipment or material handling concept.

It is important for the reader to use this collection of data, concepts, and forms as a guide. Prior to the purchase and installation of your new material handling concept or equipment, it is essential that you develop and project correct, accurate and adequate facility, inventory, SKU, transactions data, and design factors. Due to the fact that these are the design bases for your proposed equipment application or facility, it is prudent for you to gather and review vendor literature and to visit existing facilities that utilize the material handling concept and equipment application. These activities permit you to become familiar with the operational characteristics of the concept and equipment that is under consideration for implementation in your facility. The concept and performance specifications, physical design, and installation characteristics are subject to redesign, improvement, and modification, and are required to meet vendor and local governmental standards and specifications.

Each chapter of the book deals with key aspects and issues of planning and managing a warehouse distribution operation. Some of these issues are: how your facility layout and product location affects your employee productivity; when to use the 80/20 and where to locate your power SKUs; what receiving and shipping dock design and staging concept is best for your operation and product; how to route your order-pickers and organize their work for the best productivity; when to use the 'Z', 'U', or loop order-picker routing pattern, paper or paperless concept, single or batched customer order handling concept; how to control the batch release and how to identify the pick position; what is the best small item manual, mechanized or automatic order-pick, and sortation concept to handle your product and volume; how to handle hanging garments; what is the most efficient and cost effective carton manual, mechanized and automatic order-pick, and sortation concept for your business; how to pick your nonconveyable cartons; how to handle your customer returns; tips and insights on installing used equipment; what pallet load storage concept maximizes your space utilization, employee productivity, and product rotation; how to select a horizontal or vertical transportation concept to transfer your product and volume between two warehouse locations; how to maximize your lift truck investment and lift truck operator productivity; what are the alternative vehicle aisle guidance and P/D station concepts; when to unitize your product and what unit load support device do you use in your operation; when to use a fixed or variable SKU inventory allocation concept; what are the various manual and automatic identification concepts; what are the various bar codes and bar code encode devices; and what are the fundamentals and material handling concepts to an across-the-dock delivery program.

The author would like to express his thanks to all the warehouse distribution professionals with whom he has had an association at the various companies as a fellow manager, a client, and a speaker at various seminars. Special thanks are given to Mary Mulcahy, Athoney Yen, Al Wesenberg, Sam Sadler, Boles Burke, Bob Spolec, Eric Swartwood, Chester Bayko and other family members, friends, and professors.

David E. Mulcahy

CHAPTER 1

LOGISTICS AND SOME STRATEGIC CONSIDERATIONS IN FUTURE OPERATING COSTS AND CUSTOMER SERVICE

INTRODUCTION

This introductory chapter identifies the key warehouse functions, defines the terms that are used in the industry to identify warehouse and distribution operations, outlines the objective of a warehouse or distribution operation, and identifies issues and trends that are shaping present and future warehouse and distribution operations and facilities.

Warehouse and distribution operations are similar in all industry groups that have a combined product movement-storage-pick operation or facility of any size, whether it handles single items, cartons, pallet loads, or bulk materials. To some degree, each warehouse or distribution operation performs most or all basic warehouse functions. These functions include the following: (1) unloading, receiving, checking, and marking inbound merchandise; (2) internal horizontal or vertical product movement (transportation) to the storage-pick area, workstation, or outbound staging area; (3) storage (deposit, withdrawal, and replenishment); (4) order-pick (distribution) sortation and checking; (5) packing, sealing, weighing and manifesting, and shipping preparation; (6) loading and shipping; (7) handling returns, out-of-season product, and store transfers; (8) maintenance, sanitation, and loss prevention; and (9) inbound and outbound truck-yard control.

The warehouse and distribution product and information flow has a pattern that is similar to water flowing through a funnel (Fig. 1.1).

The mouth of the funnel is wide and accepts a large quantity of product and information. Over a period of time (days or weeks) a wide mix of product in various storage unit quantities from numerous vendors or from your manufacturing facility is delivered to your warehouse and distribution facility on various types of delivery vehicles.

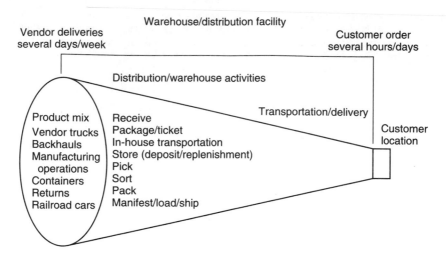

FIGURE 1.1 Warehouse product and information flow through the funnel.

Your customer information flow for these storage units (customer orders) occurs on a daily basis (more frequent than the product receipts) along with product receipt information to your warehouse or distribution operation. The time period for your warehouse or distribution operation to complete the customer order and delivery cycle is fixed. This time period is within 24 hours or 2 days, which is determined by your company's top management.

As your product flows through the funnel, various value-added activities (warehouse and distribution functions) are performed to ensure that the product satisfies your customer's needs and earns your company a profit. Therefore, with an increased number of customers and value-added activities that are handled by your operation, the allowable time to perform your warehouse and distribution functions becomes increasingly shorter and represents the small mouth (end) of the funnel.

In today's industrial magazines and other media authors tend to use the terms *distribution, warehouse, logistics,* and *material handling* to describe our industry or profession. Each term has basically the same meaning but is slightly different and is defined in this book.

Distribution. *Distribution* is defined as the function of moving various products from your vendor's facility or your manufacturing workstation (where the product was manufactured) to your company's facility for storing the product, picking the product to your customer order requirements, and delivering the product to your customer's facility or workstation in your manufacturing facility.

Warehouse. *Warehouse* is defined as the function of storing a variety of product types [stock-keeping units (SKUs)] that have a small or large quantity of storage units between the time that the product is manufactured by your facility (vendor) and the time that the product is required by your customer or workstation within your manufacturing facility.

Logistics. *Logistics* has the same meaning as distribution. But it encompasses all activities that are required to have controlled product and information flow between two locations. The first flow pattern is between your vendor location and your warehouse. The second flow pattern is between your warehouse and your customer's location. This controlled product and information flow maximizes the return on investment and the number of satisfied customers. Until the mid- and late 1980s, the term *logistics* was widely used outside of the United States; in the mid-1980s, the term became popular in the U.S. warehouse and distribution industry and media.

Storage.   *Storage,* as defined in *Webster's Dictionary,* is the activity of placing or depositing a good in a store or warehouse for safekeeping until the good is required at another location or workstation or by your customer.

Material Handling.   *Material handling* is defined by the Material Handling Institute as "the basic operation that involves the movement of bulk, packaged and individual goods in a semi-solid or solid state by means of a human or machine and within the limits of the facility."

STOCK-KEEPING UNIT (SKU)

At this time, it is best to state that a good has the same meaning as a SKU, merchandise, or product. A good, product, merchandise, or SKU is something of value that the warehouse or distribution operation receives, stores, and delivers to the customer or manufacturing department. By performing these key warehouse and distribution functions, it ensures that your company receives time-and-place value.

ECONOMIC VALUE

Your warehouse and distribution operations have an economic value to your company. The economic value assures your company that the SKU in inventory receives time-and-place value. The value is summarized in the following statement: "your warehouse and distribution product movement-storage-pick operation assures your company that the right good is in the right condition, at the right place (workstation or customer location), at the right time, in the right quantity, and at the right cost." This allows your warehouse and distribution operations to contribute to your company's bottom-line profits by reducing operating costs and satisfying your customers.

HOW YOUR DISTRIBUTION OPERATION SERVES YOUR COMPANY

To achieve company objectives, warehouse and distribution operations perform the following services for your company. The first service is to geographically consolidate your customer's demand for goods or to achieve economies of scale. With today's communication systems, this service allows your warehouse, distribution, and transportation departments to handle a greater number of customers and to reduce order-pick, handling, and transportation costs.

The second service is to provide geographic distribution of the goods to your customers. The service assures your company that your customer is receiving the best transportation cost for the goods.

The third warehouse and distribution service is to provide the means for your company to warehouse (store) goods that are produced throughout the year to accommodate your customer's seasonal demand for the goods. This service allows your company to reduce costs by purchasing large quantities of goods. This provides your customer with the lowest cost for the goods.

The fourth service is to provide the means for your company to warehouse goods which are produced from seasonal (short-time-period) production such as foods. This service allows your customer's year-round demand for the goods satisfied by your warehouse and distribution operations.

YOUR LOGISTICS RESOURCES

To achieve these company objectives, you maximize the efficient use and productive allocation of your warehouse, distribution, and transportation scarce resources. The resources that are available to you include facility layout, material handling equipment, employees (labor), land and buildings (company-owned or leased), management team (company organization), computers, vendors, consultants, and industry groups or associations.

YOUR COMPANY'S WAREHOUSE OBJECTIVES

The two major objectives of a warehouse and distribution facility are to improve profits and customer service. To achieve these objectives, your warehouse and distribution operations perform activities to (1) maximize your storage (space or cube) utilization; (2) maximize your warehouse equipment utilization; (3) maximize your labor (employee) utilization; (4) reduce your SKU handlings, maintain required SKU accessibility, and assure the designed SKU rotation or turns; (5) minimize your company's operating expenses; and (6) assure the protection of your company's assets.

IMPORTANT LOGISTICS TRENDS AND ISSUES

The important warehouse and distribution trends and issues consist of the new technologies of computer controls, automatic identification, just-in-time (JIT) replenishment and across-the-dock operations, equipment and labor flexibility, smaller inventories with material requirements planning (MRP) and distribution requirements handling (DRP), mechanized or automated machines, and contract warehouse operations. This is a list of the major factors that have had an impact on how today's warehouse and distribution facilities look and function. These factors are considered to have an increasing impact on the existing warehouse and distribution operations in the 1990s, shaping the warehouse and distribution operations of the 1990s, and affecting the warehouse and distribution facilities that are on the drawing boards which are planned for the year 2000.

Employee Training Increase.   If your company is to attain the expected results from the implementation of a new warehouse or distribution facility or from new material handling equipment or concepts in an existing facility, then you are required to train your managers and employees who work in the warehouse or distribution operation and other company departments that interact with the new facility, material handling equipment, or concept. Therefore, an important future warehouse trend is increased activity in training and motivating your managers and employees from these various groups. Managers and employees are a key ingredient in the successful implementation, start-up, and continued operation. This fact is valid for a manually operated warehouse or modern warehouse that is supported with high technology.

Impact of the Computer.   The next trend that is reshaping warehouse and distribution operations is the introduction of new computer hardware and software in almost every activity and function within the warehouse and distribution facility. The impact of the computer has allowed improvements in customer order processing inventory tracking, automatic identification, and work scheduling for warehouse and distribution equipment and labor and inbound and outbound vehicles. The implementation of computers with higher degrees of intelligence and faster processing speeds provides the opportunity to reduce operating costs, improve the flow of goods, enhance work scheduling, and improve service to customers (manufacturing workstation).

Automatic Identification.   The third important trend in the warehouse and distribution industry is the automatic identification of goods, product storage-pick positions, and assets. Today's hand-held and fixed-position scanning devices read bar-code labels, radio-frequency waves, or voice waves with network (communication) systems that transmit data. Between two warehouse locations, these concepts assure that the data is accurate and transferred on-line (time). This new technology has allowed improvements in inventory tracking; control of the product storage and merchandise flow; accurate order pick; precise order sortation; and delivery truck manifesting, loading, and tracking activities. The future trend is for more sophisticated and more extensive use of automatic identification systems in all warehouse, distribution, and transportation functions.

JIT and Across-the-Dock Operations.   The fourth important trend is the introduction of JIT replenishment and across-the-dock operations in a company's channel of distribution. These concepts schedule inbound deliveries to arrive at your warehouse or distribution facility receiving dock just in time to replenish the manufacturing workstation or warehouse ready reserve position, or to flow through the warehouse (in one door and out the other door) to your customer delivery vehicle. These philosophies have had an impact on reducing the required on-hand SKU safety stock inventory in your warehouse or distribution facility storage area levels and have redirected the warehouse or distribution emphasis or focus on the receiving and staging dock areas and operations rather than on the reserve or storage areas and functions. The across-the-dock operation has reduced the time required for the product to flow from your vendor to your retail store and has allowed the manufacturing operation to become a global activity.

MRP and DRP.   The fifth trend consists of the material requirements planning (MRP) and distribution requirements planning (DRP) philosophies. These inventory and material handling equipment planning systems are based on the marketing and sales department's forecasts. The results of these task group efforts impact your warehouse or distribution facility's on-hand SKU inventory size and quantity. These two factors affect your building's size and to some degree the mechanization or automation of the warehouse or distribution operation.

Centralized or Decentralized Operations. The sixth trend is that many companies are returning to a distribution network that has fewer distribution facilities that serve specific regions. Numerous factors affect this trend, including the degree of automation of the warehouse and distribution operations, cost of land, cost of labor, cost of transportation, and customer demand (volume or throughput). The key factors in a decentralized operation are the availability of land, availability of quality labor, customer demand (volume), and both inbound and outbound delivery costs and on-time service to the customers.

Material Handling Equipment Technology. The seventh trend is the introduction of new single-item, carton, and pallet load handling technology and equipment. The future is in the area of automatic over-the-road truck loading and unloading equipment, automatic guided vehicles (AGVs), internal transportation vehicles, mechanized order-pick vehicles, automatic or robotic order pickers or palletizers, and pallet load storage vehicles that operate faster in narrower and taller aisles. This trend has an impact on creating higher warehouse buildings that occupy a smaller-square-foot land area and have fewer employees.

Remodeling Existing Facilities. The eighth warehouse and distribution trend is the remodel (redesign or retrofit) of an old existing distribution facility. With the high capital investment requirement for a new facility (land, building, and equipment), project schedule (market entry), and sales (customer demand) fluctuations, it makes one large new distribution facility more difficult to economically justify; therefore, an alternative is to remodel an existing facility with increased mechanization or automated equipment.

Leasing Equipment and Third-Party Distribution. The ninth trend is an increase in the number of companies that are leasing equipment and buildings. This arrangement allows the company to have access to the latest technology and frees up company funds for other investments. Included in this category is the public and contract warehouse trend for companies to use an outside distribution company to handle all distribution functions within a particular market area.

Increase in Value-Added Activities. The tenth trend is toward the global or multinational company that is involved in the multilocation manufacturing and to some degree the distribution of goods. With new telecommunication technology, experience, and shift in consumer purchase patterns, warehouse and distribution facilities are increasing their value-added service (functions) to their product and their operations. The trend for warehouse and distribution segments in the channel of distribution is to increase the number of facilities and value-added activities of pick and pack, pricemark or -ticket, product repair and service, repackage, returned-goods handling, customer pickup, across-the-dock distribution, and telemarketing. These activities lessen the time that is required for product to flow through the channel of distribution to the final customer.

OVERVIEW

It is the purpose of this book to provide you with a review of equipment applications, procedures, practices, tips, and insights that you can consider for implementation in your warehouse and distribution operations. This provides you with the opportunity to maximize your company's profits by reducing operating costs and to maximize service to your customers with on-schedule and accurate deliveries.

CHAPTER 2

WAREHOUSE AND DISTRIBUTION OPERATION OBJECTIVES AND COMPANY PROFIT AND SERVICE OBJECTIVES

INTRODUCTION

This chapter defines the objective of your warehouse and distribution facility to your company and to your customers. The second objective is to list and review the various key warehouse and distribution functions. These functions include the pre-order-pick, order-pick, and post-order-pick activities in a small-item (replacement of spare parts or catalog or direct ordering) mail, carton (case), or pallet (unit-load) warehouse or distribution operation.

HANDLING CHARACTERISTICS

Small-item distribution operations receive pallet loads or cartons and send individual items of merchandise to your customers. Carton (case) distribution operations receive cartons or pallet loads and send individual cartons to your customers. Pallet load warehouse operations receive pallet loads and send pallet loads to your customers.

DISTRIBUTION FACILITY OBJECTIVE

The objective of a small-item, carton, or pallet load distribution facility is to ensure that the right SKU is in inventory, is available at the appropriate time, and in correct condition, is withdrawn in the right quantity and on schedule, is in a protective package, is properly manifested, and is delivered to the required location that satisfies your customer's order at the lowest possible operating cost.

DISTRIBUTION FACILITY ACTIVITIES

To achieve these objectives, you arrange your facility layout, material handling equipment, and product flow path to minimize handling of SKUs and to organize your employees to perform the following activities: (1) pre-order-pick activities of yard control of delivery vehicles, unloading, receiving and inspecting, identifying, inbound packaging and ticketing, internal transportation, deposit and across-the-dock (flow through) operations; (2) order-pick activities of order pick and label and storage activities in the pick area; and (3) post-order-pick activities of replenishment of fixed pick position, outbound packaging, weighing and manifesting, loading and shipping, customer returns, and out-of-season product and retail store transfers.

Pre-Order-Pick Activities

The pre-order-pick activity of packaging or ticketing individual SKUs is unique to a small-item distribution facility. The other major pre-order-pick warehouse functions are similar for a small-item, carton, and pallet load distribution operation. These activities are (1) yard control of vendor or company backhaul delivery vehicles at the dock and unloading schedule, (2) unloading the product from the delivery vehicle onto the inbound staging dock area, (3) verifying that the product quality and quantity match those specified in your company's purchase order, (4) receiving the product (entry) into inventory, (5) identifying the product (with some retail store distribution operations, this includes pricing and ticketing the individual items), (6) packaging of the product (some small-item SKUs are packaged into handling of shipping containers and cartons, or pallet SKUs are stabilized, or secured, on the unit load support device), (7) internal transportation of the product to the assigned reserve or pick position, and (8) depositing the product into the assigned reserve or pick position.

Yard Control Activity. Yard control activity of company backhauls, vendor delivery trucks, or containers unloading schedule determines what time the delivery truck is positioned at your facility unloading dock. Whenever possible, this dock location minimizes the internal transportation distance between the dock door and the storage location. Yard control includes the spotting of railcars on your rail spur to assure the shortest internal transportation distance between the rail dock location and the storage location.

Unloading Activity. The unloading activity is to unload the trolley (cart) of hanging garments, master cartons, or pallet loads of SKUs from the vendor's delivery truck, railcar, container, or company backhaul truck onto the receiving dock staging area. Chapter 4 examines the various dock layouts, dock equipment, product delivery methods, unloading equipment, and inbound product staging concepts.

Verify Product Quality and Quantity. The next pre-order-pick activity is to verify that the vendor's product quality and quantity are per your company's purchase order. This activity ensures that the quantity of product delivered to your warehouse matches your company's purchase order quantity and that the received product quality is acceptable to your company's purchase order specifications and your company's standards.

Receiving Activity. The next pre-order-pick activity is the receiving activity. The receiving department employee enters the SKU quantity into inventory and transfers the SKU from the receiving department staging area to the storage-pick staging area.

In the across-the-dock operation, the product is transferred from the receiving area to the shipping area and does not enter your inventory.

Product Identification Activity.   The fifth pre-order-pick activity is SKU identification. An employee applies (places) markings to the exterior individual SKU, master carton, or pallet load. These markings are used in other distribution facility functions to physically and discreetly distinguish one SKU from another. The markings are alphanumeric characters, a bar code, or a radio-frequency tag that serves as an instruction to your warehouse employees who handle the product. In some small-item operations, the SKU identification activity is performed after the SKU is placed into a material handling or shipping container.

Ticketing Activity.   In some retail store distribution operations, a subactivity of SKU identification is the SKU ticketing activity, in which a retail price tag is placed (ticketed) onto each individual SKU. This activity is very common in a hanging garment, flatwear, and carton (ready for retail sale) distribution operation.

In ticketing activity, a mechanical printer prints price tickets that are glued or clipped to, stitched into, or hooked onto the SKU or placed on the exterior of the SKUs.

Packaging Activity.   The next pre-order-pick activity is SKU packaging. In small-item (SKU) packaging, a warehouse employee places individual SKUs into a material handling or shipping container. These containers are plastic or paper bags or chipboard or cardboard boxes. Carton and pallet load packing activity ensures that the cartons are properly sealed and are secured to the unit load.

The objective of the SKU packing activity is to ensure in the pick position that one SKU is separate from another and during customer delivery, that the SKU is protected from damage.

Horizontal and Vertical Transportation Activity.   The next pre-order-pick activity is the internal horizontal and vertical transportation of product from the receiving area to the storage-pick staging area. Numerous methods are used to accomplish this activity, including the manual, mechanized, or automated material handling system. Because the horizontal and vertical transportation concepts are used to move units of product between two warehouse locations, a more detailed description of each method is presented in Chapter 9.

JIT and Across-the-Dock Activity.   Another pre-order-pick activity is the across-the-dock (flow-through) distribution (sortation) activity. This activity has developed as a recent trend in the retail store distribution industry to handle single-item and carton products. It is considered the retail store industry's form of a JIT (just-in-time) replenishment program. With this material handling concept, it changes the traditional sequence of activities and product flow. In this new product flow concept, the product is received and then distributed to the customer's (retail store) staging-shipping area and the residual product quantity is placed in storage. This flow concept reduces the distribution facility number of product handlings and number of days to flow from the vendor to the retail store shelf and required storage area, but places emphasis on inbound-outbound dock and sortation (distribution) activities.

This JIT operation is a manual operation or a mechanized operation that handles prelabeled or unlabeled product.

Deposit Activity.   The next pre-order-pick activity is to deposit the unit load of product in the assigned reserve or pick position. The accurate and on-schedule completion of this activity ensures that the right SKU is in the proper place, in the proper quantity, in the correct condition, and at the correct time. This allows an employee to perform on-time replenishment and order-pick activities.

Pick (Order-Pick) Activity

The SKU order-pick (withdrawal, fulfillment, shop, or selection) activity requires an employee to remove, per a customer order, the correct SKU in the correct quantity, in the correct condition, and at the correct time from the inventory (pick position) onto a picking transport device to satisfy the customer's order (demand).

The SKU order-pick activities include the following: (1) listing the SKUs that are ordered by the customers, (2) traveling and/or removing the SKUs from the pick position, (3) verifying the SKU order-pick (inventory reduction), and (4) transporting the SKU to the packing or shipping area.

Storage Activity.   The SKU storage activity provides a warehouse location to store the SKU in a reserve position until it is required at the pick position or for a customer order. Portions of Chaps. 6, 7, and 8 examine the various types of SKU storage positions (fixed vs. floating) or reserve positions. Included in each chapter is a review of the various storage concepts. Chapter 10 looks at the various mobile (lift trucks), Chap. 12 reviews the inventory control activity, and Chap. 11 reviews the methods to store the stackables and the unstackables.

Post-Order Pick Activity

Sortation Activity.   The SKU sortation activity is the first post-order-pick activity. When the single-item or carton order-pick activity is the batched mode, the sortation activity separates one of your customer's specific ordered single-item or carton SKUs from your other customer picked SKUs of the batch. It then verifies that the SKU was withdrawn from the pick position and was transported to the packaging or shipping-staging area.

The SKU sortation activities require a human or a machine to read the human or human-machine label (markings) that is on the SKU exterior surface and to transfer the SKU from the batched (grouped) picked SKUs into the specific customer temporary holding (sortation) location. This location is a bin, container, chute, or conveyor.

The various manual, active, passive, and active-passive SKU sortation methods are reviewed in Chaps. 6, 7, and 8. Chapter 13 reviews the various manual and automatic identification concepts. It includes manual identification codes, bar codes, tags, and data-entry technologies.

Replenishment Activity.   With a fixed-single-item or carton pick position concept, SKU replenishment is another post-order-pick activity. SKU replenishment ensures that the correct SKU is removed from the assigned storage (reserve) position on schedule, is in the proper quantity, and is placed into the correct SKU pick position.

SKU replenishment activities include listing of the SKU pick positions that require replenishment, withdrawal of the product from the storage (reserve) position, and transfer or placement of the SKU in the SKU pick position.

In the various replenishment methods in a small-item or carton distribution facility, a warehouse employee transfers product from a random storage (reserve) position to a fixed pick (active) position. In a pallet load operation, the putaway (from the receiving dock) of a pallet load into an assigned reserve position is the replenishment activity.

Packaging Activity. The outbound SKU packaging activity is the fourth post-order-pick activity. The objective of SKU packaging is to ensure that the SKU is protected from damage during delivery to, and is received by, your customer in satisfactory condition. In the distribution business, the exterior package condition is the first impression of your company's service that is received by your customer. This fact is especially true in the catalog and direct-mail business.

The packaging activity includes (1) verifying the order-pick accuracy (quantity and quality), (2) filling the voids in the package with protective material, (3) sealing (closing) the package or bag, and (4) placing your customer's delivery address onto the shipping container exterior.

Most SKU packaging activities involve small-item operation. The SKU packaging activities for a carton or pallet load operation include unitizing cartons onto a pallet board or cart, securing the product, and labeling (addressing) the product.

Package Sealing Activity. The next small-item warehouse activity is your customer's package sealing method. The delivery carton sealing activity ensures that the container does not open during transport to your customer and that the SKUs are in the package when it is delivered.

The type of package determines the sealing method for the package. The two basic methods are to pack multiple SKUs which are loosely packed SKUs in one large container (retail store or catalog agency industry) or as an individual SKU or a few SKUs packaged in the appropriate-sized container (catalog or direct mail industry). These sealing methods are reviewed in more detail in Chap. 6.

Package Weigh and Manifest Activities. The package weighing and manifesting activities are considered the next post-order-pick activity. The objective is to ensure that each outbound package receives the proper transportation fee (postage), lists the package number and weight, is sent by the most cost-effective transportation method, and, as required, has proper documentation. Also, you obtain the exact weight and verification (manifest) of shipment to your customer.

The weighing and manifesting activities include (1) using a scale to obtain the exact weight of each package and (2) verifying that the actual or computer-projected weight is indicated on the package and that the package identification number is listed on the transportation document.

Loading and Shipping Activities. The package loading and shipping operation function is the next direct labor function of a small-item, carton, or pallet load distribution facility. The shipping function ensures that your customer's order is placed on your customer's correct delivery vehicle. The shipping function is a direct load activity or a temporary hold for loading at a later date.

The receiving and shipping functions are reviewed in Chap. 4. The chapter reviews the various loading and unloading methods and inbound and outbound staging concepts.

Customer Returns, Out-of-Season Product, and Retail Store Transfer Activities. The activity of a warehouse and transportation department handling customer returns, out-of-season product, and retail store transfer product are considered part of the key warehouse activities.

The customer return activity is a warehouse activity that occurs in all industries. It is most evident (varying from 5 to 38 percent of the shipped volume) in the catalog and direct mail industries, but it occurs in the carton handling industry at an estimated rate of 1 to 5 percent of the volume shipped to customers.

Out-of-season product and retail store transfer activities occur in the retail store distribution industry.

Customer return activity assures your company that (1) your customer's returned-order quantity was received at your warehouse and (2) the returned merchandise physically flows in one of these patterns and, as required, is entered into inventory, placed in the SKU pick position, sent to the outlet store, donated to charity, and disposed of in the trash.

Other customer return activities include (1) sortation of the merchandise and (2) approval of credit that is issued to the customer.

The *out-of-season product* activity is a warehouse activity to temporarily hold merchandise from the retail stores that did not sell at your company's retail outlets. With your company's top management approval, the retail stores package and return the merchandise to the warehouse for temporary storage. At a later date, your top management decides how to handle the merchandise.

Retail store transfers consists of overstock merchandise from one retail store with low sales that is shipped through the warehouse or transportation system to another retail store that has high sales of the merchandise. With your top management approval, the merchandise becomes a store transfer which flows from one store to another through your distribution and transportation operations as across-the-dock merchandise with the proper paper documentation.

MAINTENANCE, SANITATION, AND SECURITY ACTIVITIES

The remaining key warehouse functions are maintenance, sanitation, and security functions. These functions satisfy two objectives: to provide protection of your company's assets and to ensure that your inventory, building, and material handling equipment are available to satisfy your customer's orders and operate at the lowest possible operating cost.

ON-SCHEDULE AND ACCURATE PERFORMANCE OF WAREHOUSE, DISTRIBUTION, AND TRANSPORTATION ACTIVITIES MEAN PROFITS AND SATISFIED CUSTOMERS

The effective and efficient completion of the pre-order-pick, order-pick, and post-order-pick warehouse, distribution, and transportation activities ensure that your company's customers are satisfied with the best service that was provided by your company. When all these activities are completed on schedule and at the lowest operating cost, then the SKU, SKU package, and documentation make a positive and lasting impression on your customer. This ensures that your warehouse and distribution facilities are profitable and have satisfied customers.

CHAPTER 3

HOW FACILITY LAYOUT AND PRODUCT FLOW PATTERN AFFECTS PRODUCTIVITY AND HANDLING COSTS

INTRODUCTION

The objective of this chapter is to familiarize you with the objective of your warehouse and distribution facility layout; the basis of a facility layout; alternative facility layout philosophies; various facility shapes; layouts that are based on SKU flow, movement, storage, and order-pick activity requirements; and methods to project design-year SKU quantity and inventory volume.

PURPOSE OF THE FACILITY

It is understood that the main purpose of your warehouse or distribution facility is to provide the housing (shelter) for your company's design-year requirements. These requirements include your material handling system, SKU pick and reserve positions to accommodate the projected inventory, and associated warehouse functions such as support and administrative activities. Some purposes of the facility layout are to (1) assure proper access to the SKUs, (2) provide proper product flow and inventory rotation, (3) assure the lowest possible operating cost, and (4) assure accurate and on-schedule customer service.

THE DESIGN VOLUME LEVEL

When you design a distribution facility and a material handling system for your new operation or a new piece of equipment in an existing operation, you design the facility, key warehouse functional areas, and equipment to handle a specific volume. Your design volume has three alternatives: average volume, some volume between the average volume and peak volume, and peak (spike) volume.

If you design your facility and key warehouse functional areas and equipment to handle your average volume and your volume exceeds the average volume, then you anticipate off-scheduled activities, outside storage, employee overtime, increased product damage, and a high potential for late customer deliveries. This design level has a small-sized building, lower capital investment, and low fixed handling cost per unit.

If you design your facility and key warehouse functional areas and equipment to handle a volume that is somewhere between your average and peak volume levels and on the occasions that your volume exceeds the design level, then your operation incurs employee overtime, congestion, increased product damage, off-schedule activities, outside storage, and some late customer deliveries. This design level has a medium-sized building, a medium capital equipment investment, and a medium fixed cost per unit.

If you design your facility and key warehouse functional areas and equipment to handle your peak volume and your volume arrives at the peak volume, then your operation will not experience additional costs or late customer deliveries. The large-sized building and extra material handling equipment capacity means a higher fixed cost per unit.

DETERMINING FACILITY SIZE

The facility size is defined by the square or cubic footage of the structure. To determine these dimensions, the facility available space is calculated by measuring the actual building or from a facility drawing with a scale to determine the building size.

Square-Foot Calculation

The building square-foot area is considered the space that houses your warehouse and distribution operations and is determined by the distance inside the building exterior wall column lines. To obtain square footage of a building, multiply the length by the width. If the building has a square or rectangular shape, then one calculation is needed to determine the building square footage. If the building is L-shaped, two calculations are required. The results of the two calculations are added to determine the building total square footage.

Cubic-Foot Calculation

If your building occupies expensive land and your warehouse or distribution operation requires that the material handling equipment reach high storage levels or have multilevels, then the building cubic footage is the available space and is more important than the building square footage. The building cubic-foot area considers the available square feet plus the clear distance between the finished floor surface and the bottom of the lowest ceiling obstruction. This is the space that houses your material handling equipment and inventory.

The typical high-rise building has a 40-ft clear ceiling height that is a square- or rectangle-shaped building. To determine the building cubic feet, multiply the building length times the width times the height. In another common building design a 40-ft-high clear ceiling high-rise building is adjacent to another building with a 20-ft-high ceiling to house the warehouse and distribution support functions. With these two buildings, you multiply each building length by the width by the height to determine the cubic feet of each building. For the total available cubic feet, you total the cubic feet of the two buildings.

Another method of calculating the cubic feet of a building is to have the total square feet area and the height as given figures. This information is obtained from the drawing or building fact sheet. To determine the cubic-foot area, you multiply the two figures. If a building has an area of 1,350,000 ft^2 and a 20-ft-high ceiling, then the volume of the building is 27,000,000 ft^3 (1,350,000 ft^2 × 20 ft^2 = 27,000,000 ft^3).

WAREHOUSE AND DISTRIBUTION ACTIVITIES

The facility layout consists of the proper arrangement with adequate space for the design year SKUs and inventory volume, warehouse functions, and the other required distribution support activities. The various key warehouse functions and activities are (1) yard control; (2) receiving and staging; (3) opening, counting, and ticketing; (4) internal transportation; (5) storage; (6) order pick and distribution; (7) packaging; (8) weighing and manifesting; (9) customer returns and out-of-season product transfer; and (10) staging and shipping.

Some of the support activities include (1) personnel; (2) restrooms and locker rooms; (3) cafeteria; (4) offices; and (5) maintenance, sanitation, and security facilities.

It has been the writer's experience that the combined square footage area for warehouse and distribution storage-pick activity represents 70 to 80 percent of the total facility area and the other key warehouse functions and support activities occupy 20 to 30 percent of the total facility area.

FACILITY LAYOUT IS A COMPLEX PROJECT

The development of a distribution facility layout is a complex project because the layout has constraints and is also required to satisfy your company objectives.

The layout constraints are (1) clear column spacing and size; (2) bay size and direction; (3) clear ceiling height; (4) door and dock locations; (5) building shape, land conditions, and shape; (6) geographic area [seismic and climatic (e.g., snow load) conditions]; and (7) local building codes.

FACILITY LAYOUT OBJECTIVES

Your company warehouse layout objectives are established by your executive management team and usually include a request or requirement to (1) maximize the space (cube) utilization or provide the maximum storage and pick positions within the building structure; (2) allow an efficient product flow from the receiving area to the storage-pick areas and from the storage-pick areas to the assembly, packing, and shipping areas; (3) provide the maximum number of, and facilitate access to, SKU pick (order-pick) positions and proper inventory rotation; (4) reduce annual operating costs; (5) improve the key warehouse function employee productivity (receiving, transportation, storage, order pick, packing, weighing and manifesting, shipping, and returns); (6) maintain the corporate philosophy and direction; (7) protect the inventory and material handling system from damage, pilferage, and infestation; (8) provide for expansion; (9) provide the employees with a safe work environment; and (10) ensure that your operation satisfies your customers.

COMPUTER SIMULATION

Today's and tomorrow's warehouse and distribution operations are considered a complex network of several sophisticated (automated or mechanized) material handling systems. All functions of a warehouse or distribution facility handle high volumes of product and are required to interface and interrelate with one another. To handle the SKU quantity and customer orders and product volumes, these key warehouse function areas require precalculated product queues and smooth, continuous product flow.

When we look at a high-volume warehouse or distribution operation, it is a complex network of product flow (transportation) paths and information (communication) transmission avenues between two warehouse locations (functions).

Computer simulation provides the warehouse and distribution designer and warehouse manager with an insight to the product and information flows through the key warehouse activity areas of an operation. These warehouse activities are the value-added activities of the warehouse or distribution operation. These value-added activities are different for each industry or type of operation.

In a conventional storage-and-hold warehouse operation these SKU flow paths are (1) inbound flow of product from delivery trucks, containers, or railcars; (2) transportation of product via AGVs (automatic guided vehicles), conveyors, or lift or pallet trucks between warehouse locations; and (3) flow of product between warehouse activity locations, from the storage area to the pick-ship area, from the pick area to the pack-ship area, and from the pack-manifest area to the ship area.

In an across-the-dock or JIT (just-in-time) operation, the SKU and information flow paths are between the receiving area, as required to the ticketing area and the shipping area. When there is residual product, then the path follows the conventional product path.

When to Use a Computer Simulation

Computer simulation is used in the design of a new warehouse or distribution facility or remodel of an existing operation that handles a high volume of product by a highly mechanized or automated material handling system. The computer simulation indicates the need to (1) design a new, highly mechanized or automated material handling system for a new operation or as an addition to an existing operation; (2) review an existing material handling system or operation; (3) determine the optimum product flow in a material handling system or operation; (4) identify the impact of adding new customers, additional volume from existing customers, or new SKUs to the inventory; and (5) ensure that the SKUs are allocated to a warehouse storage-pick location that optimizes labor and equipment.

FACILITY LAYOUT FUNDAMENTALS

Planning analysts follow numerous facility design fundamentals to design a warehouse or distribution facility. These fundamentals optimize the facility and minimize construction costs and include two steps: data collection and development of alternative layouts.

The first step of a facility layout consists of the data collection process, data analysis, establishment of design-year parameters, and consideration of alternative material handling equipment and concepts. This step includes (1) identifying and listing existing material handling equipment; (2) measuring (width, height, and weight) and cataloging all SKUs as conveyable or nonconveyable or by classification such as packaging, toxic, or edible; (3) classifying at each warehouse function, the SKU handling characteristics (per length, width, height, and weight measurements) as single items, carton, or pallet load; (4) projecting SKU inventory levels (average and peak) and at each warehouse function, the SKU volume levels (average and peak); and (5) reviewing alternative material handling concepts (manual, mechanized, or automated) for each warehouse function.

The second step is to develop alternative distribution facility layouts. These layouts include areas for key warehouse functions such as (1) yard control, truck and automobile parking, and rail spur; (2) receiving and staging; (3) open, sort, count, ticket, and packing activities; (4) returns, store transfers, and out-of-season product return to vendor; (5) internal transportation; (6) order pick and distribution; (7) sortation; (8) packing; (9) weighing and manifesting; and (10) staging and shipping.

The various secondary administrative function areas are (1) administrative and office; (2) cafeteria; (3) maintenance, security, and sanitation facilities; (4) restrooms; (5) locker (personnel) area; and (6) trash area.

MINIMIZE THE NUMBER OF SUPPORT COLUMNS FOR WAREHOUSE GROUND OR MEZZANINE FLOORS

When you design a new warehouse or a mezzanine, you have the fewest number of building or second-level support columns on the ground (first) level. This wide support column arrangement requires a greater span between two columns, higher structural support member cost, and a deeper (larger) support member. The advantages of the method are (1) fewer obstacles to decrease transportation employee productivity on the ground level, (2) increased flexibility for rearranging equipment, and (3) excellent vehicle transportation (straight) paths between two locations for on-time deliveries with minimal product, building, or equipment damage.

FACILITY LAYOUT PRESENTATION METHODS

Your project design team uses one or a combination of the following warehouse and distribution design and presentation methods to make a warehouse and distribution layout presentation to your top executive management team: (1) block layout method, (2) standard templates and layout board method, (3) drawing method, and (4) model method.

Block Layout Method

The block layout method (Fig. 3.1*a*) is a "not to scale" drawing that shows the total building size and shows each area of the building that is allocated to a specific warehouse function. These layouts allow your design employees to visualize the product flow through the facility, ensure that the size for each function has sufficient area, and estimate the total building square footage.

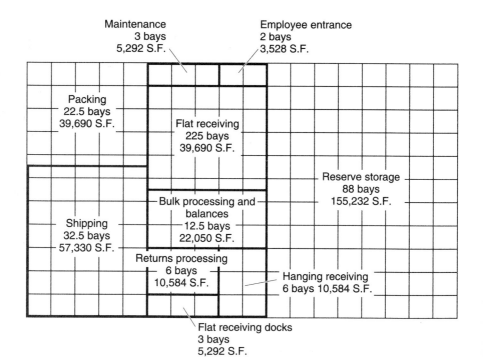

FIGURE 3.1   (*a*) Block layout.

Layout Board and Standard Templates Method (Fig. 3.1*b*)

A layout board with standard material handling equipment templates is an initial lay-out method showing a material handling concept for each function within the total facility. The layout board with templates is a two-dimensional and "to scale" repre-sentation of your designer's concept for a warehouse or distribution facility that includes the necessary equipment. The standard equipment templates are multicolored and are moved by the design employee on the planning board to show various con-cepts. These various warehouse concepts are created from standard templates in a short period of time and with low drafting expense. If mezzanines or additional floors are used in the design, then a clear plastic sheet with templates is placed on stands. This represents the floor that is above or below the main floor. After completion of the final template and layout board, your facility design is transferred to a manually or computer-produced drawing.

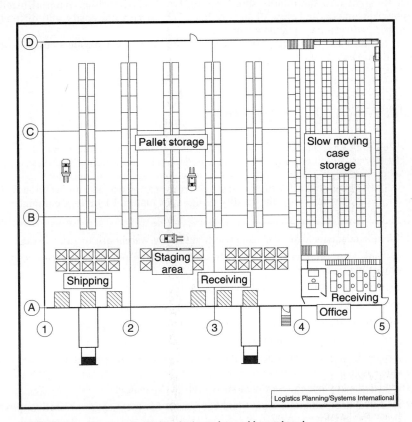

FIGURE 3.1   (*Continued*) (*b*) Standard template and layout board.

Conventional or Computer-Produced Drawing Method

The conventional drafting (human or computer)-produced drawing technique is a to-scale two-dimensional visual presentation of each function material handling concept within a distribution facility. A plan view drawing provides your design team with a view of the equipment arrangement inside the distribution facility. A detail drawing shows an exact picture of "clear" distances between pieces of equipment or between equipment and building obstacles. The conventional drawing technique is time-consuming, and changes are expensive. A computer drawing is quicker, and changes are inexpensive.

Model Method

A model consists of material handling equipment pieces and racks and is a three-dimensional presentation of a facility. Models are built to scale and provide a more exact insight into the relationship between two levels, various material handling equipment pieces, and flow of the product. The models are costly and difficult to transport between presentation locations.

If a model has additional floors in the facility, then clear plastic is used to show each floor's equipment layout.

FACILITY LAYOUT PRINCIPLES

All the warehouse, distribution, and manufacturing facility layout design and presentation methods presented in the preceding sections have two main objectives: to show how your facility will look and to describe how your facility will operate.

In development of a facility layout, numerous warehouse and distribution facility layout principles influence the facility design and material handling equipment layout:

1. Provide adequate aisles and aisle width in the key warehouse function areas.
2. Consider the product flow and volume through the reserve area, pick area, and other functional areas.
3. Provide adequate SKU accumulation prior to each workstation.
4. Provide adequate ceiling height for warehouse equipment.
5. Provide required space for fire protection and security equipment.
6. Locate all support or administrative activities.
7. Locate the building facilities on the site for excellent present utilization and future expansion.
8. Locate the key warehouse functions for future expansion.
9. Design space building columns and bay size to facilitate space utilization, product flow, and employee productivity.
10. Use gravity-propelled transportation in combination with mechanized or automated equipment.

Adequate Aisles and Aisle Width Principle

The first warehouse layout principle is to provide adequate aisles and aisle width in the key warehouse function areas. These function areas are receiving, transportation, opening, sorting, counting, ticketing, storage, order-pick (distribution), sortation, packing, and shipping functions. The proper warehouse aisle layout improves the warehouse employee transportation productivity, reduces the risk of product and equipment damage, and facilitates the transit or travel of employees and equipment between functions. With these aisle dimensions, your warehouse or distribution operation obtains satisfactory productivity, reduces product and equipment damage, becomes more profitable, and provides better service to its customers.

To achieve a warehouse layout design that follows this principle, the following features are incorporated in the equipment layout. All storage-pick aisles between racks or floor stacks have the vehicle manufacturer minimum aisle width for a right-angle (stacking) turn. This requirement is between product and product (interproduct). In a standard rack design, where the pallet boards overhang the rack load beam by 2 to 3 in into the aisle, the aisle requirement is not rack-to-rack, but product-to-product, which includes the total aisle side overhang for both pallet boards. For good operator productivity and equipment product damage it is recommended that 6 to 12 in or, with some very-narrow-aisle (VNA) storage equipment, 20 in, be added to the vehicle manufacturer minimum aisle width requirement. It is noted that this 6- to 12-in aisle width feature applies only to wide-aisle and narrow-aisle unit-load handling equipment. It does not apply to VNA equipment. When you consider VNA equipment, there are two aisle width dimensions.

Do Not Turn the Load in a Narrow Aisle.   The first aisle width is a nominal 6 in wider than an aisle width. This aisle width does not necessarily permit your operator to turn the pallet load in the aisle.

Turn the Pallet Load in a Narrow Aisle.   To turn the pallet in the aisle, consult the lift truck manufacturer for the required dimension.

1. Nominally, this aisle width dimension is 6 in wider than the diagonal dimension of your pallet load.

2. In a single-item pick aisle, 36 to 42 in is allowed between the shelves to provide a comfortable pick aisle. Some aisles have a minimum distance that permits a pick cart to travel in the aisle. This aisle width dimension is 32 to 33-in.

3. In a pallet jack operation, there is a minimum of 8 to 9 ft or a counterbalanced lift truck right-angle stacking requirement of 12 to 13 ft width, which allows for two-way pallet truck traffic in the aisle.

4. All intersecting (main traffic) aisles to the storage and pick aisles have the vehicle manufacturer minimum width to turn from one aisle to another. It is recommended that 12 in be added to this dimension.

5. All main vehicle traffic aisles between two buildings or storage structures allow two vehicles to pass each other (two-way traffic).

6. All aisle sides and floor stack pallet storage lanes have painted lines and all personnel aisles have diagonal painted lines across the equipment travel path.

7. All offices, door frames, end-of-aisle upright posts, and stationary material handling and manufacturing equipment have guard rails or posts.

8. All ramps, dock doors, and building doors (passageways) are designed to permit material handling equipment travel. The clear height from the finished floor to the passageway jam allows for the collapsed mast height of the equipment plus 6 in.

9. All personnel aisles have sufficient width (42 in to 4 ft) to handle the number of people in the work area.

Type of SKU Handled by the Operation

The first step in understanding your warehouse business is to know the type of SKU (product or merchandise) that is handled at your warehouse-distribution facility. This SKU information is determined for each key warehouse function. Each warehouse operation function has the potential to handle your product with different characteristics. The product's characteristics change as it flows through your distribution facility. This change in SKU characteristics occurs for product that is received in pallet quantities and sent to your customers in carton quantities or for product that is received in pallet or carton quantities and sent to your customers as individual items.

The first SKU category is to define the receiving, storage, pick, and shipping SKU characteristics. These characteristics are the major factors that determine how your warehouse function areas look and operate and the degree of mechanization or automation in the warehouse function. These characteristics include

- SKU type (single-item, carton, or pallet load) and volume
- Dimensions (length, width, height, and weight)
- Handling characteristics (conveyable or nonconveyable; edible or nonedible)

  Hazardous (toxic or flammable); crushable or fragile

  Security requirements (high purchase value or expensive)

  Shape and environmental conditions (whether room temperature, refrigeration, or freezing required)

Product Flow

The next basic information requirement is to identify product flow through your facility. The two alternative flow patterns are the JIT (just-in-time) replenishment concept (flow-through, break-bulk, or across-the-dock) or inventory-and-hold (storage) concept.

JIT and Across-the-Dock Product Flow Concept.   According to the flow-through (JIT) or across-the-dock product flow concept, the product arrives at the facility and is distributed (sorted) to a workstation, staging area, or your customer's delivery vehicle. In the retail store distribution industry, this flow concept means that the product is not entered in your distribution facility inventory because it has been preallocated to your individual retail store customers and is not allocated to the storage area. The merchandise arrives at your distribution facility as prelabeled or nonlabeled merchandise. The nonlabeled product requires your employees to print and place the retail store customer delivery address label on each piece of product. The concept is flexible and does permit a portion of the product (residual) placed into storage. If the small amount of product is placed in storage, then it is entered into your distribution facility inventory.

In the manufacturing industry, the JIT replenishment program is designed for direct flow of the product from the receiving dock area to the manufacturing workstations. If there is a small quantity of surplus product, then the product is assigned to the storage area and is entered into the inventory.

In both the retail store distribution and manufacturing warehouse industry, the across-the-dock and JIT replenishment programs have become more common in the channel of distribution. These programs have increased the necessity of larger and separate dock areas for the receiving and shipping activities.

Inventory-and-Hold Concept. The inventory-and-hold product flow concept is the classic product flow concept. In this concept, the distribution facility personnel unload the merchandise from the delivery vehicle, enter the product into inventory, and place the product in storage. At your customer's request, your employees withdraw the product and load it onto your customer's delivery truck or deliver the product to the workstation.

When comparing the two flow concepts, the JIT and across-the-dock flow concepts require larger dock areas. For best operational results, there are separate receiving and shipping dock areas. The inventory-and-hold concept requires a large storage area with less area allocated to the receiving and shipping functions.

Your Inventory Control and Locator Program

The next distribution facility design requirement is your distribution facility inventory control and location program. There are two types of inventory control program, in which the product has either a last-in first-out (LIFO) or first-in first-out (FIFO) product rotation.

LIFO Product Rotation. The LIFO inventory rotation criterion determines that the product rotation method ensures that your distribution facility provides your customers with the best quality product. The product rotation is a function of the product expected life cycle. In LIFO product rotation, the last product received at your distribution facility is the first product shipped to your customers. The product quality is not affected by time (examples are clothing, shoes, and basically all nonfood or nondated product); therefore, dense storage concepts are used in the distribution facility storage area. These storage concepts reduce the required storage area size.

FIFO Product Rotation. In the FIFO product rotation method the first product received at your distribution facility is the first product shipped to your customers. This method is preferred for perishable products or products with a short life (e.g., food and dated products). The inventory control criterion requires a high degree of accessibility to the first product received at the facility; therefore, it usually requires a large storage area in your facility.

Pick Position

Fixed (Permanent) SKU Pick Position. In this method the SKU is placed into a permanent pick position. All reserve cartons or pallets are placed in random reserve positions. When the SKU pick position is depleted of product, this method has a reserve

SKU pallet load, and cartons or single items are transferred from the random storage position to the fixed SKU pick position.

Variable (Floating or Random) Pick Position.   This method allocates a predetermined number (minimum of two) new storage units of a SKU to the order-pick area. These two storage units are placed into two different pick positions. When the first pick position becomes depleted of product, the computer transfers the print of new pick tickets from the first pick position to the second pick position that has the second storage unit of the SKU.

Comparison of Fixed- and Variable-Pick-Position Methods.   In most warehouse and distribution operations SKU storage positions are random (variable) positions. Comparison reveals that the fixed (slot) method requires additional labor to make replenishments and the SKU pick area requires a smaller area, which means improved employee order-pick productivity. The floating-slot method requires less labor to make replenishments, but the pick area requires an additional 20 to 25 percent pick positions, which means a larger pick area and lower employee pick productivity but a potential of fewer "stockouts."

Delivery Truck Loading Concept

The next major design factor is the outbound delivery truck loading concept. The two alternative loading concepts are immediate [direct (fluid)] concept and the stage-and-hold concept.

Direct (Fluid) Loading Concept.   The direct or fluid truck loading concept requires an employee with a pallet load handling vehicle or cartons on an extendable conveyor to transport the customer's order from the pick-pack area directly onto a delivery truck.

Stage-and-Hold Concept.   The stage-and-hold concept requires that your customer's orders are placed into a temporary holding (staging) area for consolidation and for later transfer or placement onto your customer's delivery truck.

Comparison of the Two Loading Concepts.   When we compare these concepts, the direct loading concept requires a smaller loading dock, which means a smaller building. When the concept is considered for an operation that has a carton conveyorized direct load option, then there is sufficient dock area to handle nonconveyables. In a temporary hold and manual, powered vehicle, or automatic truck loading operation, the dock space behind the dock door accommodates one truckload. The temporary hold concept requires a larger building because it requires (1) an additional temporary holding area for customer orders and (2) an aisle and transportation system to transfer cartons or pallet loads from the holding area to the loading area.

BASES FOR WAREHOUSE LAYOUT

A warehouse or distribution facility layout benefit to your company is determined by its ability to satisfy your company's warehouse objectives. These objectives are to earn a profit and to satisfy your customers. Each warehouse philosophy proposes a warehouse facility layout that includes a material handling concept and equipment and locations for the storage-pick position areas.

VARIOUS WAREHOUSE LAYOUT PHILOSOPHIES

These warehouse layout philosophies are based on the (1) type of SKU handled, (2) SKU popularity or Pareto's law (80/20 rule), (3) travel distance for the transportation vehicle, (4) family grouping, (5) SKU rotation, (6) rack row and aisle direction, (7) aisle length, (8) building height, (9) storage method, (10) storage vehicle, (11) order-pick method, (12) internal transportation method, (13) sortation method, (14) handling of returns and out-of-season product transfers, (15) receiving and shipping dock design, (16) facility construction, (17) building size and shape, and (18) SKU flow pattern.

"Type of SKU Handled" Philosophy

The first facility layout philosophy is based on the type of SKU that is handled in your warehouse operation. The various product types are (1) pallet (unit) load, (2) carton and (3) single item (hanging garment, ticketed or nonticketed; flatwear, ticketed or nonticketed; food or nonfood; aerosol cans and flammable materials).

Pallet Load Warehouse Layout.   A pallet load warehouse facility layout is a function of the pallet load storage method, inventory requirement and pallet load handling equipment requirements. Typically, the facility has a clear ceiling elevation, sufficient dock space for staging loads and floor area for turning aisles. The building is tall (25 to 40 ft) and narrow in a rectangular shape or low and wide in a square or rectangular shape.

In a normal warehouse design that has a receiving and storage area, the square-foot allocation is 20 to 30 percent of the total facility area for receiving and shipping and 70 to 80 percent of the total building for the storage-pick area.

Carton Warehouse Layout.   Two basic areas influence the layout of a carton distribution facility: the reserve (storage) area and the pick (active, primary, or forward) area.

The reserve area handles pallet loads, and the layout philosophy is influenced by the same factors as the pallet load warehouse layout.

Carton distribution facility size and ceiling height are functions of the required SKU pick positions, order-pick concept, and receiving and shipping dock areas. When compared to the pallet load facility, the carton facility does not have the high ceiling height. These buildings have a clear ceiling height of 20 to 25 ft and are square- or rectangle-shaped facilities.

Single-Item Warehouse Layout.   The single-item distribution facility is a more specialized facility. These facilities handle hanging garments or nonhanging products (spare or replacement parts, health and beauty aids, flatwear and cloths, books, and drugs). Both of these SKU types are preticketed merchandise or nonticketed merchandise. When we consider a single-item warehouse operation, the operation requires a reserve (storage) area and a primary pick area.

A facility that handles single items is designed with a clear ceiling of 25 or 36 ft. With this ceiling height, the facility is designed to accept a freestanding or equipment-supported mezzanine level (or levels) in the building area. The building shape is square or rectangular.

SKU Popularity Philosophy or Pareto's Law (80/20 Rule)

When a warehouse-distribution facility layout is based on SKU popularity, then it is based on Pareto's law (after Vilfredo Pareto, 1848–1923; Italian economist). This law states "that 85 percent of the wealth is held by 15 percent of the people." In the warehousing industry, this law indicates that 85 percent of the volume shipped to your customers is derived from 15 percent of the SKUs. Many studies have indicated that another 10 percent of the volume shipped to your customers results from another 30 percent of SKUs and that an additional 5 percent of the volume shipped to your customers is attributed to 55 percent of the SKUs. If you are in the catalog or direct mail business, then 90 to 95 percent of your business is from the 5 percent of your SKUs because two to four catalogs are introduced within a year. Each catalog has a different inventory of SKUs.

In recent studies, the results show that 95 percent of the volume shipped to your customers is obtained from 55 percent of the SKUs. This is referred to as "Pareto's law revisited."

ABC Theory. When warehouse professionals refer to the three zones of Pareto's law, their reference is to the ABC theory (Fig. 3.2). This theory simply states that the A storage-pick zone is allocated to the fast-moving SKUs. These SKUs are few in

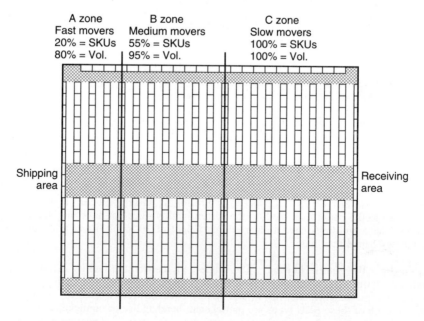

FIGURE 3.2 ABC zone theory.

number and have a large inventory quantity per SKU. The B storage-pick zone is allocated to the normally moving SKUs. These SKUs are medium in number but have a medium inventory quantity per SKU. The C storage-pick zone is allocated to the slow-moving SKUs. These SKUs are large in number and have a small inventory quantity per SKU.

If, in a warehouse layout, the receiving and shipping docks are on the front side of the building and a SKU storage-pick location is based on the ABC theory, then it locates the fast-moving SKUs at the front of the warehouse. If the receiving and shipping docks are located on opposite sides of the building, then the fast-moving SKUs are located by its unloading/loading ratio.

Unloading/Loading Ratio.   The unloading/loading ratio compares the number of trips that the unloading employees and loading employees require to handle a truckload of pallets. When the number of employee unloading trips equals the number of employee loading trips, then the SKU pallet loads are located near the shipping docks or any location in the rack row. When the employee unloading trips are more numerous than the employee loading trips, then the SKU pallet loads are located near the receiving docks. This feature reduces the employee total travel distance.

Power (Fast-Moving SKUs) in One Pick Area.   Power (fast-moving SKUs) in one pick area or zone philosophy has an inventory (SKU) allocation program that locates all fast-moving SKUs into pick positions. These pick positions are adjacent to one another. This philosophy has all your promotional, seasonal, special sale, and fast-moving SKUs in one zone. This SKU arrangement increases your order picker's hit concentration (number of hits per aisle) and hit density (number of hits per SKU). A high hit concentration and hit density means high order-picker and replenishment employee productivity due to a very short travel distance between two pick positions. However, the key to an accurate, efficient, and on-time warehouse operation is to have the replenishment merchandise available for replenishment of the pick position and a completed customer order take-away system.

Mobile Warehouse Equipment Travel Distance Philosophy

The next warehouse layout philosophy is a layout that is determined by the mobile warehouse equipment travel distances. This philosophy attempts to keep the transportation distance between two key warehouse functions as short as possible, thereby minimizing the operating costs. With this philosophy, the majority of the warehouse aisles are arranged to allow the mobile warehouse equipment to travel between the shipping docks and storage locations. Multiple load handling vehicles and automatic guided transportation vehicles (AGVs) permit an increase in the travel distance between two locations or decreases the transportation operating cost.

Inventory Characteristics Principle

The next distribution facility layout principle is to consider your design year operation projected product, SKU quantity, and inventory quantity. This principle requires your design team to establish the base-year average and peak volumes (current) and SKU inventory growth rate. The calculation determines your company's design-year SKU pick positions and inventory storage positions that are allocated to your warehouse aisles. This requires your design team to (1) identify the product flow sequence; (2) calculate the low, average, and peak SKU throughput volumes for each function (workstation)—this includes the projected impact of a JIT (just-in-time) replenishment or across-the-dock product handling program; (3) project the low, average, and peak inventory levels per handling units; (4) identify and project SKU quantity by type of packaging material, shape, whether fragile or durable, whether edible or nonedible, whether flammable paint under pressure, whether crushable, size or dimensions (length × width × height), weight (minimum, average, and maximum), and ambient storage temperature environment; (5) identify and project the frequency of product movement between the key warehouse functions and provide sufficient accumulation prior to the workstation (this answers the questions of what, when, where, how, and how much material is handled at each location); and (6) identify the material handling equipment that is used in each facility location.

"Adequate Clear Ceiling Height" Principle

The next distribution facility layout principle is to provide adequate ceiling height in the facility. The principle of clear ceiling height is to provide the required clear dimension from the finished floor surface to the lowest ceiling obstruction (joist, duct, or water pipe). This clearance has sufficient space to maximize the SKU storage stacking in the facility, allow material handling equipment movement through the facility, and permit the storage-pick vehicle to make transactions at the highest storage positions. This adequate ceiling height includes considerations for (1) a future mezzanine, (2) proper fire protection, and (3) sufficient clearance for equipment overhead guards or masts.

"Provide Proper Fire Protection" Principle

This layout design principle is to provide the required space and location for fire protection equipment. This principle requires your distribution facility design team to have knowledge of the building code requirements. These are available at the local governmental agencies, Publication NFP-231-C, and insurance underwriter publications. The best resource is to involve your company's architect, security department, insurance underwriter, local fire authorities, and building inspector. These organizations and individuals provide the (1) density and type of ceiling sprinklers, (2) location of sprinkler risers and pipes, (3) deluges or vestibules for fire wall or floor openings, (4) doors and types of windows in the door, (5) emergency lights, (6) in-rack in-sprinkler levels, (7) number and type of exits, (8) fire baffles, (9) vaults and enclosures, (10) aisle widths, (11) wire-mesh areas, (12) locks, (13) fire doors, (14) fire barriers, (15) fire walls, (16) windows, (17) exterior fencing, (18) pump station(s), (19) smoke detectors, and (20) fire pond or reserve tank.

Early Suppression, Fast Response (ESFR). The ESFR fire sprinkler system was introduced in the early 1990s. When compared to the conventional wet sprinkler system, which has multiple levels, it consists of one wet sprinkler level at the ceiling. These sprinklers have very heat-sensitive sprinkler heads that are under greater water pressure. The combination means that the sprinkler system has a greater probability to suppress a fire in its earliest stage.

Prior to the implementation of an ESFR sprinkler system in your new construction, you are required to have the sprinkler system and storage area design approved by your local authorities and your insurance underwriter. If you consider ESFR, then your storage area has warehouse ceiling height restrictions, flue space requirements (space between two unit loads) with hand-stack storage open-mesh decking, and ceiling fire baffles.

When compared to the conventional building and sprinkler system, an ESFR system means that you have an opportunity to (1) reduce construction costs from fewer regular sprinkler heads, piping, and lower building height; (2) lower sprinkler system and building cost; (3) possibly add an additional storage level to increase the number of storage units per square foot; (4) with fewer sprinkler heads, reduce the risk of an accidental discharge of sprinkler heads, which could cause product damage; and (5) reduce the time needed for sprinkler head and piping installation, which means an earlier startup.

Principle of Adequate Electric Outlets

The following principle is to provide adequate electrical outlets in your facility. These basic design requirements are (1) ensure that light and wire raceways do not interfere with the material handling equipment travel paths, (2) provide in each warehouse function area the minimum required light level (footcandles) at a specific elevation above the finished floor, (3) provide sufficient electrical outlets on columns or outlets strips hung from the ceiling for operation of processing electrical or maintenance equipment, (4) provide a clean electric (dedicated) line to the computer equipment or fixed position scanners, (5) locate the material handling equipment battery recharging area in an area with excellent ventilation, (6) allow for future additional equipment, (7) avoid installing lights that would blind the material handling equipment operators, (8) provide an uninterruptible power source (UPS) for computer equipment, and (9) locate all equipment control panels at least 10 ft from electrical panels. In new construction all electrical panels provide a 12-ft lead wire to reach the material handling equipment control panel.

Principle of Adequate Offices and Administrative Activities

The next principle of a warehouse layout is to locate administrative offices or warehouse support activities in the appropriate area. These are the indirect activities that are performed within the distribution facility and are related to your operation or employees. The completion of these activities ensure that the operation has the ability to move product and information within the facility. They ensure that the product transaction was accurate and that the update of records was on-line. Also, it ensures that the employee amenities are within easy access, with a smooth and even flow of employees between the two locations. Some of these warehouse activities include (1) maintenance tasks such as battery charging and changing, (2) tailoring and labeling special operations, (3) customer service and pickup, (4) trash handling, (5) outlet store, (6) sanitation equipment storage, (7) first aid, (8) security department, (9) conference room, (10) administrative offices, (11) salvage area, (12) returns, (13) computer system, (14) quality control, (15) trucker lounge, and (16) restrooms.

"Proper Building Column and Bay Size" Principle

The next warehouse layout principle is to obtain the proper spacing for the building columns and building bay sizes. The correct design of these building structural items assure excellent space utilization, product flow, employee productivity, minimal product damage, and future expansion.

With new construction for best results, the building is fit around the material handling system and the storage and pick positions. This means that the building columns fit in the flue space which is between the back-to-back rack positions. Another important factor is to ensure that the floor design and specifications (including wire mesh or re-bar depth) meet the rack and mobile material handling equipment requirements. If the column spacing varies, then the exterior wall column spacing is the preferred column spacing for the variance, because this area is your dock area or wide-turning aisles (open space).

"Use Gravity Transportation" Principle

The next warehouse layout principle is to use gravity transportation methods between two levels whenever possible as an internal material handling concept. In conjunction with gravity systems, your warehouse design team considers mechanical and automated material handling equipment. This combination of equipment is designed to handle the design throughput volume and SKU types. These considerations provide the lowest handling cost per unit.

SITE UTILIZATION

After the distribution facility design team has designed and sized the facility, then your team locates the distribution facility on the site. The building location on the site provides delivery vehicles and personnel easy access to the facility, allows future building expansion, and agrees with your company's building exposure to the public highway. Other important site factors include entrance to the yard, security, signage, truck maneuvering area, parking facilities for automobiles and the delivery vehicles, truck maintenance facility, landscaping, security fencing and berm, railcar access, and appearance to the local community.

EQUIPMENT LAYOUT PLANNING FOR AN EXISTING BUILDING IS A COMPLEX TASK

The distribution facility equipment layout design for an existing building is a more complex task because the rack and material handling equipment is required to fit into the building. An existing building has several constraints to your equipment layout. Some of these constraints are (1) building column spacing and size; (2) shape of shell; (3) bay direction; (4) clear ceiling height; (5) door height and location; (6) floor condition; (7) dock height and truck-yard location; (8) office and support areas; (9) availability and location of electric power, water, and other utilities; and (10) existing obstacles (pipes, ducts, and heaters).

During the equipment layout process, building columns fit in the space between the back-to-back rack sections. With some rack layouts for existing buildings, the building columns fall in the rack bay. When this situation occurs, this arrangement minimizes the loss of storage space, but it provides a rack and aisle layout with clear aisles that maximizes your employee productivity.

When your team is required to develop a distribution facility layout for a new building, then the task is less complex because the building (columns, bay spacing, shape of the shell, ceiling height, and walls are variables) fits around the material handling equipment.

VARIOUS WAREHOUSE LAYOUT PRINCIPLES

There are numerous warehouse and distribution facility layout philosophies that determine how your facility operates and looks. These layout philosophies are used for a layout in an existing or new building.

To design a warehouse-distribution facility, the designer is required to understand a company's warehouse and distribution business, operation, and product flow through the facility. If the designer does not completely understand your business, then the proposed warehouse or distribution operation does not have the greatest potential to optimize the return on investment and does not satisfy your company warehouse objectives. To understand your warehouse and distribution business, the designer must (1) visit your facility, (2) observe your warehouse and distribution operations, (3) trace the product and information flows, (4) observe your product handling methods and equipment applications, (5) interview your managers and employees, (6) review all past written reports, (7) obtain your latest annual operational statistics, and (8) obtain all your building and equipment layout drawings.

Family Group Philosophy

The fourth warehouse layout philosophy [in addition to type of SKU handled, SKU popularity (Pareto's law), and mobile warehouse equipment travel distances] is a layout that is dictated by your company's requirement that the SKUs are sorted by family group. With this philosophy, by a predetermined criterion, the SKUs are assigned to specific locations (areas) within a warehouse. This layout philosophy requires that the warehouse facility and material handling concept be designed to accommodate the SKUs that (1) have similar dimensions, weight, and SKU components; (2) have components for the same end product; (3) are located in the same aisle in the retail store; (4) require normal, refrigerated, or freezer conditions; (5) require high security; (6) include hanging wear (short and long); (7) are for toxic or nontoxic materials; (8) include shoes (one style, all sizes, or musical); (9) include edible or nonedible substances; (10) include flammable or nonflammable materials; (11) include flatwear (one style and color with all sizes); and (12) include stackable or nonstackable merchandise.

Product Rotation Philosophy

The next warehouse layout philosophy is based on the required SKU (product) rotation. Two SKU rotation methods are FIFO and LIFO rotation.

FIFO Product Rotation.   In FIFO SKU rotation the SKU that is received first in the warehouse is shipped out first from the warehouse. This indicates that the product has a predetermined life (time limit) before it spoils. After a specific date, the SKU is not withdrawn from the inventory for customer orders. A warehouse layout that is designed to store SKUs with a FIFO rotation provides access to all pallet load positions in the storage-pick area and assures that the oldest product is withdrawn first from the storage-pick position.

If there are many SKUs and few units of storage (pallet loads or cartons) for each SKU, then standard one-deep storage-pick racks or shelves are used in the warehouse. This storage concept utilizes a large-square-foot area. When there are few SKUs and many units of storage (pallet loads or cartons) for each SKU, then the flow through storage-pick racks are used in the warehouse. When compared to the single-deep rack warehouse design, this facility requires a smaller square-foot area.

LIFO Product Rotation.   In LIFO SKU rotation the unit of storage (pallet loads or cartons) that is last received in the warehouse is shipped out first from the warehouse. This type of product does not have a specific shelf life. The warehouse design does not provide access to the oldest unit of storage. This feature allows the warehouse layout to use dense storage concepts that reduce the building square-foot area.

"Rack Row and Aisle Direction of Flow" Philosophy

The next warehouse layout philosophy is based on the rack rows and aisles direction of flow to the shipping docks. The two philosophies are parallel to the receiving and shipping docks and straight to the receiving and shipping docks.

Parallel to the Docks.   In the first alternative the rack row and aisle direction of flow are parallel to the receiving and shipping docks (Fig. 3.3*a*). With the rack row and aisle direction of flow in this arrangement, at least two turning aisles at the end of the rows and a middle traffic aisle are required. These aisles lead to the receiving and shipping docks. This arrangement increases the vehicle travel time between the receiving and shipping docks and storage-pick areas.

Straight to the Receiving and Shipping Docks.   In the second rack row and aisle flow layout the rack rows and aisle direction of flow is straight from the storage-pick areas to the inbound and outbound dock areas (Fig. 3.3*b*). In this design each aisle provides access to the receiving and shipping dock areas and the main traffic aisle serves as a vehicle turning aisle. When a conventional lift truck is used in the operation and the driver performs dual cycles, then with these long aisles there is a middle cross aisle. This middle cross aisle assures good lift truck productivity to perform a transaction for a SKU that is in a different aisle. To transfer aisles, most aisle-guided vehicles must travel the full length of the aisle to the front or rear aisle exit. The number of transactions per hour of the aisle-guided lift truck are equal to or greater than those of the conventional lift truck because of a greater number of unit loads per square foot, longer travel distances at high speeds, and higher transaction speeds. With

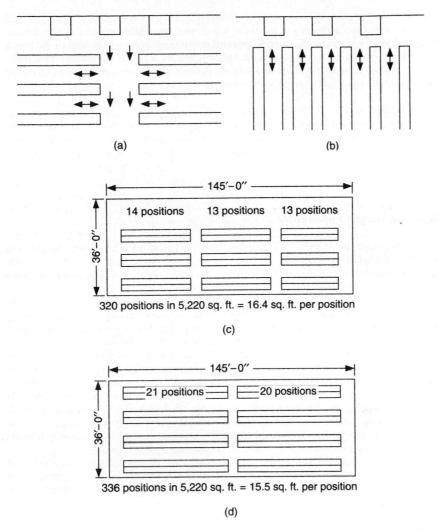

FIGURE 3.3 (a) Aisles and rows parallel to the docks; (b) aisles and rows straight to the docks; (c) short aisle length; (d) long aisle length.

multiple truck doors and proper delivery truck spotting, the straight-to-the-dock design reduces lift truck travel distances.

Aisle Length Philosophy

The next warehouse layout philosophy involves a layout that is based on the storage-pick aisle length. The two philosophies are the short-aisle concept and the long-aisle concept.

Short Aisle Length.   The short-aisle philosophy (Fig. 3.3c) specifies that the rack and shelf rows and aisles run in the short warehouse dimension or short width of a rectangle-shaped warehouse. This concept requires turning aisles at the end of each rack row. When compared to the long aisle concept and on a SKUs-per-square-foot basis, the short-aisle concept provides less storage density and lower employee productivity because of the increased nonproductive end-of-aisle turns.

Long Aisle Length.   With the long-aisle philosophy (Fig. 3.3d) the rack rows and aisles are arranged to flow in the long direction of the warehouse or rectangle. This long-aisle concept does have a cross aisle in the middle of each rack row to provide easy and quick access to other warehouse aisles. The long aisle concept does provide greater storage density and fewer nonproductive employee turning aisles.

Building Height Philosophy

The next distribution facility layout philosophy is based on the building height. The height of the building is determined by the economics (costs), available land, the local codes ("shadow" laws), and seismic, wind, and land conditions. A high-bay building is a building that has the roof at least 40 ft. high above the ground level.

An alternative building height is the medium-bay building of 30 to 40 ft height. This building is designed as a rack-supported facility, conventional storage building, multilevel building, or a building with one or two freestanding equipment-supported mezzanine structures. With minimal sprinkler and limited column fire protection, a standard constructed or equipment-supported mezzanine is allowed to occupy one-third of the building first-floor square-foot area. When your facility design exceeds the one-third criterion, then the local building code and insurance underwriter approves the design, which could require a bar-grated floor or additional fire protection. The multilevel building or a building with a mezzanine is designed for a flatwear or small-item distribution operation with powered vertical transportation concepts to move product to the second level. Powered or gravity transport concepts are used to transfer product from the mezzanine level to the lower level.

The low-bay building is no more than 30 ft high with a manually operated lift truck operation that handles unit loads or has a carton or single-item operation that does not require a large inventory. This building has the height capacity for a structural or free-standing mezzanine.

Storage Method Philosophy

This facility layout philosophy is based on the type of storage method. The type of storage method is determined by the number of SKUs and the number of storage units per SKU (cartons or pallet loads), inventory requirement, and the required product rotation. The storage method is a major factor determining the square footage of the building.

Single-Deep Storage Concept.   In this building (Fig. 3.4a), the single-deep pallet storage racks are serviced by counterbalanced or straddle lift trucks. If the cartons or single items are hand-stacked into the rack positions, then an order-picker truck is used in the facility. If the building has a clear ceiling height of at least 40 ft and has an

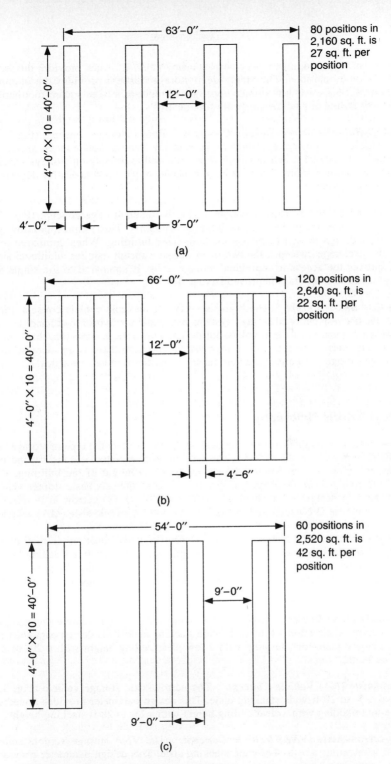

FIGURE 3.4 (a) Single-deep rack arrangement; (b) dense storage rack arrangement; (c) two-deep rack arrangement.

3.23

aisle guidance system, then a very-narrow-aisle (VNA) lift truck provides the service to the storage positions. The single-deep storage concept provides the maximum access to all SKUs, has low storage density, and requires a large-square-foot building with a low ceiling or a tall (high)-cubic-foot facility.

Low-Bay-Building Dense Storage Concepts.   Dense storage concepts (Fig. 3.4*b*); floor stack, drive-in or drive-through racks, stacking frames, mobile racks, gravity-air-flow rack, car-in racks, and dynamic flow rails with garments on trolleys require a small-square-foot building. These storage concepts require few aisles for deposit and withdrawal of product.

Two (Double)-Deep Storage Concept.   The two-deep storage concept (Fig. 3.4*c*) applies for pallet loads, cartons, or hanging garments. These concepts allow your warehouse design person to plan a medium-sized building. When compared to the other dense storage concepts, the two-deep storage concept requires additional aisles, but improves the access to individual storage units. If compared to the single-deep storage concept, it provides improved storage density.

High-Rise Rack Concept.   The high-rise very-narrow-aisle vehicle concept utilizes one of the above-mentioned storage systems but requires a hybrid load handling vehicle. When compared to the low-bay-building dense storage concepts, the high-bay concept provides an increase in storage density because there is an increase in the number of storage positions in the vertical stack and the very narrow aisles.

Storage Vehicle Philosophy

The next distribution facility planning philosophy is based on the type of storage vehicle that is planned for the facility. The type of storage vehicle is the second most important factor that determines the square foot and height of the building. If we include the manual (hand) carrying of product concept, then six basic storage vehicles affect building design: (1) manual, (2) wide-aisle (WA), (3) narrow-aisle (NA), (4) very-narrow-aisle (VNA), (5) captive aisle (CA), and (6) mobile aisle (MA) vehicle.

Manual Concept.   The manual concept is used to handle most single-item product types, but requires the greatest number of employees and restrictive stacking heights. These facts make this alternative very costly to apply and require the use of a large building. These features mean a high investment in land and construction and annual labor expense.

Wide-Aisle (WA) Vehicle Concept.   The wide-aisle storage vehicle concept (Fig. 3.5*a*) requires a minimum 10- to 13-ft-wide stacking aisle. This design parameter provides a large-square-foot building with a low clear ceiling height with a 20- to 25-ft stacking height.

Narrow-Aisle (NA) Vehicle Concept.   The narrow-aisle storage vehicle (Fig. 3.5*b*) requires a 7- to 10-ft-wide stacking aisle. This design parameter provides a medium-square-foot building with a clear ceiling height with a 25- to 28-ft stacking height.

Very-Narrow-Aisle (VNA) Vehicle Concept.   The VNA storage vehicle concept (Fig. 3.5*c*) requires a 5- to 8-ft-wide stacking aisle. This design parameter provides a

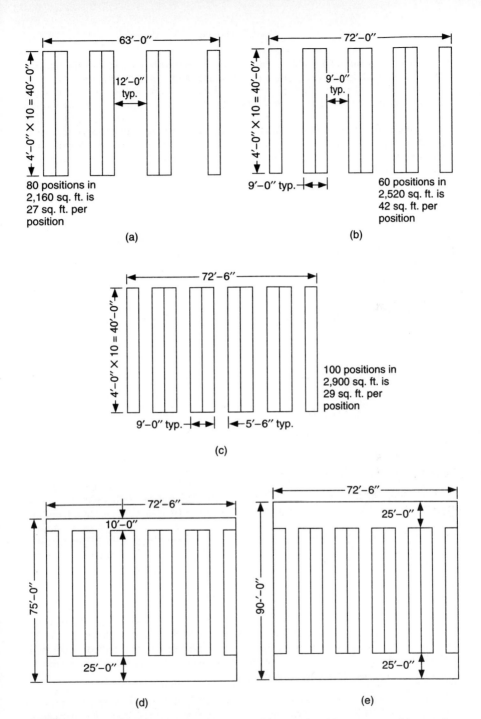

FIGURE 3.5 (*a*) Wide-aisle operation; (*b*) narrow-aisle operation; (*c*) very-narrow-aisle operation; (*d*) captive-aisle lift truck operation; (*e*) mobile-aisle lift truck operation.

small-square-foot building with a clear ceiling height with at least a 40-ft stacking height. Some buildings are designed with a stacking height of 80 ft or above.

Captive-Aisle (CA) Vehicle Concept.   With the captive-aisle storage vehicle concept (Fig. 3.5*d*) the storage vehicle remains in one storage aisle. This design feature reduces the turning aisle requirements, thus reducing the building square-foot size. But the captive-aisle design requires additional storage equipment investment, thus increasing the total facility investment cost.

Mobile-Aisle (MA) Vehicle Concept.   According to the mobile-aisle storage vehicle concept (Fig. 3.5*e*), the storage vehicle enters and exits through any of the storage aisles. This turning aisle or transfer aisle requirement results in a larger-square-foot building. But the warehouse system requires fewer pieces of storage equipment, thereby reducing the total warehouse investment cost. Some VNA storage vehicles require a transfer car (T-car) that moves the storage vehicle between aisles.

Customer Order-Pick Philosophy

The next facility layout philosophy is based on the type of order-pick (selection) method. In a pallet load operation, the vehicle stacking height and the right-angle turn (stacking) requirement are factors that determine the building height and square-foot requirement. The type of carton or single-item order-pick system determines the square-footage and height of the building. The other order-pick design factors include (1) type of order-pick concept and pick positions, (2) inventory in the pick position and number of SKUs, and (3) order-picker routing pattern.

The three order-pick concepts are (1) manual concept, (2) mechanized concept, and (3) automated concept.

Manual Order-Pick Concept.   The manual order-pick system requires all pick positions at a maximum elevation of 5 ft 6 in to 6 ft high above the finished floor. In a carton or single-item facility, this permits a maximum height of two pallet loads or four or five shelves of hand-stacked cartons or carton flow rack levels. In a hanging-garment facility, the manual order-pick system requires one long garment storage level at 56 in height and two short garment storage levels high at 36 in per garment level. If additional levels are designed per floor or in a multilevel structure, then the garments hang into a well or the aisle is raised on a platform above the finished floor. These designs permit two-high long-garment or three-high short-garment storage-pick levels.

The manual order-pick system requires a wide vehicle aisle since employees walk or ride a vehicle through the warehouse aisle. With these travel path parameters, the system requires the greatest number of order pickers, thus increasing the required support functions square footage. With these two design factors, the manual order-pick building has the largest square footage. A high-rise order-picker truck utilizes the air space and does reduce the building size.

Mechanized Order-Pick Concept.   The mechanized order-pick concept requires a medium-sized facility because it utilizes a conveyor system and permits construction of mezzanines with additional pick levels. This two- or three-level mezzanine feature and pick position replenishment that is performed with a narrow aisle vehicle increases the number of pick positions per square foot of building. Since the SKUs are separated into the mechanized (conveyable SKUs) section and the manual (nonconveyable SKUs), this concept requires a medium-square-foot floor area and a medium number of order pickers.

Automated Order-Pick Concept.   The automated order-pick concept requires a small-square-foot facility because the order-pick positions are narrow and long and there are multiple levels in a stack. These pick positions are replenished by a narrow-aisle vehicle with rack positions that are directly behind the pick area or a conveyor system. The facility has a manual order-pick section for nonconveyable SKUs. But the automated section contributes to a small-square-footprint building.

Order Sortation Philosophy

The next facility layout philosophy is used in a carton or single-item distribution facility. The type of customer order sortation method has a minor influence on the facility square footage and height. The factors that determine the size of the sortation area are the number of customers, size of customer order(s), and type of sortation concept.

The two alternative sortation concepts are the manual and the mechanical sortation concepts.

Manual Sortation Concept.   The manual sortation concept handles a low volume of product and a low number of customer orders but requires the greatest number of employees and the largest square-foot area.

Mechanical Sortation Concept.   The mechanical sortation system handles a high volume of product and a large number of customer orders but requires the fewest number of employees and a medium-square-foot area.

Internal Transportation Philosophy

The next distribution facility layout philosophy is based on the internal transportation concept, which moves product between two locations. This warehouse function design has a small impact on the square footage and height of the facility. The two basic internal transportation concepts are the horizontal and vertical product movement requirements within the facility.

The horizontal transportation concept requires a clear path, unloading and loading spurs, and 90 or 180° turning areas.

The vertical transportation concept requires an area for incline and decline conveyor paths, runouts, and a clear path between the two levels of the facility. If elevators or vertical lifts are used in the facility, then the upward runouts in the roof or floor and pit in the floor are considered in the design.

In both the horizontal and vertical transportation concepts, the travel path with loading and unloading bypass spurs and queue (accumulative linear feet) areas are key design considerations for an efficient operation.

Receiving and Shipping Philosophy

The next warehouse layout philosophy is the receiving and shipping dock (area) design. This design has an impact on the facility square footage and height. The factors that determine the area size are the location of docks, type of receiving and shipping operation, method used to perform the function, and SKU flow pattern through the facility.

The first consideration is to locate the receiving and shipping dock functions along the facility wall. The alternatives are in the middle or at the end of the facility.

The three "middle of the facility" concepts are to drive through the entire facility, have outside docks, or use an elevator.

The first alternative at the end of the facility is the truck and rail receiving and shipping functions on one wall. The alternative is the rail receiving and shipping functions on one wall and the truck receiving and shipping functions on a separate wall.

The next consideration is whether to use separate or combined receiving and shipping function locations. The separate receiving-shipping area concept requires a larger square-foot area, but is considered more efficient for a high-volume operation. Combined receiving and shipping function areas require a small area and are preferred for a low-volume operation.

The next consideration is the receiving and shipping handling method. The three alternatives are the manual (hand-carry), mechanical (carts, pallets, containers, and slip sheets with a mobile manual or powered warehouse vehicle or carton conveyor), or automated methods (pallets or dumping).

The other receiving-shipping design factor is the SKU flow pattern through the facility; alternative flow patterns are the direct (fluid) unloading-loading concept, which requires a small-square-foot building area. The second concept is the stage (temporary hold) concept, which requires floor or rack staging. This alternative requires a large-square-foot building size.

Dock design influences building size. The aspect of an enclosed, open, sawtooth, flush dock is a consideration for the receiving and shipping areas and impacts the building size. When sawtooth docks are compared to flush docks, the sawtooth dock requires the smallest truck-yard area but provides the fewest docks per linear foot of building. When the enclosed dock is compared to the open dock, the former requires the largest building with shelter from weather conditions but the latter does not provide such shelter.

Building Construction Philosophy

The next layout philosophy is determined by the architectural and construction design of the building. These architectural factors include the building exterior material ("skin"), column size, bay spacing and direction, floor type, roof type, and interior walls. The building material alternatives are (1) conventional (brick or concrete), (2) air-supported, (3) tilt-up (metal or concrete), (4) underground, (5) rack-supported, (6) rib panel, (7) wood, and (8) open air.

The rack-supported facility storage supports the roof, walls, and provides a greater number of storage positions per square foot than does the conventional building. When the rib panel or tilt-up building is used in the building, then the racks along the exterior walls fit around the building columns and project inward, thus reducing the number of storage positions.

The wood structure is a low-bay building, but typically the wood structural members do not support a conveyor system.

The air-supported building (Fig. 3.6) has a treated fabric cover supported by metal ribs and forced air. The building is quickly constructed, requires a slab, accommodates most storage rack concepts, and handles all product types. This type of facility handles product that requires separated storage or special storage conditions.

The underground warehouse is a storage system that is installed in natural or human-made caves or caverns. These caverns have a flat floor, rack storage concept, and naturally controlled temperature. Some require a vertical transport system to deliver product between the delivery and the storage areas.

Facility Shape Philosophy

The next warehouse layout philosophy is the facility shape. The facility shape has a significant impact on the arrangement of the material handling equipment, product flow, and future expansion capabilities.

The major factors that determine the building shape are (1) land shape and size and existing building; (2) inventory level, product characteristics, and pick positions; (3) product (merchandise) flow pattern; and (4) type of operation or functions performed in the facility.

FIGURE 3.6   Air-supported warehouse facility. (*Courtesy of Rubb Inc.*)

The following are alternative building configurations: (1) square-shaped, (2) L-shaped, (3) rectangle-shaped, (4) U-shaped, (5) oversized-rectangle-shaped, (6) round (circle-shaped), and (7) triangle-shaped.

Square-Shaped Building Concept.   The square-shaped building provides the best balance between wall area and floor area. This type of building design has a low wall area/floor space ratio. It creates an efficient structure for internal vehicle transportation of loads from the receiving and shipping areas to all storage and pick areas. The square building concept requires a good balance between the number of SKUs and number of storage units per SKU. It is considered a good facility shape for a remote warehouse operation because that product is in storage for a long period of time. It is normal as the warehouse operation expands that the square shape becomes a rectangle-shaped building.

Rectangle-Shaped Building Concept

The rectangle-shaped building provides an increase in wall square footage/floor square footage ratio. This type of structure provides the additional wall space for an increase in the number of dock doors. This large wall area feature permits efficient internal transportation from the dock areas to all storage-pick areas within the warehouse. The concept allows the facility to handle an inventory storage requirement that has any type of mix (number of SKUs to storage units). It is an excellent shape for a distribution facility that provides service to industrial, catalog, direct mail, and retail department store customers. As the warehouse business increases, the expansion occurs at the ends (short walls) of the rectangle-shaped building. After this expansion, an oversized-rectangle-shaped building results, with a length much greater than its width.

Oversized-Rectangle-Shaped Building Concept.   The oversized-rectangle-shaped building design has the maximum proportion of wall square foot area to floor square-foot area. This wall area feature provides the most efficient (direct) transportation path from the receiving docks to all warehouse storage areas. The oversized rectangle concept allows the facility to handle an across-the-dock product flow pattern or a break-bulk operation (freight terminal). This building shape is not recommended for a storage warehouse.

L-Shaped Building Concept.   In the L-shaped warehouse facility, the receiving and shipping docks are located at the base with the storage and distribution functions in the stem. This building shape creates an increase in the transportation costs of product between the dock areas and the other warehouse functions. This building shape is best for a processing facility that has an attached warehouse function.

U-Shaped Building Concept.   In the U-shaped warehouse facility, the shipping function is on one of the U legs and the receiving function on the other leg. The storage, distribution, and manufacturing functions are located in the remaining area between the two legs or the base of the U. The building concept increases the transportation activity and is preferred for a manufacturing facility.

Round (Circle) Shaped Building Concept.   In the round warehouse facility, the receiving and shipping functions are located along the walls or in the center of the building. The other key warehouse activities are located in the interior of the warehouse. This building concept provides an excellent transportation travel path to any interior storage location. The round building with square-shaped unit loads does not provide the best space utilization; but is an excellent building for the storage of bulk goods such as grains.

Triangle-Shaped Building Concept.   In the triangle facility the receiving and shipping functions are located on the triangle base and the other key warehouse functions and manufacturing functions in between the triangle legs. With the two legs connected at the tip, the shape does not provide the maximum space utilization with a square-or rectangle-shaped unit loads that have straight sides.

Product Flow Pattern Philosophy

The next major warehouse layout philosophy is based on the SKU (product) flow pattern through the facility. The product flow pattern has an impact on the various key warehouse function locations and productivity of the total warehouse. The two basic product flow patterns have specific product entrance and exit locations.
    The various product flow patterns are the one-way (straight) and two-way flow patterns. The two-way flow pattern has two patterns which are the U flow pattern and W flow pattern.

One-Way (Straight) Flow Pattern.   This flow pattern through the facility is also referred to as "in one side and out the other side." In this pattern the product enters the facility from one side or from the top and exits the facility from the opposite side or from the bottom. This concept requires the product to travel the entire distance between the receiving and shipping areas. This flow pattern does increase the storage-related transportation cost to move product through the facility because the operator cannot perform dual-cycle storage transactions. The pattern is preferred for a facility that handles across-the-dock product flow or product for a processing operation. On one side of the building, the operation receives palletized SKUs with mobile equipment and on the other side of the building ships SKUs as pallet loads, cartons, or single items. Examples of these warehouse and distribution operations are a freight terminal, a catalog or direct mail company, a retail store distribution across-the-dock operation, or a manufacturing JIT (just-in-time) replenishment operation.

Two-Way Flow Pattern.   In the two-way flow pattern through the facility, the product enters the warehouse from one side and exits on the same side of the warehouse. This is an excellent arrangement for a small-item, carton, or pallet load pick operation. The two-way flow pattern improves internal transportation productivity because employees make dual-cycle trips from the shipping and receiving areas and to the storage locations. When compared to the one-way product flow pattern warehouse, this facility requires less truck-yard and roadway surface, which means a reduction in land and other investment costs.

U Flow Pattern.   One of the two-way product flow patterns is the U pattern. In this pattern the inbound product is unloaded (received) on one side (right side) of the facility, transported to the storage and pick areas in the middle of the building, and loaded (shipped) on the same side (left side) of the facility. Therefore, the product movement makes a U pattern through the facility.

W Flow Pattern.   Another two-way product flow pattern is the W or double-U pattern. In this pattern all the inbound product is unloaded (received) in the middle of the building (one side), transported to the storage and distribution areas, and loaded (shipped) on the left and right side of the facility on the same side. This product movement creates a W pattern through the facility.

Material Handling System Philosophy

The next warehouse layout philosophy is a warehouse layout based on the type of material handling concept. The determining factors of the material handling concept are the SKU characteristics and numbers, throughput volume, and inventory volume.

The three major SKU characteristics are (1) pallet load, (2) carton (case), and (3) single-item.

Pallet Load Warehouse.   The pallet load warehouse has three alternative concepts that dictate the building layout. The first is the warehouse that uses a wide-aisle equipment in the operation. The wide-aisle equipment is manually operated in an aisle that has 10- to 13-ft-wide aisles with a stacking height of 20 ft. This wide aisle develops in a building in the shape of a rectangle or oversized rectangle. This wide building has aisles and rack rows that are 150 to 200 ft in length with turning aisles at both ends and a middle traffic aisle. These features create a large-square-foot building.

In the pallet load warehouse that uses narrow-aisle equipment in the operation, the equipment operates in a 7- to 10-ft-wide aisle. The aisle length is 150 to 200 ft with two turning aisles at the ends and a middle traffic aisle. The stacking height is up to 28 ft in a medium-square-foot warehouse.

In the pallet warehouse that has very-narrow-aisle pallet load handling equipment, the equipment operates in an aisle that is 5 to 7 ft wide and 200 to 300 ft long with a stacking height of 40 to 80 ft. These vehicles are moved or transferred under their own power or assisted by a transfer car (T-car) from one aisle to another. This transfer is performed in wide aisles at the ends of the rack rows. All these vehicles require end of aisle pickup and delivery stations (P/D). These P/D stations are conveyor or rack positions. Typically these facilities are narrow, long, and high buildings with small-square-foot area.

TYPES OF WAREHOUSE OPERATION

Warehouse Operation That Handles Cartons

A warehouse operation that handles cartons (cases) has different material handling concept configurations or designs that determine the building shape. The three basic order-pick concepts are the manual, mechanized, or automated concepts.

The warehouse layout that is designed for a manual order-pick concept has an aisle between two rack rows. The aisles are at least 10 to 13 ft wide. This aisle width allows two-way powered truck traffic. Typically, the SKU pick positions are two levels high in a pallet rack that has a row length of 75 to 100 ft. These features create a warehouse that has a shape of a large oversized rectangle or a smaller rectangle.

The second type of carton warehouse layout is designed for a mechanized carton order-pick concept. In a mechanized warehouse, the SKUs are separated according to either of two carton order-pick concepts: (1) manual order pick for nonconveyable SKUs and (2) mechanized order-pick concept for conveyable SKUs.

Alternative carton pick concepts are the (1) pick-to-belt, (2) pick car, and (3) cartrac. Characteristics of the pick-to-belt and pick car concepts are similar because the SKU pick positions are placed on mezzanines, the pick aisles are less than 9 ft wide, and there are separate replenishment aisles and locations. In the pick-to-belt or pick car concepts, the pick positions have several pallets deep behind the pick position.

The cartrac pick concept is a long series of platforms that travel in a closed-loop path past storage and pick areas. The pick aisle is small and is between the cartrac and the customer locations.

These design features require a warehouse that is at least 30 ft high with a square or rectangle shape.

The third type of carton warehouse layout is designed for an automated carton order-pick concept. In the automated warehouse, the SKUs are separated into the manual and automated order-pick concept areas. The manual section handles the nonconveyable SKUs, and the automated section handles the conveyable SKUs. The automated pick modules are stacked, which results in a rectangle with long sides. These features create a warehouse that has a square or rectangle shape.

Operation That Handles Single Items

A warehouse operation that handles single items (split case, small parts, replacement parts, flatwear, hanging garments, or broken cases) has a manual, mechanized, or automated pick concept. Typically, the SKUs in this type of operation are lightweight and small- to medium-cube. The product is conveyable or is placed into totes or on hangers for transportation between two levels or locations (mezzanines). This feature allows a single-item operation to fit into almost any type of building shape.

HOW TO INCREASE STORAGE SPACE

When it is necessary to increase the storage (cube) utilization in the design of a new facility or in the remodel of an existing facility, your design team has several options to consider as solutions to the project.

Use the Airspace

The first solution to improve cube utilization is to use the airspace above the floor. These alternatives are to use freestanding or equipment-supported mezzanines, taller racks, or cantilever racks, to splice onto existing racks or to use stacking frames (portable racks). Each of these storage equipment alternatives increases the number of unit loads, cartons, or single items that are vertically stacked per square foot of floor space. These storage solutions require a lift truck to reach the new elevated stacking position. This requirement could require a new base plate on the rack posts, a new lift truck, or an existing lift truck with a new mast or additional counterbalance weight.

Prior to the implementation, review the floor capacity, building codes for sprinkler requirements, upright rack capacity, and the rated capacity by the lift truck manufacturer.

Use Narrow-Aisle or Very-Narrow-Aisle Vehicles

The second solution is to use narrow-aisle or very-narrow-aisle material handling vehicles with tall racks (Fig. 3.7). In a building, these material handling vehicles and racks increase the number of rack rows for unit-load storage within the building structure. The floor thickness and re-bar depth accommodate the new vehicle and rack loadings and vehicle guidance systems.

In addition, you should determine whether the floor, the rack upright frame, and the baseplate have sufficient capacity to support the additional loads. Then review the fire code for the required sprinkler arrangement.

Use Dense Storage Concepts

The third possible solution is to use dense storage material handling rack storage equipment (Fig. 3.8). A building that uses floor stack, flow racks, drive-in or drive-through racks, car-in racks, mobile racks, and two-deep racks increases the number of unit loads and cartons within the walls of the building. These storage systems require fewer aisles than do standard racks. With the new rack design, verify that your lift trucks and fire sprinkler systems handle the new storage characteristics.

Expansion

The final solution to increase the storage capacity is to expand the existing building with the same or new material handling equipment. The expansion can be either above or below ground.

SKU UNIT LOAD POSITIONS AND OPTIMUM PICK AND STORAGE POSITIONS

SKU Positions in a Warehouse

In the facility design process and daily warehouse operation, the three SKU positions within your warehouse building and off-site building are pick positions, ready-reserve positions, remote-reserve positions, and off-site reserve positions.

SKU Pick Position.   In the SKU pick position, the order picker removes a SKU for a customer order. In a single-item or carton warehouse operation, our observations have indicated that the SKU pick-position utilization (inventory quantity) is 50 percent. In a pallet load operation, the pick position is the same as the reserve position.

Ready Reserve Position.   The ready reserve positions are carton or pallet load storage positions that are adjacent, above, or behind the SKU pick position. The type of

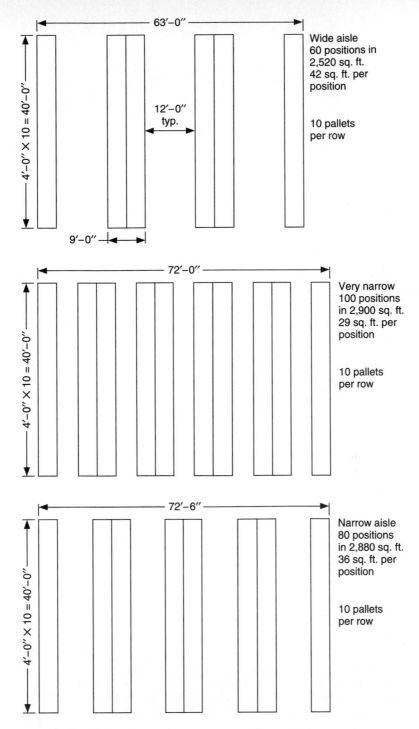

FIGURE 3.7 Wide-aisle concept (for material handling vehicles) compared to very-narrow-aisle and narrow-aisle concepts.

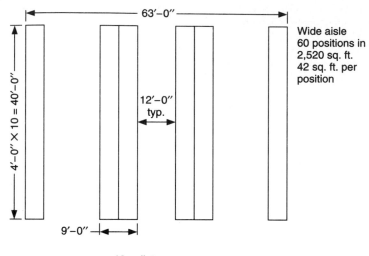

10 pallets per row

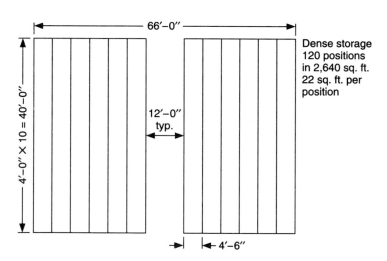

FIGURE 3.8 Wide-aisle concept compared to dense storage concept.

rack structure or floor-stack depth determines the location and number of ready-reserve positions. In the fixed-slot inventory system, the SKUs in the ready reserve positions are transferred as required to the depleted SKU pick position. The various storage-pick rack ready reserve utilization factors are reviewed in Chap. 7.

Remote Reserve Position.   The remote reserve positions are SKU storage positions that are not located in the same aisle as the pick position. These positions are located at the end of the aisle or in a separate dense storage area in the warehouse. As required, the pallet loads or cartons are transferred from the remote reserve position to the appropriate ready reserve position.

Off-Site Reserve Position. The off-site reserve positions are SKU positions that are in another facility and transported from the remote facility to the main facility. In the main facility, the SKUs are placed into ready reserve or remote reserve positions.

How to Project Your Storage and Pick Needs*

When designing a new facility or planning a facility expansion, your warehouse design team projects the SKU pick and reserve position requirements for the design year.

A new warehouse and distribution facility design effort is divided into two separate categories. In a carton or single-item operation, the first step is to project the SKU pick position requirement for the design year. In a unit load, carton, or single-item operation, the second step is to project the pallet or carton reserve position requirements. This projection requires more extensive work.

SKU Design Information

Prior to using one of the following projection methods, your design team requires data for the SKU characteristics, storage-pick position structure, and inventory movement and growth. This inventory data is part of your company's inventory management or purchasing department's computer master file and includes (1) each SKU (single item, carton, unit load) length, width, height, and weight, including stackability or crushability (when the length and width are stated, data on product overhang of the support device is also included); (2) each SKU historical velocity movement for one year; (3) project growth, throughput, and turnover rates; and (4) each SKU description, discrete number, and warehouse pick position location number.

Other important design information includes the family group, type of warehouse equipment, and environmental conditions.

SKU Pick Position Requirement Is a Multiplication Function

The method for projecting the SKU pick position requirement is a simple multiplication of the base year number of SKUs times the SKU growth factor (Fig. 3.9a). The result is the number of SKUs and SKU pick positions that are required in a carton or single-item distribution facility. In the pallet (unit) load distribution operation the storage position is the pick position.

Carton and Pallet Load Storage Projection Methods

The unit-load or carton storage position quantity projection has two alternative methods: the simple percentage increase and inventory stratification methods.

---

*This section was based on an article from *Plant Engineering Magazine,* "How to Stratify Your Storage Requirements," August 1992.

Projected number of SKUs			
(1) Year	(2) Base SKUs	(3) New SKUs	(4) Projected SKUs
Base 1	4000	10%	4400
2	4400	10%	4840
Start-up 3	4840	10%	5324
4	5324	10%	5857
5	5857	10%	6443
6	6443	10%	7087
7	7087	10%	7795
8	7795	10%	8575

(a) Base year number of SKUs × growth factor = design year number of SKUs

	Year	Annual case movement	Rotation case turns	Average weekly shipments	Average on-hand inventory	Average number of weeks on-hand
Present Year	1979	1,726,500	22	33,200	78,500	2.4
Base Year	1980	2,206,500	22	42,400	100,300	2.4
Start-Up Year	1982	2,592,300	22	49,900	117,800	2.4
Design Year	1986	3,257,300	22	62,600	148,100	2.4

(b) Base year inventory quantity × growth factor = design year inventory quantity

FIGURE 3.9 (a) SKU projection formula; (b) inventory percentage increase method. (*Courtesy of Plant Engineering Magazine.*)

Percentage Increase Method. In the simple percentage increase method (Fig. 3.9b) the warehouse design team multiplies the base-year inventory unit-load or carton quantity times the growth factor. The resulting inventory quantity is the required inventory that the storage facility accommodates within the area. To determine the pallet positions, your designer divides the inventory quantity by an average number of cartons or items per storage unit (pallet). When the distribution facility has standard single-deep rack (shelving) or when the SKUs have similar characteristics, the simple percentage increase method is an excellent SKU pallet inventory projection method.

Inventory Stratification Method. The second SKU storage projection method is the inventory stratification method (Fig. 3.10). The stratification method has four components: (1) the base-year SKUs separated according to characteristics, (2) calculations by velocity movement, (3) calculations based on growth rate, and (4) the storage concept.

After each SKU is allocated to its category, the stratification methodology projects the design year on-hand inventory.

The inventory stratification method is preferred because it takes into consideration the fact that the SKU on-hand inventory has different characteristics or is allocated to different storage structures. Also, in a carton or single-item facility, it accounts for the inventory quantity in the pick position. With this information, the facility size, storage area, and required storage equipment are based on a more exact calculation.

The inventory stratification method calculation is by velocity movement and other SKU characteristics to assign each SKU to a pick or storage rack position. The facility

(1)	(2)	(3)	2 × 3 = 4 (4)	(5)	4 × 5 = 6 (6)	4 − 6 = 7 (7)
Range	Items in inv.	Mid point	Total. pal	Type of row	Pal. in pick pos.	Pal. inv. to be put in res.
0–1.0	591	.5	296	H.O.R.S.	295	1
1.01–2.0	336	1.5	504	H.R.O.S.	168	336
0–1.0	197	.5	99	R1 × 5	98	1
1.01–2.0	98	1.5	147	R1 × 4	49	98
1.01–2.0	193	1.5	290	R1 × 5	96	194
2.01–3.0	240	2.5	600	R1 × 5	120	480
2.01–3.0	80	2.5	200	R1 × 4	40	160
3.01–4.0	168	3.5	588	R1 × 5	84	504
3.01–4.0	56	3.5	196	R1 × 4	28	1680
4.01–5.0	111	4.5	500	R1 × 5	55	445
4.01–5.0	37	4.5	167	R1 × 4	18	149)
5.01–6.0	84	5.5	462	DI2 × 5	42	420
5.01–6.0	38	5.5	209	DI2 × 4	19	190
6.01–7.0	60	6.5	390	DI2 × 5	30	360
6.01–7.0	19	6.5	124	DI2 × 4	9	115
7.01–8.0	45	7.5	338	FS3 × 4	22	316
7.01–8.0	15	7.5	113	FS3 × 3	7	106
8.01–9.0	36	8.5	306	2FS3 × 4	18	288
8.01–9.0	13	8.5	111	2FS3 × 3	6	105
9.01–10.0	35	9.5	333	2FS3 × 4	17	316
9.01–10.0	12	9.5	114	2FS3 × 3	6	108
10.1–15.0	87	12.5	1088	2FS3 × 4	43	1045
10.1–15.0	29	12.5	363	2FS3 × 3	14	349
15.01–19.0	45	17.0	765	2FS3 × 4	22	743
15.01–19.0	15	17.0	255	2FS3 × 3	7	748
19.1–Plus	55	19.0	1045	2FS4 × 4	27	1028
19.1–Plus	19	19.0	361	2FS4 × 3	9	352
	2714		9964		1349	9125

(*) The actual pallet positions provided are 544; however, due to the small vertical rack openings, all totals reflect the required pallet inventory.

(†) The actual pallet positions provided from remote reserve are 595; but, since this is drive-in rack, the availability of remote reserve is a restriction to that individual item because of the nature of the rack and in good warehousing practice, should not be used for remote reserve positions for other items.

FIGURE 3.10   Inventory stratification. (*Courtesy of Plant Engineering Magazine.*)

planner uses an inventory stratification form or computer program. The format consists of 12 columns and as many lines as are required to accommodate the SKU movement ranges.

In column 1 the planner lists the SKU unit-load movement ranges in order from the slowest range to the fastest range. Generally, each movement range has two lines. One line is for small- to medium-cube and noncrushable SKUs and the second line is for the high-cube and crushable SKUs.

In column 2, the design-year numbers of SKUs are listed in their corresponding movement range.

	IF (+)	IF (−)		
	7/UTIL = 8		8 − 9 = 10 8 − 9 = 12	
(8)	(9)	(10)	(11)	(12)
Pal. pos. for res. at a utilization factor	Rdy. res. pal. pos. provided	Pal. inv. to be stored in remote res.	Pal. pos. to be used for remote res.	Pal. pos. remote res. cumulative avail. (neg) positive
395[1]	544(*)			
1 (85%)	295		294	294
115 (85%)	294		179	473
228 (85%)	289		61	534
565 (85%)	360	205	0	329
188 (85%)	240	52	381	
593 (85%)	252	341	0	40
198 (85%)	168	30	0	10
524 (85%)	166	358	0	(348)
175 (85%)	111	64	0	(412)
560 (75%)	756		196(†)	
253 (75%)	266		13(†)	
480 (75%)	540		60(†)	
153 (75%)	133	20		(432)
421 (75%)	360	61		(493)
141 (75%)	90	51		(544)
384 (75%)	720		336	(208)
140 (75%)	195		55	(153)
421 (75%)	700		279	126
144 (75%)	180		36	162
1393 (75%)	1740		347	509
465 (75%)	435	30	479	
990 (75%)	900	90	389	
997 (75%)	225	772		(383)
1371 (75%)	1375		4	(379)
469 (75%)	399	70		(449)
11,369				

(*) The actual pallet positions provided are 544; however, due to the small vertical rack openings, all totals reflect the required pallet inventory.

(†) The actual pallet positions provided from remote reserve are 595; but, since this is drive-in rack, the availability of remote reserve is a restriction to that individual item because of the nature of the rack and in good warehousing practice, should not be used for remote reserve positions for other items.

FIGURE 3.10   (*Continued*)

Column 3 represents the midpoint for the unit-load movement range. To obtain the midpoint, the movement range line low and high figures are added, the total is divided by 2, and the figure is entered on the appropriate line under column 2.

In column 4, the SKU total unit-load on-hand inventory is entered on each movement range, which is the result of multiplying column 2 by column 3. This figure represents the movement range unit-load on-hand inventory that is to be allocated to the SKU pick and unit-load reserve positions.

Column 5 identifies the type of unit-load pick-storage rack configuration that is assigned for each SKU. The SKU velocity movement, cube, and stability characteristics are factors that determine the SKU rack configuration.

Column 6 represents the number of SKU unit loads that are allocated to the SKU pick position. These unit loads in the pick position represent a portion of the design-year on-hand inventory. Generally, in a carton order-pick operation, the unit-load pick position contains one-half unit load; therefore, the unit-load on-hand inventory in the pick position is 50 percent of one unit load.

Column 7 represents the number of the unit loads that remain to be placed in storage reserve positions. To obtain the number of unit loads, subtract the number of unit loads in the pick position (column 4) from the total unit-load on-hand inventory (column 6). The remainder represents unit loads that are to be placed into ready reserve positions or remote reserve unit-load positions. Ready reserve positions are unit-load positions above or adjacent to the pick position. Remote reserve positions are unit-load reserve positions that are not ready reserve positions but are in the same aisle or in another aisle.

Column 8 represents the number of unit loads that have been increased by the storage utilization factor and are to be placed into unit-load storage positions. This unit-load quantity is calculated by the assigned storage rack utilization factor divided into the unit loads to be placed into reserve (column 7).

Column 9 indicates the number of unit-load ready reserve positions that are provided from the storage configuration. Each SKU movement range line item has a quantity of unit-load positions that are determined by the type of unit-load rack and storage configuration. When calculating the rack configuration ready reserve positions, the following rules apply:

- Standard pallet rack (1 × 5) = two levels of pick positions and three levels of ready reserve positions. Ready reserve positions are calculated by dividing 2 into the SKU total and then multiplying by 3.

- Standard pallet rack (1 × 4) = one pick position level and three levels of ready reserve positions. Ready reserve positions are calculated by multiplying the number of SKUs by 3.

- Drive-in rack (2 × 4) = one pick position and seven ready reserve positions. Ready reserve positions are calculated by multiplying the number of SKUs by 7.

- Drive-in rack (2 × 5) = one pick position and nine ready reserve positions. Ready reserve positions are calculated by multiplying the number of SKUs by 7.

- Drive-in rack (3 × 3) = one pick position and eight ready reserve positions. Ready reserve positions are calculated by multiplying the number of SKUs by 8.

- Drive-in rack (4 × 4) = 1 pick position and 15 ready reserve positions. Ready reserve positions are calculated by multiplying the number of SKUs by 15.

- Drive-in rack (5 × 4) = 1 pick position and 19 ready reserve positions. Ready reserve positions are calculated by multiplying the number of SKUs by 19.

- Floor stack (3 × 3) = one pick position and six ready reserve positions. Ready reserve positions are calculated by multiplying the number of SKUs by 6.

- Floor stack (3 × 4) = one pick position and eight ready reserve positions. Ready reserve positions are calculated by multiplying the number of SKUs by 8.

- Floor stack (4 × 4) = 1 pick position and 12 ready reserve positions. Ready reserve positions are calculated by multiplying the number of SKUs by 12.

- Floor stack (5 × 4) = 1 pick position and 16 ready reserve positions. Ready reserve positions are calculated by multiplying the number of SKUs by 16.

If the distribution facility handles unit loads, then there are no reductions in the pick position. The planner multiplies the unit-load depth times the levels and times the number of SKUs. It should be noted that the standard rack ready reserve positions are not restricted to one SKU. The structure allows multiple SKUs per vertical stack. The structures of the drive-in, drive-through, or floor stack ready reserve positions are restricted to one SKU. When the floor-stack configuration is used in a carton order-pick facility as the pick and reserve positions, then the first unit-load vertical stack should be one unit load high.

Column 10 represents the on-hand unit-load inventory that exceeded the unit-load ready reserve positions. This figure is calculated by subtracting column 8 from column 9. When column 9 is greater than column 8, then this figure is entered in column 10 as a negative and constitutes the on-hand unit-load inventory quantity that requires remote reserve positions.

Column 11 represents the figure of the extra unit-load ready reserve positions that are not required for unit loads. This figure is calculated by subtracting column 8 from column 9. When column 9 is less than column 8, then this figure is entered in column 11 as a positive and these vacant positions are available for other movement range unit loads.

Column 12 indicates the accumulative available unit-load remote reserve positions that are available for other movement range SKU unit loads. This figure is calculated by subtracting column 10 and adding column 11 to column 12.

When these unit-load positions are standard rack positions, then they are available for other unit loads. When column 12 is negative or a drive-in or floor-stack requirement, then the facility planner is required to design additional remote unit-load reserve positions for the required unit loads.

CHAPTER 4

TRUCK AND RAIL RECEIVING AND SHIPPING DOCK AREAS

INTRODUCTION

The truck and rail receiving and shipping functions are the topics of this chapter. In this chapter, we review the primary warehouse functions of receiving and shipping activities. In all warehouse and distribution facilities, product movement and storage-pick functions start at the receiving dock by handling raw material or large quantities of product (pallet loads or cartons) and end at the shipping dock by handling pallet loads of finished product or smaller individual customer order quantities (cartons or single items).

RECEIVING OBJECTIVE

The receiving function objective is to assure you that your vendor has delivered to your company's warehouse the right product, in the right quantities, in good condition, and on schedule. Your receiving department activities include schedule of the delivery vehicle at the dock, unload the product from the delivery vehicle, accurately count the product, verify (check) the product quality, identify the product (SKU), enter the product into the inventory, and transfer the product to the storage area.

SHIPPING OBJECTIVE

The shipping function objective is to assure you that your warehouse order-pick activity has provided to your customer the correct SKU, in the correct quantity, in good condition, and on schedule. The shipping activities include scheduling the delivery vehicle at the dock; sorting, accumulating, packing, and verifying the order quantity and product; addressing (labeling) the package; manifesting the package; loading the package onto the delivery vehicle; and performing the efficient transport of the package to the customer.

RECENT TRENDS IN RECEIVING AND SHIPPING ACTIVITIES

Within the past years, there have been four major reasons for you to look more closely at your receiving and shipping functions. These reasons reflect change in the federal regulations for the over-the-road trailer lengths and width, implementation of JIT (just-in-time) replenishment and across-the-dock programs, computer technology, bar-code scanning and delivery vehicle tracking methods, and new unloading and loading dock equipment.

Longer and Wider Trailers

The federal government highway truck regulations have resulted in increased over-the-road trailer carrying capacity. This permits a trailer to carry an additional two pallets or product cartons. These revisions resulted in a trailer width increased from 96 to 102 in and length increase from 3 ft to 45 to 48 ft.

The new trailer width and length has increased the truck-yard space that is allocated for trailer parking, radius of curves, landing gear pads, and slope of pits; increased the truck dock door width and seal and shelters width; increased the receiving and shipping dock staging areas to handle two additional pallet loads or a total of 20 pallets; and improved dock leveler capacity to serve trailers with heights of 36 to 54 in.

Just-in-Time (JIT) Replenishment

Just-in-time replenishment is a system of scheduling your inbound deliveries to arrive at your receiving dock just in time to replenish your workstation or ready reserve position. This philosophy has had an impact on reducing the required on-hand safety stock inventory and focuses emphasis, with respect to warehouse and distribution activities, on the receiving and shipping dock areas (locations) and operations rather than storage-pick areas and operations.

The retail store distribution industry has implemented across-the-dock or quick-response distribution programs. When looking at these operations, these programs have the same characteristics and impact as the JIT programs.

Computers, Bar-Code Scanning, and Vehicle Tracking Networks

Computer technology and bar-code scanning have reshaped the receiving and shipping operations. The impact of the computer has allowed for improvements in document preparation, product identification, and work (employee and equipment) scheduling.

The implementation of computers with higher degrees of intelligence and faster processing speeds has improved documentation preparation, improved the flow of product and information, employee and equipment scheduling, accurate and on-line transfer of transactions, and accurate customer and product identification.

The automated identification of product with today's hand-held and fixed-position scanning devices that read bar-code labels ensure that the date is accurate and that the information transfer is on-line. The new technology has allowed improvements in inventory identification, control of product flow, efficient internal transportation, precise customer order sortation, and delivery truck document manifests.

The communication technology that has had a recent impact on delivery truck equipment and labor scheduling in a warehouse operation is the introduction of an exact delivery vehicle truck tracking network. This radio-frequency concept, which consists of a radio-frequency tag on a truck, satellite station, and central monitor station, determines the exact highway or yard location of a delivery truck. This concept improves delivery truck utilization and labor and equipment productivity.

New Loading and Unloading Equipment

The final trend is the new unloading and loading equipment, such as dock levelers, automatic unloading and loading systems, mobile warehouse equipment, and conveyor equipment. The impact of this equipment focuses on two areas: (1) it increases the flexibility of the receiving and shipping departments to handle a wide variety of trailer sizes; and (2) it improves the safety, product flow, and employee productivity of the departments.

We conclude that in the future there will be a reduction in the number of employees and building space that is allocated to the receiving and shipping functions but an increase in the number of companies that have implemented an across-the-dock operation.

The receiving and shipping department product handling methods and product flow philosophy is a subactivity of the total warehouse-distribution operation and the total product movement-storage-pick system. Many factors that determine how the dock area looks and functions, such as separate or combined dock operations, inventory control (update) philosophy, type of operation (pallet, carton, or single item), method of delivery and shipping, and product volume.

When your company is involved in a conventional warehouse storage activity, then the emphasis is placed on the storage function. If your company is involved in a JIT replenishment or an across-the-dock operation program, emphasis is on the receiving and shipping activities. If your company is a catalog or direct mail (single-item) operation, then the emphasis is on the picking and packing operations and not on the receiving and shipping operations. When your operation handles pallets or cartons at the receiving and shipping docks, then the emphasis is equal for both dock operations; but in a carton handling operation there is concentration on the storage and pick functions.

In a manually controlled pallet handling equipment operation, the dock area has sufficient space to stage unitized loads and empty pallet boards. When carts or containers are used to ship product to customers, then in the loading (shipping) dock area there is sufficient space for full or empty cart or container staging. If the operation is a conveyorized carton sortation or automatic carton pick facility, then the outbound dock area is designed with fluid (direct) load conveyors, or you would palletize or stage the product. The fluid (direct) loading of conveyorized product and manual loading of nonconveyorized product concept requires a minimal dock staging area. The second alternative is to unitize the product onto carts or pallet boards, and then load the full carts (pallet boards) onto the delivery vehicle. This concept requires a square-foot dock staging area to hold a delivery truck capacity.

RECEIVING IS WHERE IT STARTS

Since the receiving function is the start of your warehouse or distribution product handling operation, it is very important that the function be accurate and on time. If there is a failure to perform the activity in an excellent manner or there is a discrepancy in one of the activities, then the warehouse objective to satisfy customers at a profit is in jeopardy. Delivery is the first receiving activity.

Truck Delivery Methods

The various truck delivery methods are (1) common-carrier truck, (2) United Parcel Service, (3) vendor truck, (4) U.S. Postal Service, (5) company (backhaul) truck, (6) express or air freight, and (7) container.

The various truck receiving activities are determined by the type of operation. Most of these activities include yard control, vendor delivery truck or container dock scheduling, unloading product into the inbound staging area, verifying the product quality and quantity, entering the receipts into inventory, identifying the product, packaging the product, and transporting the product to the storage area.

Yard Control.   Yard control or delivery truck or container inbound dock scheduling determines the time that the delivery truck is allowed at your company's (spotted) unloading dock. Other yard control activities include use of a chock behind the trailer driver-side rear wheel or a new ICC (Interstate Commerce Commission) bar-locking dock leveler, checking the seal and opening the truck door, and visually inspecting the trailer or load condition for damage or infestation.

UNLOADING ACTIVITY

Unloading is the second receiving activity. This activity is the physical unloading (movement) of the product from the vendor's container or truck or backhaul truck onto the receiving dock staging area.

The various delivery truck unloading methods are manual, using conveyor equipment, using manual or powered mobile pallet equipment, and using automatic trailer unloading-loading equipment.

Typically, the outside (nonemployee) truck driver unloads the common-carrier or vendor delivery truck. A company employee is responsible to unload the railcar, container, backhaul, and some common-carrier delivery trucks.

These concepts are reviewed later in this chapter.

VERIFY QUALITY AND QUANTITY ACTIVITY

The third activity is to verify that the vendor's product quality and quantity are as specified in your company's purchase order. This activity ensures that the quantity of the product delivered to your warehouse matches your company's purchase specifications and that the received product quality is acceptable to your company's purchase order specifications and company standards. Excellent quality product and quantity counts from your vendors are extremely important components in your across-the-dock, JIT, or quick-response program. Quality control includes ensuring that the product exterior and interior material received in the future will be per your company's recycle specifications.

Quality and Quantity Check Methods

There are three basic methods of quality and quantity product verification, with respect to the vendor's past delivery performance and SKU characteristics: (1) 100 percent accept, random sample, and 100 percent check.

100 Percent Accept Method. The first method is to accept the vendor delivery without performing a quality and quantity check. This practice is based on the vendor's excellent past performance of complying with your company's specifications. The excellent vendors are placed on a vendor approved list, and your company accepts the vendor's quality and quantity count. This method is used for any type of product and allows for a high volume of product handled in a small warehouse square-foot space. This method does require a periodic random check of the vendor's deliveries to verify that the vendor's performance complies with your company's standards and specifications.

Random Sample (7 to 10 Percent) Method. In this method, random samples (7 to 10 percent of the vendor's delivery) are taken from the vendor's delivery. If the sample satisfies your quality standards and quantity count, then the entire inbound (vendor delivery) is accepted by your company. This method is used for any type of SKU and handles a medium volume in a medium-sized warehouse.

100 Percent Verification (Check) Method. In the 100 percent verification method, the receiving department inspects and counts each item of the inbound delivery. This method is used for vendors who have previously supplied poor-quality items or had shortages on their deliveries. This method is used for any SKU and allows a low volume to flow across the dock and requires a large-square-foot floor area.

Quantity Count Methods

When your vendor's delivery is required to have an actual count compared to the purchase order specification, one of three methods is used: manual method, mechanized method, or "count on the fly" method.

Manual Count Method. In the manual count method, the employee physically counts each individual item.

Disadvantages of the manual method are that it requires the greatest number of employees, has a greater possibility for error, handles a low volume, and requires a large-square-foot building area. Advantages of the manual count method include a low capital investment and handling of all types of SKUs.

There are several alternative methods for counting small items or flatwear merchandise, such as the stationary and flow-through methods.

The stationary (table) count concept layout is the basic count method. The components are a 3 × 6-ft flat-surface table and two 30 × 36-in flat tables. The large table is the count table, and the two small tables are the infeed and outfeed tables. As required, the product is delivered to the infeed table, and the count person opens the cartons and counts the product quantity. The product count quantity is entered onto a tally sheet. The product is placed into an empty vendor carton that was transferred from the infeed table to the outfeed table. At the completion of the count activity, the tally sheet is totaled and the total is compared to the purchase order quantity. The company's policy or procedure for a variance determines a recount or acceptance of the count. Infeed and outfeed material handling methods involve use of a two-wheeled hand truck, pallet load, or conveyor (roller or skatewheel). These components are located on both sides of the table or behind the count table. Disadvantages of this method are that it handles a low volume, increases employee fatigue, requires a large-square-foot area, requires a long setup time, and does not handle large purchase orders easily. Advantages of this method are low capital investment, easy training of employees, and capacity to handle all product types.

The *flow-through count method,* another manual count method, utilizes several conveyors, work shelves, and conveyable containers.

The product flow starts at the dock as the product is delivered in vendor-supplied cartons to the off-loading station. Employees transfer these cartons onto the infeed end of the opening conveyor section. Each conveyor section handles one purchase order, and whenever possible each conveyor lane with a section (opening) handles one SKU (same size and color). The conveyor on the left side contains the smallest-sized SKU. On the outfeed end of the opening conveyor, an opening employee opens and removes the product from the vendor-supplied cartons. The product is counted and placed into a conveyable container. The arrangement of the product in the tote is vertical, with the vendor label facing up and toward the lead end of the tote. Until the entire purchase order is processed, the full totes of counted product are transferred to holding conveyors. When the employee has completed the purchase order count, the entire purchase order is released onto the take-away conveyor. The take-away conveyor transports the totes to the ticketing station. The product counts are entered onto a tally sheet. Then the trash (empty cartons, paper, and wrapping) is placed into a trash conveyor or trash container. Disadvantages of the conveyorized method are that it requires an additional investment and requires management control and discipline. Advantages of the method are that it handles a large volume, requires no downtime for setup, handles large quantities and purchase orders and allows queues to be established prior to workstations.

In the *"ti × hi" physical manual count method,* the receiving employee determines the ti (cases per layer) and hi (number of layers) per pallet load. After determining these facts, the employee multiplies the ti by the hi, which provides the number of cartons per pallet load, and then counts the pallet loads. An alternative procedure is to physically count each carton of the pallet load.

These manual methods require additional staging dock area and allow only a low product volume to flow through the receiving area, and empty or partially full cartons can be counted in the total. The count methods are used on all types of SKUs.

Mechanized Count Methods. The mechanized count option is the scale count method. This concept is used to count small items, cartons, and truckloads. The method requires the receiving employee to place an individual item, several items, cartons, or truckloads onto a scale that registers the weight. Then with a full container on the scale, the employee determines the count for the entire box, pallet, or truck.

When *small items* are counted, the receiving employee places an individual item or several items onto scales that registers the weight. In the next step, the receiving employee allows for a container (tare) weight and places the entire carton quantity onto the scale. The scale indicates the carton count on its display screen. Disadvantages of the method are that it does not handle all SKU types and increases capital investment. Some advantages of the scale count method are capacity to handle a high volume, accurate count, and requirement for a small-square-foot building area and the fewest number of employees.

When *cartons* are counted, there are two alternatives: the pallet load scale and the truck scale. In the *pallet load scale count* method, the receiving employee performs the activity on the receiving dock. To determine the carton delivery quantity, the receiving employee weighs an individual carton, weighs the pallet load, reduces the pallet board weight from the pallet load weight, and performs the math calculation (division) that provides the carton quantity per pallet load. To verify that this weight matches the previous pallet board weight, an employee weighs each pallet load. If each pallet weight matches the sample pallet scale weight, then the number of cartons is the number of cartons per pallet load for all the pallets on the delivery.

The *truck scale count* method involves (1) weighing the delivery vehicle prior to unloading, (2) unloading the delivery vehicle, (3) weighing the empty delivery vehicle, (4) determining the weight difference between the two delivery vehicle weighings, and (5) comparing the truck weight difference to the purchase weight.

If there is no discrepancy between the actual truck weight and the purchase order weight, then the actual weight (count) is accurate. If the actual truck weight difference does not match the purchase order weight, then the employee uses the pallet count method or physically counts the cartons.

The disadvantages are that it requires additional capital investment and the vendor is required to indicate the delivery weight of the carton or truckload.

The scale count concept is used for all types of SKUs. It provides an accurate count, requires a small staging area, handles a high volume, and requires few employees.

Count on the Fly.   In the "count on the fly" method (Fig. 4.1), the vendor standardizes all carton shipments for each SKU. Each carton contains a predetermined number of SKUs and displays the SKU label on the exterior. As the carton travels on the conveyor system, the label is read by an automatic scanning device and is weighed by a scale. The computer matches the actual carton weight for each SKU to the SKU predetermined weight (including tare weight) that is in the file. If the weights match, then the carton continues travel on the receiving conveyor. If the weights do not match, then the carton is diverted to the quality control area for a detail check.

Disadvantages of this method are need to update the computer files, requirement for a computer or conveyor investment, and requirement for vendor compliance. Advantages are that it handles a very high volume, requires few employees, provides an accurate count, requires the fewest SKU handlings, and requires a small-square-foot building area.

HOW TO HANDLE PRODUCT DAMAGE ON YOUR DELIVERY VEHICLE

When your receiving employee discovers product damage on a delivery vehicle (railcar, truck, or container), then your company receiving procedure is to verify the prod-

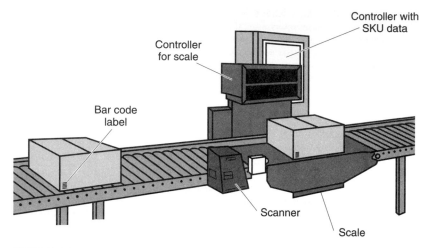

FIGURE 4.1   "Count on the fly" quantity count method. (*Courtesy of Accu Scan.*)

uct damage on the delivery vehicle. The two methods to verify delivery vehicle product damage is the written document (report) and a combination of the written document (report with illustrations).

The written document (report) is your employee's description of the delivery vehicle interior condition and the number of damaged cartons on the delivery vehicle. If
the document is not cosigned by the transportation (truck driver or railcar or container
inspector) company employee, then it is your employee's statement against the transportation company opinion.

In the written document (report) with illustrations concept, your receiving employee completes the written document and has the transportation employee cosign it.
Then employee photographs the delivery vehicle interior condition and damage cartons. An instant camera (Polaroid) is preferred because your employee verifies the
quality of the photographs. To assure a good view of the damage, several photographs
from various angles are taken of the delivery vehicle interior condition and the damaged cartons.

The advantage of this method is that a "picture is worth a thousand words."

RETURN TO VENDOR OR HOLD FOR QUALITY CONTROL AREA

During your receiving activity as your receiving department discovers product that
does not meet your company's specifications or standards, then by your company policy the product is held for the purchasing department to make the final decision on the
product. The purchasing department has two decisions and typically, this decision
process takes 1 to 3 days.

The first, is to temporarily hold the product and return it to the vendor or wait for
the vendor to pick it up. This decision requires storage of the product for 4 to 5 days.

The second decision is to perform a 100 percent inspection of the product that separates the good-quality product from poor-quality product. The good-quality product

is sent to the storage area. The poor-quality product is held in the separate holding area and waits for vendor pickup or authorization for shipment to the vendor. This activity requires storage for 4 to 5 days.

With these decisions, the receiving process is slow and requires temporary storage positions. These storage positions are occupied for 1 to 2 weeks. With these criteria, the storage positions are dense storage types and are located in a remote area. This area is in the receiving department or in the storage area. Past experience determines the required storage positions.

VARIOUS INVENTORY UPDATE METHODS

The next receiving activity is to update the product on-hand inventory files and internal transportation of the product from the receiving area to the storage area. The two inventory update methods are (1) delayed-entry method and (2) on-line receiving method.

Delayed-Entry Method. In the delayed-entry method there is physical movement of the product from the receiving area to the storage area; the on-hand inventory book entry is made at a later time. The method could cause a "no stock" occurrence with the inventory physically in the pick position because the on-hand book inventory is not updated to show the new receipt. This situation means that the computer does not create a pick instruction for the customer order.

On-Line Entry Method. In the on-line receiving inventory entry (adjustment) method the product on-hand book inventory is updated as the product is moved from the receiving area to the storage area. This method could cause a "stockout" occurrence because the book inventory is increased for the receipt quantity but the product is physically not in the pick position. With this situation, the computer creates a pick instruction for the customer order.

PRODUCT IDENTIFICATION

The next receiving activity is product identification. The employee applies alphabetic or numeric markings to the product exterior. These markings serve as instructions to the warehouse employee who handles the product.

Various Product Identification Methods

Your company can use any of several methods for product identification: (1) no identification method that relies on the vendor markings on the product exterior, (2) manual handwritten label method, where a receiving employee handwrites the product number on a self-adhesive label that is placed onto the product exterior; (3) machine-printed human-readable label method, in which a computer-controlled machine-printed label is placed onto the product exterior; and (4) machine-printed human- and machine-readable label method, in which alphanumeric characters and a bar code appear on the surface. The label is placed onto the product exterior.

Ticketing Activity

In some single-item distribution facilities, a subactivity of the product identification activity is price ticketing. An employee places a retail price tag onto the individual product package. See Chap. 6 for details regarding ticketing activities.

PRODUCT PACKAGING

The next receiving activity is to place the product into a container or cart or to secure the product onto the pallet load. The activity ensures that the product is efficiently handled in the warehouse transportation and storage activities with minimal product damage. A more detailed review of the packaging is presented in Chaps. 6 and 7.

RECEIVING AND SHIPPING DOCK LOCATIONS

The location of a warehouse and distribution facility or terminal receiving and shipping dock locations directly affects product flow and employee productivity. The main objective of the receiving and shipping dock locations are to reduce the in-house transportation distance between the dock area and the storage-pick area. Therefore, the best dock location is determined by the product storage aisle or position in the storage area. The incremental cost of an additional dock is a small investment when compared to the increase in the additional in-house transportation investment, low employee productivity, and poor product flow.

The three basic truck receiving and shipping dock concepts are (1) combination receiving and shipping docks, (2) separate receiving and shipping docks, and (3) scattered receiving and shipping docks.

Combination Docks Concept

In the combination (shipping and receiving) docks concept (Fig. 4.2*a*) the receiving and shipping activities are performed in one building area, which means that few dock positions are used. Receiving and shipping activities utilize the same building area, equipment, and employees. For best results, the combination concept requires a truck dock schedule in which the inbound product is delivered in the morning and outbound product loaded in the afternoon.

This concept is best for a small facility that handles a low volume, a small quantity of SKUs, and product characteristics that are small or large. With the receiving and shipping docks on the same wall, the product flows through the facility in a horseshoe or U pattern.

Disadvantages of this method are that it increases in-house transportation and requires exact scheduling of inbound and outbound trucks, and with this method it is difficult to compensate for product delivery problems and business fluctuations.

Advantages include maximum use of building area, equipment and employees; also, with this method less supervision is required and it is easy to assign trucks and provide security.

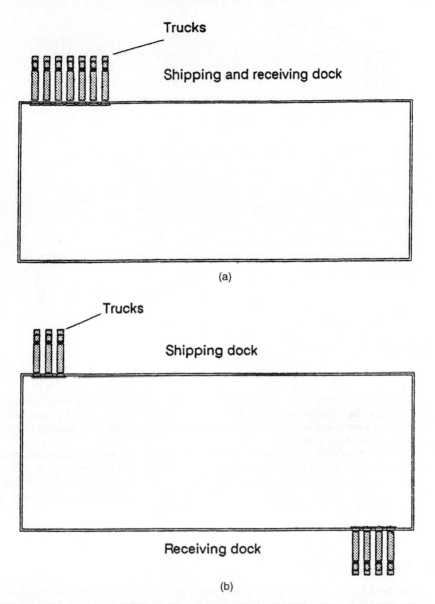

FIGURE 4.2 (*a*) Combination dock concept; (*b*) separated dock method; (*c*) scattered dock method. (*Courtesy of Rite Hite.*)

Separated Docks Concept

The second alternative is to have separated dock locations for the receiving and shipping activities (Fig. 4.2*b*). Receiving activities and shipping activities are performed in separate building areas and with separate equipment and employees and separate supervision. This concept reduces in-house transportation requirements and activities.

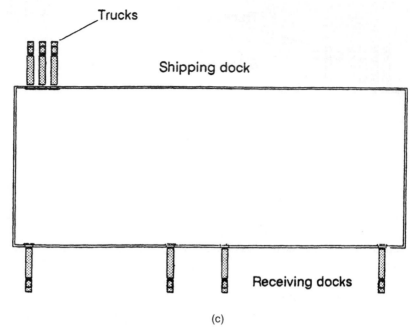

FIGURE 4.2   (*Continued*)

The separated dock location concept is best for a large operation that handles a high volume, has a large number of SKUs (inventory), and has a large product mix. With the receiving and shipping docks in separate locations, the dock operations handle any product flow patterns.

Disadvantages of this method are that it increases investment in building and dock equipment and number of employees needed. Advantages include flexibility in scheduling of trucks; increased capacity to handle problems, business fluctuations, a large volume, and an inventory with a large product mix; easier tracking and control of product flow; reduced need for in-house transportation.

Scattered Docks Concept

The scattered dock concept (Fig. 4.2c) is a variation of the separated dock concept. The receiving docks are located along one entire building wall. This area is assigned as the inbound side of the building. Each dock is located directly across from the product storage area or manufacturing (processing) area. This dock arrangement permits the product to flow directly (shortest distance) in a straight line from the delivery truck dock area to the assigned storage area or production line.

The shipping docks are located along the opposite building wall, thus allowing the product to flow from the storage areas or production lines into the shipping dock area.

The scattered docks concept is best suited for a company that has implemented a

JIT (just-in-time) or across-the-dock product flow concept. The concept dictates a one-way straight product flow pattern.

Disadvantages are increased design and planning activity and increased need for management discipline and control of the operation.

When compared to the separated docks concept, the scattered docks concept has an additional advantage of providing a continuous and uninterrupted product flow.

TRUCK ACCESS TO THE DOCKS

The second important factor for the receiving and shipping function is the truck (access) entrance and exit to your company facility. This is a factor that determines the receiving and shipping dock locations.

In a small facility that is located in the urban city area, the inbound and outbound truck access is direct from the city street. This restrictive land area requires the delivery truck to use the city street as its access road to your receiving and shipping docks.

With this situation, the time required to position (spot) a truck at the dock increases and often a company employee must assist the driver.

At a large facility located on a site with sufficient land, the inbound and outbound truck access is designed to improve vehicular flow and reduce the risk of accidents.

In general, a well-planned truck access road allows over-the-road trucks driven forward from the public road onto the truck yard safely, rapidly, and with minimal maneuvering of the truck. Likewise, the exit road permits the delivery truck driven forward from the truck yard onto the public road. To permit good truck flow, the access-exit road is, at least, twice the length of the longest vehicle (tractor and trailer combination). Most facilities have separate traffic lanes for truck-yard entrance and exit.

Between the public road and truck yard the truck passes through a security gate and a security station (guardhouse). The guard ensures that authorized vehicles are allowed on your property and that the truck has arrived at the correct address and contains product that is assigned to the facility, verifies that the seal is not broken, and logs (notes) the truck arrival time. The guard notifies the receiving department of the truck arrival. A receiving employee notifies the guard to direct the truck driver to drive the vehicle to a preassigned dock location or truck-yard holding spot.

Since truck (and other motor-vehicle) drivers are on the left side of the tractor cab in the United States, the truck traffic flow around the facility is counterclockwise in pattern. A counterclockwise pattern provides the driver with clear visibility while driving through the yard or backing the truck up to the dock. When a driver backs up to a dock from a clockwise traffic flow pattern and sits on the left side of the tractor, there is low productivity because of the "blind side" or difficulty to see the truck side and path.

If the facility is in a country where the driver sits on the right-hand side of the tractor cab, then the traffic flow pattern around the facility is in a clockwise pattern. This pattern allows clear visibility while driving through the yard or backing the truck to the dock.

In accordance with the site configuration, local codes, or architectural requirements, normally the truck docks are located in the rear or side of the building; therefore, the truck is driven to the rear or to the appropriate side of the building.

With a vehicle load of at least 40,000 lb traveling on the service road, the road has a minimum of 9 in of crushed gravel and 5 in of asphalt.

Truck Traffic Flow Patterns

The two alternative vehicle traffic flow patterns in the truck yard or service roads are (1) one-way service road and (2) two-way service road.

One-Way Pattern.   In this pattern the facility is engulfed by the road, which is at least 13 ft wide from the building, or is of a dimension that allows for the longest parked tractor-trailer combination plus 13 ft and the maneuvering area that extends outward from the truck dock. For an employee walkway adjacent to the truck lane, the width is wider than this combined road and tractor-trailer dimension by at least 4 ft.

Disadvantages of the one-way service road are necessity for increased road construction and additional land investment. Advantages are (1) better vehicle flow and (2) improved yard safety.

Two-Way Pattern.   The two-way service road allows the delivery truck to travel in both directions between the guard station and truck dock. The concept does not require the road to engulf the building but requires, at least, the length of the longest tractor-trailer combination, maneuvering area, and a 26-ft-wide truck lane. An employee walkway requires an additional 4 ft to the width of the road, which becomes 30 ft wide.

Disadvantages of this method are decreased yard safety and vehicle control. An advantage is improved vehicle spotting (docking) time.

Delivery Truck Holding Area

The truck temporary holding (waiting) area, an important truck-yard feature, is located in a secured (fenced) area. When a truck arrives ahead of its scheduled truck dock time, then the truck is allowed to exit the public highway and is assigned to a temporary parking area (buffer zone).

In the truck yard, the three trailer parking patterns are (1) block (square) pattern, (2) 45°-angle pattern, and (3) back-to-back and side-to-side arrangement.

Block Concept.   In the block (square) trailer parking pattern (Fig. 4.3a), trailers are parked on the perimeter straight back against the security fence and trailers that are parked in the interior are in a single row back to back. An option is to place wheel bumpers on the pavement in specific locations. These bumpers stop the trailer wheels, preventing the trailer rear from hitting the fence. The block concept provides the maximum number of trailer parking positions but requires the widest truck turning aisle (maneuvering area). Moreover, the trailers with rear doors that face the perimeter are considered a potential security problem. The security (rear door) problem is reduced with the interior trailers because they are back to back.

45°-Angle Concept.   In the 45°-angle parking pattern (Fig. 4.3b), trailers are parked on the perimeter and in the interior at a 45° angle to the security fence and to the middle of the yard. When compared to the block pattern, the 45°-angle arrangement provides fewer trailer parking positions but a narrower truck turning (maneuvering) area.

Back-to-Back and Side-to-Side Concept.   In the back-to-back and side-to-side trailer parking arrangement (Fig. 4.3c) all trailers are parked in the yard arranged in the block or 45° pattern on the interior of the truck yard. According to this concept, each

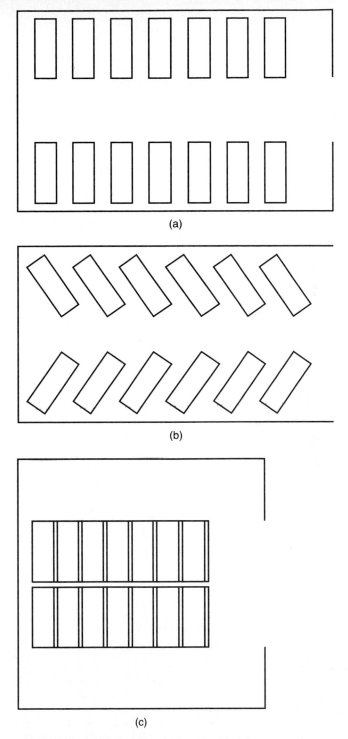

FIGURE 4.3 (*a*) Block trailer concept; (*b*) 45°-angle truck parking concept; (*c*) back-to-back and side-to-side truck parking concept.

trailer is backed up to the rear door of another trailer and the trailer side is adjacent to the next trailer. This trailer concept does reduce the door security problem because most of the trailer doors are blocked by another trailer.

Landing Gear Pad

When trailers are parked (dropped) in the waiting (parking) area without being hooked to a tractor, then the trailer landing gear rests on a landing gear pad. This pad is a concrete strip that is 48 in wide and provides a solid base for the landing gear. If this pad is not provided in the yard and during hot or warm weather, then a fully loaded trailer has the potential to sink below the tractor fifth wheel and make it difficult to move the trailer.

Other Important Truck and Truck-Yard Features

Other important truck-yard features are (1) return building, (2) maintenance building, (3) trailer-tractor wash rack, (4) dispatch station, (5) fuel island, (6) lift truck ramp, (7) engine heaters, and (8) truck canopy.

Truck-Yard Security.   The next important truck-yard feature is security.
    The first security feature is a security fence and berm located on the truck-yard perimeter. This fence-berm combination restricts unauthorized personnel from entering the property, reduces the outside view of the truck yard, and reduces the noise traveling from the truck yard to neighbors in the area. The second security feature is a series of cameras that view the parked trailers and dock locations. The view is transmitted to screens in a guardhouse. The third security feature is a guard station and a guard in a mobile vehicle to patrol the truck yard. The guard maintains the company's truck-yard policies and procedures, stops all unauthorized vehicles or personnel from entering; and notes any damage to company trucks, building, road surface, lighting, or security fence. The fourth security feature is to provide adequate lighting.
    Other truck-yard factors that render the truck yard more efficient and safe include identification of dock doors and stripping of truck parking spots, traffic arrows, speed and stop signs, maps and facility identification (signage), and guardrails and bumpers.

Truck Loading and Maneuvering Areas.   The truck-loading and maneuvering yard area is located in front of the docks. The loading area is directly in front of each dock location. During loading and unloading, the tractor-trailer is parked in this area. The loading area is designed for the overall length of the longest tractor-trailer combination. The minimum length is 65 ft from the dock. The width is designed for the overall width of the widest trailer plus 3 ft on each side of the tractor. This dimension compensates for the two side mirrors on the truck. If the loading area is paved with asphalt and trailers are spotted on their landing gear (without the tractor), then a 4-ft-wide concrete landing gear pad is installed at required distance in front of and parallel to the docks. The concrete pad has sufficient strength to prevent the trailer landing gear from sinking below surface. The pad supports a 40,000-lb load on 6-ft centers. The exact landing gear location is determined by the trailer length. The normal location for the landing gear on an over-the-road trailer is 10 ft behind the trailer nose.
    The maneuvering space is the space that is required to back up to the dock the longest overall tractor-trailer combination. With a counterclockwise vehicular traffic

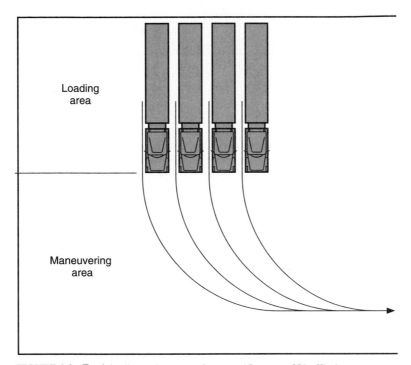

FIGURE 4.4   Truck loading and maneuvering area. (*Courtesy of Rite Hite.*)

pattern, the maneuvering area is a minimum of 70 ft extending outward from the loading area. For the normal tractor-trailer length of 65 ft, the total loading and maneuvering area that extends outward from the dock edge is 135 ft (Fig. 4.4). If shorter tractor-trailer combinations are used, the area is decreased in length.

For a clockwise vehicular traffic pattern, the maneuvering area is doubled in length, and the maneuvering length is 140 ft with a total loading and maneuvering area of 205 ft from the dock edge.

The loading area of the truck yard has a slight grade or slope from the maneuvering area to the dock. The degree of slope is a minimum of 3 percent to a maximum of 10 percent, with a preferred slope of 6 percent in geographic areas with snow and ice conditions.

To assist with water drainage, the loading area that is adjacent to the dock (building) is slightly sloped from the dock edge toward the maneuvering area to a drain. This outward slope is not extended beyond 3 ft from the dock edge. If the extension is greater, then the slope has an effect on the elevation of the truck rear to the dock and the top of the truck to the overhead curtain.

Since the maneuvering, loading, and dock areas have the sole purpose of facilitating the loading and unloading of trucks, these areas are designed to accommodate the widest range of truck sizes. At some companies, separate docks are designed to handle oversized trailers.

Truck Dimensions for Docks.   In designing the truck dock height, the height of the trailer bed is the most important dimension. This dimension is determined for both receiving and shipping trucks. If your company owns or leases its shipping and deliv-

Dock height, in		Truck overall dimensions, ft		
Truck type		Length	Width	Height
Flatbed	52	55–70	8–$8\frac{1}{2}$	12–14
Over-the-road	48	55–70	8–$8\frac{1}{2}$	12–14
City trailer	46	55–70	8–$8\frac{1}{2}$	12–14
Container	55	40–55	8–$8\frac{1}{2}$	12–14
Refrigerated	52	30–35	8–$8\frac{1}{2}$	12–14
High-cube	40	55–70	8–$8\frac{1}{2}$	12–14
Straight truck	44	15–35	8–$8\frac{1}{2}$	12–14

FIGURE 4.5   Delivery truck dimensions. (*Courtesy of Rite Hite.*)

ery trucks, then this is an easy task. If your company uses a common-carrier service, then an employee is assigned to obtain the actual delivery truck dimensions or contact the delivery company (common carrier) for the dimensions.

During this truck information collection process, other required dock equipment and tractor-trailer information dimensions are (1) overall length and overall height (including wind deflectors) of tractor-trailer; (2) overall width, height, and length of trailer; (3) overall width and length of tractor, including side mirrors; (4) landing gear location from the front (nose) of the trailer; and (5) centerline of rear axle (wheels) from the end of the trailer.

Dimensions of various types of trucks are shown in Fig. 4.5.

The best dock height provides the smallest variance between the dock edge and rear of the trailer bed. To accommodate a wide variety of truck heights to the dock height, the height differential between the dock edge and the trailer bed is bridged by a loading ramp (dock plate) or dock leveler. As the height differential between the dock edge and truck or trailer bed increases, the incline-decline slope of the bridge device increases. If this slope increases to a high degree, then, as a result of mobile warehouse pallet handling equipment problems on the dock plate or leveler, there is a decrease in dock employee productivity, increased need for mobile dock equipment maintenance, and a decrease in dock safety.

As a general guide, the dock height for most delivery trucks ranges between 46 and 52 in. The most common dock heights for the various truck types are listed in Fig. 4.5.

If there is a wide variety of truck bed heights, then alternative truck dock designs include allowance for separate docks for specific truck bed heights, using an extralong dock leveler (up to 12 ft) with an 18-in up-down (vertical) travel from the dock edge, installing truck leveler in the loading area to raise or lower the entire trailer, installing a dock lift, and using portable rear wheel risers to elevate the rear bed of the low trucks.

HOW TO PROJECT THE REQUIRED NUMBER OF DOCKS

The number of truck or rail receiving or shipping docks that are required at a facility are determined by one of the three following methods: manual calculation, manual simulation, and computer simulation.

Manual Calculation Method

The number of truck docks is determined by several factors that are easily obtained or calculated by your department. These factors include (1) tabulating for one week the number of daily truck deliveries (peak and average) plus growth; (2) time of day and frequency of truck delivery arrivals; (3) delivery percentage (floor stack, slip sheet, and pallet load); (4) time required to unload or load and stage product on the dock per type of delivery (common carrier, backhaul, container, Federal Express, United Parcel Service, or U.S. Postal Service); (5) number of mobile material handling equipment vehicles or vehicles with attachments or conveyors that are used in the dock area; (6) including 30 to 45 min for employee coffee or lunch breaks, truck maneuvering time, and document signing time; and (7) determining a reasonable safety factor such as 20 or 25 percent.

A general rule of thumb is that there are a sufficient number of dock positions to unload and load the maximum number of trucks that arrive or are scheduled for unloading and loading at your facility. A simple formula for determining the number of truck doors is the number of trucks per year multiplied by the hours it takes to load or unload a truck divided by the work hours of a year. Figure 4.6 shows a hypothetical calculation of the required docks for a facility.

When designing the number of docks, add at least one dock space for each of the following activities: ramp or dock for lift truck travel to the ground from the dock, maintenance dock with a high door, dock for trash, dock for paper bales, and one dock for United Parcel Service or U.S. Postal Service trucks.

For the example given in Fig. 4.6, the facility requires 16 dock positions.

Manual Dock Simulation

A second method to determine your number of dock doors that are required at your distribution facility is the manual dock simulation method. This method projects the number of docks that are required for the various types of vehicles that deliver or ship product to your facility. Data required for a manual truck dock simulation consists of the following: (1) for an average day, the number of trucks that arrive at your facility, type of load (floor-stack or unitized), and rear of truck height; (2) the time of day that each truck arrived at your facility; (3) the number of hours that each truck was at the dock position (including spot and departure time); (4) the number of standard and special docks; and (5) your company's anticipated growth rate.

Given:

- Trucks per year = 7000
- Safety factor = 25%
- Turnaround time = 2.5 h
- Work hours per year = 2080 h

$$\frac{7000 \times 2.5}{2{,}080} = 8.4 \times 1.25 = 10.5 \text{ or } 11 \text{ dock positions}$$

FIGURE 4.6 Example of a dock calculation.

With this dock data, you have the information to perform a manual dock simulation for your existing or proposed dock area and the results show the following: (1) number of dock utilizations per hour; (2) the total available work hours for one day; and (3) the number of docks required to handle your projected truck activity, which is based on the previous truck receiving and shipping experience and policy of your company. This policy is a truck dock for every truck as it arrives at your facility or with a fixed number of dock positions. With the latter arrangement, your operation has a number of trucks that are sent to the temporary holding area.

A dock simulation shows you how to improve your dock assignment or delivery schedule for better dock utilization and determines the number of docks that are required for your warehouse.

Manual Dock Simulation Steps. To perform a manual dock simulation, you require the truck information mentioned previously plus (1) a dock simulation form showing the number of dock positions along the top, number of trucks that arrived per time slot, and hours per day (in 15-min intervals) as lines on the side of the form and (2) the number and time lengths of employee coffee and lunch breaks.

The required dock simulation steps are (1) for each truck arrival time under the "truck arrivals" column, enter the number of trucks; (2) for each truck under the "Dock position" column, enter the time that the truck was assigned to the dock in the appropriate square; (3) fill in the squares under the dock position that represents the unloading time (remember that coffee and lunch breaks represent dock occupancy time but unloading activity does not occur); (4) as other trucks arrive at the facility, then the other dock positions become occupied with trucks and (5) if a (new) truck arrives at the facility and existing trucks occupy the docks, then the truck is assigned to the holding area until a dock becomes available. When a dock becomes available, the new truck from the holding area is assigned to the dock and you follow the previously mentioned steps.

Computer Simulation of Dock Operations

When you use a computer simulation to design your truck and rail dock operation, then you provide the same truck and rail delivery and shipping activity, inbound and outbound volume, productivity, and dock door and other design information that was mentioned in the section on manual dock calculation. After the data is entered into the computer, the computer program projects each dock door utilization, required number of doors, and other dock area design information.

TRUCK DOCK DESIGN FACTORS

The truck dock type and location that is designed for your warehouse and distribution facility is determined by several factors, including climate or weather conditions (e.g., prevailing winds), available land, security, delivery truck traffic flow and control, employee comfort, safety, available investment funds, type of delivery vehicle, and type of delivery load.

The various dock concepts are flush dock with a slight incline, flush dock with a depressed driveway, open dock, open dock with a canopy, enclosed dock with a side entry, enclosed dock with a straight-in entry, staggered dock (sawtooth), side-loading dock ("finger" dock), dock extension, warehouse yard with fixed ramp or mobile dock plate, flatbed trailer inside the warehouse, pier dock, and free-standing dock (dock house).

Flush Dock Design

There are two designs for the flush dock, which has a slight incline: the cantilever and vestibule types.

Cantilever Flush Dock.   The cantilever flush dock (Fig. 4.7a) is basically a hole in the building wall with a dock door and dock seal along the outside of the door frame. This concept provides excellent weather protection and security. For unloading, the delivery truck is backed up to the building wall and is secured against the dock seal. The trailer remains outside the facility, and the truck driver enters the building through a personnel entrance. Inside the building, the dock staging area is open to the warehouse storage area.

A cantilever flush dock requires a dock leveler with a truck ICC bar-restraining device (dock plate or chock block). It also requires an excellent communication between the truck driver and dock employees to prevent unexpected truck movement during the loading or unloading activity. An alternative communication method is the stop/go dock light system. The building wall is set back to provide at least 8 in of clearance between the rear of the truck and the building wall at an elevation of 6 ft above the dock and a minimum of 6 in of clearance between the top of the trailer and the building wall.

This clearance is achieved by a dock bumper that extends outward from the building wall. The bumper absorbs the trailer impact from backing up to the dock. The bumper dimensions are determined by the opening size, the loading area shape, the distance between the rear of the truck bed, and the rear wheel center. The bumper is also used with a yard jockey (mule) tractor, where the slope of the trailer is increased as the yard tractor lifts the front of the trailer higher than the normal over-the-road tractor. With this angle and height, the tractor-trailer could potentially hit the building.

With a flush dock, to prevent the risk of inclement weather conditions (e.g., rain or snow) destroying the top of the seal, a small canopy or strut is installed over the seal top.

When your warehouse facility is on grade level (the ground floor is level to the land), a flush dock with a depressed driveway is designed to provide the dock positions. The driveway (loading area) is depressed to create the proper height between the edge of the dock and the rear of the truck bed. Special consideration is given to dock bumpers, water or snow removal or drainage, hand railings, and windows in the door. The window in the door allows the dock employee to observe the status of the truck yard in front of the dock door without opening the door.

Vestibule Flush Dock.   The vestibule dock design (Fig. 4.7b) is a variation of the flush dock type. The difference between the cantilever flush dock and the vestibule dock is inside the building. With the vestibule dock, there is an open area between the building exterior wall and a second interior wall, which is the storage area wall. This dock design is extremely practical as a staging area and useful in the perishable food and cold storage distribution industry, where storage temperatures are closely controlled to protect the product. Lift truck and personnel doors in the interior wall or

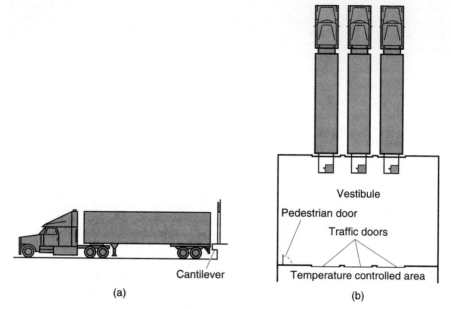

FIGURE 4.7   (*a*) Cantilivered dock design; (*b*) vestibule dock design; (*c*) open dock concept; (*d*) dock curtains or sliding panels. (*Courtesy of Rite Hite.*)

through a tunnel permits access between the storage area and truck dock area. With two required lift truck aisles (one for the dock area and one for the storage area), the vestibule dock area requires a large-square-foot area.

Open Dock Design

The open dock design (Fig. 4.7*c*) is one of the least expensive dock designs for a warehouse facility. The open dock consists of a concrete platform that extends outward from the exterior building wall. The dock distance (depth) between the building side of the dock plate and the building wall provides sufficient open space for unloading and loading equipment to travel onto the dock plate and two-way vehicular traffic on the dock. The dock width and length provide access to all building doors.

Since the open dock is exposed to the weather, safety is a concern in bad weather. When the open dock design is considered for a warehouse, a canopy is extended at least 4 ft beyond the edge of dock. The canopy prevents rain from making the dock area slippery. A wet dock is a hazardous condition for operating lift trucks. An alternative is to install sliding panels or dock curtains on the dock perimeter, thus reducing the negative impact (of rainy or snowy weather) on the dock operations. Additional open dock safety features are yellow-coated concrete posts and chains on the perimeter.

Dock Curtains or Sliding Panels.   To convert an open dock into a semiclosed dock facility, sliding panels or dock curtains (Fig. 4.7*d*) are installed on the dock perimeter to create a solid barrier between the dock area and the truck well.

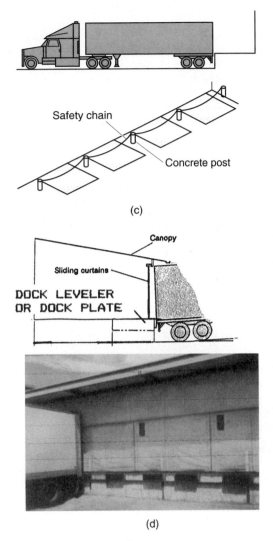

Safety chain

Concrete post

(c)

Canopy

Sliding curtains

DOCK LEVELER
OR DOCK PLATE

(d)

FIGURE 4.7 (*Continued*)

When required to perform dock activities, these sliding panels or dock curtains are opened to sufficient width to create a passageway for a lift or pallet truck between the dock area and the delivery truck. The fixed portion of the dock panels or curtains maintains a solid barrier on the dock edge and the doors of the delivery truck.

When not required to perform dock loading or unloading activities, the panels or curtains are closed and create a solid barrier between the dock and the truck well. When the external building doors are opened for delivery truck entry into the truck well, this barrier helps keep the cold air or dust in the truck well and from entering the dock area.

Enclosed Dock Design

In the enclosed dock design, the truck-loading area (combined length of tractor and trailer) is inside the building or sheltered area. At the exterior of the building is a set of doors that control the delivery truck entry to the loading and dock area. The interior enclosed dock area is an open or flush dock design.

In the enclosed dock area, adequate ventilation is required to prevent truck engine fumes from building up in the area. Other features are adequate water drainage, lighting, dry or wet sprinkler system that is protected with heat tape or wire, and markings on the floor or dock wall that identify the dock space. These features improve the dock safety and driver truck stopping efficiency.

Side-Entrance Enclosed Dock. The first enclosed dock design is the side-entrance design (Fig. 4.8*a*); however, construction costs for this design are high. In this design there are two sets of doors (one at each side of the building) and the truck-loading and maneuvering areas are inside the building. This feature dictates one-way delivery vehicular traffic flow through the building ("in one door on the right-hand side of the building and out the second door on the left-hand side of the building"). The side-entrance enclosed dock design accommodates the staggered (sawtooth) dock design.

Straight-in-Entrance Enclosed Dock. In the straight-in enclosed dock design (Fig. 4.8*b*) there is one exterior door per truck dock position or one extralarge wide exterior door for two truck dock positions. The delivery truck maneuvering area is outside the building. Since the delivery truck loading area is inside the building, the delivery truck backs straight up in the loading area to the dock edge. The straight-in enclosed dock design provides a shorter truck-loading lane that permits the trailer to be dropped inside the enclosed area or with a long truck-loading lane to accommodate both tractor and trailer.

Side-Loading or Finger Dock. The side-loading dock or finger dock design (Fig. 4.8*c*) is used primarily for flatbed trucks and open-sided vans (side-opening vans with side door open for loading and unloading). This dock design is a cutout in the building interior floor. The cutout is directed inward from the building exterior wall into the dock staging area. The finger length and width is sufficient in size to permit a flatbed trailer open-sided van to back into the opening. Concrete platforms on both sides of the cutout permits lift trucks to unload and load the trailer. To improve dock safety, yellow-coated posts and chains surround the cutout.

Drive-through-the Facility Design. This drive-through design (Fig. 4.8*d*) is used for flatbed trucks or open-sided vans. The flatbed truck or open-sided van drives into the building, and lift trucks or an overhead crane unloads or loads the product from or onto the flatbed trailer. This dock design requires a one-way flow of vehicles through the facility, or the trucks are backed up into the facility. It is a very common truck-unloading design in certain countries such as Australia or in the beverage distribution industry.

Staggered (Sawtooth) Dock Design. When there is limited truck maneuvering area, the staggered (sawtooth) dock design (Fig. 4.9*a*) is preferred. This design requires a one-way flow of vehicles in the truck yard that matches the approach to the angle of the staggered dock positions. This dock arrangement requires either a counterclockwise or clockwise truck flow pattern. In addition, the loading area must be level at the dock, or the trailer bed will not line up with the edge of dock.

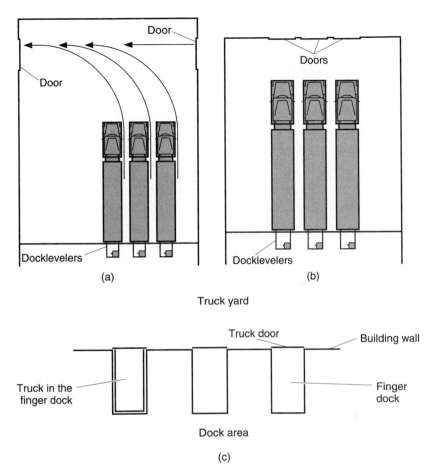

FIGURE 4.8 (*a*) Side-entrance enclosed dock (*Courtesy of Rite Hite.*); (*b*) straight-in-entrance enclosed dock (*Courtesy of Rite Hite.*); (*c*) Side-loading or finger dock.; (*d*) drive-through concept.

The major disadvantages of the sawtooth dock design are that it requires approximately twice the building dock space for fewer docks and an increase in the construction requirements and costs.

Pier Dock Design.   If the dock side of the building does not have sufficient wall space for the required number of dock positions or the building interior layout does not permit construction of a dock area, then the pier dock design (Fig. 4.9*b*) has the potential to provide the necessary dock positions. In this design a section of the building extends outward from the building. Dock positions are located on each side of the extension. The width of the extension permits a lift truck turning aisle for turning onto the dock levelers and two-way lift truck traffic on the dock. With the pier dock design, there is a tractor-trailer maneuvering area on both sides of the pier dock.

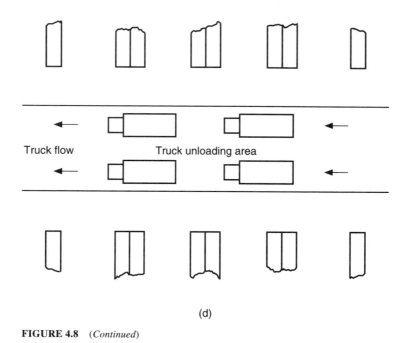

(d)

FIGURE 4.8   (*Continued*)

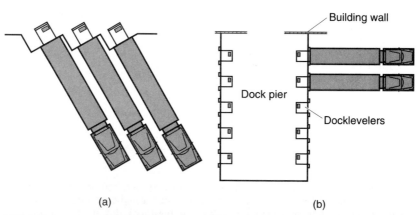

(a)   (b)

FIGURE 4.9   (*a*) Sawtooth dock concept (*Courtesy of Rite Hite.*); (*b*) pier dock concept (*Courtesy of Rite Hite.*); (*c*) freestanding (dock house) concept (*Courtesy of Rite Hite.*); (*d*) mobile yard ramp (*Courtesy of AAR Brooks and Perkins.*).

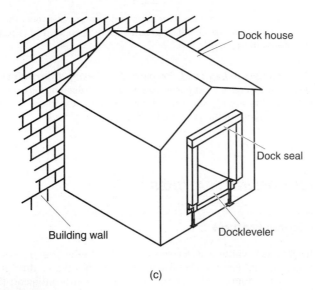

(c)

(d)

FIGURE 4.9   (*Continued*)

Freestanding Dock (Dock House) Design.   To increase dock positions in a low- or small-volume operation with limited interior space inside the building, a freestanding dock or dock house extension is constructed on the building wall (Fig. 4.9*c*). The freestanding dock is a platform with a dock leveler that extends outward from the building and is enclosed with metal or plastic panels. If the dock extension is an open platform, then to improve safety, chains and posts are on the dock perimeter.

Mobile Yard Ramp Design.   The mobile ramp design (Fig. 4.9*d*) is used outside the building. This dock operation uses a mobile ramp that permits a lift truck to unload or load a trailer and to have the mobile warehouse equipment transport the product into the warehouse. If the warehouse is not at ground level, then a ramp is required for lift truck entrance into the building.

Summary of Alternatives When Dock Property Is a Design Constraint.   In summary, when building wall area or property size is a dock design constraint, alternative dock designs for facility are (1) staggered (sawtooth), (2) extension (dock house), (3) pier dock, (4) mobile ramp in the yard, and (5) freestanding dock.

OTHER DOCK DESIGN FEATURES

Dock Door Design Features

Dock door heights and widths are determined by the trailer rear dimensions, seal or shelter type, and internal environmental conditions within the facility. It is a general rule of thumb that the doors are an energy loss and security problem; therefore, the number of dock doors is kept as small as possible.

For average warehouse operations, door widths range from 8 to 9 ft to agree with the corresponding rear trailer dimensions of 8 and 8½ ft. When a truck is backed up to the dock door, this arrangement permits the trailer to have a slight variation from an exact center.

With the trend in the trailer design of 102-in-wide trucks, permitting an increase in the pallet load-carrying capacity, the 8½- or 9-ft-wide door is preferred for trailer dock doors on new buildings.

For regular over-the-road trailer loading and unloading activity with a 4-ft-high dock, the dock door clear heights range from 8 to 10 ft high. The 8-ft-high door permits a 7- to 7½-in-high pallet load to travel across the dock plate or leveler. A 9-ft-high door with 3 in of top clearance permits an 8¾-ft pallet load height to be handled across the dock plate or leveler. The 10-ft-high door handles a high-cube trailer.

One important door height principle is that at least one dock door have a 13½- to 14-ft clear height to accommodate the height of the tallest warehouse material handling equipment with a collapsed mast. Therefore, a 14-ft-high door accommodates most warehouse equipment deliveries in the United States, where the highway bridge clearance is 13½ to 14 ft high.

How to Bridge the Gap between Dock Edge and Trailer Bed

At the edge of dock height, the dock plate or leveler bridges the gap (open space) between the dock edge and the rear of the trailer bed. The dock plate or leveler is the factor that assures high employee productivity and safety. The height variation between the dock edge and rear of the trailer bed is the major factor determining the type of mobile pallet handling equipment that is used in the dock function. The other dock plate or leveler selection factors include the frequency of deliveries, unit-load handling equipment, type of dock concept, and available capital.

Dock equipment needed to bridge the gap includes (1) portable walk ramps, (2) portable dock plates with a T-bar, (3) portable dock boards, (4) mobile dock, (5) recessed (pit) dock levelers (manual or hydraulic), (6) vertical stored dock levelers (manual or hydraulic), front of dock, (7) edge of dock, (8) hydraulic jack in the floor, (9) scissors lift, (10) pit installed scissor lift, (11) dock lift lever, and (12) truck trail gate.

Portable Walk Ramp. The portable walk ramp (Fig. 4.10*a*) is an aluminum ramp that is designed with a skid-resistant deck and is used at a truck dock or a grade-level facility. The ramp is up to 3 ft in length to bridge a 9-in height variance. The ramp is up to 16 ft long to bridge a 5-ft height variance and handles loads of up to 1300 lb. Beveled ends and a 11-in lip permit an even passage for an employee with a two-wheel hand truck between the truck bed and the floor or ground. Side curbs on the walk ramp reduces the overall usable width by $2\frac{1}{2}$ in, but improves safety. The ramp handles a low volume, has a low capital investment, permits side delivery truck door loading and unloading, and enables a two-wheel hand truck or employee to move the product.

Portable Dock Plate. The portable dock plate (Fig. 4.10*b*) with an equalizing bend is an aluminum plate with a length of 2 to 5 ft and width of 5 to 6 ft and handles a height differential from 3 to 10 in. At the truck dock, a short plate handles a low height differential and a long plate handles a high height differential. To reduce mobile warehouse pallet handling equipment skids, the dock plate has a diamond-pattern surface. To secure the plate position in the gap, the plate has a locking leg (T-bar) that is attached to the underside and fits into the gap. This lightweight and low-cost dock plate with handles on both sides is easily moved by an employee to the required dock position. Safety yellow side strips along both sides of the dock plate surface minimize pallet truck runoff. The dock plate has sufficient structural strength and width to handle a low-lift pallet truck (jack). When a high height differential exists between the truck and the dock edge, then the rear two pallet loads of the delivery truck are removed by a lift truck. Disadvantages of the dock plate are that it limits the use of warehouse equipment, it requires a lift truck to handle the rear two loads, and mobile equipment can slide off the plate and potentially increase the risk of damage to delivery truck sides.

Advantages of the dock plate are low cost, mobility between dock positions, and capacity to handle a low volume.

Portable Dock Board. The portable truck dock board (Fig. 4.10*c*) with an equalizing bend is an aluminum device that permits an electric pallet truck or a rider fork-lift truck to unload and load trailers. To reduce skids, the dock board has a diamond-pattern surface, and to secure the dock board in the gap, locking legs (T-bars) are attached to the underside and are inserted in the gap. Yellow-painted safety curbs help eliminate equipment runoffs. The dock board weight requires powered mobile dock equipment to move and position the dock board in the required dock position. When fork-lift trucks are used in the dock area, some dock plates have fold-down loops or lifting chains to assist in relocation and positioning of the dock board.

When compared to the portable dock plate, the portable dock board has the same disadvantages, except it is moved to the required dock positions by a lift truck. The advantages are the same, except it permits a lift truck to load or unload the delivery truck.

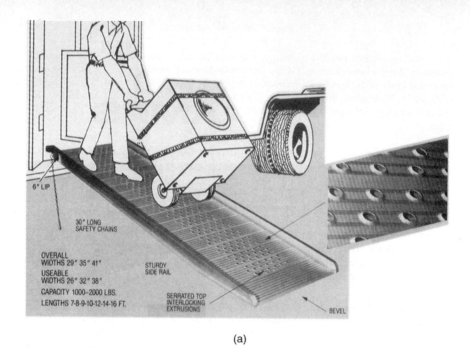

(a)

(b)

FIGURE 4.10 (*a*) Portable walk ramp; (*b*) portable dock plate; (*c*) portable dock board. (*Courtesy of MTD Product.*)

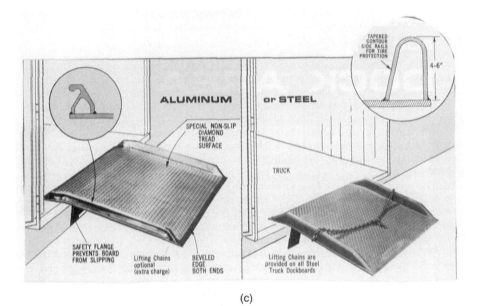

(c)

FIGURE 4.10 (*Continued*)

Mobile Yard Ramp. For facilities at grade (ground) level, a mobile yard ramp, ranging from 30 to 36 ft in length, permits a lift truck to unload or load a trailer. The ramp has aluminum or steel grating platform and a usable width that ranges between 4 ft 5 in and 7 ft 4 in. Side guards on both sides reduces powered equipment runoff. Dual wheels are located under the center, and a tow bar permits a lift truck to tow and position the ramp. At the required location, some mobile ramp models have a hydraulic system for easy ramp positioning onto the truck bed and a level-off top for lift truck maneuvering and removing pallets that are at the rear of the delivery trailer.

Recessed Dock Leveler

For facilities that have a truck dock edge at 4 ft height, the recessed dock leveler method is most commonly used to bridge the gap. Each dock leveler is installed in a pit or in a cutout of the warehouse floor. It has a diamond-pattern surface with a lip. As required, the dock leveler surface is raised up with a lip that extends outward and lowers to rest on the rear of the delivery trailer bed. The structural strength of the dock leveler permits powered mobile warehouse vehicles to travel between the trailer and the dock. With electric low lift trucks (jacks), the dock leveler is 6 ft long and handles a 7- to 8-in height differential. A 12-ft-long dock leveler handles a 15- to 16-in height differential. A gas-powered lift truck travels over a 12-ft-long dock leveler, then the dock leveler compensates a 17- to 18-in height differential. A pallet jack travels over a

12-ft-long dock leveler, then the dock leveler compensates a 13- to 14-in height differential. The most common recessed dock leveler length is 8 ft, which handles a height differential of 10 in.

The width of the recessed dock leveler has a range from 6 to 7 ft. The 7-ft-wide leveler is best for 8-in-wide trailer and has a tapered lip on each side to compensate for a delivery trailer that is spotted (positioned) off center, or for a narrow delivery trailer. When a tapered dock leveler is used at the dock position, it creates a material handling equipment drop-off zone as a result of the space that is created by the tapered ends.

When a 12-ft-long dock leveler is required at the dock area, the dock leveler extends further into the warehouse. This feature compensates for the extreme height differential, but increases the building area requirement and the cost or reduces the dock staging area.

The standard recessed dock leveler lip is 16 in long, permitting a 12-in extension into the rear of the delivery trailer. Another important lip feature is that the lip or edge permits the lip to stay in firm contact with the delivery trailer bed. To reduce pallet jack wheel hang-up in the floor recess (drain) of refrigerated delivery trucks, the lip projection length is a very critical factor.

When compared to the other dock equipment that bridges the dock edge and trailer gap, disadvantages of the recessed dock leveler are higher capital investment and the fact that each dock position requires the use of a dock leveler. Advantages include compensation for a wide height differential, ability to handle a lift truck with a low grade range, improved flexibility to service a wide range of trucks, capacity to handle a heavy load, long life, allowance for high employee productivity, greater dock safety, and ability to handle all loading and unloading methods.

The recessed dock leveler options (Fig. 4.11a) that improve dock area safety, energy conservation, and sanitation are full-range toe guards on both sides of the leveler, ICC bar trailer restraints to prevent delivery trailer rollaway, dock restraint lip to prevent lift trucks from running off an open dock, weather seals between the dock leveler and curb channel to improve sanitation and reduce loss of refrigerated air, and dock occupancy lights.

The two types of recessed dock levelers are (1) hydraulic (automatically operated) and (2) mechanical (manually operated).

Hydraulic Recessed Dock Leveler.   The hydraulic recessed dock leveler (Fig. 4.11b) has a hydraulic pump and motor in the pit that moves the dock plate to the desired elevation. The dock leveler controls are push buttons. The buttons are located on the interior of the building wall that is adjacent to the dock position. After the delivery truck is positioned at the dock, the employee presses the button that causes the dock leveler to automatically rise and lower to a flat-level position with the lip inside on the trailer bed.

With few mechanical parts and structural members in the pit, the hydraulic recessed dock leveler is very reliable, requires little maintenance, and permits cleaning of the pit. This last feature is very important in the food and hospital industries or for any industry that has very high sanitation or hygiene standards. However, the hydraulic recessed dock leveler requires a higher capital investment.

Mechanical Recessed Dock Leveler.   The mechanical recessed dock leveler (Fig. 4.11c) has a diamond-pattern surface and upwardly biased ramp with a spring mechanism that is held down by a releasable ratchet device. To operate, the employee walks onto the ramp and pulls on the release chain that is connected to the ratchet mechanism. This pulling of the ratchet releases the ramp, and the ramp rises to its uppermost position. During the upward movement, the ramp lip is automatically extended for-

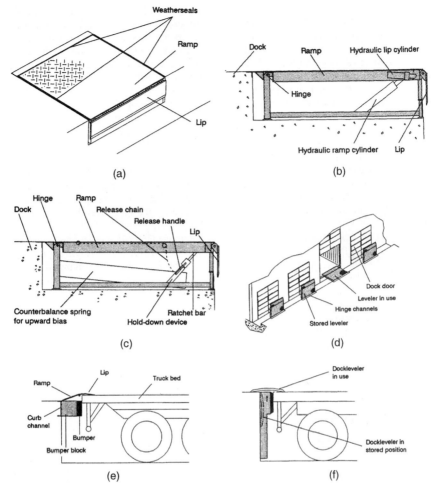

FIGURE 4.11 (*a*) Dock leveler options; (*b*) hydraulic dock leveler; (*c*) mechanical dock leveler; (*d*) vertically storing dock leveler; (*e*) edge-of-dock leveler; (*f*) front-of-dock leveler. (*Courtesy of Rite Hite.*)

ward. The employee walks forward onto the extended ramp, which is forced down onto the trailer bed, and the dock leveler is ready for the loading or unloading activity.

The main advantage of the mechanical recessed dock leveler is that it costs less than the hydraulic model. With a greater number of components in the pit, however, there is increased expense on labor to clean the pit and for maintenance expense.

Vertically Storing Dock Leveler

The vertically storing dock leveler (Fig. 4.11*d*) consists of a ramp that is withdrawn in the vertical position inside the building. The standard vertical storing ramp is available in 5-ft length with a 6- to 7-ft-wide plate. The 6-ft-wide leveler is best for a delivery trailer that is less than 96 in wide. The lip projection and bend have a standard lip length of 16 to 18 in and a grade break of 12.5 percent. This allows the vertically storing dock leveler to handle a dock edge − truck bed height differential of 6 in. The vertically storing ramp pivots on hinges that are installed on the interior side of the wall in a step-down. The step-down is a continuous pit that runs on a steel channel along the entire wall face that parallels the dock doors.

Since the vertically storing dock leveler does not have a pit, it reduces the sanitation problem and loss of refrigerated air. Other energy control features are a tighter seal on a flush dock and tighter dock seal on a delivery trailer that is backed up to the dock.

The vertically storing dock leveler is available in hydraulic (automatic) or mechanical (manual) models.

With the hydraulic dock leveler, the dock employee controls the lip and ramp with push buttons from a control panel that is located on the interior of the building wall. After use of the dock leveler, the vertically storing dock leveler ramp is locked in the vertical position. At this point, the leveler is pushed by the employee to another dock position.

The mechanical vertically storing dock leveler is manually moved between dock positions. Generally, the lip is in the extended position; however, if required, the lip is manually retracted. At the required dock position, the dock employee manually lowers the ramp onto the trailer bed. Counterbalanced springs assist in the lowering effort. When not in use the ramp is returned to the vertical position.

When compared to the recessed dock leveler design, the vertically storing dock leveler design requires a higher initial capital investment per each dock leveler and the step-down collects debris. Advantages of this design are lower maintenance costs, energy control improvements, and lower total dock investment due to increased ramp mobility.

Edge-of-Dock Leveler

The edge-of-dock (EOD) leveler (Fig. 4.11e) is a short ramp with a lip that is mounted to a steel channel that is embedded to the front of the dock. The EOD leveler is available in ramp lengths of 27 and 30 in and widths of 66 and 72 in. The EOD leveler has a standard 15-in lip with an 11½-in lip projection. For EODs with a 17-in lip, a 13½-in lip projection into the truck is provided. Edge-of-dock leveler devices are available in both standard-profile and low-profile types. The low-profile type is best used for dock operations using a lift truck and pallet truck with a low clearance. A generally accepted rule of thumb is that the EOD leveler is not used with high-speed mobile dock equipment or vehicles with a low underclearance.

If powered belt, roller, or skatewheel extendible conveyors are used in the unloading or loading activity, then the EOD leveler reduces the total dock area investment.

The EOD leveler is a hydraulic or manual device. An employee controls the hydraulic dock leveler with push buttons located on the interior wall of the building. When activated, the hydraulic model pushes the ramp upward with the lip in the extended position. After reaching full extension, the lip is lowered onto the rear bed of trailer.

With the manually operated model, counterbalanced springs assist in raising and lowering the dock leveler.

After the trailer departs from the dock, the EOD leveler automatically returns to the stored position.

The EOD leveler requires a low capital investment but handles a trailer bed − dock edge height differential of only 5 in above or 5 in below the level.

Front-of-Dock Leveler

The front-of-dock (FOD) leveler (Fig. 4.11f) is a bridge device that has its own built-in bumpers and is bolted to the concrete edge of the dock. The device is available with a ramp length of 30 in and width of 72 in. The dock leveler is available in both high- and low-profile types with a standard projection of 11 in. A special grocery industry model has a 14-in lip projection for placement in a trailer that delivers perishable food items. The high-profile bridge device serves a height differential of 6 in above and 4 in below; and the low-profile device, 4 in above and $1\frac{1}{2}$ in below.

If powered belt, roller, or skatewheel extendible conveyors are used in the unloading or loading activity, then the EOD leveler reduces the total dock area investment.

The dock employee manually lifts the ramp with the help of counterweights, flips the ramp into the horizontal position, and lowers the dock leveler onto the rear of the delivery trailer.

The FOD leveler is not recommended for loading or unloading activity that uses high-speed warehouse vehicles (lift trucks) or vehicles with low underclearances.

After the trailer departs from the dock, the FOD leveler automatically lowers to the stored position.

Disadvantages of the device are that it cannot service all vehicles and accommodates only limited height differentials. Advantages are low capital investment and ease in relocating.

Lift and Bridge Devices

In-Floor Hydraulic Lift.   The in-floor hydraulic (ram) lift (Fig. 4.12a) is an outdoor device that has an elevating platform. The elevating platform is pit-installed in the delivery truck-loading yard area and is driven by a hydraulic pump that permits the lift to travel in a vertical direction. The in-floor hydraulic lift controls are located on an interior wall of the building or on the platform. The 7 × 10-ft platform has a diamond-pattern surface, and two sides have kickplates and handrails. One open end of the platform faces the warehouse door, and the second open end, with a lip, faces the rear of the delivery trailer.

As required, the ramp lift permits employees to move the pallet load and warehouse pallet load handling equipment between the delivery trailer and the warehouse. The rear of the delivery trailer − dock height differential does not impact the lift to provide the bridge between the trailer and the warehouse. When the lift is not in use, then the platform side guards (kickplates or handrails) are removed and the surface is at grade level.

The hydraulic pump is electrically operated, and in cold environments attention is given to keeping the hydraulic fluid warm.

To operate the lift, an employee plugs in the controls and raises or lowers the platform to the desired transaction level. At the desired level, the employee performs the transaction and then raises or lowers the platform to the next required level.

To improve the safety of the operation, a protective shroud engulfs the three or

(a)

(b)

FIGURE 4.12 (*a*) Dock leveler lift (*Courtesy of Rite Hite.*); (*b*) scissors lift (*Courtesy of Advance Lifts.*); (*c*) bascule bridge (*Courtesy of Magline, Inc.*); (*d*) tailgate truck or trailer (*Courtesy of Rite Hite.*); (*e*) manual wheel lift (*Courtesy of MTD Products.*); (*f*) hydraulic wheel lift (*Courtesy of Kelley Co.*).

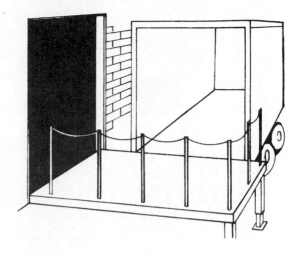

(c)

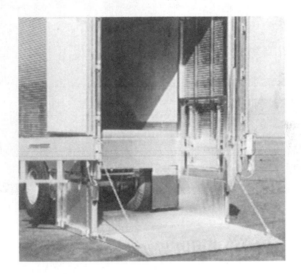

(d)

FIGURE 4.12 (*Continued*)

four sides of the lift. The expanded length of the shroud is from the ground level to the maximum platform elevation. This feature reduces the risk of injury to people and animals becoming trapped under the lift.

Scissors Lift.   The scissors lift (Fig. 4.12*b*) is an outdoor bridge device that is an elevating or declining platform. The scissors lift has a diamond-pattern surface that has the ability to move up or down. The surface area has two sides with kickplates and

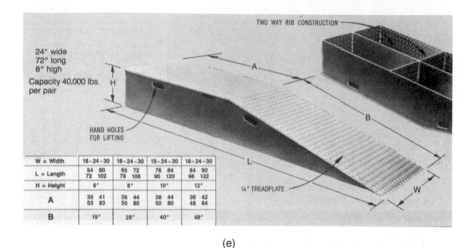

24" wide
72" long
8" high
Capacity 40,000 lbs.
per pair

HAND HOLES
FOR LIFTING

TWO WAY RIB CONSTRUCTION

¼" TREADPLATE

W = Width	16-24-30	18-24-30	18-24-30	18-24-30
L = Length	54 60 72 102	66 72 78 108	78 84 90 120	84 90 96 132
H = Height	6"	8"	10"	12"
A	35 41 53 83	38 44 50 80	38 44 50 80	36 42 48 84
B	19"	28"	40"	48"

(e)

(f)

FIGURE 4.12   *(Continued)*

handrails. Removable gates are located onto the two open sides. One open side faces the warehouse door, and the second open side faces the delivery truck. The delivery trailer end has a lip to assure a smooth transition of the pallet load handling equipment between the delivery trailer and the scissor lift.

The scissors lift has a normal lift height differential of 48 in from a lowered minimum height of 10 in above the grade level. Some heavy-duty models have a lowered elevation of 20 in. With these lowered elevations above the grade, the lift is installed in a pit, which has a lip projection or a slight ramp to compensate for the minimal height differential. The 6 × 8-ft or 8 × 10-ft diamond-pattern surface platform size handles a manual pallet truck, a standard pallet load, and two employees.

The scissors lift consists of a dual set of legs (scissors) that are moved by an electrically controlled hydraulic pump. As required, the pump pressure moves the legs upward. The top of the legs have rollers that move on the underside of the platform. The legs moving upward and the rollers moving horizontally cause the platform to lower.

The electric controls are attached to the lift or on the interior building wall. During operation of a scissors lift, a skirt (shroud) around the platform improves the safety of the area.

The scissors lift is available in three designs. The first design is the lift that is installed on the grade level. This design has a baseframe that rests on the ground and has the platform lowered to a height that is above the ground.

The second design is the pit scissor lift design that has the baseframe resting in a pit. This concept does permit the platform to remain level with the floor or ground.

The third scissors lift design is the portable lift that handles a 3000-lb load on a 6 × 6-ft platform. The mobile lift with lockable wheels is easily located by the employee at the required location to perform the dock activity.

To operate the scissors lift, an employee adjusts or aligns the platform elevation to the rear of the trailer and transfers a manually operated pallet truck with a pallet load from the trailer onto the platform. Next, the platform is lowered (declined) to a level that corresponds with the warehouse dock. At this elevation, the pallet load and vehicle are moved into the warehouse.

Bascule Bridge Dock. The bascule bridge dock (Fig. 4.12*c*) is another outdoor bridge device. The bridge has a diamond-pattern-surface platform that is attached to the side of the building in front of the dock door. All controls are located on an interior wall of the building.

When not in operation, the bridge is in the raised position against the truck door. As required, the electronically controlled hydraulic powered unit lowers the bridge in a 90° outward arch from the building. In the lowered extended position, legs on each corner support the bridge.

In the lowered position, the platform (6 × 20 ft or 8 × 20 ft) has a dock lip that sets on the rear of the delivery trailer, and along the two sides the safety rails, chains, and kickplates are inserted into the platform holes.

For safety when lowering the bridge, a second employee outside in the yard ensures that there is no person, animal, or equipment in the bridge lowering path.

Tailgate Trailer. The next bridge device is the tailgate delivery trailer (Fig. 4.12*d*). When the delivery trailer is on the highway, the tailgate is stored under the delivery trailer. At the unloading and loading location, the truck electronically controlled hydraulic system unfolds the two sections of the unloading tailgate onto the ground. The tailgate has an 8 × 8-ft diamond-pattern surface with a safety plate across the entire width.

The hydraulic pump as required uses the truck electrically powered system or uses the building electrical outlet.

To operate, the tailgate is raised to the rear of the trailer and the cart or pallet is placed on the tailgate. After the employee lowers the tailgate to the ground or dock, the unit load is transferred to the warehouse.

Wheel Lift. The next outdoor device to raise the delivery truck bed is the wheel lift. The two basic types are the manually and the hydraulically operated wheel lifts. Both wheel lifts are outdoor devices that are located next to the building and extend outward in the loading area.

The manual wheel lift (Fig. 4.12*e*) (two per truck) consists of wood or metal structures that have the width and length surface to support the delivery truck rear wheels. These devices are placed against the building wall under the dock edge and are located in the delivery truck (two rear wheels) path. As the truck backs up to the dock, it backs up on the wheel lifts. With the rear wheels on the two wheel lifts, the rear bed section of the delivery trailer is raised to a height differential that is handled by a dock board or dock leveler.

The hydraulic rear wheel lift (Fig. 4.12*f*) (leveler) has a diamond-pattern surface that is pit-installed in the loading area against the building wall under the dock edge. The surface is 10 ft wide by 14 ft long. A 5-in-high wheel locator is an elevated middle portion of the platform. When the truck is backed up to the dock, the wheel locator is between the truck rear wheels.

The hydraulic device has controls on the inside of the building wall and has the ability to raise or lower the rear delivery truck bed to the required height for a dock board or leveler.

When using the wheel lift devices, attention is given to the truck new slope and the height of the trailer or dock. The new slope has the delivery truck bed that is higher at the dock edge and lower at the truck nose. To unload a delivery truck, this new slope creates an upgrade pull for an employee with a pallet truck. To reduce truck or product damage, the loading employee is required to be extremely careful because of the decline slope that is created inside the delivery truck.

Hi-Way Lift.   The next device to reduce the height differential between the rear bed of the delivery truck and the dock is the "hi-way" lift. The hi-way lift is a hydraulic electronically controlled device that has a diamond-pattern surface. The surface is 10 ft wide by 40 ft long and has the capacity to lift and lower a fully loaded delivery trailer (without a tractor) to the required dock elevation.

The device is pit installed in the loading area with rams that are controlled from the inside of the building. The device requires a tractor to drop the trailer onto the surface. With the delivery trailer secured on the platform and the rear doors unlocked, the front of the lift is raised to a specific elevation. At this elevation and slope, the product slides from the truck into the appropriate warehouse receiving trough. This device is used to handle granular or bulk materials.

Dock Leveler Lift.   The dock leveler lift (Fig. 4.12*a*) is an indoor device that compensates for the elevation change between the ground level and the dock. This device is a lift plate with a diamond-pattern surface that travels the distance between the ground level and the dock. The capacity is to hold a normal counterbalanced lift truck with an operator and pallet load.

The dock lift is equipped with a protective shroud. When the dock lift is not in use, a sensing device automatically ensures that the lift returns to the warehouse floor level. This feature, and yellow-coated chains and posts around the perimeter, improve safety.

When compared to the ramp design, one disadvantage of the lift design is that it handles only one vehicle per trip and requires greater maintenance; however, it requires only a small area.

OPTIONS TO IMPROVE DOCK SAFETY AND EFFICIENCY

Your design for an efficient and safe dock operation requires additional devices other than the dock type or bridge device to reduce the height differential between the dock and the truck. These devices and areas are in front of and behind the dock door. These include (1) wheel chocks, (2) exterior guard posts, (3) dock bumpers, (4) interior door guard posts, (5) dock leveler lip, (6) truck canopy, (7) dock ladders, (8) fifth wheel lock jack, (9) guide lines, and (10) dock identification.

Dock Doors Design Features

The dock door is a building device that permits passage through a wall that provides security and assures energy conservation. The three most common door types are vertical lift, vertical and horizontal movement, and rollup.

A small window installed in the door is a design advantage as it reduces an employee's need to open the door to verify that a delivery truck is at the dock position.

A second feature is that each door is designed with a device to lock it from the interior of the building with an industrial paddle lock. This feature, as well as the dock door window, improve the security of the facility.

The third feature in cold and windy climates is weather stripping along the door, frame, and bottom to reduce energy loss.

The fourth dock door feature in warm climates is a folding gate that covers the entire door passageway. During warm and sunny days, the gate is in a closed and locked in position, preventing personnel from passing through the door. This arrangement maintains good security and does permit a cool breeze to enter the building.

Another door feature in a warm climate with a flush dock to reduce energy loss is a short dock house that extends outward from the exterior of the building at each dock location. This house engulfs the rear of the trailer, and the feature reduces warm air movement into the building.

Dock Seals and Shelters

Dock seals and shelters are exterior door frame and jam features that extend outward from the building and permit the truck rear sides and top flush against this extension. These devices ensure that there is no gap on the two sides and top of the delivery truck and building door frame and jam. With no gap, it improves dock security and energy conservation.

Dock Seals.　The dock seal (Fig. 4.13*a*) consists of two pads or an air pad (Fig. 4.13*b*) on the two side frames of the door and one head curtain along the top of the door jam. Some models have movable head curtains that are manually adjusted for the trailer height. Yellow strips or pads with yellow strips are options on the seal exterior sides. These yellow strips help the truck driver to properly locate the delivery trailer at the dock position. The pads extend the life of the seal by reducing wear from the delivery truck sides movement, which rises and lowers as the delivery truck is loaded or unloaded by the warehouse employees.

Dock Shelters.　Dock shelters (Fig. 4.13*c*) conserve energy and improve dock security. The dock shelter is a fixed-frame structure or flexible-frame structure with a head curtain that extends outward from the building. The two sides of the frame hang onto the delivery trailer sides and roof and conform to the rear door of the delivery trailer or truck. Both types of shelters have steel channel protectors to reduce trailer damage as the trailer is backed up to the dock. Yellow strips on the pads or plain pads serve the same purpose as the yellow seal pads.

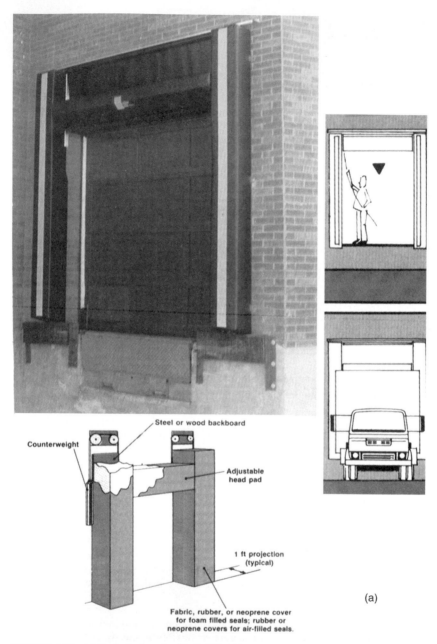

Counterweight

Steel or wood backboard

Adjustable head pad

1 ft projection (typical)

Fabric, rubber, or neoprene cover for foam filled seals; rubber or neoprene covers for air-filled seals.

(a)

FIGURE 4.13 (*a*) Foam dock seals; (*b*) air-inflatable dock seals; (*c*) dock shelter. (*Courtesy of Airlocke Dock Seal and Material Handling Engineering Magazine.*)

(b)

FIGURE 4.13   (*Continued*)

Dock Lights

Dock lights are adjustable, movable devices that provide light into the interior of the delivery truck. The dock light consists of a single lamp that has a flexible arm which reaches to the side and upper portion of the trailer rear door. Most dock light arms are located between two truck doors and service both dock doors. Some dock light arms provide light to a single dock door. Lockable screens protect the lamp from damage and pilferage.

An option is a wire prong that extends in the door path. As the door is opened by an employee, the wire prong automatically turns the light on as the door is moved into the open position. As the dock door is closed by the employee, the wire prong comes in contact with the dock door and turns off the light.

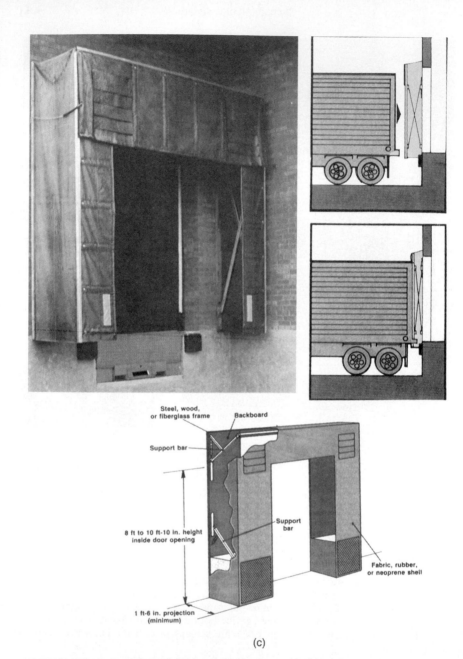

Steel, wood, or fiberglass frame

Backboard

Support bar

8 ft to 10 ft-10 in. height inside door opening

Support bar

Fabric, rubber, or neoprene shell

1 ft-6 in. projection (minimum)

(c)

FIGURE 4.13 (*Continued*)

4.44

ICC Bar Lock (Hook) Device

The ICC bar lock (hook) device (Fig. 4.14*a*) is a dock leveler option that improves lift truck safety in the dock area. The lock (hook) is a device in the dock leveler pit area and is controlled from the dock to extend forward, upward, and beyond the dock leveler lip. When a delivery truck is spotted (backed up) at the dock, the hook motion and new position are in front of the delivery truck or trailer ICC bar. The hook in this position restricts the delivery truck forward movement, thus preventing trailer rollaway or unexpected delivery truck departure.

Stop and Go Lights

Stop and go lights (Fig. 4.14*b*) consist of red and green lights on the inside and outside of each dock door location. When a delivery trailer or truck is at the dock location, the light is red and signals to both the lift truck operator and truck driver that the delivery truck is not ready for departure. When there is no delivery truck in the dock position or the delivery truck is ready for departure, the light is green. This green color signals to the lift truck driver and delivery truck driver that the truck is departing from the dock position and it is not safe to enter with the delivery truck or trailer.

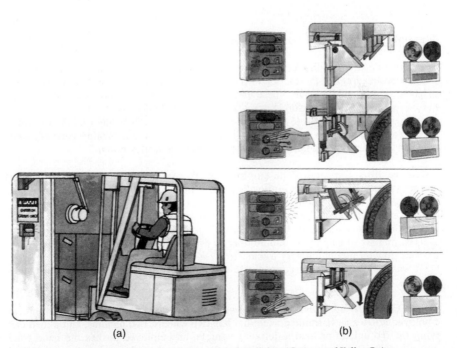

(a)   (b)

FIGURE 4.14   (*a*) ICC bar lock (hook); (*b*) stop-and-go lights. (*Courtesy of Kelley Co.*)

RECEIVING AND SHIPPING OFFICE

An important dock area feature is the receiving and shipping office complex. The office complex consists of the supervisor and clerk office area for document control. Other portions of the office complex are restrooms for employees and separate toilets for truckers. The truckers' area has pay phones, lobby, vending machines, and a sliding window on the office wall for document transfer. Some companies, for company truck drivers, provide a sleeping area with full toilet facilities.

UNLOADING AND LOADING CONCEPTS

When you consider the unloading and loading activity, you are considering the physical movement of the product between the delivery vehicle and the receiving-shipping dock. Three methods are used to move product: (1) manual method, (2) mechanized method, and (3) automated method.

For effective operation and good employee productivity, each method is matched with its product, type, and volume and your product movement (in-house transportation) system.

Manual Unloading and Loading Concepts

The first unloading-loading concept group consists of manual methods. An employee is required to carry the product or to pull or push a carrier with product between the delivery vehicle and the dock. These various concepts are applied in any type of distribution facility that handles single items, cartons, or pallet loads.

These methods include (1) employee carry, (2) two-wheel hand truck, (3) roller pallet, (4) manual pallet truck, (5) semilive skid, (6) garment trolley cart, (7) four-wheeled and hanging garment carts, and (8) extendible trolley boom.

Employee Walk Concept. The employee walk concept is considered the simplest and most basic in loading and unloading delivery vehicles. This method requires an employee to carry the product between the delivery truck and the dock. However, this concept requires a large dock area and a maximum amount of time to handle a trailer load. Other disadvantages include increased risk of employee injury, requirement for greatest number of employees and number of dock positions, low employee productivity, and slow turning of delivery trucks.

Advantages of employee carry are low capital investment and capacity to handle all types of SKUs.

Two-Wheeled Hand Truck. In the two-wheeled hand truck manual method, an employee utilizes an H-frame device that has two wheels, two handles, and a load carrying lip. To unload or load a delivery vehicle, an employee pushes the hand truck into the delivery vehicle, loads product onto the hand truck, pushes the hand truck from the delivery vehicle, and unloads the product onto the staging area. This method requires 6 to 7 h to unload and load a delivery truck.

Disadvantages of this method are slow delivery truck turning, requirement for stackable product, and difficulty in handling a high volume and traveling up grades.

Advantages of the two-wheeled hand truck include low cost, improved employee productivity, and reduced risk of employee injury.

Roller Dolly or Pallet Roller Method. The roller dolly or pallet roller (Fig. 4.15*a*) is the next manual loading-unloading method, which consists of a structural metal or wood platform with at least four wheels to roll across the truck floor. The metal pallet rollers have additional center wheels that are rigid, and the wood carton dollies have swivel casters and wheels.

MANUAL PALLET TRUCK

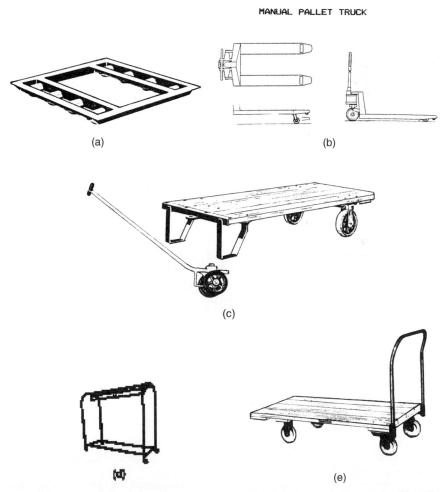

(a)   (b)

(c)

(d)   (e)

FIGURE 4.15 (*a*) Roller dolly or pallet roller (*Courtesy of Magline Inc.*); (*b*) manual pallet truck (*Courtesy of Crown Lift Truck.*); (*c*) semilive skid (*Courtesy of Hamilton Caster Mfg. Co.*); (*d*) garment trolley cart (*Courtesy of SDI.*); (*e*) four-wheeled platform truck (*Courtesy of Hamilton Caster Mfg. Co.*).

To unload or load a delivery vehicle, the employee carries, pulls, and pushes the empty dolly or roller into the delivery truck. After a predetermined number of cartons are placed onto the dolly or pallet board that is on the roller pallet, the device is pushed to the door of the delivery truck. At the dock end of the delivery truck, a lift truck removes the pallet board from the roller pallet.

This method requires 6 to 7 h for loading and unloading a delivery truck that has a floor-stacked load.

Disadvantages of this method are requirement for a smooth truck floor and a great number of employees, slow turning of trailers, and low employee productivity.

Advantages of the roller pallet concept include low capital investment, ability to handle all types of product, and less physical effort and little employee training needed.

Manual Low-Lift Pallet Truck Method.   In the manual low-lift pallet truck (jack) unloading-loading method, an employee pushes a manual low-lift pallet truck that consists of an operator's handle, which has two wheels and two load-carrying forks with wheels. These two forks extend forward from the vehicle base. To lift a pallet load, the employee pushes the forks into the pallet board openings and operates the hydraulic system, which lifts the pallet boards above the finished floor surface. To deposit the load, the employee releases the hydraulic system and pulls the forks from the pallet board opening.

To unload a delivery vehicle, an employee pushes or pulls the low-lift pallet truck into the delivery vehicle, picks up a pallet load, and pushes or pulls the pallet load to the dock staging area for deposit. With a manual pallet truck, the unloading of a delivery truck ranges from 2 to 6 h. Unloading of palletized deliveries takes less time than unloading of floor-stacked deliveries.

Disadvantages of this method are that it handles only a medium volume, product is palletized, travel up grades is difficult, and a backrest is required to reduce product damage.

Advantages of this method are reduced physical effort and risk of injury, low capital investment, improved employee productivity and delivery vehicle turning, reduced number of dock positions required, and efficient handling of product in other warehouse functions.

Semilive Skid Method.   The semilive skid (Fig. 4.15c) consists of two components. The first is a solid surface platform that has two rigid rear wheels and two front skid legs. The front of the platform has a tongue with an eyehole. The second component is a removable elevating jack handle that has two wheels and a stud.

To operate, the employee pulls the semilive skid into the delivery trailer and removes the jack handle. With this action, the legs rest on the floor to stabilize the skid. After the cartons are placed onto the semilive skid platform, the jack handle stud is inserted into the tongue. When the handle is lowered to the proper level, it elevates the skid legs and the skid is moved from the delivery truck. This method requires 4 to 6 h for unloading a trailer.

Disadvantages of this method are similar to those of the manual pallet truck. Additional disadvantages are that the semilive skid is difficult to handle in a multilevel storage system and is difficult to push.

Advantages of the semilive skid are low capital investment, less physical effort and used in a low-bay facility.

Garment Trolley Cart.   The next unloading-loading device is the garment trolley cart

(Fig. 4.15*d*), a four-wheeled vehicle with two structural members and a trolley load bar and two adjustable end stops. The load bar has the capacity to carry one full garment trolley. The trolley cart is used to unload hanging garments from a hanging-garment (rope) delivery trailer onto the dock staging area. In the dock staging area, the garment trolley cart is positioned at the trolley rail infeed section. An employee adjusts the end stop to the open position, thereby releasing the trolley to the warehouse gravity rail system.

To operate, the employee places an empty trolley on the cart load trolley bar and sets the adjustable end stops to the desired position. The cart is pushed or pulled into the delivery truck, and individual garments are removed from the delivery vehicle hang ropes and are placed onto the trolley. When the trolley is at capacity, the cart with the full trolley is pushed onto the dock area. On the dock area, the cart is placed in the staging area or aligned with the warehouse trolley rail system. The employee releases the end stop, and the trolley is pushed onto the warehouse trolley rail system.

With this design, a full garment rope trailer is handled in 6 to 8 h.

Disadvantages of the garment trolley cart are requirement for a dock plate at the dock position, difficulty in handling a large volume, and requirement for a trolley rail system in the warehouse. Advantages include low capital investment and efficiency in handling a low volume of hanging garments that are delivered on a rope vehicle.

Four-Wheel Cart or Platform Truck.   The next manual unloading-loading device, the four-wheeled cart or platform truck (Fig. 4.15*e*), has a load-carrying surface. The cart caster-wheel arrangement has two swivel casters and two rigid wheels for a heavy-duty platform truck and four swivel wheels and casters for a cart. The load-carrying surface is a solid deck for pallets, shelves for cartons or single items, or a hang bar for hanging garments.

To operate, an employee pushes or pulls the cart into the delivery vehicle and transfers the product from the delivery vehicle onto the cart. The full cart is pushed or pulled onto the dock staging area. With this method, a full trailer is unloaded in 6 to 8 h.

Disadvantages and advantages of this method are the same as those of the semilive skid method.

Extendible Hanging Garment Trolley Boom.   The extendible trolley boom is a specialized manually operated hanging-garment unloading-loading device. This method is used in a hanging-garment facility that handles a high volume of garments on trolleys. The extendible boom consists of a series of extendible channels that have headers and support devices with a trolley rail that has an end stop. When the delivery truck (rope hanging garments) is at the dock, the extendible boom is extended into the delivery truck and is supported from the delivery truck roof or with a series of frame structures.

To unload, an employee places an empty trolley on the rail and then transfers hanging garments from the delivery truck rope hooks (loops) onto the trolley. When the trolley is full or after the accumulation of a group of trolleys, the trolleys are gently pushed on the extendible boom rail to accumulate against an end stop. As required, the trolleys are released to the warehouse trolley rail system. With the extendible trolley boom concept, a full rope trailer is unloaded in 3 to 4 h.

Disadvantages of this concept are higher capital investment, requirement for a trolley rail system in the warehouse, and ability to service only one dock position. Advantages include high employee productivity, requirement for the fewest dock positions, ability to handle a high volume, and compactness (the extendible boom does not require floor space when not in use).

Mechanical Unloading and Loading Methods

The second unloading-loading group consists of the mechanical methods. In these methods gravity-, electricity-, and fuel-powered vehicles or conveying surfaces move cartons or unitized product between the warehouse dock and the delivery truck. These methods are used in any type of distribution facility.

The various mechanical methods include (1) electric pallet truck (jack), (2) electric pallet truck with a slip sheet attachment, (3) powered lift truck, (4) powered lift truck with a slip sheet attachment, (5) powered lift truck with pallet grabs (claws), and (6) conveyor (skatewheel- or roller-powered conveyor).

Electric Powered Pallet Truck. The first mechanical unloading-loading device is the electric powered (electricity-powered) pallet truck (Fig. 4.16*a*), which consists of a three-wheeled vehicle. One wheel is the drive and steer wheel, and the other wheels are under the two load-carrying forks that elevate or lower the unit load. These vehicles are walk-behind or rider models. Optional full-height load backrests and fork entry wheels improve pallet handling productivity and reduce product damage.

To unload, an employee drives the pallet truck into the delivery vehicle, picks up a pallet load, and drives with the pallet load from the delivery vehicle onto the warehouse dock staging area.

For maximum efficiency and less damage, if the delivery vehicle has palletized and secured loads, a full trailer is unloaded in 1 to 2 h.

Disadvantages of the electric pallet truck are that it requires greater employee training, a separate battery-charging area, and unitized product and it is difficult to handle stacked pallets. Advantages are mainly that it requires only moderate capital investment, handles a high volume, travels up grades, reduces employee injuries, requires fewest dock positions, and can be used in other warehouse activities.

Electric Powered Pallet Truck with Slip Sheet Device. The second mechanical unloading-loading device is the electric powered pallet truck with a slip sheet attachment (Fig. 4.16*b*). In this method, a powered pallet jack with a special platen and lip clamp device handles a slip sheet unitized load. The slip sheet method is an alternative to the floor stack or palletized method, which allows unitized product handled as a unit load rather than as individual cartons.

To operate, an employee drives the pallet truck into the delivery vehicle and uses the slip sheet lip clamp device to extend outward and grab and hold the slip sheet lip. After the load is completely on the platens, the employee drives to the dock area and deposits the slip sheet unit load onto a pallet board. With this method, a delivery truck is unloaded in 2 to $2\frac{1}{2}$ h.

When compared to the powered pallet truck concept, the slip sheet method has an additional disadvantage in that the slip sheet vehicle handles only slip sheet unit loads. An additional advantage is, however, that the product is shipped without pallet boards, which means that the delivery truck carries the maximum capacity in product weight.

Lift Truck. The electric or fuel-powered lift truck (Fig. 4.16*c*) is the next mechanized unloading-loading device. The lift truck is an electric battery-, gas-, liquefied petroleum gas-, or diesel-powered vehicle that is a sit-down or stand-up model. These models have three or four wheels and a set of forks that extend outward from the mast. These forks elevate and lower palletized loads. With the full free lift option, the forks elevate without moving the mast. The mast height, overhead guard and vehicle underclearance permits the lift truck to enter delivery vehicles. For maximum employee

(a)

FIGURE 4.16 (*a*) Electrically powered pallet truck; (*b*) electrically powered pallet truck with slip sheet (*Courtesy of Crown Lift Truck.*); (*c*) lift truck (*Courtesy of Caterpillar Lift Truck.*); (*d*) lift truck with slip sheet (*Courtesy of Cascade Corp.*).

productivity and reduced product damage, fork side shift and tilt, and spotlights on the mast options are considered for the lift truck.

To operate, the employee drives the lift truck into the delivery vehicle, picks up a pallet load, and drives with the pallet load to the dock area. With the lift truck concept, a delivery truck is unloaded in 1 to 2 h.

FIGURE 4.16   (*Continued*)

When compared to the electric pallet truck concept, the lift truck has additional disadvantages of increased need for operator training and a higher capital investment.

Additional advantages are that the lift truck handles stacked pallet loads and is used in other product movement and storage functions.

Lift Truck with Slip Sheet Device.   The next mechanical unloading-loading device is the lift truck with a slip sheet attachment (Fig. 4.16*d*) that permits the lift truck to handle slip sheet unit loads.

When compared to the electric powered pallet truck (jack) slip sheet, the lift truck with slip sheet attachment has additional disadvantages of requiring a higher capital investment and more extensive operator training. An additional advantage is that it handles a unit load from a two-high stack.

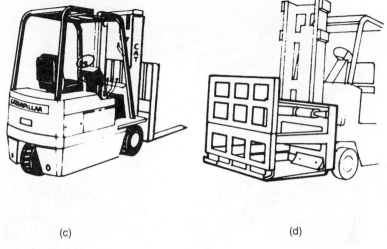

(c)   (d)

FIGURE 4.16   *(Continued)*

Gravity Conveyor.   The next mechanical unloading-loading device handles cartons. This device is the gravity skatewheel or roller conveyor (Fig. 4.17a), which consists of gravity conveyor sections. These conveyor sections have plastic or metal wheels. The sections are extended from the appropriate dock location to the nose of the delivery vehicle.

To unload a floor-stacked delivery trailer, the unloading employees in the delivery trailer place the cartons onto the conveyor. With a push on the carton and gravity force, the carton flows to the end of the conveyor onto the dock area or for loading activity to the nose of the delivery truck. At this location, an employee removes the carton from the conveyor and stacks the carton onto the floor or onto a pallet board.

This method has three alternative concepts: fixed conveyor in the delivery vehicle, tripod-supported conveyor, and nestable or extendible conveyor.

The *fixed conveyor in the delivery vehicle* is a carton conveyorized loading-unloading concept that is used with a captive truck delivery fleet. These delivery vehicles have gravity conveyor sections that are permanently mounted on the floor and to one of the sides of the delivery vehicle interior. In the United States, the preferred side in the delivery truck is the left side as you enter the truck because all soft shoulders are on the right side of the road or highway and the road surface is lower on the right side. The disadvantages of this method are increased weight and loss of cube (space) in the delivery vehicle, use limited to floor-stacked loads, and additional capital investment in the delivery vehicle. Advantages include reduced setup time, less investment in extendible conveyors, no warehouse space requirement for conveyors, and use at all locations.

In the *tripod conveyor on the dock* loading-unloading method, employees set up portable tripods (conveyor support stands) on the delivery truck floor and warehouse dock area floor. With this arrangement, the conveyor is extended from the dock to the trailer nose. In comparison to the fixed conveyor, the portable tripod requires longer setup time and there is no conveyor at the delivery location. Advantages are full use of the delivery vehicle and less conveyor investment.

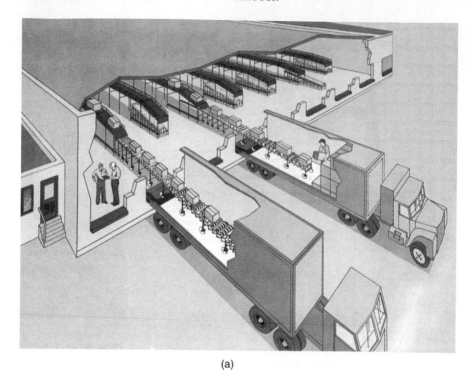

(a)

(b)

FIGURE 4.17 (*a*) Extendible or nestable manually powered conveyor (*Courtesy of Flexible Material Handling.*); (*b*) manually powered extendible belt conveyor (*Courtesy of Stewart Glapat Corp.*); (*c*) manually powered extendible skatewheel conveyor (*Courtesy of Best Diversified Products.*).

(c)

FIGURE 4.17 (*Continued*)

In the extendible-nestable conveyor, which has rollers or skatewheels at the conveyor surface, an employee, as required, pulls the conveyor forward from the dock area into the delivery vehicle. The conveyor sections are on casters or wheels and a rear brake that permits easy movement. In some mobile models the rear conveyor stands are fixed in a channel. Each conveyor section overlaps the next section or is at a slope (angle) which permits an easy, uninterrupted flow (transport) of product. When the nestable conveyor is fixed in the dock area, to allow the handling of nonconveyable product across the dock leveler, the conveyor path is designed to extend along the delivery truck wall and to one side of the dock leveler. Disadvantages of the gravity conveyor concept are that employees must lift and push cartons and product is conveyable and is unitized inside the warehouse. Advantages are low capital investment, only moderate capital investment, turning of the dock position twice every 8 h, and flexibility.

Comparison of the various skatewheel and roller conveyor concepts reveals that the skatewheel has lighter weight and the roller handles heavier loads. If small cartons (length) are handled, then the skatewheels are meshed or rollers are on close centers to assure carton movement across the conveyor. If you compare metal skatewheels to plastic skatewheels, you see that the metal skatewheel permits an easier flow of cardboard cartons.

Powered Extendible Belt or Roller Conveyor. The next mechanical loading-unloading method is the powered extendible belt roller conveyor, which is the most sophisticated of the mechanical carton dock handling methods. This method consists of an electric powered belt-roller conveyor that is set on the dock. The conveyor is a series of belt-roller conveyor sections that extend outward from the dock to the nose of the trailer. Some models are in a stationary position on the dock, and other models have electric motors that move the conveyor between dock positions. All conveyor controls and emergency stop devices are located on the front of the belt-roller conveyor unit. As required, a series of controls and safety stops are located on the side of traversing models. The conveyor belt-roller speed is 60 to 85 ft/min.

To operate, an employee activates the extendible belt-roller conveyor bed to extend forward into the trailer. At the desired length, an employee transfers cartons between the conveyor and the delivery vehicle. The conveyor automatically propels the cartons between the dock area and the delivery vehicle.

The extendible conveyor has six alternative designs: (1) fixed (stationary) position, (2) portable powered skatewheel, (3) portable with attached decline-incline, (4) portable with detached decline-incline, (5) hinged conveyor with an electric powered hoist, and (6) pallet handling device (PHD).

With the *fixed (stationary) extendible belt conveyor* design, the stationary housing (base) to the extendible conveyor is located to the side of the dock position. This arrangement permits the handling of nonconveyables between the dock and the delivery truck. With this method, the capital investment is lower per each extendible conveyor but the conveyor provides service to only one door.

The *powered extendible skatewheel conveyor* (Fig. 4.17c) consists of extendible skatewheel conveyor sections that are driven by an endless chain. The chain is looped over a series of cams, which turn the skatewheels in the middle of the conveyor. This cam arrangement does not cause employee injuries. The chain is turned by an electric-motor-driven sprocket. As the chain turns the skatewheels, its motion and the force of gravity move the carton or tote forward, across a distance of as much as 40 to 60 ft. For maximum efficiency, the conveyor charge end is slightly higher than the discharge end. Disadvantages are that an employee must pull or push the conveyor into position and that this device is available in only 10- to 12-ft-long sections to make up the required extended length. Advantages are low capital investment and mobile equipment.

The *portable (mobile) extendible belt conveyor* is installed on a set of tracks that permits the extendible belt conveyor housing (base) to travel on four wheels between dock positions. This concept requires that the following features be designed for the system: overhead electric power source, embedded channel track in the rear in the floor to guide the base as it moves between dock positions, front track which must be flat for the front wheels, and safety stop devices in all movement directions (left, right, and forward). A disadvantage of the extendible conveyor is higher capital investment. Advantages are less employee fatigue, lower risk of injury, and ability to handle a high volume.

The portable (mobile) extendible powered conveyor has two alternative designs. The first option is the *mounted incline or decline design,* according to which the incline-decline belt conveyor is mounted to the extendible conveyor base. This conveyor extends from the mobile conveyor base to the warehouse in-house transport conveyor merge. When compared to the other portable conveyors, this feature reduces the total conveyor investment, but requires a wider clear path for the extendible conveyor to move product between dock doors, which means a larger dock area.

In the second option, there is an *incline-decline conveyor section at each required location* for access to the dock door. This conveyor section extends down from the transport conveyor merge to the mobile conveyor. This design increases the total conveyor investment but requires a narrow, clear path for the extendible conveyor to move between dock doors, which means a smaller dock area.

When the shipping transport conveyor has a fixed-decline conveyor to the gravity extendible conveyor, then a *hinged conveyor section attached to the transport conveyor or as a bridge across the receiving-shipping dock aisle* can be applied in the dock area. One end of the hinged conveyor is attached to the discharge end of the decline transport conveyor. The discharge end of the hinged conveyor is attached to a chain that raises or lowers the conveyor section. The upward-downward (vertical) movement is controlled by a manually operated electric powered hoist. In the elevated position, the hinged conveyor design permits a clear lift truck path between the dock leveler and staging area. In the lowered position, the hinged conveyor permits the maximum extension of the gravity extendible conveyor into the delivery vehicle. Another benefit of the hinged conveyor concept is that the decline conveyor section has the maximum slope to assure a smooth flow of cartons into the delivery truck.

The *pallet handling device* (PHD) (Fig. 4.18a) consists of a stationary or mobile fully automatic device that is used to load and unload delivery trucks that have palletized or container loads. With the base of the PHD on the dock and live roller pallet

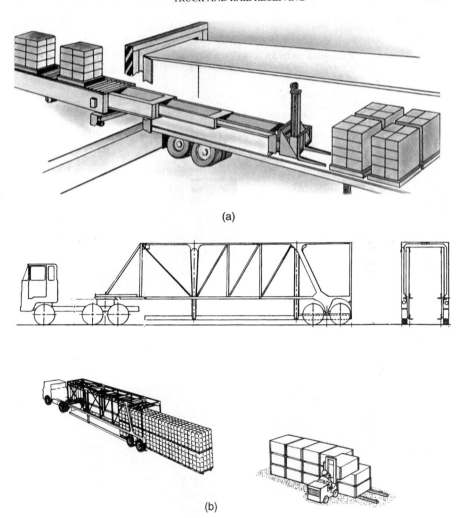

(a)

(b)

FIGURE 4.18 (*a*) Pallet handling device (*Courtesy of Stewart Glapat Corp.*); (*b*) strad-o-lift loading-unloading device (*Courtesy of Kornylak Corp.*); (*c*) automatic pallet flow loading-unloading apparatus (*Courtesy of Alevy Corp.*).

conveyor fixed to the top of the base, this design provides the staging area for the pallet loads. In the loading process with the delivery truck positioned at the dock, the PHD with its sensor devices determines the truck length and is ready to receive pallet loads. The PHD forks raise and pick up a pallet load from the live roller conveyor. With the load on the forks, the PHD rotates and lowers for entry into the delivery truck. After entry (cantilever extension of the forks) into the delivery truck, it deposits the pallet load in the proper position on the delivery truck floor. The microcomputer program determines the other pallet drop locations on the delivery truck floor. In the unloading process, the PHD uses fork sensors to determine the pallet openings of the loads on the delivery truck floor. With the pallet load on the forks, the PHD returns to

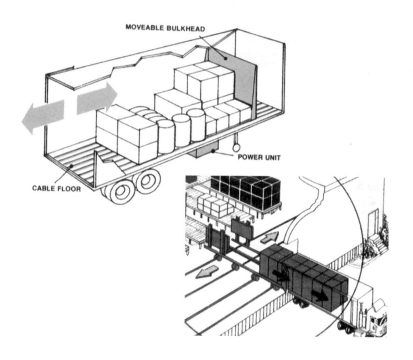

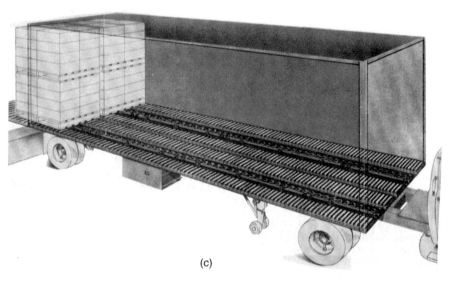

(c)

FIGURE 4.18 (*Continued*)

the dock position and rotates and raises the load for placement onto the live roller conveyor that is on the top of the base.

The single-fork PHD device handles a 3500-lb pallet load and a full delivery truck of 18 pallets in 30 min. If the truck spotting is required, then add 15 to 20 min to the 30 min. A double-fork device handles a total of 3500 lb for two pallet loads. When two pallet loads are handled per trip, the activity time is reduced to 20 to 25 min for each delivery truckload.

When used with the fork sensors, the PHD handles pinwheel loading patterns or four-way pallet loads or containers, is capable of handling stacked pallet loads or containers, has a complete safety stop device plus a safety stop light curtain that extends on both sides of the dock area from the dock door to the PHD base, and handles all delivery truck lengths. Disadvantages of the PHD are high capital investment, large dock area requirement (37 to 40 ft), and, to achieve the objective rate, requirement that the loads be handled in a fluid manner (without interruption). Advantages of the PHD are no labor expense, no dock leveler investment or security guarding for doors or walls, minimal dock area and truck dock lights, reduced building and mobile warehouse equipment investment, and reduced product, building, and equipment damage and maintenance.

The various *powered extendible conveyor options* are designed to improve dock employee productivity, reduce product damage, and improve employee safety. These options are (1) angle guides, (2) motorized hydraulic pump to direct the extendible conveyor to a new angle (left, right, up, or down), and (3) bulkhead lights.

Pallet Claw Design. The last mechanical unloading design is the pallet gripper or "claw," which is a set of clamps and chain. The clamps are connected to a chain with a loop that is attached to a lift truck.

To operate this system, an employee sets the claws onto the pallet board center stringer and attaches the loop to a lift truck. As the lift truck travels from the pallet board or truck, the grip tightens on the pallet load stringer and moves the pallet board to the rear of the delivery truck.

This design is applied when the trailer rear end is not equal to the dock. Disadvantages of the pallet claw design are that it can be considered dangerous and handles only a low volume.

Automatic Unloading and Loading Concepts

The final receiving-shipping concept group consists of automatic unloading and loading methods. The various designs in this group utilize an electric powered conveyor system, hydraulic platform, or a specially designed trailer to move product between the dock and the trailer. All these automatic devices move product between the truck and the dock area with very little labor.

Truck Lift Design. The automatic truck lift (tilt) is designed to unload bulk or granular product from a delivery truck into a receiving trough. The design consists of a platform, truck securing clamps, and a hydraulic cylinder that tilts the front of the delivery truck upward. The elevation and gravity force cause the product to slide from the delivery truck into the receiving trough.

Disadvantages of this design are that it handles only bulk or granular product and entails a high capital investment. The advantage is reduced product handling.

Strad-o-Lift. The automatic strad-o-lift device (Fig. 4.18*b*) picks up, transports, and deposits its cargo onto a warehouse floor or in the truck-yard loading area. The device consists of a specially designed trailer that has a set of lifting shoes. These shoes run the full length of the truck and hydraulic mechanisms to clamp the shoes together, thereby lifting the entire load. After the load is in the raised position, the trailer travels to the next distribution facility. At the new distribution facility, the delivery trailer load is lowered onto the floor and the shoes release their hold on the load. To facilitate loading and unloading, the product is set on two full-length stands.

Disadvantages of this design are that the delivery vehicle has a low clearance, requires secured product, and both receiving and shipping locations use the method. The advantage is that load is dropped and picked up in the yard or on the warehouse floor. This represents a short time to unload and load the vehicle.

Pallet Flow Device. The automatic pallet loading-unloading device (Fig. 4.18*c*) handles pallet loads between the truck dock and the delivery vehicle. This concept consists of air pressure, powered chain, and conveyors that move pallet loads between the delivery truck and the dock. These specially designed delivery trucks have specially designed floors, and the warehouse dock space has a specially designed push-pull device and conveyors.

To operate, after the delivery truck is located at the dock position, an employee activates the dock area section to push the palletized product from the dock area onto the conveyorized truck floor. The entire load is pushed into the delivery truck. The delivery truck driver ensures that the load is secured in the delivery truck and the building power cable is disconnected from the dock.

Disadvantages of this method include higher capital investment in the truck, high dock area investment, increased maintenance, and both receiving and shipping locations designed to handle the load. Advantages of the automatic pallet loading-unloading method are less dock time, fewer dock positions, and fewer mobile dock equipment and employees.

Railroad Car Unloading and Loading Concepts

The second major warehouse receiving and shipping area is the rail(road)car dock area. Since over-the-road trucks normally have easy access to the highways, the over-the-road transportation method handles, in some industries, the majority of or all the product deliveries to the warehouse (catalog or direct mail). In other industries, delivery trucks handle the majority of the perishable products, nonperishable products ranging in size from small- to medium-cube and lightweight. In these industries, the railcar transportation method handles nonperishable product and some perishable products that are promotional or seasonal-demand SKUs or large-cube, low-weight SKUs.

To provide railcar access to the distribution facility without interruption to the delivery truck deliveries, a rail spur and rail tracks run parallel to the building on the opposite side of the building from the truck dock area.

When you consider a railcar dock area, the most critical design factor is to determine the state and local codes for the centerline rail clearances from your warehouse building. The second important design factor is the railcar length, door length, capacity in pallet loads, and number of railcars (average and peak) that are handled at your present facility. This information is obtained from your railcar company that services your facility, your company's railcar receiving records, and company growth rate.

Height, ft	Length, ft	Wheel height, in	Door width, ft
22	40	41–45	6–8
	50	41–46	$16\frac{1}{2}$
	80	43–46	20

FIGURE 4.19 Typical railcar dimensions. (*Courtesy of Rite Hite.*)

With this information, your rail dock is designed for your new distribution facility. The most common railcar dimensions found in the United States are shown in Fig. 4.19

The rail track section that is between the railroad main line and your distribution facility dock is your rail spur. The rail spur permits the railroad company to transfer your railcars from the main line to your dock area. Usually the maintenance of your rail spur is the responsibility of your distribution facility.

The four rail dock designs are (1) flush dock, (2) rail platform, (3) inside rail dock, and (4) portable dock.

Flush Dock. The first rail dock type is the flush dock design (Fig. 4.20*a*). In the flush dock design the railroad spur track runs parallel along the side of your distribution facility. The distance between the centerline of the rail spur and the furthest extension of the building (dock) is $8\frac{1}{2}$ to $9\frac{1}{2}$ ft. This clearance permits unobstructed railcar travel on the track.

The flush dock design minimizes the building construction cost, maximizes the building utilization of the site, and maximizes the square-foot utilization of the building. With the flush dock design, the dock doors have at least the same width as the widest railcar door. To facilitate railcar spotting and possible wider railcar doors, these dock doors are as wide as possible. To improve safety and employee productivity during bad weather, a canopy is extended outward from the side of the building to a distance at least to the center of the track and at least 23 ft above the tracks. Lighting under the canopy improves dock safety. This canopy protects the dock area from becoming wet. A second energy-conservation feature is an inside the building shelter or vestibule.

To provide maximum energy conservation and safety, a dock shelter is added to the dock door frame. The shelter is a fabric-covered frame-constructed device. After the railcar is secured at its dock position, the shelter is extended outward from the building to cover the railcar door opening. This extension forms a protected tunnel for the movement of product across the dock plate and between the railcar and the warehouse.

Disadvantages of the flush dock design are that it requires exact railcar spot, unloading could require additional internal transportation, and the number of doors determines the number of railcars spotted or unloaded at the facility.

Advantages of the flush dock include minimum construction costs, maximum energy conservation, maximum building utilization, and improved dock security and safety.

Rail Platform Dock. The second rail dock design is the rail platform dock (Fig. 4.20*b*). The rail platform dock extends outward from the building side toward the center of the track. The concrete platform is constructed along the exterior building wall

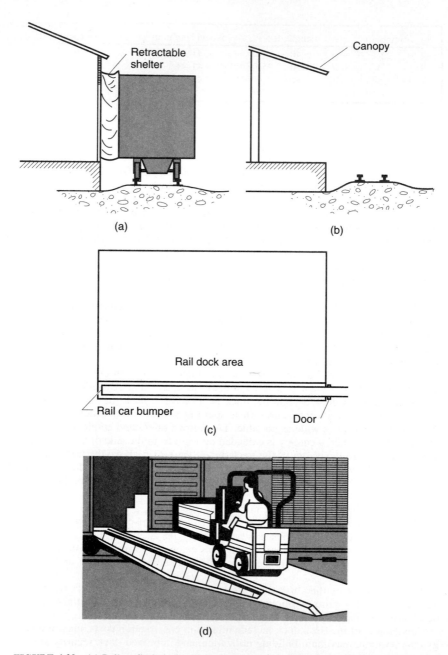

FIGURE 4.20 (*a*) Railcar flush dock with shelter; (*b*) rail platform dock with canopy (*Courtesy of Rite Hite.*); (*c*) inside rail dock; (*d*) mobile (remote) dock ramp (*Courtesy of AAR Brooks and Perkins.*).

that parallels the rail tracks. This feature permits railcar dock activity at any spot along the platform. The platform has sufficient width that pallet loads can be stacked along the exterior wall and pallet load handling equipment can make right-angle turns from the railcar. This dimension is approximately 14 to 20 ft between the building wall and the dock edge.

This wide platform adds to the building land utilization and construction costs. To facilitate dock activity in bad weather, a canopy extends outward from the building wall to the centerline of the tracks and at least 23 ft above the tracks. Lighting under the canopy improves the safety of the dock activity.

Disadvantages of the rail platform include increased construction cost, hazardous conditions during rainy weather, and reduced dock security and safety. Advantages are provision of maximum unloading spots and reduced internal transportation.

Inside Rail Dock. The third rail dock design is inside the building rail dock (Fig. 4.20c). A rail dock inside the building provides weather protection and permits day and night dock activity. It also permits railcars to unload at any dock position along the track. With the track between the building rear wall and the dock, the design adds to the building construction costs and building utilization of the land and requires a door for railcar entry and exit from the building.

With the tracks inside the building and portable dock boards, the distance between the edge of dock and the centerline of the track is approximately 8 ft.

Disadvantages include potential rodent infestation; if there are rodents in the car, they may enter your facility. Building and site (land) utilization and construction costs are also increased with the inside rail dock design. Advantages include dock activity in any weather, unloading at any dock position, improved security and safety, and reduced internal transportation.

Mobile (Remote) Dock Ramp. The next rail dock design is the mobile dock ramp (Fig. 4.20d). When your building and site size cannot have a rail spur adjacent to the building and the railcar volume is very low, then the railcar siding is located remote from the building. The remote rail spur is serviced by a dock ramp that is the same as the truck dock ramp.

Disadvantages of the remote dock ramp are not used in bad weather and requires setup time. Advantages are low cost and maximum site utilization.

Railcar Unloading

Single-Railcar Unloading Strategy. The previously described railcar dock designs are parallel to the rear or the to the side of the building. Whenever possible, the railcar track is a single track because it permits railcars to be spotted at any location for easy access to the product storage location.

When the land or building shape and spur does not permit sufficient space for a single track to handle the design-year daily railcar requirement, there are two alternative strategies.

Controlled (Selective) Railcar Unloading. The first strategy is to sequence the railcar spotting on a daily required basis by first unloading the oldest railcars or railcars that contain priority product. This strategy potentially incurs detention charges or inventory flow problems.

Double-Railcar Unloading Strategy. The second strategy is a railcar dock area design concept in which two rail tracks are parallel to the dock edge. This design increases the number of railcars handled by your receiving operation in a day. To operate, this design requires a bridge dock board that bridges the gap between the two railcars.

To unload or load the railcars, a specially designed railcar dock plate bridges the gap between the two railcars and a second dock plate bridges the gap between the first railcar and the dock edge. The bridge between the two railcars supports fully loaded mobile pallet load handling equipment as it travels across the gap. Side guards on the dock plate improve the safety of the railcar unloading or loading activity.

Various Rail Dock Board Designs to Bridge the Gap

When you consider the various dock boards that bridge the gap between the railcar and the dock or between two railcars, options are a portable dock board (lift truck moved or vertical stored) or a bridge between two railcar dock boards.

Portable Dock Board. The first type of railcar dock board is the portable dock board (Fig. 4.21a). The dock board is designed with a top diamond-pattern surface with structural reinforced members on the underside. These members serve a second purpose of reducing the dock board movement. Lips on both ends provide a smooth transition from the dock board to the railcar or warehouse floor. Two lift truck fork handles (loops) provide openings for lift truck forks that allow the lift truck to move the dock board. Securing pins on the railcar end of the dock board secure the dock board to the railcar, thus reducing dock board movement. Side curbs reduce mobile equipment runoffs.

Disadvantages of the portable dock board method are that it requires a lift truck to move the dock plate and two employees to position the dock plate. Advantages of the portable dock board are low capital investment and mobility to all dock positions.

Vertically Stored Dock Board. The second portable rail dock board is the vertically stored dock board (Fig. 4.21b). It has the same design and operational characteristics as the truck vertical design and the same operational characteristics as the truck vertically stored dock board. The railcar vertically stored dock board is available in both manual and hydraulic types. Both types are attached to a channel that is mounted to the side of the rail dock platform or building and requires an exact dimension from the dock edge to the rail track centerline. Therefore, the dock boards are best installed after construction of the rail dock and rail spur.

A disadvantage of the vertically stored dock board is high capital investment, and advantages are lower maintenance costs and improved employee productivity.

Portable Railcar-to-Railcar Dock Board. The bridge between two railcars is the next portable railcar dock board (Fig. 4.21c). As described in the section on double railcar unloading, this dock board bridges the gap between two railcars. After the inside railcar (adjacent to the dock) is unloaded, a lift truck places the dock board in the door opening of each railcar and thus bridges the gap between the two railcars.

The two-railcar dock board bridge has two lips. One lip (end) rests on the outside car door opening and the other, on the inside railcar door opening. The surface has a diamond pattern, and the underside has support members that prevent movement of the dock board. Each lip end has securing pins to secure the bridge to the railcar and prevent movement.

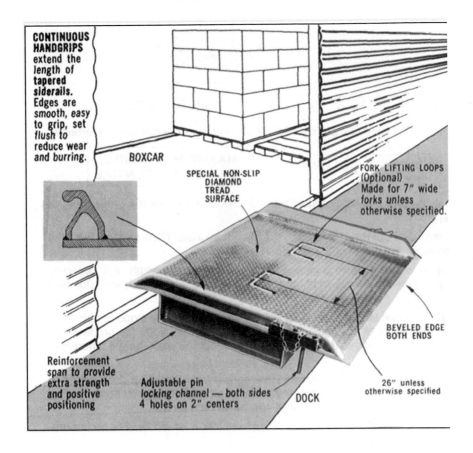

CONTINUOUS HANDGRIPS extend the length of tapered siderails. Edges are smooth, easy to grip, set flush to reduce wear and burring.

BOXCAR

SPECIAL NON-SLIP DIAMOND TREAD SURFACE

FORK LIFTING LOOPS (Optional) Made for 7" wide forks unless otherwise specified.

BEVELED EDGE BOTH ENDS

Reinforcement span to provide extra strength and positive positioning

Adjustable pin locking channel — both sides 4 holes on 2" centers

DOCK

26" unless otherwise specified

(a)

FIGURE 4.21   (*a*) Portable dock board (*Courtesy of MTD Products.*);

With the bridge between the two-railcar arrangement, product from the railcar on the outside track is unloaded through the railcar on the inside track onto the warehouse floor.

If your building or site size and configuration prevents a single-track expansion, the railcar-to-railcar unloading design is recommended to handle the projected volume.

Bascule Bridge.   As a result of building expansion or railroad company spur design requirements, the rail spur is designed inside the building with dock space and warehouse space on both sides of the track. This rail spur design separates the building into two warehouse zones. To have mobile warehouse equipment travel the entire warehouse or between the two zones, a bascule bridge (drawbridge) (Fig. 4.21*d*) spans the track and permits travel between the two areas.

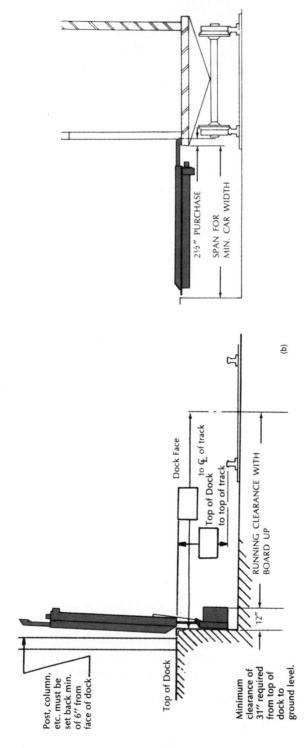

FIGURE 4.21 (b) vertically stored dock plate (*Courtesy of Serco Co.*);

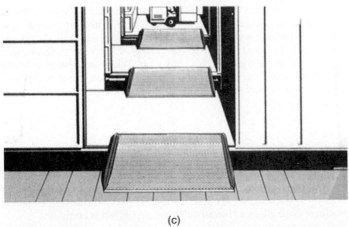

(c)

(d)

FIGURE 4.21 (c) portable railcar-to-railcar dock board (Courtesy of Roll Rite Corp.); (d) bascule bridge (*Courtesy of AAR Brooks and Perkins.*).

The bridge is an electrically, mechanically, or hydraulically powered device that consists of a diamond-pattern surface with underside structural members, safety sidings, and ability to raise or lower one end. The bridge device is attached to one side of the dock edge, and the other side is in the raised or lowered position. In the raised position, it permits railcar traffic on the rail spur. When the bridge is in the lowered position, the bridge spans the gap between the two warehouse zones and mobile warehouse equipment travels between the two warehouse zones.

Rail Dock Area Options

When we consider the rail dock area, there are several pieces of equipment that improve the rail dock productivity and safety. These items are (1) wheel chocks, (2) railcar bumper, (3) dock lights, (4) hose outlets, (5) sealed rail spur or siding, and (6) on open platform dock edges to install yellow-coated chains and posts.

Methods of Transferring Product between Delivery Vehicles and Docks

One of the main receiving and shipping objectives is to transfer product between the delivery vehicle and dock area. The method of supporting (handling) the product in the delivery vehicle has a direct influence on the dock employee's productivity. There are five basic methods of shipping and receiving product at a warehouse or distribution operation: (1) floor-stacked product, (2) slip sheet product, (3) pallet boards with product, (4) carts with product, and (5) containers with product.

Floor-Stack Method.   In the floor-stacked method the product is set directly onto the delivery vehicle floor. To receive or ship the product, the employee must lift and place the product onto a cart, pallet board, or conveyor for transfer between the delivery vehicle and the dock. An alternative is for your employee to hand-carry the product. Loading or unloading of a floor-stacked delivery truck takes 6 to 8 h.

The disadvantages of this method are increased risk of employee injury, low employee productivity, and requirement for the greatest number of employees and the greatest number of dock positions. Advantages of the floor-stack method include excellent cube utilization of the delivery vehicle, no extra weight on the delivery vehicle, no storage requirement, and no additional capital investment in equipment.

Pallet Board Method.   The second method is to have the product palletized on a pallet board. The pallet board is a support device with top and bottom deck boards. Mobile fork-lift equipment forks are inserted in the pallet board fork openings and lifts the pallet load. This feature permits the mobile equipment to transport the pallet load between the dock area and the delivery truck.

The normal delivery vehicle has the capacity of 18 to 20 pallets, and the concept handles a delivery truck in 1 to 2 h to handle the load. To minimize product damage and assure high employee productivity, the product is secured to the pallet board.

Disadvantages of this method are that it requires storage area for empty pallets and additional expense for pallets, adds weight to the delivery vehicle, and entails additional capital investment in pallet handling equipment and loss of some cube (space) in the delivery vehicle.

The advantages of the pallet load method include high employee productivity, fewer employees and dock positions needed, turning of a trailer within 2 h, and reduced risk of employee injury.

Slip Sheet Method. The next method is the slip sheet method. The product is unitized onto a corrugated, plastic, or other material sheet that has a lip. The lip extends forward from the load and permits mobile equipment with a slip sheet attachment to clamp onto the lip and lift the unit load. After the unit load is secured onto the mobile equipment, it is transferred from the delivery vehicle to the dock staging area.

To handle a delivery of 18 to 20 unit loads, the slip sheet method requires 2 to $2\frac{1}{2}$ h. When you use a slip sheet in a truck, you need a single-lip (-tab) on the slip sheet; when you use a slip sheet in a railcar, you require a multilip (-tab) slip sheet.

Additional disadvantages of this method, in comparison to the pallet load method, are increased capital investment in slip sheet handling equipment and requirement for a slip sheet device.

Cart Method. The four-wheeled cart is the next product handling method. With this delivery method, the product is unitized onto a four-wheeled cart for transfer of product between the delivery vehicle and the warehouse dock. The cart transfer is made by an employee manually moving the carts or an employee controlling a powered vehicle that moves the cart.

To reduce product and equipment damage, the carts in the delivery truck are secured with load locks to prevent uncontrolled movement of carts inside the delivery truck. A full trailer is loaded or unloaded within 1 to 2 h.

The disadvantages of the cart method are additional investment in carts, additional weight on the delivery vehicle, and loss of cube in the delivery vehicle. Also, warehouse space is required for empty carts and carts are returned to the warehouse. Advantages of the cart method are that the cart can be used at the other facility and manually moved in the dock area.

Container Method. The next shipping-loading method is the container method. The container is a device that has a product support surface, two fork openings, four sides, and legs or sets on a pallet board. To transfer product between the warehouse dock and delivery truck, an employee uses pallet load handling equipment.

Disadvantages of the method are (1) powered pallet load handling equipment required to move the product and (2) additional cost in the container.

Advantages of the container method are similar to those of the cart method except that the container is used in the storage area and handles liquid, granular, and temperature-controlled products. Also, for special products, the container is captive between your facility and the vendor facility.

SHIPPING FUNCTION

The various shipping dock activities for your warehouse or distribution operation includes (1) checking, (2) packing, (3) sealing, (4) securing, (5) manifesting, and (6) loading.

Checking Activity

The checking activity ensures that the correct product, in the right quantity, and in good condition is delivered to the appropriate area. With a pallet load or carton distribution facility, the checking activity is performed in the staging area. In a single-item distribution operation, the checking activity is performed at the packing station.

This checking activity is performed by the manual method or bar-code method. In the manual method a checking employee visually verifies that the actual product that appears on the customer order form is in the package and places a mark on a copy of the customer's order form. With the bar-code method, the checking employee directs a hand-held scanning device light beam onto each product bar-code label. On a conveyorized sortation system, a fixed-position scanning device reads the product bar code, and the information triggers the sortation device to divert the product onto the appropriate shipping lane. This bar-code information is held in memory or is transmitted on-line to the computer. Some hand-held devices indicate that the product belongs to the customer.

After the checking activity in a single-item (catalog or direct mail or retail store) distribution facility, packing is done; this is a more extensive activity. This activity ensures that the product is protected against damage and that the customer's address appears on the package exterior.

In a pallet load or carton distribution facility, the packing activity ensures that the cartons are secured onto the shipping device.

Manifest Activity

The next shipping activity is the manifest activity, which ensures that the customer's order was handled and processed by the distribution facility. This activity is a manual operation or a bar-code scanning operation.

PRODUCT STAGING CONCEPTS

One of the critical activities in the receiving and shipping area is the inbound-outbound product staging activity. To have productive employees with minimal product damage and accurate inventory control, you design sufficient space directly behind the dock doors to accommodate the inbound and outbound capacity for a full delivery truck.

Other storage space requirements in the dock area include space for empty shipping and receiving devices (pallet and carts and return-to-vendor and trash containers).

It has been the writer's experience that when a company reduces the square-foot area that is allocated to the dock staging activity, then the company discovers that the congested staging area lowers employee productivity; increases product, equipment, and building damage; increases frequency of loading and unloading errors; and creates off-scheduled activities.

The best dock area design allows mobile warehouse equipment to travel between the dock plate and the staging area and between the staging area and the storage area. The staging area requires vehicular traffic aisles that connect the dock area to the storage area.

VARIOUS RECEIVING AND SHIPPING STAGING CONCEPTS

The alternative dock staging designs are (1) floor stack, (2) standard rack, (3) drive-in or drive-through rack, (4) flow rack, (5) push-back rack, (6) stacking frames or portable racks, and (7) conveyor accumulation.

Floor-Stack Design.   The floor-stack dock staging design is the first, and the most common, design used in all types of warehouse, distribution, and manufacturing facilities. In the floor-stack method an employee places the product on the floor or onto a warehouse vehicle that is located directly behind the dock door. With this staging concept, the product is handled by any of the product transfer methods. The floor-stack dock area requires square-foot area sufficient for 18 to 20 pallet loads.

With the typical dock door width of $12\frac{1}{2}$ ft and with a 6-in space between the pallet boards, the standard dock door width handles three pallet loads. The length of the floor staging area is seven to eight pallet loads deep with a 6-in space between loads. This floor staging area has $31\frac{1}{2}$ to $43\frac{1}{2}$ ft in length. In this method, the product flow is a straight line from the storage area to the dock area. In the dock area, a yellow 6-in painted line on the floor guides the employee's pallet load transactions in the dock area.

Disadvantages of this method are that it (1) requires a large floor area; (2) increases building construction costs; and (3) if ceiling height is not used, then it has poor space utilization. Advantages of the floor staging method include no capital investment in additional equipment and ability to handle a high volume and all types of shipping devices.

Standard Single-Pallet Rack Design.   The second staging method is the single-deep standard pallet rack. The standard rack staging design requires that the standard rack rows be placed in the dock area between the dock aisle and the storage aisle. The number of rack bays that are required to hold a delivery truck quantity is three to four. This configuration has an overall length of 28 to 36 ft. This type of staging requires that the storage area and the dock area vehicle make transactions in the staging positions and in the delivery truck. The vehicle lift height reaches the top rack level in the staging area rack and permits entry into the delivery vehicle. This top rack level has a nominal elevation of 10 ft above the floor.

In this procedure a product flow makes a right or left turn prior to being placed onto the delivery vehicle and is used in a small or limited-space facility.

Disadvantages are (1) increased capital cost in rack, (2) dock equipment needed to lift and retrieve the pallet loads, (3) all product placed on pallet boards or in containers, and (4) small increase in truck loading and unloading time.

Advantages of the standard rack are that it (1) requires less or smaller square-foot area, (2) reduces building construction costs, and (3) permits access to all pallet boards and containers.

Two-Deep Rack Configuration.   The two-deep rack configuration requires a two-deep straddle reach truck and two-deep rack. The two-deep straddle reach truck has difficulty entering delivery vehicles. This restriction increases equipment costs and labor needs, which makes it less economical and operationally feasible than the alternative (the single-pallet rack), but when you floor-stack the delivery, the two-deep rack concept becomes very viable.

Cart Staging Method.   If additional storage is required at a distribution facility that handles carts, then back-to-back standard pallet racks with a high first load beam level provide additional storage locations for supplies, records, or reserve positions for very-slow-moving SKUs. When this concept is used, the rack is anchored to the floor, all posts have protection devices, a minimum of $7\frac{1}{2}$ ft is required between the finished floor and the bottom of the first load beam, and a minimum of two load beam levels and sufficient aisle width between the two rack rows are required to turn a fully loaded cart into the rack position.

Carton Conveyor Method.   The standard storage rack with walkways and conveyors is the next staging method. It is used in a distribution facility that handles cartons on a conveyor system. In this system, prior to the shipping area, the customer order is prelabeled (selected) and transported via a conveyor system to a temporary holding area. In this holding area, the cartons are accumulated on conveyors. At the required shipping time, these cartons are released from these holding conveyor lanes and are transported to the loading area on the shipping dock.

Manual Palletized Load-and-Hold Method.   This temporary hold method has two alternative designs: the standard rack and the pallet flow rack. Both require that the order-pick function be coordinated with the staging function. In the staging area, these two designs have walkway rack positions and transport conveyors. The rack positions are your customer holding position.

Standard Rack Carton Conveyor Hold Method.   When you consider the standard rack temporary hold method, each rack bay has a solid bottom with netting or wire mesh on the two sides and rear. In this rack bay, the arrangement permits carton stacking from the deck to the maximum reach of an employee.

Before the carton arrives in the holding zone, an employee sets an end stop at the conveyor discharge end. As the cartons travel on the conveyor, the employee matches the carton identification number to the rack bay number. If two identification numbers match, the employee transfers the carton from the conveyor to the rack position.

When required to ship cartons, the employee is given a release instruction document and adjusts the end stop to permit cartons to travel on the conveyor from the holding area. As required, the employee matches the release document number to the rack position number and transfers cartons from the rack position to the conveyor. The release document sequence matches the delivery truck loading sequence.

An alternative conveyor arrangement is a double-stacked conveyor network which has inbound cartons traveling on the first level and shipping cartons traveling on the second level.

Pallet Flow Rack Hold Method.   When you consider the pallet flow rack temporary hold method, you envision walkways, an inbound conveyor with a fixed end stop, pallet flow lanes, and outbound conveyor. Each pallet flow lane is a customer holding position that is three to four pallets deep and requires "slave" pallet boards. Each pallet flow lane has a slight slope and at the charge end, a device (restricted wheels) to hold the pallet at the palletizing station.

As the inbound conveyor transports cartons to the temporary hold area, the employee matches the carton number to the pallet flow lane number. When the numbers match, the carton is transferred from the conveyor to the pallet flow lane. After the pallet load attains a predetermined height, the employee pushes it toward the discharge end and places a new slave pallet on the palletizing station.

When required to ship customer cartons, an employee is given the release instruction form and walks to the holding area, matches the instruction customer and pallet flow lane numbers, and transfers the cartons from the pallet flow lane onto the take-away conveyor. As required, the empty slave pallet board is placed onto a return lane. The additional pallets in the customer holding lane are pulled or gravity indexes (moves) the next pallet forward to the depalletizing station.

Drive-in or Drive-through Rack Concept. Drive-in and drive-through racks with post protectors are designed to hold product in the dock staging area. This type of staging concept is similar to standard rack, except that it provides increased density (number of pallet boards) per aisle. To hold a truckload, it requires (1) a three-deep rack (12 to 13 ft deep) 15 to 16 ft long and (2) a two-deep rack configuration of 8 to 9 ft depth with a staging area of 29 to 30 ft length.

The drive-in concept requires a turning aisle, and the rack rows are perpendicular to the dock. This provides a right- or left-angle product flow path. The drive-through rack provides a straight product flow path. This design requires a three-deep rack.

Disadvantages of this method are (1) additional rack investment, (2) requirement for a lift truck, and (3) slower truck loading and unloading. Advantages of the drive-in or drive-through rack method are that it requires less warehouse space and lower building construction costs and, with the drive-through rack, the product flow is straight.

Flow Rack. The flow rack is the next dock staging design. Each upright post requires post protectors and lift truck guards. This configuration is three pallet flow lanes high and three deep. The flow rack directs the product flow from the storage aisle to the dock aisle. In this design the product flow remains in a straight line and requires a lift truck. To hold 27 pallet loads, it requires a rack depth of 13 to 15 ft and a width of 15 to 16 ft. Within this configuration one flow lane is designed to flow empty pallet boards from the dock area to the storage area. This design feature permits empty pallets to be handled with little effort.

Disadvantages of the pallet flow design are increased rack investment and requirement for good pallet boards and a lift truck. Advantages include ability to handle pallets or containers, less building space requirement, and lower construction costs.

Push-Back Rack. The push-back rack is a dock staging apparatus that requires upright post protectors and lift truck guards. There are two alternate push-back rack designs: (1) the standard pallet flow rack with end stops and (2) the telescoping flow rack with nestable carriages. This configuration is three pallets high and three or four pallets deep. The slope depends on the weight of the unit load. The rack is placed along the dock walls or in a back-to-back arrangement. To handle a delivery truck, a rack depth of 13 to 16 ft is required.

Disadvantages and advantages of the push-back rack are similar to those of the pallet flow rack.

Stacking Frame or Portable Rack. The next dock staging method consists of stacking frames or portable racks. This design requires product to be placed into portable stacking frames that are placed in the floor-stack positions in the staging area. As additional stacking frames are required in the dock staging area, these frames are stacked two to three high onto the floor-level frames. The two-high design requires a 10-ft-wide by 30-ft-long area; the three-high, a 15-ft-wide by 24-ft-long area. The stacking frame design provides a straight product flow path between the storage and dock areas. This design handles pallet loads, or employees hand-stack directly into the stacking frame. Some models are nestable when not in use; these frames reduce the storage area.

Disadvantages of the stacking frame method are that a lift truck is required and

employee productivity is only moderate. Advantages are only moderate capital investment, large storage space not required by the nestable type, requirement for only moderate building size area, and handling of hand-stacked or palletized loads.

OTHER DOCK AREA CONSIDERATIONS

In a conveyorized carton sortation loading and unloading system design for inbound
or outbound product that has a direct (fluid) activity or unitizing (palletizing) activity
on the dock, a sufficient accumulation conveyor is provided prior to each workstation.
The accumulation conveyor provides the queue area to handle surges or the stops or
starts of handling product. This accumulation conveyor is rack- or ceiling-supported in
the warehouse facility.

In all building designs, the distance between the finished floor and the ceiling
accommodates the dock staging method. If the dock method involves the floor-stack
design in a high-ceiling warehouse, then an additional floor level or mezzanine for
offices is considered for this area. Prior to the design of the additional floor level, the
fire protection and the other codes are reviewed for its design.

CHAPTER 5

ROUTING AND ORGANIZING THE WORK OF ORDER PICKERS

INTRODUCTION

The objective of this chapter is to develop an understanding of the various order-picker routing patterns and the operating characteristics of each pattern. Each order-picker routing pattern, concept, and path through the aisles is considered individually and is matched to the type of SKU (pallet load, carton, or single item). Disadvantages and advantages of each order-pick concept are also discussed.

The second portion of the chapter analyzes the various order-picker instruction methods. These methods indicate to the order picker each SKU and SKU quantity that appears on a customer order form. This section identifies the disadvantages and advantages of each method and matches the instruction method to an appropriate order picker.

The next section of the chapter reviews the alternative customer order handling methods. These methods determine the number of order pickers that are in a warehouse aisle and the various batch control methods. The section examines the disadvantages and advantages of each order handling method and indicates the appropriate order-pick concept.

Finally, the chapter examines the theory that determines the product allocation to the pick position. This includes the preferred elevation for high-, medium-, and low-volume SKUs, heavy- and lightweight SKUs, or large- and small-cube SKUs.

REASONS FOR IMPROVING ORDER-PICK ACTIVITY

There are several reasons for an accurate, timely, and highly productive order-pick labor force. First, when an order picker withdraws SKUs that are on your customer order list accurately and on time, then your distribution operation achieves the following:

- Maintains a satisfied customer group

- Allows your manufacturing department to maintain its production schedule
- Allows your retail store to have the product on the retail shelves
- Achieves your budget cost

In addition, a review of the annual expense budget of most distribution facilities indicates that the order-picker line item has the highest budget dollar value and correspondingly the greatest number of employees. This is true for the manual and mechanized order-pick systems. If you consider the replenishment activity in an automated facility as an order-pick activity, then in the automated facility the order-pick labor force has the greatest number of employees.

FUNDAMENTAL CRITERION FOR ANY ORDER-PICKER ROUTING PATTERN

The fundamental rule of a successful order-picker routing pattern is that the pattern follow the rack, shelf, row, and aisle layout. The routing pattern guides the order picker down-aisle to the required pick position or directs the order picker's hand to the appropriate pick position in a container.

THE BASIC ORDER-PICK METHODS

In the warehouse and distribution industry, the three basic order-pick methods (Fig. 5.1) are:

- Order picker walks to the product location.
- Order picker rides to the product location.
- The product is transferred from the storage location to an order picker at a workstation.

All three methods require an order-picker routing pattern to direct the order picker to the SKU physical pick location in a warehouse aisle or to direct the employee's hand to the pick position of the SKUs in a container. The order-picker routing pattern objective is to reduce or minimize the order-picker's nonproductive time of walking, traveling between two SKU pick positions, or hand movement between two SKU pick positions within a container.

To achieve the maximum employee productivity, the appropriate order-picker routing pattern is implemented in conjunction with the fundamentals of good warehouse practices. Some of these practices are 80/20 or family grouping principle, clear aisles, pick position identification, and clear instructions.

VARIOUS ORDER-PICKER ROUTING PATTERNS

Many types of order-picker routing patterns can be implemented in your distribution facility. You realize that some patterns are the preferred patterns in a carton or single-

Manual method (picker to SKU)
Order picker ➡➡➡➡ SKU ➡➡➡➡ packer/shipping area

FIGURE 5.1   Three basic order-pick methods. (*Courtesy of White Storage and Retrieval Systems and Electro Com Automation L.P.*)

item operation, but are not preferred in a pallet load operation. In this chapter the order-picker routing pattern is matched to the type of operation. These order-picker routing patterns are

- Nonrouting method
- Sequential routing method

  Single side: one or two order pickers

  Loop

  Horseshoe, U

  Z

  Block

  Stitch

  Multilevel: one-way or two-way travel; four or six levels

  Vertical (up and down) movement

  Lateral or horizontal (front and rear) movement

  Zone: fixed or variable

  When your order-picker routing pattern matches the SKU characteristics, through-

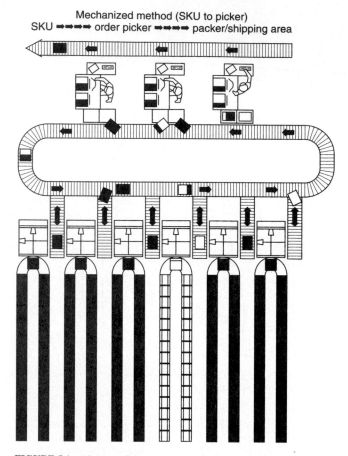

Mechanized method (SKU to picker)
SKU ➡➡➡➡ order picker ➡➡➡➡ packer/shipping area

FIGURE 5.1   *(Continued)*

put volume, and material handling concept, then the implemented order-picker routing pattern advantages are obtained by your operation. This match of the order-picker routing pattern with these other factors provides you with the opportunity to obtain the best order-picker productivity, accurate picks, on-schedule activity, and on-budget activity.

Nonrouting (Nonsequential) Pattern

When you consider all available order-picker routing patterns, then you consider implementation of the nonrouting (nonsequential) pattern, where order pickers determine their own pick path through the warehouse aisle or aisles. The nonrouting pattern is not considered for implementation in a warehouse because it offers no advantages. The disadvantage of this method is that it provides extremely low employee productivity because

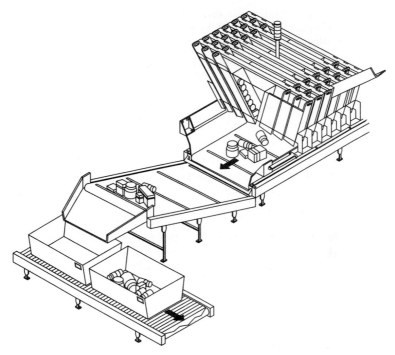

FIGURE 5.1 *(Continued)*

- The employee may walk the same path twice.
- The employee fatigues from increased walking or hand movement.
- The employee spends nonproductive time trying to locate the SKU aisle.

Sequential Order-Pick Patterns

In the warehouse and distribution industry, the sequential order-picker routing patterns are the preferred and most commonly used. The fundamental characteristic of the sequential order-picker routing pattern is that there is an arithmetic progression to the pick position numbers through each warehouse aisle. This means that the lowest SKU pick position number (1 or 0) is at the entrance to the aisle and the highest pick position number (99 or 100) is at the exit of the aisle.

In these patterns the order picker starts at the first required SKU pick position in the warehouse aisle, and as the order picker travels (progresses) down-aisle to the end of the aisle, the next required SKU pick position is as close as possible to the previous SKU pick position. In your order-pick operation, any one of the sequential routing patterns provides you with an efficient and productive order-picker group.

Advantages of the sequential routing method include reduced employee nonproductive time (two or more trips down an aisle, etc.), employee fatigue, and employee confusion and increased employee productivity.

Several elements (criteria) are part of an order-picker routing pattern and warehouse aisle condition to assure optimum results:

- Pick position numbers that end with an even digit on the right side of the aisle as you travel down-aisle; pick position numbers that end with an odd digit on the left side of the aisle as you travel down-aisle.
- Use arithmetic progression through the aisle.
- Keep the picker in the aisle (work area) as long as possible.
- Improve the SKU hit concentration and hit density.
- Start the pickers in the fast-moving section.
- With single-item orders, whenever possible pick and pack.
- Cube out (divide) the pick activity.
- Keep the aisles clear (maintain good housekeeping) and well illuminated.

Evens on the Right and Odds on the Left.　　To maximize order-picker productivity, whenever possible all order-picker routing patterns have pick position numbers that end with even digits on the right side of the aisle and pick position numbers that end with odd digits on the left side of the aisle (Fig. 5.2a). During the picking activity, this pattern reduces confusion because several routine, everyday activities have this numerical arrangement.

Arithmetic Progression through the Aisle.　　The second element of a sequential order-pick system is that the method routes the order picker to remove single items or cartons from a divided rack or shelf opening or pallets from a storage position. The divided rack or shelf opening pick position number sequence within the opening as the order picker walks (in the aisle) is in an arithmetic progression from the first pick position (lowest number) to last pick position (highest number).

Keep the Order Picker in the Aisle.　　The third element of the sequential order-picker system is that all the routing patterns keep the order-picker in the aisle as long as possible. This feature reduces the nonproductive travel time and increases your order picker's SKU hit concentration and hit density.

Cube the Pick Activity.　　The fourth element for a good order-picker system is to reduce or eliminate unnecessary conversation between order pickers while they are in the warehouse aisles and to ensure that the order picker is capable of handling the work without additional trips.

If your distribution operation order-pick instruction method "cubes out" (divides) the order picker's activity, then the computer program and the total cubic feet for one of your customer's ordered SKUs determines the number of order pickers per aisle. Also, the order picker has the ability to fill the container or vehicle load-carrying surface to capacity each trip, thus optimizing the trip. This feature reduces the order picker's nonproductive travel time and reduces product damage.

If your distribution operation order-picker instruction method does not cube your order picker's activity, then the order control clerk estimates and issues to each order picker a quantity of pick tickets (instructions) that balances the load-carrying capacity of the vehicle and maintains one order picker per aisle.

When there is no match of the SKU pick quantity to the vehicle load-carrying capacity, then there is a high probability that the order picker's portion of the cus-

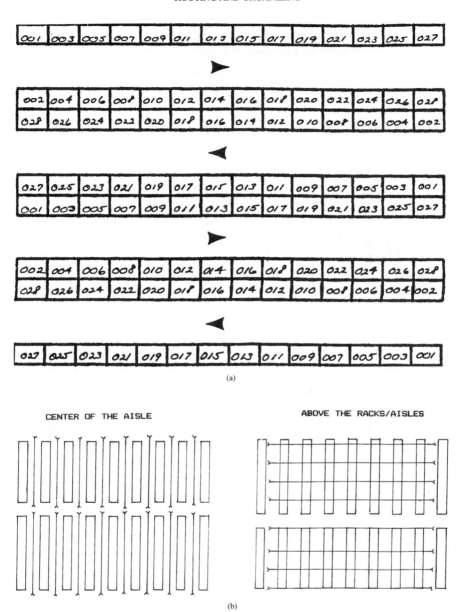

FIGURE 5.2 (*a*) Order-picker routing using even numbers on the right side of the aisle and odd numbers on the left side of the aisle; (*b*) alternative light fixture hanging arrangements.

tomer order exceeds the vehicle load-carrying capacity. This situation has a high potential to cause the order pickers to make unnecessary, nonproductive travel trips to complete their portion of the customer order.

Where to Start the Single-Item Order Pickers. When your company's single-item order-pick area layout has fast-moving SKUs and slow-moving SKUs in separate areas, then to obtain the highest possible employee productivity in the single or batched customer pick and pack operation, where do you start (in what section) your order pickers? In the single-item order-pick area design, the fast-moving SKUs represent 80 percent of the business from 20 percent of the SKUs and the slow-moving SKUs represent 20 percent of the business from 80 percent of the SKUs.

The two possible order-picker start location options are:

- Start the order pickers in the fast-moving section.
- Start the order pickers in the slow-moving section.

If the order pickers start in the fast-moving section, then the results of the start location are

- This option improves pick position replenishment productivity.
- This option assures a higher volume handled by the pick operation and the flow of product from the pick area to the packing area.
- With slow-moving SKUs, it is easier to top off customer order cartons from the fast-moving section.
- This option decreases the number of customer order carton handlings.
- This option improves knock-down (KD) carton replenishment (retrieval of empty cartons) because a large number of tall or partially filled cartons may be used in the fast-moving section and short cartons may be used in the slow-moving section.
- This option assures high order-picker productivity because the fast-moving SKUs with high hit density and concentration are easily (first) read on the order-pick instruction form.

If the order-pick operation handles each container of multiple tote or container customer order as an individual container and your transportation system has a bypass transportation conveyor spur for full containers from the fast-moving section, then the preferred start location is the fast-moving section.

If you start the order pickers in the slow-moving area, then the results of the start location are

- There is lower picker productivity because the order pickers in the fast-moving area handle two cartons for fast-moving SKUs (one container that is partially full of slow-moving SKUs which is topped off with fast-moving SKUs and a second container for fast-moving SKUs).
- It is more difficult to read the order-pick instruction document because the fast-moving SKUs are located at the bottom of the document.
- It is more difficult to transport a partially full carton of slow-moving SKUs (because of lighter carton weight) on the transport system.
- There is increased KD carton handling and all types of KD cartons in the fast-moving area.

- To keep replenishment activities coordinated with the order-pick activities, the slow-moving SKU replenishment is handled prior to the fast-moving SKU replenishment, which does not represent the majority of your business.

Single Item and Single Customer Means Pick-Pack Activity.   The pick-pack activity is a single-item activity that occurs when your distribution operation has a high volume of a single (sometimes two SKUs) SKU per customer order. The pick-pack activity consists of one or two additional steps in the normal order-pick activity. When compared to the normal pick activity, there is a slight decrease in order-picker productivity because the picker packs, seals, and labels the package. This dramatically improves the total facility productivity and product flow because the merchandise bypasses the packing station. If the SKU is picked en masse and transferred to a specific pick-pack station, then the order-picker productivity is very high.

With the pick-pack activity, the order picker picks the SKU, packs the invoice and SKU into a shipping container, seals and labels the shipping container, and places the container onto a shipping take-away container.

Improve Order-Picker SKU Hit Concentration and SKU Hit Density.   The next element in improving order picker productivity is to improve the SKU hit concentration and SKU hit density. To do this, you allocate your SKUs according to Pareto's law (80/20), which states that 80 percent of the volume is derived from 20 percent of the SKUs.

The SKU *hit concentration* means the number of SKUs that are ordered by your customers or lines (stops or hits) within a particular warehouse aisle or the number of pick positions within one aisle to withdraw for a customer order. The SKU *hit density* is the number of times (quantity) that a customer's order has a particular SKU (one pick position) or the number (quantity or hits) for that one SKU to complete a line on a customer's order.

A high SKU hit concentration and SKU hit density dramatically improves order picker productivity because the travel time (distance) between picks (hits) is reduced. For best results in a single-item pick to conveyor or tote (or carton pick to conveyor) warehouse operation, the order pickers are assigned pick zones. In these operations, to assure on-time arrival at the appropriate sortation or packing location, each pick zone conveyor section requires a different travel speed. The most distant zone has the speed that is the minimum to transport the merchandise to the final (packing) station. The next zone is 5 to 10 ft/min faster than the previous zone.

How to Light (Illuminate) Your Warehouse Storage-Pick Aisle.   There are two alternative light fixture arrangements for warehouse storage and pick aisles. With both alternatives, the light fixture program requires you to specify the desired lighting (lumen) level 30 in above the finished floor. The two arrangements (shown in Fig. 5.2 *b*) are

1. Fixtures hung directly above the center of the aisle (Fig. 5.2*b*)
2. Fixtures hung perpendicular to the aisle and above the racks or shelves

In arrangement 1, with the light fixture directly above the center of the aisle, the light fixtures hang from the joists or second-level platform (walkway) support members. This arrangement includes a series of fixtures that illuminate the entire length of the aisle. When required to replace the light fixtures or tubes or bulbs, maintenance employees have easy access to the fixture with this arrangement. Also, warehouse employees are generally more familiar with this arrangement in their daily lives.

Arrangement 2, where the light fixture is perpendicular to the aisle and above the racks or shelves, includes a series of light fixtures hung from the joist of the second-level structural members. In this arrangement a row of fixtures begins at your storage-pick area first aisle, strung above the other storage-pick structures and aisles, and ends at your last storage-pick aisle. The center spacing between the light fixture rows ensures that the light (lumen) level is as desired. When required to replace light fixtures or tube or bulbs, maintenance employees have a more difficult task with arrangement 2 as the light fixture is less accessible. Also, warehouse employees are generally unlikely to have this type of arrangement in their homes and will be less familiar with it.

Maintain Clear Aisles and Good Housekeeping. When you consider that an order picker travels through a warehouse aisle to various pick positions, then to attain the proper travel speed without unintentional stops, the aisles must be free of obstacles and clear. Many professionals have stated that good housekeeping in the warehouse enhances employee productivity.

Single-Side Order-Picker Routing Patterns

The single-side order-picker routing pattern directs the order picker through the warehouse aisles. These aisles are between racks or shelves that have the SKU pick positions. These SKU pick positions contain SKUs that appear on a customer's order document.

Single-Side Order-Picker Pattern with One Picker. The single-side order-picker routing pattern has two alternative designs. In the first design one order picker travels down the aisle and withdraws SKUs from pick positions on the right side of the aisles, to turn at the end of the row and to travel down the same aisle on the left side and withdraw SKUs from aisle pick positions (Fig. 5.3a).

Single-Side Order-Picker Pattern with Two Order Pickers. The other alternative involves two order pickers (Fig. 5.3b). The first order picker travels down-aisle and withdraws SKUs from the right side of aisle A pick positions and then transfers to aisle B. In aisle B, the first order picker withdraws SKUs from the right side of aisle B pick positions. The second order picker travels down aisle A and withdraws SKUs from the left side of aisle A pick positions, transfers to aisle B, and withdraws SKUs from the left side of aisle B pick positions.

Single-Side Order-Picker Routing Advantages and Disadvantages. The single-side order-picker routing pattern is used to withdraw SKUs from floor or rack pick positions that contain pallet loads (especially SKUs with high cube and length), cartons, and single items from a rack or shelf bay.

The pattern is commonly used in a distribution facility that has a "human walk," "human ride a truck," or "human walk pick to belt" material handling concept. The single-side order-selector routing pattern is implemented in a pallet load, carton, single-item, or hanging-garment distribution facility.

When your fast- and slow-moving SKUs are randomly located in the aisle SKU pick positions, then the single-side order-picker routing pattern has the disadvantage of low order-picker productivity. This is due to a high probability of nonproductive travel (walking distance) between two SKU pick locations. In addition, to complete the customer order, an order picker must travel twice, or two order pickers must travel once, down the same aisle to withdraw product from both sides of the aisle.

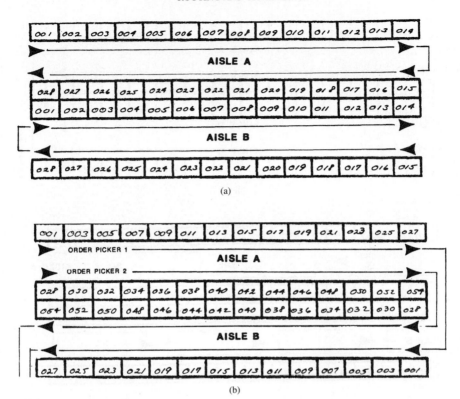

FIGURE 5.3 (*a*) Single-side order-picker pattern with one order picker; (*b*) same pattern with two order pickers.

There are several advantages, however, as single-side order-picker routing is easy to implement in a new or existing distribution facility. Moreover, it requires very little employee training, and when the order picker withdraws one pallet load (a large number of cartons or single items to complete the customer order), the single-side routing pattern provides the opportunity for excellent employee productivity. Also, when the order picker is picking onto a series of towed carts, the single-side routing pattern reduces the employee's nonproductive time of walking across the aisle to select and place SKUs onto the cart. Final advantage is when the rack-and-aisle layout dead ends into a wall, then the single side order picker routing pattern productivity is equal to the productivity that is achieved from the other order picker routing patterns.

Other Order-Picker Routing Patterns

Loop Order-Picker Routing Pattern. In the loop order-picker routing pattern the order picker travels down the aisle to withdraw SKUs from aisle A right-side pick positions (Fig. 5.3*c*). At the end of aisle A, the order picker transfers to aisle B. In

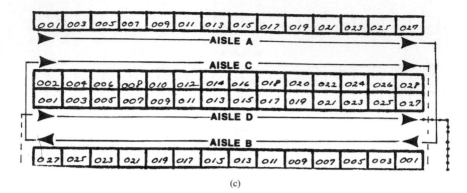

(c)

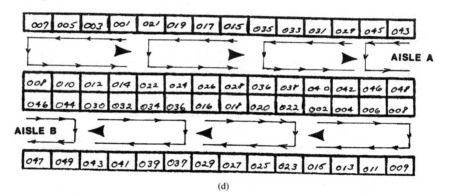

(d)

FIGURE 5.3 (*Continued*) (*c*) Loop order-picker routing pattern; (*d*) U or horseshoe order-picker routing pattern.

aisle B, the order picker withdraws SKUs from aisle B left-side pick positions. After completing travel in aisle B, the order picker returns to aisle A and withdraws SKUs from aisle A left-side pick positions. After completing aisle A travel, the order picker travels to aisle C and withdraws SKUs from the left side of that aisle. This pattern is repeated until completion of the entire customer order or the order picker completes the cube or portion of your customer's order.

The loop order-picker routing pattern is used to withdraw cartons or single items from pick positions that are in floor or rack pick positions. It is best used in a carton or single-item distribution facility that handles orders with numerous different SKUs that are located in numerous aisle locations. The loop routing pattern is best used in a human (employee) walk, pallet jack, or train of carts with a tugger order-picker system. With a conveyor in the middle of the aisle, the loop pattern is not used in a pick-to-belt system.

The loop order-picker routing pattern is not the preferred pattern for a pallet load operation. In a pallet load warehouse, it is very difficult to obtain high order-picker productivity because of the loop pattern pick position number sequence.

If you locate fast- and slow-moving SKUs randomly in the aisle pick positions, the loop order-picker routing pattern will present many disadvantages. Low order-picker productivity may result because of a very high probability of nonproductive travel time (walking distance) between two SKU pick positions. To complete your customer's order, the order picker must double travel in the same aisle to withdraw SKUs from both sides of one aisle. This feature requires the order picker to pass the previously selected pick positions. When compared to other order-picker routing patterns, the loop order-picker routing pattern requires extensive employee training.

If you have a high-volume carton or single-item distribution facility, then excellent order-picker productivity is achievable with the following parameters. If you locate fast-moving SKUs on the right side of the aisle and slow-moving SKUs on the left side, there is an excellent opportunity for high order-picker productivity. If the order picker is picking SKUs onto a series of towed carts, then the loop order-picker routing pattern reduces the employee's nonproductive time spent walking across the aisle to select and place SKUs onto the carts.

U or Horseshoe Order-Picker Routing Pattern.   In the U or horseshoe order-picker routing pattern (Fig. 5.3*d*), the order picker travels down the aisle and at predetermined pick locations halts in the middle of the aisle. At each halt location, the order picker has the opportunity to withdraw SKUs from the four pick positions on both right and left sides of aisle A (i.e., from a total of eight pick positions). Until the customer order is completed, completion of the cube (portion of the customer's order), or the vehicle SKU carrying device reaches a predetermined height, the pattern is followed through the warehouse aisles.

The U or horseshoe order-picker routing pattern is best used to withdraw single items or cartons from floor or rack pick positions and in a distribution facility that handles customer orders that have numerous SKUs in various aisles. The U pattern is not considered for implementation in a pallet load warehouse. The pattern is best used for a human (employee)-walk or human-ride order-pick system. Since the conveyor is in the middle of the aisle, the horseshoe pattern is not implemented in a pick-to-belt order-pick system.

Disadvantages and advantages of this pattern are similar to those of the loop order-pick routing pattern. An additional disadvantage of the U pattern is low productivity or counterproductive time spent by the order picker stopping at predetermined aisle halt locations and passing pick positions that have SKUs that are not on your customer's order.

Z Order-Picker Routing Pattern.   In this routing pattern the order picker travels once down-aisle between the aisle pick positions (Fig. 5.4*a* and *b*). The pattern directs the order picker to withdraw SKUs from the first four down-aisle pick positions on the left side of aisle A, withdraw SKUs from the next eight down-aisle pick positions on the right side of aisle A, and withdraw SKUs from the next eight down-aisle pick positions on the left side of aisle A. Until the customer order is completed or the container or pallet load reaches a predetermined height, this pattern is followed through the distribution facility aisles.

The Z order-picker routing pattern is used in a distribution facility that handles cartons or single items. It is not used in a pallet load warehouse. The pattern is used in a human walk or human ride on a powered pallet truck order-pick system. It is not implemented in a pick-to-belt or high-rise order-pick system.

One disadvantage of the Z order-pick routing pattern is that it requires additional employee training. Without an excellent SKU locator system, the employee's nonproductive walk time between SKU pick positions increases to complete your customer's order.

Advantages of the Z order-picker routing pattern are many. With an excellent SKU locator system, improved order picker productivity is realized due to one down-aisle

LARGE 'Z' PATTERN

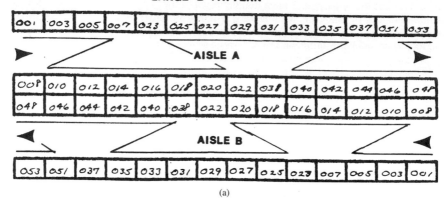

(a)

SMALL 'Z' PATTERN

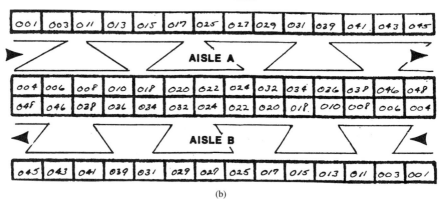

(b)

FIGURE 5.4 (*a*) and (*b*) Z order-picker routing pattern; (*c*) block order-picker routing pattern; (*d*) stitch order-picker routing pattern.

trip to withdraw SKUs from the pick positions that are on both sides of the aisle. With individual pick positions in an arithmetic progression through the warehouse aisle, there is a reduction in employee nonproductive time to locate the SKU pick position. This nonproductive time of walking between SKU pick positions is reduced by allocating fast-moving SKUs to the aisle right-side pick positions and slow-moving SKUs to the aisle left-side pick positions.

Block Order Picker Routing Pattern. In this routing pattern the order picker travels once down-aisle between the aisle pick positions (Fig. 5.4*c*). The order picker then withdraws SKUs from the first two pick positions on the right side of the aisle and SKUs from two pick positions on the left side of the aisle. Until completion of the customer order or the pallet load or container is full, the pattern is followed through the warehouse aisles.

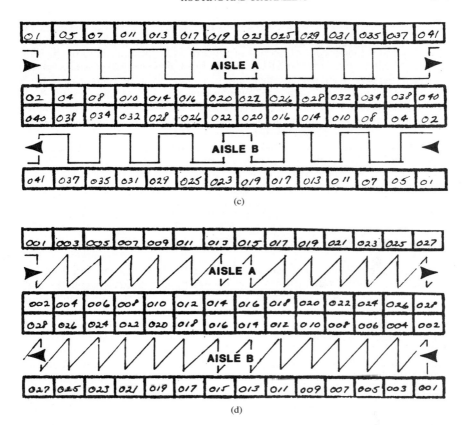

(c)

(d)

FIGURE 5.4 (*Continued*)

The block order-picker routing pattern is best used in a carton or single-item distribution facility. It is very rarely used in a pallet load distribution facility. The pattern is best used in a human walk or human ride on a vehicle order-pick system. The block pattern is not implemented in a pick-to-belt conveyor system.

If your carton distribution facility has a large number of slow-moving SKUs, one block pattern disadvantage is an increased number of nonproductive walking trips across the aisle to withdraw SKUs from the aisle right- and left-side pick positions. Furthermore, in this pattern the down-aisle travel distance between halt locations has the potential to increase the order picker's nonproductive starts and stops. The pattern requires additional employee training and a good stock locator system.

If your distribution facility handles fast-moving cartons of single-item SKUs, then the block order-picker routing pattern offers several advantages. First is improved order-picker productivity because there is only one down-aisle trip to withdraw SKUs from pick positions on both sides of the aisle. Second, if you locate fast-moving SKUs on the right side of the aisle and slow-moving SKUs on the left side of the aisle, order-picker productivity is further improved because there are fewer nonproductive cross-aisle trips.

Stitch Order-Picker Routing Pattern. In this routing pattern the order picker travels down-aisle between the aisle pick positions (Fig. 5.4*d*). The pattern then directs the order picker to withdraw SKUs from one rack or shelf bay (pick position) on the right side of the aisle and to withdraw SKUs from the rack or shelf (pick position) on the left side of the aisle. Until the customer order is completed or the container reaches a predetermined height, this pattern is followed through the aisles of your facility.

The stitch order-picker routing pattern is best used in a single-item distribution facility. The pattern is not preferred for a unit-load or carton distribution facility and is not used in a human-walk pick-to-belt conveyor system.

Disadvantages and advantages of the stitch order-picker routing pattern are similar to those indicated for the block order-picker routing pattern. In a single-item distribution facility that has a high SKU hit concentration and hit density, the stitch routing pattern improves order-picker productivity.

Multilevel Order-Picker Routing Patterns

Multilevel order picker routing patterns are specialized patterns that are used in a machine-ride multilevel (high-rise) case or single-item order-picker system. The multilevel order-picker routing pattern directs the order picker who rides on board a vehicle to the required pick position. The vehicle travels in vertical and horizontal down-aisle directions between two rack or shelf row pick positions. If the vehicle travels from the aisle first pick position to the aisle last pick position or from the last pick position to the first pick position, then this travel is considered as one down-aisle trip. If it is important to obtain maximum order-picker efficiency with low product or equipment damage, then the order-pick vehicle travels in a rail- or wire-guided aisle. A picking cage, which basically is a specially engineered device consisting of an empty pallet board at the bottom, with two high sides and a high back, may also be used to reduce product damage.

There are two fundamental high-rise order-selector (HROS) routing patterns: one-way vehicle (vehicular) traffic through the aisle and two-way (forward-reverse direction) traffic within the aisle. Other multilevel order-picker routing patterns are also discussed in this section.

One-Way High-Rise Truck Traffic. One-way high-rise vehicle traffic through the aisle directs the order picker to make one trip through the aisle (Fig. 5.5*a*). With this pattern, as the order picker enters the aisle, the first pick position (lowest number) is at the bottom of the first rack or shelf bay and the last pick position (highest number) is at the top of the last rack or shelf bay.

This routing pattern guides the order picker to enter the aisle at one end and exit at the opposite end. Each bottom rack or shelf position of the next rack or shelf bay has the lowest number for the bay (stack). At the bottom of the bay, the pick position number sequence is in the vertical direction of the vertical bays (stack). With this pattern the bottom pick position is numbered 0010 and the next pick position is numbered 0011, which is in the same rack bay on the second level, and so on. In the rack bay after the pick positions reach the maximum number of the rack positions, then the next arithmetic pick position is the bottom bay position of the opposite of the aisle rack or shelf position. With the first bottom bay pick position number 0010, the next bottom bay pick position number is 0020. With this order-picker routing pattern, the order picker picks from both rack positions in the aisle and makes one trip in the aisle and transfers to the next aisle to complete your customer's order.

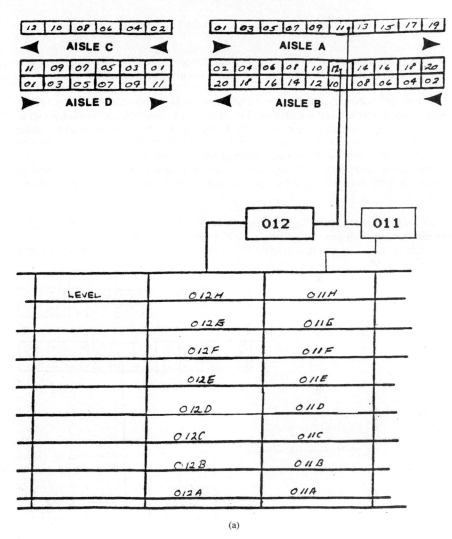

	12	10	08	06	04	02

◄ **AISLE C** ◄

	11	09	07	05	03	01
	01	03	05	07	09	11

► **AISLE D** ►

01	03	05	07	09	11	13	15	17	19

► **AISLE A** ►

02	04	06	08	10	12	14	16	18	20
20	18	16	14	12	10	08	06	04	02

◄ **AISLE B** ◄

LEVEL	012H	011H
	012G	011G
	012F	011F
	012E	011E
	012D	011D
	012C	011C
	012B	011B
	012A	011A

(a)

FIGURE 5.5 (*a*) One-way high-rise order-selector (HROS) truck traffic pattern.

The one-way high-rise order-picker routing pattern is implemented in a carton or single-item distribution facility. In a specialized pallet load distribution facility, the one-way high-rise truck routing pattern is used for order-picker control.

Disadvantages of the pattern are

- Lower productivity due to increased up-down vehicle movement and double travel with full loads
- Requirement for turning aisles at each end of the aisle

- Increased product damage resulting from product falling from full loads at high levels

Two advantages of the one-way high-rise order-picker routing pattern are that it is easy to implement and all trips are in one direction through the aisle.

Two-Way High-Rise Truck Traffic.   The two-way high-rise vehicle routing system (Fig. 5.5*b*) directs the order picker to start at the highest-level pick positions, and at the end of the aisle the operator lowers to the bottom-level pick position of the same aisle. This is a second trip back to the aisle entrance. Therefore, the two-way routing system has an aisle with one entrance and exit for the vehicle.

Two basic order-picker routing patterns are used in a two-way high-rise vehicle routing system: four-level and six-level systems, described in the next two subsections.

Four-Pick-Level Order-Picker Routing Pattern.   In a four-vertical-rack-bay pick position system with a maximum height of 36 in per opening, the order picker's down-aisle trip is to elevate and to withdraw SKUs from the third and fourth (highest) pick

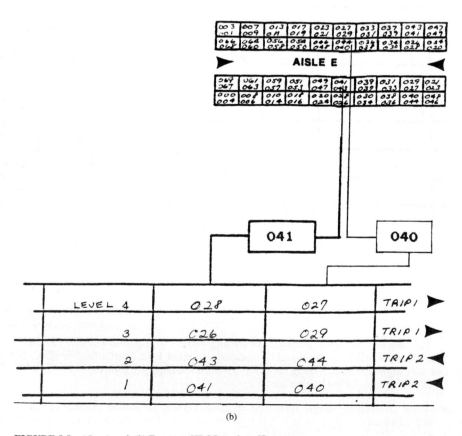

(b)

FIGURE 5.5   (*Continued*) (*b*) Two-way HROS truck traffic pattern.

levels of the rack (Fig. 5.6a). At the end of the aisle, the operator lowers the machine to the floor and travels (reverse direction) down-aisle to the aisle entrance. During this trip, the order picker withdraws SKUs from the first- and second-high rack pick positions.

Six-Pick-Level Order-Picker Routing Pattern.   If the vertical rack structure consists of six rack bay levels (Fig. 5.6b), then to complete the down-aisle trip, one trip must be made down-aisle in the elevated position, thus permitting the order picker to withdraw SKUs from the fourth (4th), fifth (5th), or sixth (6th) rack level. At the end of the aisle, the machine lowers to the floor level and makes the reverse down-aisle trip that permits the order picker to withdraw SKUs from the first (1st), second (2d), and third (3d) rack levels.

8- or 10-Pick-Level Order-Picker Routing Pattern.   In an 8- or 10-vertical-shelf pick position system with a maximum of 30-in-high openings, the order picker's down-aisle travel is to elevate and withdraw SKUs from the highest four or five levels of the shelving bay. In one trip, the order-picker picks from levels 10, 9, and 8 and in the second trip, from levels 7 and 8. The third aisle trip permits the order picker to withdraw SKUs from the levels 5, 4, and 3. In the final trip the picker picks SKUs from levels 2 and 1 or the lowest pick positions in the rack bay.

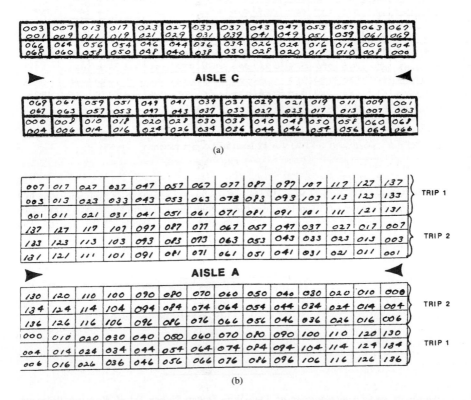

(a)

(b)

FIGURE 5.6   (a) Two-way four-level HROS truck traffic pattern; (b) two-way six-level HROS pick pattern.

007	017	027	037	097	057	067	077	087	097	107	117	127	139	TRIP 1
003	013	023	033	043	053	063	073	083	093	103	113	123	133	
001	011	021	031	041	051	061	071	081	091	101	111	121	131	
137	127	117	107	097	087	077	067	057	047	037	027	017	007	TRIP 2
133	123	113	103	093	083	073	063	053	043	033	023	013	003	
131	121	111	101	091	081	071	061	051	041	031	021	011	001	
007	017	027	037	047	057	067	077	087	097	107	117	127	137	TRIP 3
003	013	023	033	043	053	063	073	083	093	103	113	123	133	
001	011	021	031	041	051	061	071	081	091	101	111	121	131	
137	127	117	107	097	087	077	067	057	047	037	027	017	007	TRIP 4
133	123	113	103	093	083	073	063	053	043	033	023	013	003	
131	121	111	101	091	081	071	061	051	041	031	021	011	001	

▶   **AISLE C**   ◀

130	120	110	100	090	080	070	060	050	040	030	020	010	000	TRIP 1
134	124	114	104	094	084	074	064	054	044	034	024	014	004	
136	126	116	106	096	086	076	066	056	046	036	026	016	006	
000	010	020	030	040	050	060	070	080	090	100	110	120	130	TRIP 2
004	014	024	034	044	054	064	074	084	094	104	114	124	134	
006	016	026	036	046	056	066	076	086	096	106	116	126	136	
130	120	110	100	090	080	070	060	050	040	030	020	010	000	TRIP 3
134	124	114	104	094	084	074	064	054	044	084	024	014	004	
136	126	116	106	096	086	076	066	056	046	036	026	016	006	
000	010	020	030	040	050	060	070	080	090	100	110	120	130	TRIP 4
004	014	024	034	044	054	064	074	084	094	104	114	124	134	
006	016	026	036	046	056	066	076	086	096	106	116	126	136	

(c)

FIGURE 5.6 (*Continued*) (*c*) Two-way 12-level HROS pick pattern.

If the shelf pick positions have 12 (Fig. 5.6*c*) to 20 vertical shelf pick positions in a bay with a maximum of 30-in-high openings, then the order picker makes eight trips through the aisle. The first trip is at the highest three pick levels 20, 19, and 18. The second trip is at pick levels 17, 16, and 15. The third trip is at the pick levels 14, 13, and 12. The fourth trip is the pick levels 11 and 10. The other trips are

Trip	Pick levels
5	9, 8, 7
6	6, 5
7	4, 3
8	2,1

Advantages and Disadvantages of Two-Way Order-Picker Truck Routing Patterns.
The two-way high-rise order-picker routing patterns are best used in a carton or sin-

gle-item distribution facility. A combination of stitch and one-way high-rise order-picker routing patterns is used in any type of distribution facility.

One disadvantage of the two-way high-rise order-picker routing pattern is that the order picker is restricted to withdraw product from the pick positions in the aisle pick positions on each side of the vehicle. When the order picker moves the vehicle up or down for maximum reach, then it is nonproductive time.

Advantages of the two-way high-rise order-picker routing patterns are many. The order picker at each stop has access to the maximum number of pick positions without moving the vehicle in the vertical or horizontal directions. The down-aisle pick position arithmetic progression reduces the order picker's nonproductive time spent searching for the required pick position. All full-load travel is at the floor level, which reduces product damage because the vehicle is more stable at the lower levels. When the SKUs are allocated by the power rule (ABC zone theory) theory, then there is less nonproductive full-aisle-length travel.

Vertical, Front-to-Rear, Zone, and Other Order-Picker Routing

Vertical (Up-and-Down) or Vertical Up Order-Picker Routing Pattern. The vertical up-down or vertical up order-picker pattern is a specialized single-item or carton order-picker pattern that is used in a carousel or carton (case) flow rack order-pick system (Fig. 5.7a and b). The order-picker pattern requires two steps. First is one of the preceding order-picker routing patterns which directs the order picker to the aisle SKU pick position or bin that is delivered to the order-pick station.

The vertical up-down order-picker pattern has many advantages. With even-numbered pick position digits on the upper levels and odd-numbered pick position digits on the lower levels, the picker spends less nonproductive time locating the required pick position. An alternative pick number pattern is to have the lowest number (1) one at the lowest level with an arithmetic progression upward to the highest level and number. With a proper stock locator system, there is improved replenishment and order-picker productivity.

Front-to-Rear Order-Picker Pattern. The front-to-rear order-picker routing pattern (Fig. 5.7c) is a specialized pattern designed for a drawer pick system. The drawer or container is a static drawer that it manually pulled outward from a cabinet (larger container) or is mechanically extracted from a miniload storage system for delivery to a pick station. The front-to-rear pick pattern is used for single items.

This order-picker routing system consists of two steps. The first is to use one of the preceding order picker routing patterns that directs the order picker to the cabinet or the drawer that is delivered to a workstation. At this location, the front-to-rear order-picker routing pattern directs the order picker's hand to withdraw a single item from one of the drawer or container compartments.

One advantage of the front-to-rear order-picker routing pattern is that it reduces search time since the high pick position numbers are in the rear and low pick position numbers are in the front of a drawer or container.

Zone Order-Picker Routing Pattern. The zone order-picker routing pattern directs the order picker to travel in a predetermined warehouse area or within a specific area of an aisle. This feature permits the order picker access to a limited number of SKUs (pick positions). The pattern is used in a single-item, carton, or pallet load warehouse facility.

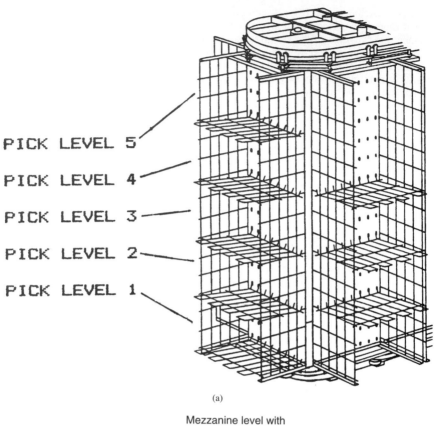

PICK LEVEL 5

PICK LEVEL 4

PICK LEVEL 3

PICK LEVEL 2

PICK LEVEL 1

(a)

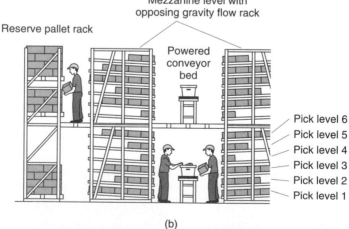

Mezzanine level with
opposing gravity flow rack

Reserve pallet rack

Powered
conveyor
bed

Pick level 6
Pick level 5
Pick level 4
Pick level 3
Pick level 2
Pick level 1

(b)

FIGURE 5.7 (*a*) Vertical up and down pick pattern (*Courtesy of ACCO Systems*); (*b*) vertical up pick pattern. (*Courtesy of Material Handling Engineering Magazine.*)

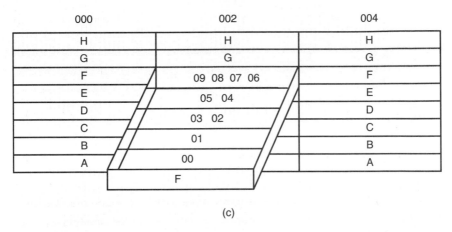

(c)

FIGURE 5.7   (*Continued*) (*c*) Front-to-rear pick pattern.

There are two zone order-picker routing patterns: the fixed pattern and the variable pattern.

The *fixed-zone order-picker routing pattern* requires the order picker to travel past a predetermined (fixed) number of aisles or within an aisle (Fig. 5.8*a*). Within this area, the order picker withdraws SKUs from a specific number of pick positions.

The *variable-zone order-picker routing pattern* (Fig. 5.8*b*) directs the order picker to travel within a warehouse aisle that does not have rigid (predetermined) aisle limits or does not have (rigid) predetermined start and finish pick positions. This variable-zone pattern is best used in an operation that issues batched or single customer orders per wave.

The zone order-picker routing pattern is best used in a distribution facility that handles cartons or single items and in a manual push-the-container or human-walk pick-to-belt conveyor system.

Disadvantages of zone order-picker routing are:

- It requires a high volume.
- It requires an increase in management control of the zones.
- Queues tend to slow down other workers.

The advantages of the pattern include

- It establishes good work standards.
- It assures accurate employee productivity.
- It reduces product damage and improves container utilization because employees are familiar with the zones and SKUs.
- It reduces replenishment and order-pick errors because there are fewer SKU changes and the employees are familiar with the SKUs.

Order-Picker Routing Pattern with an Overhead Trolley.   If your garment or soft-goods distribution facility has an overhead trolley or floor-mounted conveyor in the main aisle, then the order-picker routing system does not direct the order picker to travel across the main traffic aisle. The preferred scheme is to turn the order picker

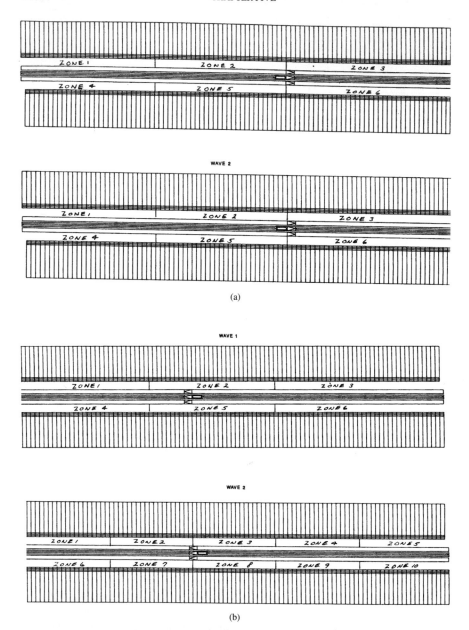

FIGURE 5.8 (*a*) Fixed-zone order-picker routing pattern; (*b*) variable-zone order-picker routing pattern.

from one aisle into an adjacent aisle. This pattern involves increased rail and switch cost but decreases employee nonproductive walking time.

The second alternative is to have the order picker walk from the main traffic aisle into the pick aisle to the pick position and return with the garment to the trolley in the main traffic aisle. This pattern increases your employee walking time and lowers capital investment.

Pick-to-Belt Conveyor Order-Picker Routing Pattern. If your distribution facility has a pick-to-belt conveyor with pick aisles on both sides of the conveyor, then the order-picker routing path does not direct the order picker to cross the conveyor. The preferred order-selector path is to travel the entire length in one pick aisle, then cross over or under the conveyor, to a lower level or to the other side of the conveyor for travel in the second pick aisle.

ORDER-PICKER INSTRUCTION METHODS

The fundamental component of any order-pick system is the order-picker instruction method. This directs the order picker to the SKU pick position, describes the SKU, indicates the SKU quantity, and states the customer's name and address or workstation.

Keep It Simple and Clear

The basic philosophy of any order-pick instruction method is to keep order-picker instructions as simple as possible. This allows the order picker to read the instructions and clearly understand them. These instructions are on a paper document, label or lighted display in the form of alphabetic characters, numeric digits, or a combination of both. Each individual or series of characters and digits identifies a specific aisle and SKU pick position in your distribution facility.

Information on the Instruction

The typical order-picker instruction identifies the following:

- Warehouse or warehouse location
- Warehouse aisle
- Pick position

Two order-picker instruction formats are shown in Fig. 5.9a and b.

SKU pick positions are digits (numbers). You may note, that both order-picker instruction examples in Fig. 5.9 have a combination of alphabetic characters and numeric digits in their formats. This combination helps reduce replenishment and order-picker errors which may result from an employee mentally confusing the numbers.

Digits are generally easier for order pickers to read and understand because most

Type of unit	Warehouse or location	Warehouse aisle	Bay	Level	Position (location)
Pallet	A	BB	120	—	—
Carton	A	CC	100	B	1
Single item	A	DDN	010	C	N

Key: A = alphabetic characters; B, C, and D = shelf level; N = numeric digits.

(a)

Warehouse location	Format digit
Building	2
Floor	4
Zone or section	5
Aisle	00–99
Bin	1236
Shelf	2

(b)

FIGURE 5.9   Order-picker instruction formats (a) and (b).

order pickers, in their everyday life, more frequently deal in numbers rather than alphabetic characters. Also, numbers (digits) are progressive and unlimited in quantity, but alphabetic characters are limited to the 26 letters of the alphabet, and thus to provide more than 26 pick position identifiers, you double or triple the letters (e.g., BB) for the pick position identification. It is noted that there is potential confusion between the letter o and the digit 0 (zero) and the letter I and the digit 1 (one).

Pick Position.   The pick position is the most significant component (field) of the order picker instruction format because once in your warehouse aisle, the order picker frequently (each pick) refers to the pick position. Thus, the pick position print must have the largest and boldest characters or digits feasible. The subdivided characters and digits of a divided shelf or rack have the same impact as the pick position.

Pick Level.   If the pick area has shelf, hand-stack, or pallet load pick positions and one-way truck traffic flow pattern, then an additional identification component (field) is required for the picker instruction format. When this additional component is required to identify the pick position level, then the component is an alphabetic character, for the following reasons:

- Typically the number of levels in all previously mentioned order-picker routing patterns is less than 26.
- With an alphabetic character between two numeric digits, an employee is less likely to become confused than with three consecutive numeric components.

Warehouse Aisle. The next most important instruction format component (field) is the warehouse aisle. This character or digit identifies your warehouse aisle that contains the pick position for the SKU that appears on the instruction (customer order). This format component is an alphabetic character.

Warehouse Location. The order-picker instruction component (field) with the lowest significance is the warehouse location. It is used if the distribution facility has a mezzanine or different subwarehouse areas such as dry-goods grocery, or perishable food or freezer areas.

If you use an order-picker instruction format for the manual sortation or checking function instructions, then on the instruction document sortation instructions should have the same degree of emphasis as the pick position but with a different character and different-colored background and in a different location.

VARIOUS PICK INSTRUCTIONS

The pick instruction method that is used in your picking operation is determined by the type of order-pick (manual, mechanized, or automated) and customer handling (single or batched) method used. The various pick instruction methods are

- Manually or computer-printed document
- Machine-printed labels
- Paperless picking

Manually Printed Order-Pick Document Method

The first order-picker instruction method is the manual (typed or preprinted) or computer-printed paper document (Fig. 5.10*a*). The paper document method is the simplest of the order-picker instruction methods. This method is used in a single-customer order-pick system, in an operation with a limited number of SKUs and a low volume.

The manual paper document is a preprinted page (single sheet or multicarbon sheets) that lists all available pick positions and SKU description. Each pick position and SKU description has a corresponding empty block for your customer-ordered quantity and an empty block for actual order-picked quantity. The manually printed document method requires an office clerk or assembly clerk to enter (write or type) in the appropriate block the number of SKUs that were ordered by your customer. When the order picker is given the document, then the picker reads the first page pick position and travels to its location. After the order picker removes the SKU quantity from the pick position, then the order picker places a mark in the appropriate empty block.

Disadvantages of the manual paper document order-picker instruction method are the following:

- It does not cube the work.
- SKU changes are difficult.
- Transposition errors occur.
- It creates errors in filling (customer orders).

(a)

FIGURE 5.10   (*a*) Machine-printed pick document (*Courtesy of Cornell Home Study Course Warehouse and Transportation*).

- This method requires additional clerks.
- It is difficult to batch-pick with this method.

Advantages are

- This method does not require a capital investment.
- Order entry is flexible.
- This method requires little employee training.

```
------------------------------------------
|  07/20-1          48907-08N   -51       |
|                                         |
|  TIME       LOCN ZONE ROW   BIN         |
|  07:05       PL   AA   01   0A01        |
|                                         |
|  1-19495064-00-01-3-731001              |
|                                         |
|  00000   99999          BBBBBB          |
|   0   0  9   9          B    B          |
|   0   0  99999          BBBBB           |
|   0   0      9          B    B          |
|   0   0      9          B    B          |
|  00000       9          BBBBBB          |
|                                         |
|   Bin: 125        # PIECES: 99          |
------------------------------------------
```

(b)

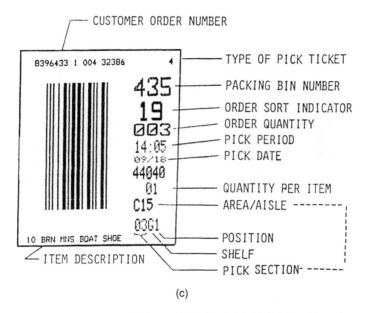

(c)

FIGURE 5.10 (*Continued*) (*b*) Machine-printed pick label; (*c*) machine-printed pick label with sort location.

Machine-Printed Pick Document Method. The computer-printed document has the same format as the manually printed document (Fig. 5.10*b*). The significant additional advantages of the computer-printed document are

- It facilitates SKU changes.
- It cubes the order selector work.

- It reduces transposition errors.

  The disadvantages of the method are

- It requires greater capital investment.
- It requires print time.
- It requires a standard customer order entry time.

Machine-Printed Label Method. The next order instruction method employs the machine (computer)-printed label. This method is implemented in a pallet load, carton, or single-item operation. The distribution operation is a single customer or batched customer order handling system. The label order-picker instruction method has the SKU pick position, SKU description, customer name and address, SKU customer quantity, and other required company information on the label front.

The computer software and printer create labels in a sequence that routes the order picker through the warehouse aisles. These labels have pressure-sensitive self-adhesive backs for easy attachment to the SKU exterior. After removing the SKU from the pick position, the order picker removes the label from the backing, places the label onto the SKU, and then places the labeled SKU on the take-away device. As required, the backing paper is discarded into a strategically located trash container. The trash container is in an apron worn by the order picker or a container on the mobile vehicle.

Disadvantages of the computer-printed label method are that it requires a higher capital investment in label print software and hardware. Unless a label holder is used by the order picker, the label stack is difficult to handle.

Order-Pick and Sortation Instructions on the Label. A frequent disadvantage with the label print method is that many companies forget that the order-picker instructions and sortation instructions (characters and digits) are the most important items on the label (Fig. 5.10c). Many companies have this information in small characters and digits. When this situation occurs, the order-pick and the sortation instructions are difficult to read, thus reducing the employee productivity and increasing errors. The preferred method is that both instructions have equal emphasis with a different-style character or digit, with different-colored background, or in a different location.

Advantages of the machine-printed label are

- Use in a single or batch system
- Clear and accurate instructions
- Quick and easy to make SKU changes
- Cubed order selector activity
- Use in a high-volume operation with a large number of SKUs
- Retail price information can appear on the label
- Allows time-motion (time vs. motion) standards applied to the activity

Paperless Picking Method. The next order-picker instruction method is the paperless picking method, which is used in a single-item, carton, or pallet load operation. The paperless picking method has three alternatives: the manually controlled paperless picking method, the computer-controlled paperless picking method, and the voice-directed pick instruction method.

The *manually controlled paperless picking method* is the digital display method. In this method, the order picker, by a fixed scanner reading a label on a tote at the entrance to the pick zone or aisle, enters your customer's order in the computer (Fig. 5.11*a*). The computer lights (activates) the individual digital display screens at the required pick position. On arriving at the pick position, the order picker removes the required SKU quantity and presses the order-pick button that registers a pick. The computer then reduces the quantity on the digital display by one. If another pick is required at the same pick position, then the order picker repeats the activity until the digital display indicates zero. The SKUs are placed onto a belt conveyor or into a con-

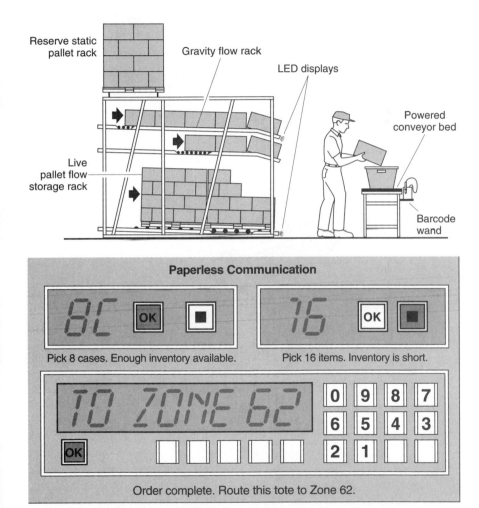

(a)

FIGURE 5.11   (*a*) Manually controlled paperless pick method. (*Courtesy of Material Handling Engineering Magazine.*)

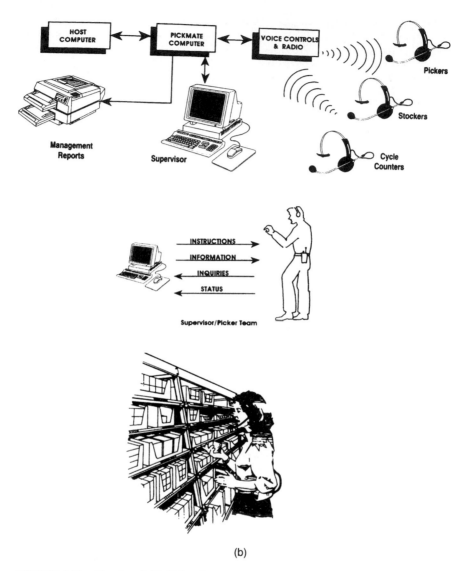

(b)

FIGURE 5.11 (*Continued*) (*b*) Voice-directed pick instruction method. (*Courtesy of Interlake Material Handling.*)

tainer. If a pick is required at another location, then the order picker travels to the required location and performs the pick activity. These pick activities are performed by the order picker until your customer's order is completed. An alternative paperless pick system requires a conveyor and bar-code scanning system. It has fixed order-picker pick zones and a customer-labeled container that travels on a powered accumulation conveyor past each pick zone. As the customer's container enters the pick zone, the customer's discreet label is read by a scanner device that sends the customer's number to the host computer. The host computer indicates on each pick position digital display the appropriate number of SKUs required to complete the customer order. After the first order picker completes all the required picks in the zone, the tote is passed by conveyor or cart onto the second zone. This arrangement has the potential to create tote or car queues. An option is to divert the customer tote with a discreet label on a pick spur. While a slug of totes are on the spur, the other customer order totes are diverted to other pick spurs (zones). After the order picker completes the necessary picks, the tote is transferred to the main traffic line for transfer to another pick spur or to the packing station. The computer remembers that the tote (discreet label) was transferred to a particular pick spur and so does not transfer that particular tote a second time to that spur. Therefore, a particular customer tote with a discreet label travels the main spur until it has passed through each pick spur.

The *computer-controlled paperless picking method* is used in an automated order-pick system that handles single items, cartons, or pallet loads. In these single-item or carton automated systems, the computer transmits an order impulse to the SKU pick mechanism (device) to release one SKU from the pick position onto a take-away conveyor system. In some computer-controlled systems an order-picker device (pick head) moves vertically or horizontally to the required pick position and releases a SKU onto a take-away conveyor.

Disadvantages of these systems are

- High capital investment
- On-time orders required
- Off-system handling of nonconveyables
- On-schedule replenishments
- Multiple lanes for high-volume SKUs

Advantages of the computer-controlled order-pick system include

- Accurate order selection
- Throughput rates projected easily
- Less product or equipment damage
- No labels required
- Fewest order selectors required

Voice-Directed Pick Instruction. The voice-directed pick instruction method consists of a computer system and an order-picker microphone or earphones (headset). With this voice method (voice recognition and speech synthesis), the order picker has on-line communication with the computer. In this method each operator talks to the computer through a microphone and receives verbal order-pick instructions from the computer through the headset (Fig. 5.11b). The order-pick instruction is achieved by a radio transmission which permits the order picker to have free hands to transfer merchandise from the pick position to the vehicle load-carrying surface or container.

The voice-directed pick instruction is considered a paperless pick method that handles both single and batched customer orders. It is used in single-item, carton, or pallet load operation.

Disadvantages of the method are

- High capital investment
- Additional employee training required

The advantages of the method are

- On-line transaction
- Reduced pick instruction print expense
- Order picker's hands less occupied

Automated Storage and Retrieval System.   Another automated order pick system is the automated storage and retrieval system, in which a computer-controlled storage and retrieval vehicle performs all deposit and withdrawal transactions between storage positions and pickup and delivery stations. Automated systems handle small items in containers, cartons, or pallet loads.

CUSTOMER ORDER HANDLING METHODS

The next major order-pick factor of a distribution operation is how customer orders are handled in the order-pick area. This factor determines the following:

- Number of order pickers per aisle
- How the order picker proceeds through the aisle
- SKU hit concentration (number of picks or lines per aisle) and SKU hit density (number of picks per SKU)

The various customer order handling methods are

- Single customer order handled by one order picker
- Single customer order handled by multiple order pickers

Zone: fixed or variable
Batched (grouped)

One Customer Order, One Order Picker.   The first customer order handling method is to have one order picker handle (pick all SKUs of) one customer's order. In the single customer order-picker method one order picker travels the entire warehouse (all aisles) and picks all items for your customer's order. This customer order handling method is used in a manual or mechanized order-pick system and is used in a pallet load, carton, or single-item distribution operation.

The disadvantages of this method include

- Low volume handled
- Reduced employee productivity due to long travel distances

Advantages of the single order picker for one customer order method are:

- Employee familiarity with all SKUs
- Employee errors easy to identify
- No sortation area or labor required
- Order integrity maintained

Multiple Order Pickers, For One Customer Order. The next customer order h
dling method has multiple order pickers handling one of your customer's order. In this
method your customer order is separated into random or predetermined portions. Each
portion is given to an employee who performs all order transactions. The random portion separates your customer's order into a cube that is a pallet load or a container,
which is a very common manual carton handling or single-item-to-a-container operation. Predetermined portions separate your customer's order per a pick zone requirement. This method is very common in a mechanized order pick-to-belt system and is
used in a pallet load, carton, or single-item warehouse and distribution operation.

The advantages of this multiple order picker per one customer method includes

- Flexibility of employees to pick anywhere in the warehouse
- Large volume and order size handled
- Different SKU types handled
- Completion of order in shorter time

Zone Area Picking. Zone area picking is the next customer order handling method.
The zone area picking method is a well-defined single customer multiple order-pick
method. This method separates one or several of your customer's orders into fixed
zone or variable zones (areas) within your warehouse or within a warehouse aisle.
This method is implemented in a pallet load, carton, or single-item warehouse operation.

Batched (Grouped) Customer Orders. The next customer order handling method is
the batched (grouped) customer or SKU method. The batched customer order handling
method separates your daily customers into random number of your customer orders
(waves) that are combined by the computer. Therefore, it is possible for your operation to have a number of waves each day. The number of waves and number of customers per wave are determined by the SKU volume, the number of final workstations
(docks or packing stations), or the number and productivity of the order pickers. The
computer-controlled printer prints an individual label for each SKU on all of your customer's batched orders. With these labels the order picker makes one trip through the
warehouse aisle and stops at each pick position that appears on each label. At the pick
position, for each label, the order picker removes a SKU, then labels and places the
labeled SKU onto the takeaway conveyor or into a container. The most common
batched customer order-pick method requires the batched picked SKUs to be sorted
(manually or mechanically) by your customer order into individual bins, containers,
chutes, or lanes. The batched (grouped) customer order handling method is used in a
single-item or carton warehouse operation.

Disadvantages of this method are that it

- Requires your customer orders on time and to be reviewed by a computer program
- Requires a sortation activity and area
- Requires an investment into a computer program
- Requires a human- or machine-readable label with pick and sortation instruction

Advantages are

- Improved order-picker productivity due to increased SKU hit concentration and
  SKU hit density
- Large number of SKUs handled

VARIOUS BATCHED (GROUPED) PICKER CONTROL CONCEPTS

The batched customer order-pick concept requires a high degree of discipline and control of the order picker's activity. You must ensure that all order pickers complete their portion of the batch on time. If your completed customer orders are not released on time, then tremendous confusion and errors in the sortation area may result.

The batch control concept is a technique that is used to instruct order pickers regarding the time when they can place their picked SKUs onto the take-away conveyor system. This feature is very critical because there are several batches (waves) that are picked and sorted within a day. The various concepts ensure that the order picker's activity is in coordination with the sortation activity and require communication between the sortation area and the pick area. The sortation area notifies the pick area regarding the receipt of last tote, item, or label of the batch. To reduce batch overlaps, these guidelines should be followed:

- Ensure that the total pick productivity is equal to your packing and sortation productivity.

- When your picking activity is in advance of the sortation activity, provide a temporary setdown area for the advance picked volume. This holding area is adjacent to the take-away conveyor.

- In the sortation area, ensure that there are sufficient sort lanes to handle two complete batches.

- In a single-item packing area, there should be at least three chutes.

Various Batch Pick Release Methods

The basic batch pick release methods are

- Control desk and clerk
- Printing on the label or ticket with clock(s) in each pick area
- Scoreboard in each pick area

With any of these methods, if the order pickers are picking in advance of the batch release time, then they are not placing the merchandise onto the take-away conveyor. The advance picked merchandise is held in a temporary holding area or in containers for later release at the appropriate time.

Batch Control Desk with a Clerk.   With the control desk method, in the pick area a clerk sits at a control desk and issues to each order picker their pick instructions (labels or documents) (Fig. 5.12a). The control clerk issues batch 1 pick labels to all order pickers. After issuing all batch 1 pick labels, then the control clerk issues batch 2 pick labels. To assure proper control, the clerk lists the names of the order pickers who received the labels and their return time.

Batch Control Printed on the Label and Clock in the Pick Area.   With this method, the computer prints each pick zone release time on each pick label(Fig. 5.12b). This release time indicates to the order picker the expected time that batch 1 merchandise is scheduled for transfer onto the take-away conveyor. Each zone has a clock and has a different release time because the travel time from the pick area to the sortation area is different for each zone.

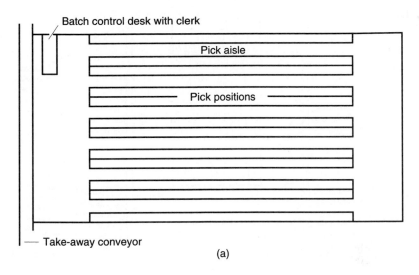

(a)

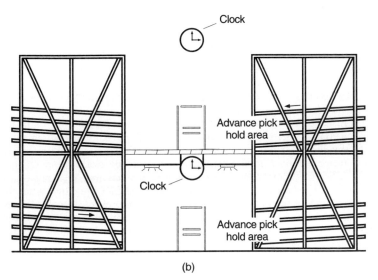

(b)

FIGURE 5.12   (a) Batch control desk with clerk concept; (b) clock in the pick area batch release concept.

Batch Control with the Scoreboard.   With the scoreboard method, each pick area of the warehouse has a scoreboard or signal light that has a symbol for each batch (Fig. 5.12c). In the pick area to the pickers, it indicates the appropriate batch that is scheduled for pick and placement onto the take-away conveyor.

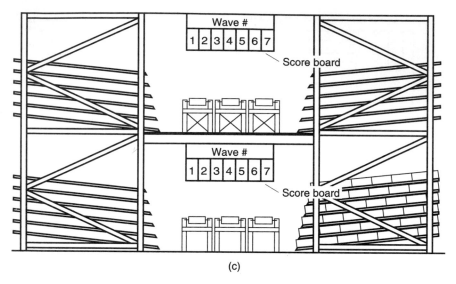

(c)

FIGURE 5.12   (*Continued*) (*c*) Scoreboard batch control concept.

Single SKU, High Volume Means Pick and Pack

The next batched customer order pick method handles multiple customer orders that have a high volume for one, two, or three SKUs. With this method the computer collects all your customer orders that have a single particular SKU or two SKUs and prints the customer order-pick instruction forms. The SKU total requirement is recapitulated (recapped) for all your customer orders. Your individual customer order instruction forms are given to one employee. With all the pick instruction forms and each SKU recap total, an order picker travels to the required pick position and en masse (pallet loads or cartons) withdraws the indicated SKU quantity. After the SKU quantity is secured on the order-picker vehicle, the order picker travels to a predetermined location. At the location, an employee with the individual customer order instruction forms picks, invoices, packs, seals, and labels the shipping container and places the container onto the shipping container.

How to Determine the Number of Items or Cartons per Batch

With the batched (group) customer order handling method, the entire material handling, order-pick, sortation system is designed according to the SKU quantity that is handled at each workstation. Either of two concepts is used to determine the quantity of items or cartons per batch: the picker-driven and the final workstation-driven (packer-fluid loader-unitizing) concept.

Picker-Driven Concept.   With the picker-driven concept, your batch (customer quantity) is determined by your order picker's productivity. This concept relies on the computer program to distribute the customer-ordered SKU quantity by a predetermined pick productivity rate (number of pickers), and this SKU quantity is evenly distributed to all final workstations.

If your business is the catalog or direct mail business, then the picker-driven method is an acceptable alternative because the computer allocates approximately 50 pieces per packing station and the average customer order size is 2 to 5 pieces. But the batch quantity cannot exceed the number of pieces per packing station and number of active packing stations. The design parameters for the picker-driven small item batch quantity concept are as follows:

- Number of orders per packing station
- Number of SKUs (pieces) per order [do not exceed 50 SKUs (pieces) per packing station]
- Anticipated packing productivity
- Total average daily pieces
- Work hours per day
- Anticipated order-picker productivity
- Anticipated induction productivity

Final Workstation (Fluid Loader) Method.   With the final workstation (packer-fluid loader-unitizing)-driven concept, the batched quantity is determined by the number of SKUs that are handled by each station. With this method your computer program ensures that the batched quantity has a sufficient SKU quantity to keep each workstation active. It also calculates the number of SKUs to assure a clean cutoff for loading of your customer delivery truck.

If you operate a carton handling facility, then the final workstation batch quantity concept is preferred because it assures a constant work quantity for all warehouse functions. The design parameters for a final workstation batch quantity concept are

- Number of palletizing stations or truck docks or fluid loaders that are based on an anticipated productivity
- Average daily conveyable pieces and nonconveyable pieces
- Time to set up fluid loaders or palletizing stations
- Time to load nonconveyable pieces
- Number of customers per day
- Average number of pieces per order

ELEVATION AND LOCATION OF THE PICK POSITION

The final section of the chapter reviews the SKU pick position elevation and the physical location for the shelf or rack levels.

The physical elevation and location in a hand-stack rack, case flow rack, or shelf pick position concept is a very important factor that contributes to the order picker's productivity and replenishment employee's productivity. To achieve maximum order-picker (replenishment) productivity, the SKU pick position elevation reduces the reaching and bending of the employees. This means that the top and bottom levels (pigeonhole) are the least desirable pick positions and that the pick level between the 20-in and $5\frac{1}{2}$-ft elevations are the preferred positions.

When SKUs are placed in a pick position concept, allocation is according to the following parameters.

Pick position elevation level and location	SKU weight	SKU velocity and movement	SKU volume, ft^3	Vertical space opening height, in
		Example of shelf pick positions (Fig. 5.13*a*)		
5, top	Light	Slowest, reversed	Small	16
4	Heavy	Fast	Large	17 (golden zone)*
3	Heavy	Medium	Large	18 (golden zone)
2	Medium	Slow	Medium	16
1, bottom	Light	Slowest	Small	16
				3 (structure)
		Example of flow rack and hand-stack pick positions (Fig. 5.13*b*)		
4, top	Light	Slowest	Small	20
3	Medium		Medium	20 (golden zone)
2	Heavy	Fast	Large	21 (golden zone)
1, bottom	Light	Slow	Small	20
				3 (structure)
		Example of pallet load pick positions (Fig. 5.13*c*)		
2, top	Medium, light	Medium, slow	Medium, small	44 (Noncrush-able)
1, bottom	Heavy	Fast	Large	0 (floor crush-able or fragile)
		Example of hanging-garment pick positions (Fig. 5.13*d*)		
Description	Length, in	Position		
Long	61	One (top)		
Short	78	Top		
Short	39	Bottom		

*The golden zone is the pick position elevation that is easy for the order picker to perform pick transactions.

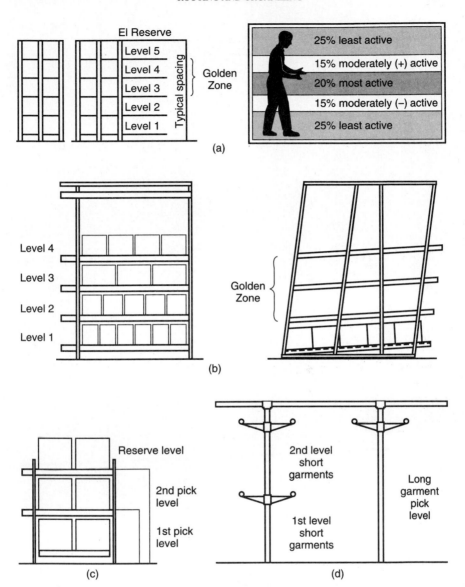

FIGURE 5.13 (*a*) Shelf pick levels and the golden zone (*Courtesy of Material Handling Engineering Magazine.*); (*b*) flow rack and hand-stack pick levels (*Courtesy of UNARCO Material Handling.*); (*c*) pallet load pick levels; (*d*) hanging-garment pick levels.

CHAPTER 6
ORGANIZING SMALL-ITEM FACILITY HANDLING ACTIVITIES

INTRODUCTION

The main objective of this chapter is to identify and evaluate material handling and storage systems and technologies that make your small-parts distribution operation effectively and efficiently service your company's present and future customers (workstations) at a profit. The second objective is to review the activities, methods, techniques, and equipment applications that you can implement in your distribution facility. The last objective is to analyze and evaluate these concepts and techniques.

The first warehouse function is the unloading of the product from the delivery vehicle. The receiving function was reviewed in Chap. 4. The most common small-item delivery methods are (1) United Parcel Service, (2) common carrier or container, (3) U.S. Postal Service Express Mail, (4) vendor truck, and (5) backhaul truck.

The various carton and box unloading methods are (1) hand-carry, (2) gravity-powered conveyor, (3) two-wheeled hand truck, (4) powered lift truck, and (5) manually or electrically powered pallet truck.

To unload hanging garments, the various unloading methods are (1) hand carry–rolling rack (cart) or (2) extendible trolley boom.

The type (size and weight) and volume of the product determine the material handling equipment that is used to unload the delivery vehicle.

UNLOADING ACTIVITY

During the unloading process, the following procedures assure high employee productivity at the next warehouse activity: (1) separate the delivery by purchase order, (2) separate the delivery by purchase order and style (SKU), (3) separate the delivery by purchase order, style, and color; and (4) separate the delivery by purchase order, style, color, and size.

In the future, retail store distribution industry vendors will package the product in quantity lots by size, style, and color and require merchandise ticketing with retail price tags. This vendor activity will increase as companies become involved with across-the-dock activities. Across-the-dock operation reduces overhead warehouse and distribution labor, material handling equipment, and building area.

QUANTITY AND QUALITY CONTROL ACTIVITY

The next key warehouse activities are product quality and quantity check to verify that the vendor delivery quantity and quality are per your company's purchase order specifications and standards. The three receiving options are 100 percent accept, random sample, and 100 percent check.

PACKAGING ACTIVITY

In the next receiving function, packaging, an employee places individual SKUs (product) into a package or container or sends the product loose in a large container to the reserve or pick location. The three packaging concepts are loose items, material handling container, or shipping container.

Product Handled Loose in a Large Container

The first packaging method is to place the individual items loose into a transportation container or into the vendor's carton. In the container, the product is transported to the warehouse reserve or pick position aisle. At the position an employee transfers the product from the transportation container into the warehouse reserve or pick position container.

Some disadvantages of the loose method are that empty containers are returned to the receiving area, there is increased risk of damage or lost items, and SKUs are difficult to stack in the pick position and are difficult to count and identify during inventory.

Advantages are no package material or labor cost and no space for requirement package material storage.

Product Placed in Material Handling Container

When the product is placed into a material handling container, the individual item is placed into a chipboard box, plastic bag, or paper bag. The material handling container provides the means to keep an individual item separate during transport to the position and separate in the pick position. The exterior of the container has a SKU identifier attached to it.

Disadvantages of the method include increased square-foot building area needed for package material storage and increased expense. Advantages of the material handling container are that it identifies the SKUs, reduces damaged inventory, is easy to count during inventory, and SKU is separated in the position and easy to handle.

Various Small-Parts Handling Containers

When loose small parts are transported or handled in your distribution operation, then you place these small parts into a container to ensure that your operation is efficient and effective. The various small-part containers are cardboard or chipboard box, coated corrugated box, or plastic tote or container.

Cardboard/Chipboard Box. The cardboard or chipboard box is the basic small-parts container; this can be a specially designed box or your vendor delivery box.

Disadvantages of the cardboard or chipboard box are that it is disposed of in the storage-pick area and has a limited durability, especially in humid environments; it is also difficult to maximize storage-pick area space, and if the boxes are not of uniform size, they are difficult to convey or stack. Also your vendor box has other markings on its exterior and usually contains only one SKU. Advantages of this method are reduced expense, and with vendor boxes, there is no requirement for storage space or temporary storage at the workstation, and the vendor box is reused as a shipping box.

Coated Corrugated Container. The coated corrugated container is designed to your specifications and is coated with a resin substance that hardens and increases its rigidity and durability. Specially designed dividers can be installed inside the container to handle multiple SKUs. When used in a conveyor system, a rough wood or pressedboard bottom exterior surface increases durability of the container. Use of nestable types reduces the required storage area at the workstation.

Disadvantages of the coated container are increased expense, storage area requirement, and repeated SKU handling; also the coated container is difficult to clean and to use in humid areas.

Advantages of the coated containers are increased durability; handling of heavier SKUs and with dividers, handling of multiple SKUs per container; standard size easy to convey; increased stacking height and maximum use of storage area; and handling of trash in the processing area.

Plastic Container. The plastic container has four sides and a bottom. It is manufactured from plastic according to your specifications. As required, openings or handle grips are designed on the sides to improve handling by your employees. As required by the operation, removable or lockable tops are part of the container. If future operation requires a dividable container, then slots are designed in the sides to accept dividers.

Your SKU and material handling system determines whether the bottom surface is of the open or closed type. The open type has a basket weave pattern on the bottom which permits easy removal of dirt or dust. When the plastic container is used on a conveyor system, the bottom exterior surface has a checker-square pattern to assure positive traction. With this pattern, the container is easily moved by your conveying surface. To reduce the required storage space in the workstation area, stacked empty plastic totes are placed in a nested or crisscrossed pattern. The desired pattern is part of your plastic tote specification.

If employees move containers across your warehouse floor, then prior to the placement of the next tote on top of another tote, have them place paper sheets on top of the merchandise in the lower totes. This practice reduces the possibility of dirt or dust rubbing off onto your product.

When your tote is used on a bar-code scanning system, then your tote exterior has a label pocket that presents the label to the scanner.

Disadvantages of the plastic container are higher cost, and solid-bottom types required cleaning and if stacked or dragged across the floor, dirt or dust accumulates on the bottom surface or inside.

Advantages of the plastic container include longer life, handling of heavier SKUs and with dividers, handling of multiple SKUs per container, easy conveyability because of standard size, maximum stacking height, use in humid environments, and handling of trash in the processing area.

Product Placed into Shipping Container.   When a SKU is placed into a shipping container, the individual SKU is placed into a cardboard box, corrugated bag, or plastic bag. The SKU identifier is placed on the exterior of the container; therefore, in the pick position the SKU is separated from the other SKUs and is protected from damage during shipment.

An additional disadvantage is that if the container is placed into a larger shipping container, then the second shipping container adds more expense to the shipment.

Advantages of the shipping container packaging method are the same as those of the material handling container packaging method. Also, if the SKU is picked and packed by the employee, there is lower outbound packing expense.

PRODUCT IDENTIFICATION ACTIVITY

The final receiving activity is product identification. The product (SKU) identification activity is the process that you use to discreetly identify one SKU from another. In many distribution facilities that service retail customers, the SKU identification process includes price-ticketing the SKU that has the SKU identification label, which includes the retail store price for the SKU.

Identification Must Be Easy to See

When a SKU identification label is placed onto the SKU exterior, the label is placed in a location that is easily recognized by the internal transport, order-pick, sortation, and outbound packaging employees. The various product identification methods are self-adhesive label, paper label with a plastic self-locking band, paper label with a staple, or paper label with string.

Handling Hanging Garments

If your distribution facility handles hanging garments, then the garments are received either in boxes or on hangers.

Handling Garments in Boxes.   When the garments are received in the boxes, they are placed onto hangers. When the garments are in boxes, to assure a high productive opening activity, the boxes are delivered to the opening station. There are two alternative garment handling methods: (1) opening and hanging the garments onto a rolling rack and (2) opening and hanging garments onto a trolley.

Garments Transferred to a Rolling Rack.   In the rolling-rack concept (Fig. 6.1*a*), the boxes are stacked from the pallet load onto a table or gravity conveyor section that feeds to the opening employee. The opening employee removes the garment from the box, places it onto a hanger, and places it onto a rolling rack. The garments are placed onto a rolling rack with the left arm hanging to the left side of the rolling rack's lead end. In a retail store this arrangement permits the customer to easily see the price tag. After the rack is full or at the completion of the purchase order, the rolling rack is transferred to the sort-and-count staging area. The trash (empty boxes and paper) is placed into a trash container or conveyor.

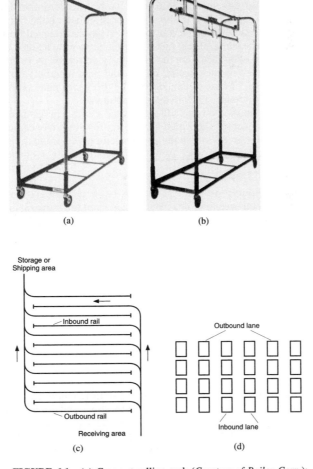

FIGURE 6.1   (*a*) Garment rolling rack (*Courtesy of Railex Corp.*); (*b*) hanging-garment trolley rolling rack (*Courtesy of Railex Corp.*); (*c*) rolling-rack sort and count concept; (*d*) trolley-rail sort and count concept.

Disadvantages of this method include low box opening and transport employee productivity, difficult handling of a large number or volume of SKUs, and requirement for a large-square-foot building area.

Rolling-rack method advantages are portable workstation, handling of a small volume, and low capital investment.

Garments Transferred to a Trolley. The second method is to place the hanging garments onto a trolley (Fig. 6.1b). This is the most productive method, in which the SKUs are separated by color and size. In this method, the cartons are placed onto a gravity conveyor, which provides carton accumulation prior to the opening station. A cart adjacent to or an empty trolley above the carton conveyor supplies the workstation with empty trolleys. This empty trolley conveyor has a slight pitch toward the employee and an end stop. The top of rail (TOR) is at an elevation of 30 in to 3 ft above the carton conveyor. The full trolley take-away rail is directly behind the opening station. After an empty trolley is placed on the take-away trolley line, the hanging garment is placed onto the trolley with the left-hand sleeve facing the left side of the trolley lead end. This arrangement in the retail store allows proper location of the price tag. When the trolley is full, it is allowed to accumulate on the rail.

Both trolley rails are at a nominal $6\frac{1}{3}$ ft to TOR from the finished floor.

After the boxes are opened, the garment is placed onto a hanger. The hanger is placed onto the empty trolley that is on the take-away conveyor. A fixed end stop prevents the trolley from passing the opening station and permits empty trolley accumulation. An adjustable (open or closed) end stop at the discharge end controls the flow of full trolleys onto the sort-count or steaming staging area.

Disadvantages of this method include additional capital investment and requirement for a trolley rail system in the warehouse; also, this system occupies a fixed area.

Advantages are high box opening and transport employee productivity, handling of a high volume, accumulation of work prior to the station, and sequentially arranged workstations.

Hanging-Garment Steaming Activity. The next hanging-garment receiving activity is garment steaming to remove the folds or wrinkles from the garments. The two alternative garment steaming methods are the manual and automatic methods.

Manual Steaming Method. In this method an employee uses an electric portable steaming machine to steam a garment that is not in a bag. As required, the garment is steamed at a workstation. The individual garment is removed from the trolley or rolling rack, steamed, and returned to the transport device.

Disadvantages of the garment steaming method are low volume handling and low employee productivity. Advantages are low capital investment and no requirement for a fixed-square-foot building area.

Automatic Steaming Method. In this method, unbagged individual garments are transferred to the steaming device infeed conveyor. The individual garment conveyor system consists of slide rail and powered screw conveyor sections that transport the garment on a hanger through the steam machine. The steaming machine requires an electric or gas boiler.

Disadvantages of this method are additional maintenance expense and capital investment, and some models cannot handle both short and long garments or both plastic and wood hangers and require the use of an electric or gas boiler. Advantages of automatic steaming include high employee productivity and removal of all folds and wrinkles.

Hanging-Garment Bagging Activity. The next hanging garment receiving activity is the packaging (bagging) operation. The hanging garment is placed into a protective container that protects it from dirt and dust.

When considering the bagging of hanging garments, there are two alternatives: the manual or the automatic method.

Manual Bagging Method. The manual bagging method is considered the most basic method. The garment is placed onto a bagging device that consists of a hook that is supported from a stand and has a roll of plastic bags at the top. After the hanging garment is on the hook, the bagging employee pulls the plastic bag over the garment and separates the bag material at the precut location. When the plastic bag is positioned squarely over the garment on the hanger, the garment is inside the plastic bag and is transferred to the rolling rack or trolley.

Disadvantages of the method are low employee productivity and low volume handling. Advantages of the manual method include low capital investment, portable workstation, little employee training required, all sizes handled, and no requirement for installation time or a fixed-square-foot warehouse area.

Automatic Bagging Method. In this method an automatic bagging machine rotates the garment to four different stations. The automatic bagging machine requires that an individual garment be manually or mechanically placed onto the infeed conveyor at the infeed station.

After the hanging garment is placed on the hook, the machine rotates it to the bagging station. At this station, the plastic bag is automatically (mechanically) pulled down over the hanging garment. The automatic bagging device is returned to the upright position and separates the bag, and then rotates the hanging garment to the outfeed station. At the outfeed station, an employee transfers the hanging garment from the bagging machine to a take-away conveyor. At the end of the conveyor, another employee transfers the garment onto a rolling rack or trolley. Disadvantages of the automatic bagging machine are that it requires a fixed-square-foot building area, uses one length of bag, and increases maintenance expense. Advantages are that it handles a large volume of both long and short garments.

Hanging-Garment Sort-and-Count Activity. The final hanging-garment receiving activity is sorting and counting. The sort-count activity ensures the company that the individual hanging-garment quantity matches your purchase order quantity. The garments are identified and grouped by style, color, and size, which means that the garments are ready for transfer to the ticketing operation. The two sort and count methods are the rolling-rack and trolley or rail method.

Rolling-Rack Sort-and-Count Method. The rolling-rack sort and count system (Fig. 6.1c) consists of three rows or rolling racks. The middle (interior) rack row contains the inbound (mixed garments) racks. The two exterior (outside) rolling-rack rows are empty. As the employee sorts the garments, the individual garment (SKU) is transferred from the middle rolling rack onto the outside rolling racks; when all the garments are sorted and distributed to the rolling racks, then the employee counts them by SKU.

Disadvantages of the rolling-rack method are lower employee productivity, setup time requirement, and low-volume handling. Advantages include low capital investment and no requirement for a specific building square-foot area.

Trolley Rail Sort-and-Count Method. The trolley-rail sort-and-count area consists of a three-trolley rail network. The middle (interior) rail contains the inbound (mixed garments) trolleys and has a fixed end stop. The two exterior (outside) rails contain empty trolleys. As the employee sorts the garments, the individual garment (SKU) is transferred from the middle trolley to a trolley on the outside rail. After all the garments are sorted and placed on trolleys, the employee counts them by SKU.

Disadvantages of the trolley-rail method are that it requires a fixed-squre-foot building area and increased capital investment. Advantages include controlled garment flow and high employee productivity.

ORDER-PICKER INSTRUCTIONS

The fundamental component of any single-item order-pick system is the order-pick instruction method, as mentioned in Chap. 5. The order-pick instruction method directs the order picker to the SKU pick position and indicates the quantity that is desired by the customer or workstation.

The order-pick instruction method is matched with the order-pick system. These order-pick instruction methods are considered for implementation in a new or existing facility. This chapter briefly lists the order-pick methods; in Chap. 5 each order-pick instruction method is reviewed in more detail. These methods are (1) printed paper document (manually or computer-printed), (2) printed label (by computer), (3) paperless, and (4) voice directed pick instruction methods.

HANDLING CUSTOMER ORDERS

The next major factor in a single-item order-pick system is how your customer's order is handled in the order-pick area. This factor determines the number of order pickers in your aisles. The various customer order handling methods are (1) single customer order, single order picker; (2) single customer order, multiple order pickers; (3) zone (variable or fixed); and (4) batched group.

A more detailed analysis of each customer order handling method is presented in Chap. 5.

ORDER-PICKER ROUTING PATTERNS

The next major factor that determines a small-item picker's productivity is the order-picker routing pattern.

The objective of an order-picker routing pattern is to optimize the order picker's walk, travel, or hand movement between two pick positions. These pick positions contain SKUs that appear on a customer's order.

To achieve maximum order-picker productivity, the appropriate order-picker routing pattern is implemented with the appropriate customer order handling system.

Any one of 10 basic order-picker routing patterns can be implemented in a small-item distribution facility:

- Nonrouting pattern
- Sequential routing patterns

  Single-side order-picker pattern: one or two order pickers per aisle

  Loop pattern

  Horseshoe or U pattern

  Z pattern

  Multilevel [high-rise order-selector (HROS)] pattern: one- or two-way aisle, with either four or six levels

Block pattern

Stitch pattern

Zone pattern: fixed or variable

Front-to-rear pattern

A detailed description of each order-picker routing pattern is presented in Chap. 5.

INVENTORY CONTROL ACTIVITY

The type of inventory SKU location system that is used to allocate product to the SKU reserve and pick position is the next major small-item pick productivity factor. This factor partially determines employee productivity. The two alternative SKU location methods are the fixed- and floating-slot methods.

These systems are reviewed in more detail in Chap. 12.

The primary objective of this section is to develop an understanding of the various small-item order-pick concepts. Looking at each concept, this section describes the layout, operational characteristics, disadvantages, and advantages of each concept.

Basic Order-Pick Concepts and Their Design Parameters and Operational Guidelines

Small-item order-pick concepts are separated into three basic types: (1) employee to stock, (2) stock to employee, and (3) automated system.

The implementation of one of these concepts in a distribution operation is based on numerous factors, such as customer order (information) transmission method, paper flow, pick instruction method, labor productivity and availability, SKU characteristics, SKUs per customer order (SKU mix and hit concentration), number of customer orders, throughput volume (customer demand), material handling equipment, storage and replenishment methods and procedures, facility layout and shape, customer order-delivery cycle, and available capital for investment.

Methods for Protecting Employees from Workstation Edges.   When you design a work (receiving, ticketing, order-picking, and packing) station to handle your product and volume, you are required to protect your employees from sharp metal edges as they handle your product between the workstation and conveyor. The type of protection is determined by the type of workstation construction. The most frequently used workstation material is formed sheet metal and conveyor bed.

When you use formed sheet metal on the workstation, it has a sharp metal edge and the best protective material is a hard plastic strip. The hard plastic strip has a front and two sides that slip over the sheet-metal edge. These features secure the hard plastic to the sheet metal and provide protection to the metal edge.

The second workstation material is a conveyor that has C channel beds which is used on a pick-to-belt conveyor system, in a manual sortation area, and at an inspection station.

To reduce employee injury or clothing damage from the conveyor bed edge, the conveyor bed must have some form of protection. The various protection materials are hard plastic strips or a hard rubber or hard plastic tube that is attached to the conveyor bed.

The hard plastic strip is attached full length to each conveyor bed section. This strip is slipped over the bed edges and secures itself to the bed (channel). This strip becomes a protective cover to the conveyor bed.

Using a hard rubber or hard plastic tube is the second protective method. The tube is cut in half and is inserted between the two conveyor bed channels. The diameter of the half tube secures the tube in the bed channel cavity and extends outward to provide protection. Plastic straps or tape provide a means to secure the half tube in the conveyor bed channel.

Employee to Stock (Manual) Pick Concepts.   The first group of small-item order-pick concepts is the employee-to-stock group. An employee walks or rides to the pick position and places the required SKUs onto the take-away conveyor or vehicle load-carrying surface. The employee either (1) walks to the stock (hand-carry, retail store shopping cart, pushcart, or rolling rack), rolling ladder, and trolley or (2) rides to the stock (manually places SKUs on a load-carrying surface or mechanized take-away vehicle).

Manual Carry Concept.   The manual hand-carry order-pick concept is the most basic order-pick concept. The employee walks with the picking instruction form to the SKU pick position, withdraws the appropriate SKU quantity from the pick position, and verifies the order pick. This activity is repeated for other customer-ordered SKUs until the employee cannot handle additional SKUs or completion of the customer's order. At this time, the employee walks and deposits these SKUs into the assigned staging area or onto the take-away conveyor.

The manual hand-carry method handles a single customer order. When an outbound sortation system is part of the total system, then the hand-carry concept is used in the batched order-pick method. With the batched customer order system, the employee walks from the pick position to a cart or conveyor that is in the main aisle and performs a sortation-and-dump activity. For a detailed review of the pick-to-cart (sort) arrangement, see the "Sortation" section of this chapter.

Disadvantages of the concept include (1) increased walk time, (2) low order-picker productivity, (3) limited pick height, (4) requirement for the greatest number of employees, and (5) low volume handling. Advantages of the manual hand-carry concept are (1) applicability for hanging garments and nonhanging SKUs, (2) capital investment requirement, (3) minimal employee training requirement, (4) low maintenance, (5) handling of all types of SKUs, and (6) variable-path system.

Methods to Increase Productivity

Use of an Apron.   To improve your employee's ability to carry a larger quantity of SKUs, an apron (Fig. 6.2a) is used by the order picker. The specially designed apron forms a pouch that is attached to the employee's waist and to the wrist of one arm.

Hamper or Cart with a Second Load-Carrying Surface with Springs.   When a hamper or cart that has a deep cavity is used to carry merchandise between two warehouse work locations, then at these locations your employee is required to remove the merchandise from the cavity. This deep reach into the cavity for the lowest merchandise on the bottom creates a long reach for the employee. This long reach means low employee productivity.

An option for the hamper or cart is to provide an additional load-carrying surface (Fig. 6.2b) inside the hamper or hanger cart cavity that is supported with four springs. These springs are attached either (1) together to the underside of the second load-carrying surface and to the top of the regular load-carrying surface (base) or (2) individu-

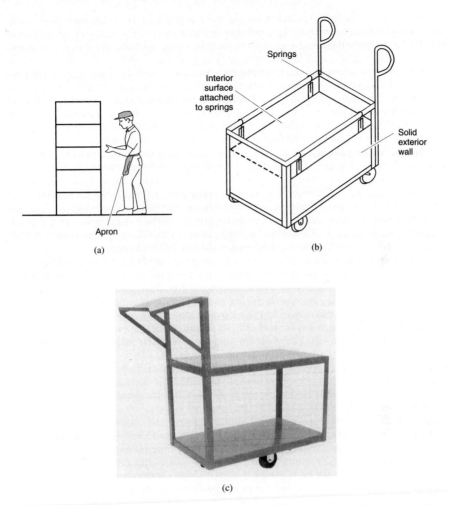

FIGURE 6.2 (*a*) Order picker wearing an apron; (*b*) additional carrying surface in a cart; (*c*) manual pushcart.

ally (each spring separately) to the four corners of the load-carrying surface, with the other end of each spring attached to the four corners of the second load-carrying surface edges.

With both methods, as the merchandise is transferred to or from the cart, the springs maintain the merchandise at an accessible level for the employee. This feature reduces the employee's nonproductive and physical effort to bend or reach for the merchandise from the lower levels of the hamper or cart cavity.

Retail Store Shopping Cart.   The next small-item manual controlled order-pick vehicle is the retail store shopping cart. This device has four wheels (the two front wheels are swivel casters or wheels, and the two rear ones are rigid casters or wheels), a push handle, and formed wire strands that create the merchandise carrying cavity.

As required, the order picker transfers the merchandise from the pick position to the cart. After completion of the customer order or with a full cart, the order picker manually transfers the merchandise from the cart to the take-away conveyor or workstation.

When compared to the other order-pick carts, the retail shopping cart is not preferred for a dynamic small-item distribution operation because of low order-picker productivity resulting from additional employee movements (greater number of trips) and smaller load-carrying capacity. If the merchandise is not in a container, it can hang up on the strands; it is difficult to handle batched orders with no dividers inside the load-carrying cavity or if the pick instruction holder is inadequate.

The advantages of the cart are that it is easy to push and steer and is nestable and thus requires less storage space, and most employees are familiar with the cart.

Manual Pushcart.   The second small-item order-pick vehicle is the manual pushcart (Fig. 6.2c). An employee pushes a four-wheeled cart with a load-carrying surface through the warehouse aisles. The cart load-carrying surface is designed with sides to contain a large quantity of loose SKUs, with a flat surface to carry a container or with pigeonholes (bins or slots) to handle batched picked SKUs or orders.

When compared to the employee carry concept, the pushcart offers advantages and disadvantages similar to those of the hand-carry method. An additional disadvantage with the use of carts is capital investment. The additional advantages are that it is used in a single or batched customer handling method, carries a larger cube quantity of SKUs, travels a greater distance, and is a variable-path system.

The pushcart concept has several options (accessories or optional equipment) that can be attached to the cart. These options increase the order picker's productivity and are added to the cart as a single option or in combination.

Stepladder Option.   The first feature is a stepladder (Fig. 6.3a) that is attached to the order picker's (rear) side of the cart. The stepladder increases the order picker's ability to withdraw SKUs from elevated pick positions that are above normal employee reach.

Self-Dumping Option.   The other option is a loose SKU (small-item) carrying surface that is a self-dumping container (Fig. 6.3b). This self-dumping container has four side walls and a bottom surface. The four side walls slide forward across the carrying surface, and the picked loose merchandise in the cart cavity falls onto the belt takeaway conveyor. When the cart is used in a pick-to-belt conveyor system, then the self-dumping feature improves the order-picker transfer productivity.

Swivel Casters or Wheels in the Front and Rigid Casters or Wheels in the Rear. The third option is to have two rigid casters or wheels in the rear and two swivel casters or wheels in the front of the cart. This feature improves cart steering and end-of-aisle cart turning by employees.

Document, Ticket, and Tool Holder.   The next cart option is an order-picker document, ticket, and tool holder (Fig. 6.3c). This device provides a location on the cart to hold all necessary order-pick tools and items that are within easy reach of the order picker. This location permits the order picker to perform the order-pick transaction productively. Some of these items are order-pick labels or documents, packing bags, stapler and staples, trash container, and stuffers. The design of the holder permits unobstructed depositing of the picked merchandise onto or into the cart load-carrying surface or cavity.

Top-Attached Rolling-Ladder Method.   The next order-pick method is similar to the employee carry order-pick method. This concept employs a rolling ladder (Fig. 6.4a). When the pick positions are elevated above normal employee reach height, the employee pushes a portable ladder with a safety bottom step to the required pick posi-

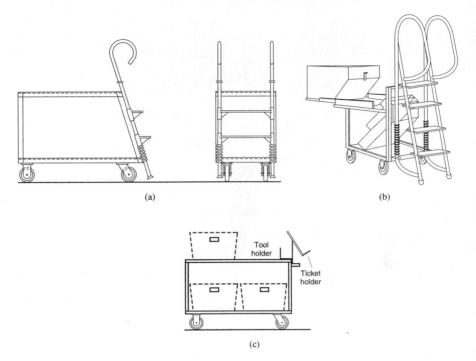

(a) (b)

(c)

FIGURE 6.3 (*a*) Pushcart stepladder option; (*b*) self-dumping cart; (*c*) document, ticket, and tool holder.

tion. This ladder has two wheels that are attached to the top guiderail, to provide guidance as the ladder travels down-aisle. Spring-loaded safety steps reduce risk of employee injury. The ladder has a slight slope that extends into the aisle. This factor is added to the aisle width.

Four-Wheeled Rolling Ladder. The four-wheeled rolling ladder (Fig. 6.4*b*) is used in a distribution facility that handles small-cube or lightweight SKUs that are in elevated pick positions. With the additional employee time that is required to move and climb the ladder with the product, use of this method lowers employee productivity. The method is considered for a low-volume facility, to access very-slow-moving SKUs or for access to reserve positions. Use of the ladder increases the number of SKUs per square foot.

Collapsible Platform on a Four-Wheeled Rolling-Ladder. To reach garment positions that are in elevated positions, a collapsible top guardrail platform ladder is used in a trolley rail system warehouse (Fig. 6.4*c*). The disadvantages and advantages are similar to the four-wheeled rolling ladder.

Regular or Z-Frame-Bottom Cart for Hanging Garments. When the distribution facility handles hanging garments, then a rolling rack is used in the human (employee)-push method. The rolling rack is a four-wheeled cart that is designed with four swivel wheels and metal support structure members that provide rigidity and a hang

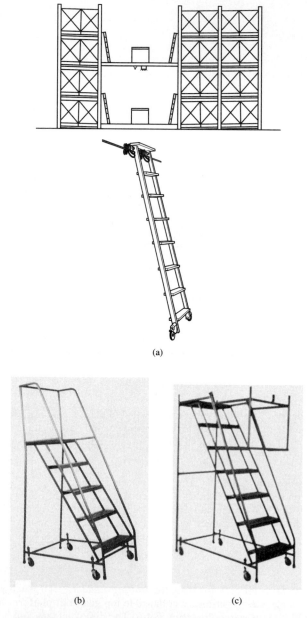

(a)

(b)   (c)

FIGURE 6.4   (a) Top-attached rolling ladder (*Courtesy of Cotterman Co.*); (b) four-wheeled rolling ladder (*Courtesy of Railex Corp.*); (c) four-wheeled rolling ladder with collapsible platform (*Courtesy of Railex Corp.*).

bar for garments. Some carts have a series of bins above the hang bar for flatwear items. The Z-frame bottom support member of the rolling rack (Fig. 6.5a) permits the employee to hook two carts together and to pull two carts.

Manually Pushed Trolley. The next order-pick method is the manual push trolley method. In this method for handling flatwear or single items, the employee pushes a basket hung from a trolley. For hanging garments (Fig. 6.5b), a trolley with pegs (stays) is pushed on an elevated trolley rail through the warehouse aisles. The operational characteristics, disadvantages, and advantages of this method are similar to those of the pushcart method.

Trolley with a Basket. The trolley basket is a wire-mesh basket (Fig. 6.6a) with dividers or multiple levels, circular levels (Fig. 6.6b), or a canvas bag (Fig. 6.6c) that permits small items and flatwear items to be handled. With the basket or carrier divided into small compartments, the order picker can batch-pick small-sized customer orders.

Disadvantages of the trolley basket are that it is a fixed-path vehicle, and some baskets have only one side fill and are difficult to sequentially enter in all aisles.

Advantages of the trolley basket are that it requires little employee training and travels on the same system as a hanging-garment trolley.

Three to Seven Pegs Per Trolley. The hanging-garment trolley is designed with three or up to seven pegs (stays) (Fig. 6.6d), which reduces the hanger pressure on one end of the trolley. Also, when there are only a few SKUs per customer order, these pegs enable the order picker to batch-pick the hanging garments. When the distribution facility handles a high volume of hanging garments, the trolley is preferred.

Disadvantages are a fixed-path system and difficulty in sequentially entering all aisles. Advantages of the hanging-garment trolley are that it handles a greater number of garments and can be used in the batch-pick mode, and the employee handles at least two trolleys.

Employee Riding on a Vehicle. In the next major employee-to-stock order-pick methods the employee rides on a vehicle to the pick position. This method is divided into five order-pick submethods: (1) powered [by electricity, gas, or LP (liquefied petroleum) gas] (employee) burden carrier, (2) order-picker cart or powered pallet truck with shelves, (3) Data Mobile, (4) tractor train of carts, and (5) high-rise order picker (multilevel or HROS).

Powered Burden Carrier. In this method an employee rides on an electric, gas, or LP-gas powered burden carrier with a steering device that guides the vehicle through the warehouse aisles to the required pick position. At the pick position, the operator leaves the operator's area, removes the SKU from the pick position, and places the SKU into the burden carrier small-load-carrying platform. The carrying capacity is increased if a divided container or a shelf device is used.

Electric Powered Order Picker Truck or Pallet Truck. This employee ride-to-pick position method involves the use of an electric powered order-picker cart (Fig. 6.7a) or electric powered pallet truck. This cart has a steering device in the operator's platform area and a large load-carrying surface that easily carries totes. The operational characteristics, disadvantages, and advantages of the cart are similar to those of the burden carrier except for the additional advantage of increased employee productivity. Increased employee productivity is due to an increase in the load-carrying capacity, as an employee stands on the vehicle while picking the SKUs. With some models the employee with a remote-control device or a movable operator's handle directs the vehicle in the aisle while walking beside the vehicle.

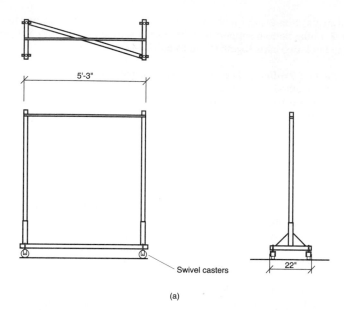

5'-3"

Swivel casters

22"

(a)

(b)

FIGURE 6.5 (a) Z-framed cart; (b) tubular rail (*Courtesy of Railex Corp.*).

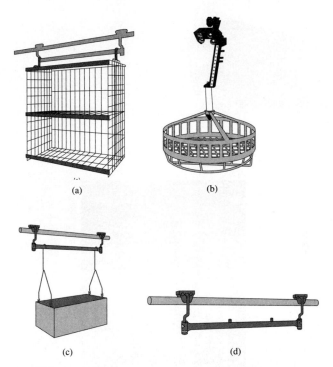

(a)　　　　　　　　(b)

(c)　　　　　　　　(d)

FIGURE 6.6 (*a*) Wire basket (*Courtesy of SDI.*); (*b*) circular basket (*Courtesy of Boyson, Inc.*); (*c*) canvas bags (*Courtesy of Railex Corp.*); (*d*) trolley with pegs (*Courtesy of SDI.*).

Tractor or Train of Carts. This employee ride-to-pick position concept involves the use of an electric or gas powered tractor train (Fig. 6.7*b*). The tugger (tractor) has a steering device, tow pins, and an operator platform. A tractor pulls one or a series of carts through the warehouse aisles. With some models, the employee is able to walk or ride the vehicle through the aisles to the required positions. As the vehicle is stopped at a required pick position, the SKUs are removed from the pick position and placed onto the cart load-carrying surface. If this surface is divided into shelves or compartments, the vehicle can be used for a batch-picking method. Some models are available with remote controls for forward, left, or right directional control. This feature permits the picker from the aisle to steer the tractor through the warehouse aisle.

Disadvantages of this method are that it increases capital investment and requires additional employee training and a wider aisle. Advantages are that the vehicle carries larger quantities, travels greater distances, handles most SKUs, and improves employee productivity.

Data Mobile Pick. The Data Mobile pick (Fig. 6.7*c*) is an electric powered single-item manual order-picking cart that consists of an on-board computer and printer, with controls for tote, container, and operator carrying platforms; and has three to four wheels, top guiderail (fixed path), and shelf or rack pick positions. The Data Mobile cart receives its electric power from a DC (direct-current) overhead rail and its customer order-pick requirements via a wireless infrared interface from the host computer. The Data Mobile pick cart can handle a single customer order or 8 to 12 batched

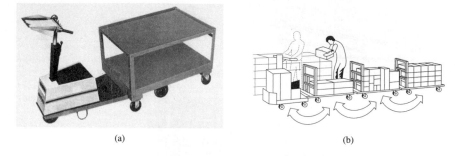

(a)   (b)

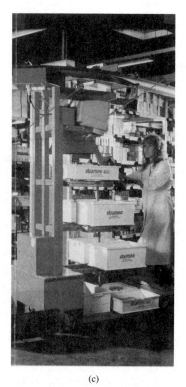

(c)

FIGURE 6.7 (a) Electrically powered order-pickcart (*Courtesy of Lyon Metal Products.*); (b) tractor train (*Courtesy of Sims Consulting.*); (c) data mobile order-pick cart (*Courtesy of Distribution Tech, Inc.*).

customer orders. Each vehicle makes one pass through a 3-ft-wide aisle from the start location through the entire pick aisles to the end pick location.

In this manual pick method, each on-board container (customer order) is equipped with a digital display screen that directs order-picker activity. Also, for each SKU on any customer order, the computer controls automatically stop the vehicle in the pick aisle at the appropriate pick position. At this position, it allows the order picker to transfer the required number of SKUs from the pick position to the appropriate customer container. If more than one customer order has the same SKU, then the appropriate digital display screen indicates this requirement. With these pick instructions, for this particular SKU the order picker completes all pick activities for the batch.

After the order picker obtains the assigned Data Mobile vehicle, the Data Mobile CRT (cathode-ray-tube) screen indicates the appropriate-sized tote for each customer order. Each tote receives a human- or machine-readable label and is placed onto its assigned position on the Data Mobile vehicle. Next, the order picker prepares (signs on) the vehicle for travel through the order-pick area and to complete the order-pick activity.

As the order-pick vehicle travels through the aisles and approaches the required pick position, per each customer order, the on-board computer, as required, illuminates the digital display for each customer order. The computer stops the vehicle travel in the pick aisle at the pick position. At this aisle location, the order picker has access to the pick position and transfers the appropriate number of SKUs from the pick position to the customer tote on the Data Mobile vehicle. This activity is repeated for each customer order of the batch that requires this SKU.

After completion of all picks at this pick position and with the proper setting of the safety controls, the vehicle travels in the pick aisles to the next required pick location that contains a SKU which is required on one customer order of the batch.

At the completion of all pick requirements, each Data Mobile vehicle sends the pick information to the host computer. This permits the preparation of an accurate packing slip for each customer tote. The completed customer order totes are transferred from the Data Mobile vehicle to the take-away vehicle or conveyor.

With the Data Mobile, when a customer order has more than one tote, the computer system "cubes out" (divides) the customer order into the appropriate number of totes. With this concept for a large order, there are multiple totes for one customer order but each portion of the order is an individual delivery package with an associated packing slip.

Disadvantages of the Data Mobile concept are that it is a fixed-path system, and if the vehicle is not dispatched properly (within the single-vehicle path), then there is a potential for queues in the aisles. Also there is greater requirement for management control and disciplines and for customer order or SKU cube (dimensional and weight) information, and a high capital investment.

Advantages of the concept include accurate pick activity and packing list, batched or single-customer order handling, high picker productivity, reduced product or equipment damage, and handling of all types of SKUs and on-line inventory adjustments.

Multilevel Order-Picker Truck Concepts. Next is the high-rise or multilevel order-pick concept, which is considered one of the most sophisticated concepts of the group. Four different types of vehicles can be used to order-pick SKUs: (1) order-picker or platform truck, (2) decombe, (3) pick car, and (4) MS/RS ("man-on-board" storage and retrieval system).

Order-Picker or Platform Truck Concept. The order picker or platform truck (Fig. 6.8*a* and *b*) is a manually operated and controlled vehicle. In a guided aisle, the vehicle elevates as it travels horizontally or vertically down-aisle between pick positions. These positions are located on both sides of the aisle and are up to 35 ft high. The operational characteristics are similar to those of other employee-to-stock concepts, except that the vehicle travel is guided in the aisle and elevates the operator.

Disadvantages of the vehicle are increased capital investment and employee training and requirement for a vehicle guidance system.

The advantages of the concept are that it increases space utilization, increases security of SKUs, improves employee productivity, provides easy access to all SKUs,

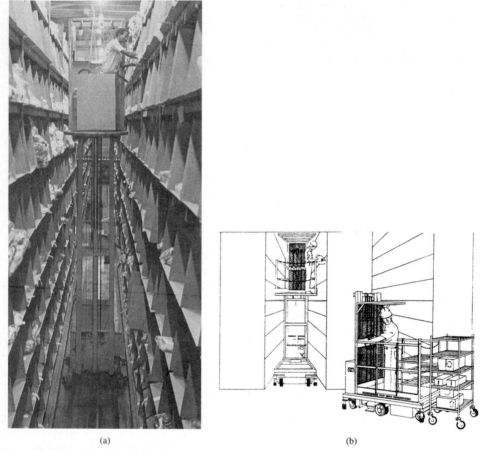

(a) (b)

FIGURE 6.8 (*a*) High-rise order-selector truck (*Courtesy of Barrett Indl Trucks, Inc.*); (*b*) platform truck (*Courtesy of West Bend Equipment Co.*).

has lower risk of product or equipment damage, and can be used as a hand-stack storage vehicle.

When comparing the individual vehicle operational characteristics, the decombe and pick car have on-board SKU take-away conveyors: therefore, these vehicles are reviewed in the next subsection.

Employee On-Board Storage and Retrieval Vehicle System. The next order-pick method is the employee on-board vehicle and computer-controlled storage-retrieval vehicle system (Fig. 6.8c). The unique characteristics of these order-pick vehicles are that some are captive-aisle and others are mobile-aisle vehicles. They have versatile mobility as they can travel either vertically or horizontally in the aisle to the required pick position. This travel is either computer-controlled or manually controlled. When not used as an order-pick vehicle, some vehicles are used for pallet load storage.

The operational characteristics, disadvantages, and advantages of these vehicles are similar to those of the manually controlled order-pick vehicles.

(c)

FIGURE 6.8 (*Continued*) (*c*) Employee-on-board aisle storage and retrieval vehicle. (*Courtesy of Rapistan Demag Corp.*)

Mechanized Order-Pick Methods

The next group of employee-to-stock order-pick methods consists of the mechanized group. In these procedures the order picker walks or rides to the SKU pick position, removes the SKUs from the pick position, places an identification label on the SKU, and places the SKU into a tote or directly onto a mechanical transport take-away conveyor system.

The take-away conveyor system transports the SKUs to a customer order sortation station. At the sortation station, the mixed (batch-picked) SKUs are separated into the appropriate customer order bin by a mechanized sortation system or by a sortation employee who sorts the SKUs.

Mechanized Employee Walk Method. The mechanized walk-to-stock methods of this group have very similar operational characteristics, disadvantages, and advantages. The two alternatives for this concept are the pick-to-belt or pick-to-tote on a conveyor.

Alternative Employee Pick-to-Conveyor Concepts.   The pick-to-conveyor alternative has two basic designs: (1) belt or roller take-away conveyor in the main aisle with perpendicular pick modules on one or both sides of the pick conveyor and (2) belt or roller take-away conveyor with aisles and pick positions that are parallel on both sides of the pick conveyor.

Three Basic Pick-to-Tote on a Conveyor Layouts.   The pick-to-tote on powered roller conveyor has three basic layouts:

1. The first layout design consists of a powered roller take-away conveyor in the main aisle with perpendicular pick modules on one or both sides (Fig. 6.9*a*). The order picker picks to a tote on a cart. The cart load-carrying surface has an elevation that is slightly higher than that of the take-away conveyor. The cart also has a removable side for easy tote discharge onto the take-away conveyor. The cart has a load-carrying surface that permits the full tote to be easily pushed onto the take-away conveyor.

2. The second layout includes sloped pick-to-tote conveyors that are attached full length, as required, to the roller take-away conveyor in the main aisle (Fig. 6.9*b*). The pick conveyor is divided into separate pick zones, which requires the order picker to pass the partially completed order (tote) on the roller conveyor to the next pick zone. The perpendicular pick modules are on one or both sides of the take-away conveyor. The employee walks from the pick conveyor to the pick position, returns to the pick conveyor, and places the merchandise into the appropriate tote. A gap plate is full length between the two conveyors, or the take-away conveyor is set low to assist with the transfer of totes and travel of totes on the main take-away conveyor.

3. The layout consists of sloped pick-to-tote conveyors that are attached full length to the sides, or a flat conveyor that is above the roller take-away conveyor in the main aisle with pick positions parallel on both sides of the take-away conveyor (Fig. 6.9*c*). During the pick activity, the employee turns, removes the SKUs from the pick positions, then turns and places the merchandise into the appropriate tote or directly onto the conveyor.

In the first two methods all full totes are transferred to the take-away conveyor. In the first method, all full totes with proper customer identification are transferred from the pick conveyor to the take-away conveyor for transport to the sortation-holding area.

In the third method an empty tote with a customer identification is placed onto the pick conveyor in front of the full tote. When the customer order is completed by the order picker(s) from all zones, all customer totes are transferred from the pick conveyor to the take-away conveyor. This transfer concept maintains customer order integrity and reduces the sortation effort.

Disadvantages of the walk pick-to-belt/tote concept are higher capital investment and increased management control requirement; also, all SKUs are conveyable or are placed in totes, and the order-picker routing pattern agrees with the layout.

The advantages of the concept include high volume handling, high employee productivity, reduced vehicle damage, and availability of zone picking.

Design a Bypass Take-Away Conveyor Path in Your Small-Item Pick Area

Some small-item pick methods do not transport the picked product en masse, but are required to transport your batched customer orders in individual totes as a single cus-

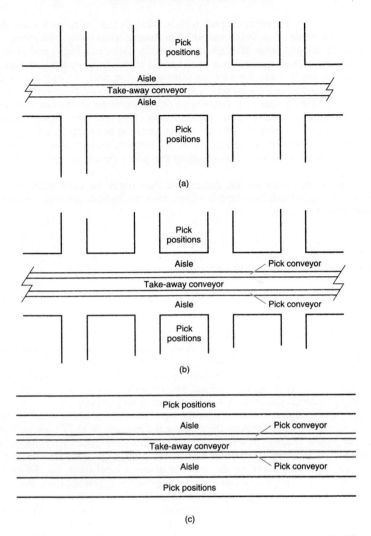

FIGURE 6.9 (*a*) Pick aisle perpendicular to the take-away conveyor; (*b*) pick aisle perpendicular to the take-away conveyor that has a pick conveyor; (*c*) pick aisle parallel to the take-away conveyor.

tomer order because the operation does not have a sortation activity or area. These customer order-pick concepts are (1) zone-controlled single customer order-pick concept and (2) batched customer order-pick concepts (pick-packed and picked into a divided container, tote, or cart).

When you use one of these customer order-pick concepts, the order-pick container is transferred from zone to zone or handled by order pickers until the container is full or until completion of the customer order. For a high order-picker productivity, you design a full tote or completed order take-away conveyor path. The take-away conveyor path transports these completed orders or full containers or totes to the packing area.

If the take-away conveyor or path travels through the entire pick area, then your tote, container, or cart that has a completed customer order travels the entire conveyor system. With the pick area divided into zones for fast-, medium-, and slow-moving SKUs, the one take-away conveyor that travels through the entire pick area for all three zones creates a delay for packing and completion time. To optimize your packing employee productivity, your take-away conveyor network design has a divert spur that directs the appropriate tote, containers, or cartons from the end of your fast-moving SKU pick zone directly to the packing area. This conveyor layout means that a greater number of completed customer orders arrive at the packing area in less travel time. At the beginning of the packing shift, this means a greater amount of work available at your packing stations, thus reducing the peaks (levels of daily activity) of the packing activity.

Since most operations do not stagger the start times for each warehouse activity (picking, packing, manifesting, and loading), then this bypass conveyor reduces surges on the product flow path.

Mechanized Employee-Ride Concepts

The mechanized employee ride-to-stock concepts are the pick car and decombe vehicles.

Pick Car Concept. The pick car (Fig. 6.10a) is considered as a single-item order-pick vehicle. It is a manually operated vehicle that is guided in a captive aisle, has a direct electric powered rail, and travels on four wheels that are above the belt take-away conveyor. The vehicle travels between two pick positions of racks or shelves, bridges the take-away conveyor, and has the ability to elevate the order picker on the order-picker (control) platform to the elevated high pick positions.

As the vehicle is driven down-aisle, the operator stops it at the required pick position. The pick position is between the floor or the top rack or shelf pick position level. In the batch-pick order-pick mode, the pick car has two alternative layout options. First is the "pick direct to the belt take-away conveyor" concept. As the SKU is removed from the pick position, the operator places a label on it. The labeled SKU is then placed on the take-away conveyor. After the SKU travels from the order-pick aisle, it arrives at the sortation conveyor area for customer sortation.

In the second layout option, the order picker picks merchandise into a tote that is on the pick car. When the tote is full, it is transferred to the take-away conveyor for transport to the sortation station. A further batch-pick option is to have a series of totes on a divided shelf. Each tote represents a customer order.

Disadvantages of the pick car concept are captive aisle and fixed-path vehicle, high capital investment, requirement for a separate replenishment aisle or shift, and a limit of one employee per car (aisle). Advantages are increased pick height, reduced employee walking and reduced product or equipment damage.

Decombe. The decombe (Fig. 6.10b) is an employee ride-to-stock multilevel vehicle. The vehicle provides the order picker with the ability to reach pick levels that are 20 ft high. The decombe method is considered a single-item order-pick method. The decombe is a manually controlled order-pick vehicle that is guided, and is a powered vehicle. The decombe has a spiral chute that is in the front of the operator's platform and extends from the belt take-away conveyor to the highest pick position. The belt take-away conveyor is at floor level and is located on the right side of the decombe

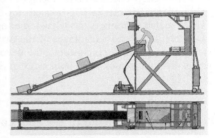

(a)

(b)

FIGURE 6.10 (a) Pick car order-pick vehicle (*Courtesy of Rapistan Demag Corp.*); (b) decombe or multilevel order-pick vehicle (*Courtesy of Rapistan Demag Corp.*).

travel aisle. This feature allows the order picker to pick SKUs from the rack positions on the left side of the aisle and makes it difficult for the operator to pick from the right side of the aisle.

While driving the decombe down-aisle, the operator stops the vehicle at the required pick position that is on the left side of the aisle. The pick position is between the floor level or at the top rack or shelf position level. In the batch pick concept, as the SKU is removed from the pick position, the operator places the label on the SKU. The labeled SKU is placed onto the spiral chute, which allows it to flow to the floor-level conveyor, or the SKU is placed into a tote. The conveyor transports the loose SKUs or sealed containers to the customer order sortation or packing station.

Disadvantages and advantages of the decombe are similar to those of the pick car. When comparing the two concepts, the decombe has a higher vertical travel capability but it is difficult to pick from both sides of the aisle.

"Product Is Transferred to the Pick Location" Methods Group

In the second major small-item order-pick group of procedures the SKUs come to the order picker. The main characteristic of this group is that the SKUs in their storage containers are transported from the storage area to the order-pick station. At the order-pick station, the required SKU is removed from the pick position. The SKU is labeled, or the customer order-pick document is marked by the employee. The employee places the SKU onto a take-away conveyor or into a tote. The SKU pick position container is returned to the storage location. The stock-to-employee order-pick group includes (1) miniload or full-load S.I. Cartrac with shelves, (2) miniload, (3) carousel, and (4) rapid pack.

Cartrac System.   The first mechanized stock-to-employee order-pick system is the miniload or full-load shelved S.I. Cartrac system (Fig. 6.11*a*). In this method the SKUs in pick position containers are placed onto one platform (car) of a series of an electric powered rail platforms. These platforms travel on a fixed path which is a closed loop. In the closed loop each car travels past a SKU replenishment station and a series of order-pick stations.

Each car is assigned a group of SKUs. After the SKUs are placed onto the car, the car travels to the order-pick station. The order picker stops the desired car by pressing the car stop bar. If the order picker is in batch-pick mode, the employee removes the SKUs, identifies the SKU, and places the SKUs onto the take-away system or into a tote. The system transports the labeled SKUs or totes to the sortation area for sortation to the appropriate customer staging container. In a single customer order handling system, the SKUs are placed into the appropriate customer container. When completion of the customer order, the full container is sent to the packing station.

After the car SKU pick position is depleted, the car is stopped in the replenishment area. In this area, a new SKU quantity is transferred from the storage area to the pick car.

Disadvantages of the S.I. Cartrac concept are fixed-path equipment and high capital investment. Some advantages of the concept are that it reduces employee walking time, reduces product or equipment damage, and handles batched customer orders.

Miniload System.   The second mechanized stock-to-employee order-pick system is the miniload. Individual large SKUs are placed into single containers, and small SKUs are placed into smaller compartments within a larger container. These smaller compartments are equipped with dividers that are placed in the large container interior.

Operationally, the miniload is a captive-aisle vehicle that travels to the end-of-aisle input station, takes on-board the container, travels down-aisle to the assigned location, and deposits the container in the storage position. To withdraw, the machine travels to another location, withdraws the container that has the required SKU from the storage position, travels down-aisle to the output station, and places the container onto the

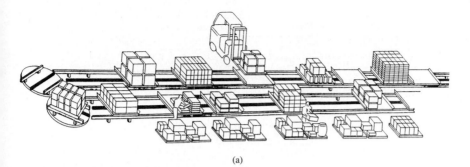

(a)

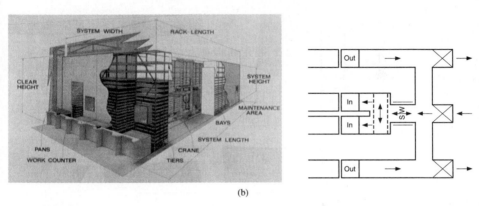

(b)

FIGURE 6.11   (a) Cartrac system (*Courtesy of S.I. Handling Systems.*); (b) front-end miniload (*Courtesy of Litton Indl. Automation.*).

take-away conveyor system. The containers are transported via conveyor or placed onto carts for transport to the order-pick station.

The ministacker is controlled by either of two mechanisms: an employee-controlled vehicle, with the operator at the end-of-aisle control station (the operator enters in the controller each input/output transaction, and the controller directs the vehicle to perform the transaction) or (2) a computer, which controls the entire vehicle activities. The control concept is based on the economic savings from the labor and the on-time, accurate transactions.

A ministacker order-pick station has three design options: the front-end pick location, front-to-rear pick locations, and remote pick location.

Front-End Pick Area Miniload.   The front-end pick location has a series of conveyors that are placed in the front of the ministacker aisle (Fig. 6.11b). One conveyor direction of travel is to the stacker aisle and provides the infeed (input) station. A second conveyor travel direction is from the ministacker aisle. This conveyor is the ministacker outfeed (output) path. The third conveyor section is the order-pick station that has the direction of travel from the output conveyor to the input conveyor and is across the front of the ministacker aisle. Stop-start controls and accumulation order-pick conveyors halt the container transport for the order-pick activity.

Front-to-Rear Pick Area Miniload.   The front-to-rear miniload (Fig. 6.11c) has the same operational characteristics, disadvantages, and advantages as the front-end

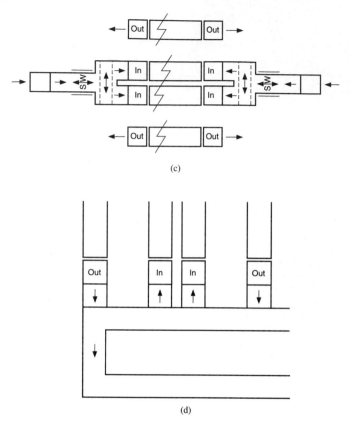

FIGURE 6.11 (*Continued*) (*c*) Front-to-rear miniload; (*d*) order-picker stations remote from miniload.

miniload except that both ends of the ministacker aisle have a pick conveyor network that allows the ministacker to handle a higher volume.

Remote Pick Area Miniload.   In this design alternative the order-picker stations are located in an area remote from the miniload stacker (Fig. 6.11*d*). The ministacker has an infeed and outfeed conveyor. When compared to the other two layouts, the unique characteristics are more extensive accumulation conveyor transport systems with divert devices and an increase in the number of pick station.

The infeed-outfeed conveyor system of the ministacker with remote order-pick stations is more extensive than the other conveyor systems. These conveyors transport totes from the output station to the appropriate order-pick station. These conveyors allow accumulation of totes prior to each workstation. Tote divert devices direct (divert) the totes from the outfeed conveyor to the appropriate order-pick station. At the station, the order picker removes the appropriate SKUs from the tote and places them into a customer's container and the ministacker tote is placed onto the infeed conveyor. Each completed customer order is placed onto the outbound take-away transport device or conveyor.

Disadvantages of this miniload concept are high capital investment, queuing prior to workstations, and increased need for management control. Some advantages are that the ministacker handles a medium volume, increases order picker productivity, and reduces product and equipment damage.

Carousel System.   The third stock-to-employee order-pick method is the carousel order-pick system. The carousel consists of a series of bins (containers) or holes or hooks for individual hanging garments that are attached to a chain which is an endless loop. The forward-reverse movement of the carousel is controlled by a control panel that receives its instructions from an operator who makes a key entry or by a mini-computer that makes the entry. These instructions command the carousel to rotate (travel) until the required bin is halted at the pick station. The order picker removes the required SKU quantity from the bin, places the SKU into the customer's container or onto a take-away conveyor, and programs the carousel to advance to the next required bin.

There are several alternative layouts for the vertical and horizontal carousel designs. The type of carousel used determines the chain or bins direction of travel.

Vertical Carousel.   The vertical carousel (Fig. 6.12*a*) has an endless chain of bins that travel in a vertical direction. To access the bins, the carousel has one workstation or two stations, one on each floor. When there are two workstations, then there are controls to prevent the carousel movement as an employee is performing a transaction. In the one floor design, the order-pick and replenishment activity occurs at one station. In the two-floor design, the order-pick and replenishment activities occur on separate floors. The major advantage of the vertical carousel is that is requires a small-square-foot floor space with a high ceiling.

Horizontal Carousel.   The horizontal carousel (Fig. 6.12*b*) has an endless chain that is attached to a top or bottom drive. With the use of mezzanines and available building space, the horizontal carousel is one or two carousels high. An alternative one floor carousel layout provides one position for both order-pick and replenishment activities or two positions with one position for the pick activity and the second position for the replenishment activity.

There is also a *multiple horizontal carousel.* When two carousels are designed side by side, an order-pick station has access to both carousels (Fig. 6.12*c*). When four carousels are designed to service one order pick station, the order picker's productivity improves because there is less waiting time and an increased number of SKUs.

When the horizontal carousel is compared to the vertical carousel, the horizontal carousel requires more square-foot floor space but a lower clear ceiling.

Comparison.   The disadvantages of the carousel method in comparison to the other methods are that it limits pick positions, limits SKU quantity in the pick position, is slow-moving (rotating), and requires a replenishment cycle.

Advantages of the carousel are that it does not require a sortation area, reduces the number of aisles, reduces order-picker walking or traveling time, and is an excellent method for handling small parts.

Rapid Pick-Pack Method.   The next stock-to-employee order-pick method is the rapid pick and pack (pick-pack) system (Fig. 6.13*a*). The rapid pick-pack order-pick is a small-item (flatwear) system that consists of a series of order-pick stations that are located between a SKU pick position and an order-pick conveyor with a take-away conveyor. The SKU pick positions are gravity flow lanes or trolley lanes.

Empty individual discreetly identified store shipping containers are placed onto the order-pick conveyor. At each order-pick station or zone a group of store shipping containers are diverted from the store shipping conveyor to the appropriate order-pick area. In the order-pick area, an order picker picks the required SKUs from the SKU

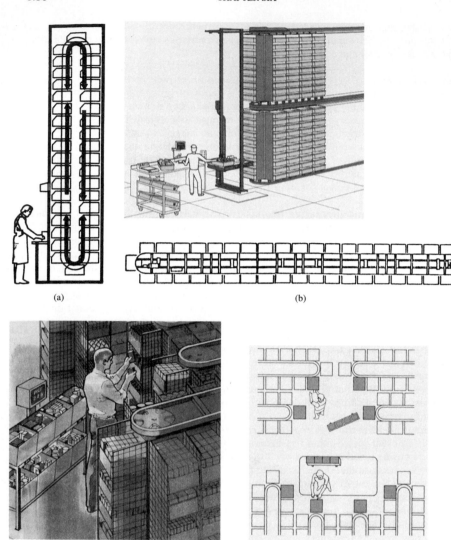

(a)   (b)

(c)

FIGURE 6.12   (*a*) Vertical carousel (*Courtesy of Remstar Int'l.*); (*b*) horizontal carousel (*Courtesy of White Storage and Retrieval System, Inc.*); (*c*) multiple horizontal carousels (*Courtesy of White Storage and Retrieval System, Inc.*).

pick position. The employee places the SKU into the shipping container. Each order-pick area (zone) has a specific number of SKUs that are assigned to the pick area; therefore, to complete a customer order, the customer's shipping container must travel on the order-pick conveyor through several order-pick zones.

The unique feature of the rapid pick-pack concept is that the SKUs are moved from the reserve area to pick position. When the customer orders are completed by the order picker, then the partial customer order container is sent to the temporary reserve area or sent to the packing-shipping area.

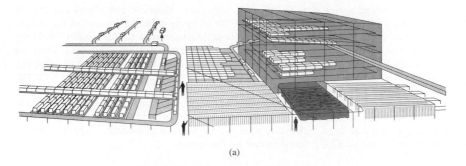

(a)

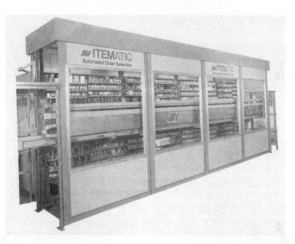

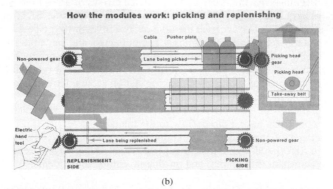

(b)

FIGURE 6.13 (*a*) Rapid pick and pack (*Courtesy of SDI.*); (*b*) Itematic order-pick machine (*Courtesy of S.I. Handling Systems, Inc.*).

During the order-pick day all full customer order shipping containers are placed onto a take-away conveyor for transport to the shipping area. At the end of the order-pick day activity, all partial full shipping containers are sent to a temporary holding area or sent to the shipping area. If sent to the temporary holding area, then on demand or on the next customer-assigned order-pick day, these partially full shipping containers are released to the order-pick area for completion of the container with the customer's next-day ordered product.

Disadvantages of the rapid pick concept are higher equipment investment and multiple pickers per customer order, and some product is held for next-day delivery. Advantages are that rapid pick reduces employee walking and handles a high volume, zone picking, and all SKUs (hanging and nonhanging).

Automatic Order-Pick Methods

The third major small-item SKU order-pick methods group is the automatic concept group. In these order-pick methods the required SKUs are automatically (by computer control) withdrawn from the pick position and transported by a belt conveyor or slide to the customer packing station.

This group includes the Itematic, automated order-pick system (A Frame), Flex Pick, and automatic item dispenser (Robo Pick).

Itematic Concept.   The itematic is considered as a stand-alone SKU order-pick machine (Fig. 6.13b). Each Itematic machine is a computer-controlled single order-pick concept that handles a specific size range and number of SKUs. In the Itematic machine, individual SKUs are assigned to specific replenishment or order-pick lanes. After an employee places the SKU into its assigned replenishment lane that is on the back side of the machine and is horizontal to the floor, then the SKU is ready for order-pick activity. During the order-pick process, a computer-controlled picking device that has a single customer order requirement travels along the order-picking side of the SKU lane. As the order-picking device arrives at a required SKU lane, the computer triggers the picking device mechanism to release one SKU onto its take-away conveyor. The SKU travels on the take-away conveyor to the packing station. At the packing station, the individual items are collected and packaged into the customer's shipping container. To optimize the packing station productivity, a divert blade on the take-away conveyor is used to divert an individual (slug of SKUs) order to accumulate prior to the appropriate packing station accumulation slide.

The Itematic is located on a mezzanine or adjacent to other Itematic machines. The design requirements are that there is sufficient area to handle replenishment and the SKU reserve positions have a direct flow to the Itematic.

Disadvantages of the Itematic are single order-pick activity, high capital investment, limited SKU handling, and slow replenishment. Advantages are accurate order-pick activity, constant throughput, and on-line inventory adjustments.

Automatic Order-Picking Machine (A Frame).   The second automated single-item order-pick method is the automated order (A Frame) (Fig. 6.14a) picking machine that is a computer controlled concept. The automatic picking concept consists of at least 64 product dispensers that are arranged in a A-framed structure over a belt take-away conveyor. The product dispensers or sleeves are considered one unit; therefore, each (module) unit handles SKUs with common characteristics.

The automatic picking machine concept computer program handles all order picks of SKUs for one customer order. All the customer orders are held in the computer. As required, to satisfy the delivery schedule or as a coded tote passes a scanner, the com-

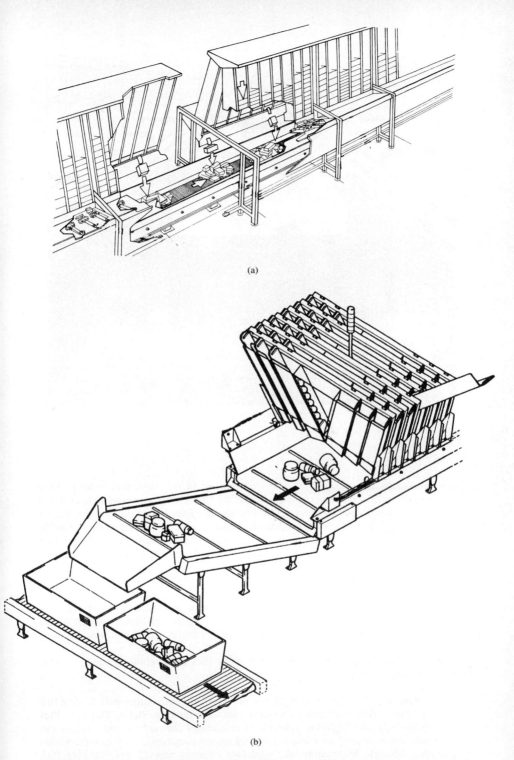

(a)

(b)

FIGURE 6.14 (a) A frame order-picking machine (*Courtesy of Electro Com Automation L.P.*); (b) Flex Pick (*Courtesy of Electro Com Automation L.P.*); (c) Robo Pick (*Courtesy of Versa Ferguson.*).

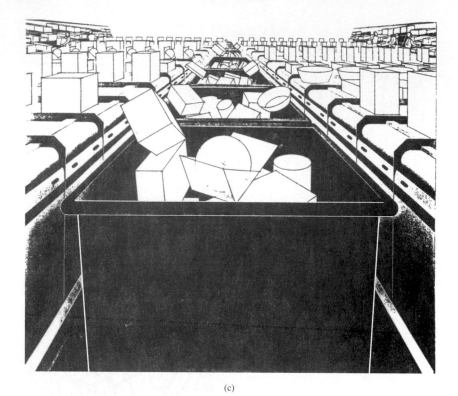

(c)

FIGURE 6.14   (*Continued*)

puter releases the appropriate customer requirement to the automatic order-picking machine module. The customer order release is in a random or sequential arrangement. The computer sends an order-select impulse to module, and the impulse travels along the order release side of the product dispensers. After the dispenser that contains the SKU which appears on the customer order receives the impulse, then the dispenser releases the SKU at a rate of 5 to 7 items per second onto the take-away conveyor. The take-away conveyor transports the SKUs to the outbound packing station or into a container at the end of the pick belt. The concept fills 1,400 totes per hour.

Pick position (SKU dispenser sleeve) or magazine replenishment takes place during order-pick activity. Replenishment lights in the pick area identify the sleeves that require replenishment. The replenishment employee adjusts the sleeve holder, places the SKUs into the sleeve, and replaces the full sleeve holder to the automatic picking machine.

Advantages and disadvantages of the automatic picking machine method are similar to those of the Itematic.

Flex Pick System.   The Flex Pick (Fig. 6.14*b*) is an automated order-pick system that is computer-controlled and handles medium- to slow-moving SKUs. The Flex Pick concept consists of SKU (product) dispenser magazines and support rails, take-away conveyor belt, tote or container conveyor, and required computer and electric wiring and sensing devices. As required, the customer order is entered into the Flex Pick computer system that sends a pick impulse to the appropriate product dispenser to release one unit of product from the sleeve onto the belt conveyor.

The Flex Pick concept handles product that has a typical dimension of $6 \times 6 \times 6$ in, and the SKU is solid and stackable in any shape (rectangular, round, or oblong). On the average, there is 24 in to the product magazine and the magazine dispenses product at five (5) units per second. It has 4050 dispenser sleeves per belt, which is 250 ft long and handles up to 5000 orders per day.

Operation of the Flex Pick occurs as the customer order is entered into the computer. As the customer container, tote, or box with the customer discreet label approaches the fill position or the discharge end of the gathering belt (take-away belt conveyor full length under the dispenser magazine), a sensing device identifies the customer label and communicates this information to the computer. The computer activates the pick activity. As required by the customer order, the computer controls the appropriate dispenser magazine to release one unit of product per impulse from the dispenser magazine onto the gathering belt. As the unit of product passes a photo-eye, it verifies the pick to the computer, which updates the inventory level in the SKU magazine and stores the pick information for manifest preparation. All released product travels on the gathering belt to the discharge end for deposit into the customer container.

After all the product for one customer order (tote) has been released onto the gathering belt, the next customer order (tote) approaches the fill position and passes the sensing device and activates the computer.

This pick arrangement divides (cubes out) a large customer order into the appropriate number of totes that are required to handle the number of SKUs. Each tote of a multitote customer order is handled as an individual package with its individual manifest.

During the pick process or at specific time periods, product is replenished to the product dispenser magazines. To reduce "stockouts," a large low product light or lights are located in the dispenser magazine area and a smaller light indicates the exact location.

Disadvantages of the Flex Pick concept are high capital investment, requirement for on-time replenishment, double handling of units if the dispenser magazine does not hold a master carton SKU quantity, and requirement for accurate SKU cube and weight information. Advantages of the concept are accurate picks, on-line inventory update, high pick rate, and replenishment occurring during the pick activity.

Automatic Item Dispenser (Robo Pick) Concept. The Robo Pick is another automatic order-pick system that is manually replenished and computer-controlled. The major components are a central container belt conveyor with banks or item dispensers on each side. These banks are individual belt conveyors.

As the empty container travels on the belt conveyor between the item dispensers, per the customer's order the computer activates the appropriate item dispenser lane to index (move) forward. This forward movement permits one SKU to fall into the container. The forward action automatically causes the item to pass a sensor and fall into the container. The sensor transmits a signal to the computer to verify the pick.

All partial or full containers travel on the conveyor to a packing station. At the packing station, the items are prepared for customer delivery.

Manual replenishment to the SKU dispenser lane is made from an aisle by an employee placing the SKU onto the empty belt lane position. On the average, the pick belt holds a small quantity of SKUs.

The Robo Pick Concept handles loose or individual packaged SKUs that are medium- to small-cube, medium-weight to lightweight, medium- to slow-moving, and with a small inventory. The concept provides a FIFO (first-in first-out) product rotation and medium facings.

The Robo Pick has the capacity to handle a batch of 54 orders and 20,000 pieces per hour. Disadvantages of the Robo Pick are higher capital investment, requirement for increased management control and discipline, and slow replenishment. Advantages of the Robo Pick are that it handles most SKUs, has a constant throughput, handles a large number of orders, improves inventory control, and improves employee productivity.

Shipping Carton (KD) and Tote Replenishment to the Pick Area

A unique feature of the small-item pick activity is that most conveyor or pick-pack methods require that empty shipping cartons or totes be replenished to the picker in the pick area.

The empty carton or tote replenishment method is based on the following design parameters: (1) the order-pick method, (2) SKU characteristics and number of pieces per order; (3) number of required cartons or totes per minute (average and peak); (4) size and other physical characteristics of the carton or tote such as handles, shape, and lips; (5) available space in the pick area; (6) carton or tote identification requirement; (7) tabbed or opened carton; (8) whether the carton is placed into a container; (9) whether the carrier handles trash or outbound cartons; and (10) whether the master cartons are reused as the pick carton.

Three basic types of cartons can be used in your pick operation: your vendor carton, the tabbed carton, and the open carton. If you consider using totes, then your alternatives are a tote either with or without handle locations.

Vendor Cartons.   Some distribution companies use the vendor cartons to pack their customer orders. When a company uses the vendor carton, there is a savings of carton expense but an increase in packing labor to form the carton and place the proper protective material inside it. Also, the carton exterior appearance is not professional to the customer; therefore, in a dynamic operation the vendor carton is not considered as an alternative.

Tabbed Carton.   A tabbed carton is a specially designed carton with four flaps that are not entirely cut. When assembled, the flaps are held together on the exterior of the carton with a small portion of each flap. This portion is a nominal $\frac{1}{16}$- to $\frac{1}{8}$-in-wide piece of cardboard. After the carton is assembled at the workstation, the bottom is sealed by an employee and placed onto the transport device. If a destination label is required on the outbound carton, then consideration is given to the flap position because the length could eliminate a scanner view of the label. With the tabbed carton, the clear path of the transport device is 1 in wider than the carton. Prior to a carton seal station, the tabs are broken and the flaps are folded in.

Disadvantages of the tabbed carton are that it is more expensive on a per unit basis and some cartons cannot handle destination labels. Advantages of the tabbed carton are that it requires a narrow transportation path and reduces hang-ups, and with no flaps extending outward, additional cartons are accumulated on a conveyor.

Open Carton.   An open carton is a regularly designed carton with four flaps that are entirely cut. When assembled, the four flaps are loose on the exterior of the carton. After the carton is assembled, the bottom is sealed by an employee and is placed onto the transport device. The transport device clear path width compensates for the loose flaps, which means you allow the carton transportation path an additional 4 to 6 in width than the carton side-to-side dimension.

Disadvantages of the open carton are that it is difficult to handle, occupies additional space on a conveyor, can cause hang-ups on a transport path, and requires a wider transport path. Advantages are that it is less expensive and reduces destination label location problems.

Various Empty Shipping Carton and Tote Replenishment Methods

When you consider the various empty carton and tote replenishment methods, there are three alternatives: (1) at the end of the aisle, in an assigned pick position or above the take-away conveyor; (2) gravity-powered roller conveyor above or below the pick conveyor or pick position; and (3) powered trolley conveyor.

Manual Method. The manual method utilizes empty shipping cartons or totes that are placed at the end of the pick aisle, in an assigned pick position, or above or below the take-away conveyor. An employee with a two-wheeled hand truck transports empty cartons to the assigned location and deposits the cartons or totes in the appropriate position. As required, order pickers remove an empty carton or tote from the temporary storage position to the pick conveyor. With this arrangement an employee makes up the carton, or the cartons are made up in another location and nested in the pick area.

Disadvantages of this method are that it increases the number of handlings, represents an additional order-picker activity and potential out-of-stock situations, requires the order picker to walk and transfer the empty carton or tote, and occupies a pick position. An advantage of the manual concept is that there is no capital investment.

Mechanized Methods

Roller Conveyor Above or Below the Pick Conveyor or Pick Position. In this conveyor method the empty shipping cartons or totes travel on a conveyor (Fig. 6.15a). The conveyor path permits both replenishment and order-picker employees easy access to the carton or tote. The conveyor path is past the pick position, and the conveyor design is a recirculation type or one-way dead-end conveyor.

At the replenishment or KD carton make-up location, an employee places the empty bottom-sealed cartons or totes onto the conveyor, and as required in the pick area, the order picker transfers the carton or tote to the pick conveyor.

Disadvantages of this method are higher investment and that it is designed in the pick area. Advantages are consistent supply of empty cartons or totes and reduced picker nonproductive walking time.

Powered Overhead Trolley Conveyor. The powered overhead monorail trolley (Fig. 6.15b) consists of a powered endless chain with a pendant every 6 to 5 ft on centers, which means approximately 12 to 10 carriers per minute. These pendants extend downward and have a hook or flat carrier. The path of these carriers is over the take-away conveyor, permitting the order picker easy access for the replenishment employee. In the pick area, the bottom of the carrier permits a tote or carton to be handed downward and not to interfere with the travel of completed customer orders on the take-away conveyor. As required, antisway and antiswing devices are added to the pendants to stabilize the carrier. When lift gates are installed to cross the conveyor, then at the gate locations the bottom of the tray is at 7½-ft elevation above the floor. This elevation with a 36-in-high conveyor provides a clear passageway through the conveyor.

Flat Carrier. The flat carrier (Fig. 6.15c) transports two corrugated cartons. The cartons are placed on the carrier with its length or width in the direction of travel. The

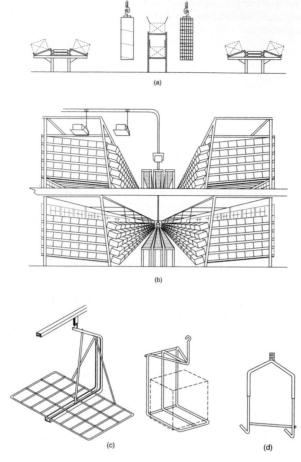

FIGURE 6.15 (*a*) Conveyor above the take-away conveyor; (*b*) overhead trolley (*Courtesy of Sims Consulting Group.*); (*c*) flat carrier; (*d*) hook carrier.

carton placement arrangement depends on the relationship between pick conveyor and carrier paths. Plastic totes are difficult to transport on the metal carriers because plastic slides easily on metal. But with a higher lip on the carrier and the tote turned upside down (the tote lip resting on the carrier), the sliding is reduced and the plastic tote rides on the carrier.

Hook Carrier. The hook carrier (Fig. 6.15*d*) transports one or two cartons or plastic totes. The carton or tote is placed onto the hook and extends downward. To reduce the carton or tote from accidentally falling from the hook, the hook is turned in an upward direction and lodged in one of the handle grips of the tote.

Employees place empty cartons or totes onto the flat carrier or hook that transports the carton or tote through the pick area. As required in the pick area, the order picker transfers the carton or tote from the carrier or hook onto the pick conveyor.

Disadvantages of this method are that it requires a higher elevation in the pick area, requires guarding in open areas, handles at a maximum of 24 to 28 cartons per minute, and is designed in the layout of the pick area. Advantages of the trolley method are that it reduces the order picker's nonproductive walking time, provides a constant supply, and when compared to the powered roller conveyor, is less expensive.

Determining the Best Storage-Pick Position

The information presented in this section helps you to determine the best small-item or replacement-parts storage-pick concept for your facility.

The definition of a small-item facility pick position is a location that contains a sufficient SKU inventory to satisfy at least one day of customer demand. In most small-item or replacement-parts facilities, the inventory is allocated to a pick (forward or active) position, a ready reserve storage position, or remote storage position. The ready and remote reserve positions hold inventory that does not fit into the pick position.

The six major factors involved to determine the best storage-pick concept for small-item or replacement parts are (1) storage-pick position type, (2) material handling system, (3) inventory control method, (4) customer order requirements, (5) building design, and (6) SKU characteristics.

The other factors that influence the storage-pick position selection are the type of operation, fire protection and seismic conditions, and order-pick and replenishment procedures.

Storage-Pick Position Objectives.   The small-item or replenishment-parts storage-pick objectives are to satisfy the throughput (volume) requirement, assure the proper product rotation, provide the best storage-pick density per aisle, provide the maximum SKU openings per aisle, and assure lowest operating costs. These objectives are similar regardless of the facility size or whether it has a manual (employee-walk or -ride), mechanized, or automated (product delivered to a pick area) concept.

To satisfy the storage-pick objective, numerous storage-pick concepts are available for use in a distribution facility. These concepts are reviewed in this section.

Design Parameters.   Prior to selection, purchase, and implementation of the storage-pick system, your design parameters are most important and are clearly defined to your design team. These factors are item (SKU) dimensions (length, width, height, and weight), storage-pick area size (clear height and bay spacing), floor surface and condition, on-hand inventory level (average and peak), withdrawal (pick) characteristics (single or multipacks), SKU velocity movement, replenishment method, packing characteristics (loose, individual packages, crushable, fragile), existing or proposed material handling equipment, labor productivity and availability, inventory rotation requirements, degree of security for inventory and safety, and fire protection and seismic requirements.

The concept that best satisfies the majority of your objectives is designed for your facility. Whenever possible, these design parameters are uniform. If these design parameters are not uniform, then SKUs with similar characteristics are grouped in one area within the facility.

For best results, the storage-pick concept design is made in conjunction with the fundamentals of good warehouse practice that dictates the SKU location within the distribution facility storage area.

General Rule of Thumb for Carton or Large Container or Tote of SKUs. A general rule of thumb for small-item (cases that contain 12 SKUs) storage-pick position criteria is illustrated in Fig. 6.16.

General Rule of Thumb for Bin or Container SKUs. If the SKUs are very small parts that have the size of $1 \times 1 \times 3$ in, then the storage-pick position criteria are as listed in Fig. 6.17.

The first objective of this section is to develop an understanding of the various small-item storage-pick positions. The second is to indicate their alternative designs and operating characteristics and to look at the ability of each storage-pick position to satisfy your distribution facility objective of effective use of space, efficient use of labor, and proper product rotation.

Case movement	Storage-pick concept
0–1 case	Shelving
1.1–3 cases	5-ft flow rack
3.1–5 cases	10-ft flow rack
5.1–10 cases	20-ft flow rack
10.1–40 cases	Standard rack or hand stack in racks
40.1–80 cases	Floor stack

*Case dimensions are 11 in width $\times$ 18 in length $\times$ 11 in height and 25 lb, and movement is cases per day.

FIGURE 6.16   General rule of thumb for storage-pick position for small-item SKUs.

Product movement	Storage-pick concept
0–4 items	$2 \times 18 \times 5$-in bin
4.1–8 items	$4 \times 18 \times 5$-in bin
8.1–12 items	$6 \times 18 \times 5$-in bin
12.1–14 items	$8 \times 18 \times 5$-in bin
14.1–18 items	$10 \times 18 \times 5$-in bin
18.1–24 items	$12 \times 18 \times 5$-in bin
24.1–48 items	Divided shelving
48.1–144 items	Shelving
3.1–5 cases	Flow rack
5.2–10 cases	Hand stack
10.1–40 cases	Standard pallet rack

FIGURE 6.17   General rule of thumb for storage-pick position for SKUs containing small parts.

Various Small-Item Storage-Pick Position and Single-Item Pick Position Replenishment Methods

The small-item (SKU) storage-pick position concepts are divided into the nonhanging and hanging types. The nonhanging type is reviewed first, and these include (1) floor stack, (2) wood racks or cardboard bins, (3) pallet rack, (4) hand stack in pallet rack, (5) versa (multilevel) shelving, (6) shelving, (7) mobile shelving, (8) case or tote flow rack, (9) push-back rack, (10) chute or slide, (11) bins on racks or board, (12) drawers, (13) kit, (14) carousel, (15) miniload, and (16) cantilever shelving.

One of the critical factors in a single-item pick system is the on-time replenishment of the SKU pick position. During the pick activity of customer orders, a vacant pick position indicates a stockout, nonproductive employee, and an unsatisfied customer. The various methods to replenish vacant pick positions include mobile equipment, roller conveyor, or tilt-tray conveyor methods.

When you use mobile equipment (Fig. 6.18a) to replenish your single-item pick positions, a unitized (pallet or cart) load of cartons is transferred from the reserve area to the replenishment side of the pick position. The pallet load contains one SKU or multiple SKUs, and each carton has a human- and machine-readable label.

From the location in the replenishment aisle, your replenishment employee transfers the carton from the pallet load directly to the replenishment side of the pick position or onto a manual pushcart for transportation or transfer to the pick position. At the completion of the transfer the employee records the replenishment transaction on a paper document or by a bar-code scanner.

Some disadvantages of this method are that it requires a large number of employees and mobile equipment pieces, has increased risk of product or equipment damage and employee injury, handles a low volume, requires a human- or machine-readable label, and has an increased potential for replenishment errors. Advantages of the method are that is handles all SKU types, shapes, and sizes; is inexpensive, and does not require management control and discipline.

The second carton replenishment method is the carton conveyor method (Fig. 6.18b). This method consists of a carton conveyor system that transports cartons from the reserve area to the replenishment side of the pick area. Each carton has a human- or machine-readable label on the carton. After the carton is labeled, it is placed onto a conveyor system for transport to the pick area. An alternative label method is to pick the carton which has a machine-readable label on the carton and have an automatic label application machine or jet-ink spray machine apply a human-readable code to the carton.

In the pick area, the replenishment conveyor transports the carton on the replenishment side of the pick position. As the replenishment employee reads the human-readable code, the carton is transferred from the conveyor to the pick position. The replenishment transaction is recorded on a paper document or by a hand-held bar-code scanner.

The replenishment transport conveyor has three alternative designs: (1) for a short length, run on a dead-end conveyor; (2) for a medium length, a recirculation loop; and (3) for a long length, a combination of sortation and transport conveyor for divert and transport to the specific zone.

Disadvantages of the method are that it requires an employee to handle the cartons and carries an increased risk of injury, requires a medium investment, and has increased potential for replenishment errors. Advantages of the method are that it handles a medium investment and requires fewer employees.

The third replenishment method is a hybrid method using a tilt-tray or Novasort

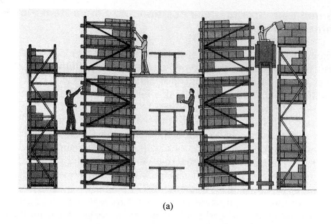

(a)

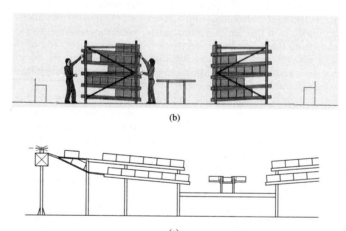

(b)

(c)

FIGURE 6.18 (*a*) Manual or mobile equipment replenishment (*Courtesy of Kingston-Warren Corp.*); (*b*) conveyor replenishment (*Courtesy of Kingston-Warren Corp.*); (*c*) tilt-tray replenishment.

system (Fig. 6.18*c*). In addition to handling the sortation activities, it handles the replenishment of SKUs. The tilt-tray or Novasort sortation system requires a conveyor network that transports the cartons from the reserve area to the sortation induction area and that each carton has a human- and machine-readable label.

After the carton is inducted onto the tilt-tray or Novasort sortation system, it is tilted (tipped) directly onto the replenishment side of the flow lane. An alternative transfer method is to have the carton tipped onto a conveyor system that transports the carton to the replenishment lane. This conveyor method was described in the preceding paragraphs.

Disadvantages of the method are that is requires a standard-sized and -shaped carton, a high investment, a replenishment carton accumulation or queue area, management control and discipline, and a human- or machine-readable label. Advantages of the method are that it requires few employees, handles a high volume, and assures accurate replenishment.

Floor-Stack Method.   The first storage-pick method is the floor-stack method, which is considered the most basic concept and requires a large floor area. This concept requires the lowest economic investment because it has an employee stack containers (master cartons) of SKUs on the facility floor, onto pallet boards, or in a wire-mesh container. This concept requires that the container be stackable, noncrushable, and able to support the stack weight. As the containers are stacked on the floor, the stack height should not exceed $5\frac{1}{2}$ ft, and each lane is one SKU. The exterior position is the pick position and the interior positions are the ready reserve positions. This concept provides a LIFO (last-in first-out) product rotation and few facings per aisle. It handles SKUs that have a large inventory, fast movement, high cube, and heavy weight.

The disadvantage of the concept are low employee productivity, product damages, poor access to the SKUs, and LIFO product rotation. Advantages are low investment, good storage density, and heavy or large-cube SKU handling.

Wood Rack and Carton or Tote Method.   The next storage-pick method utilizes the wood rack and carton system (Fig. 6.19*a*). This arrangement consists of 2 × 2-in wood top and bottom members and end upright members that are nailed together. For rigidity, several 1 × 11-in members are placed at the top and bottom. If required, additional 1 × 11-in members are attached to the end member as end panels. The wood members form a rectangle frame.

The SKU pick positions are created from the cardboard boxes that are placed inside the wood frame. To assure rigidity and uniform openings, each box flap is stapled to the wood members, stapled, and taped to one another, to form one unit. With the small pick position opening which holds a small inventory quantity, the preferred inventory allocation design is the floating-slot design.

Disadvantages of the wood frame and cardboard method are that some components are difficult to relocate, difficult to use in humid locations, handle only lightweight SKUs, limit the height to 7 to 8 ft, and have fixed opening size.

Advantages are low capital cost, best used with a floating-pick location design, easy to install, frame divided into pick positions, and excellent SKU hit density and hit concentration.

Standard Pallet Rack.   Next is the standard pallet rack, which consists of two upright frames and a pair of load beams for each storage level. There are at least two load beam levels in a rack opening to prevent rack sway. In the typical storage design, there are two to three pallet loads wide per opening and two openings high. The additional rack openings are the ready reserve positions. The first level is on the floor, and the second level is a nominal 40 in high above the first level. To reduce employee reach distance into the rack opening, the pallet load is placed with the 40-in dimension into the rack opening, or the pallet load is a 40 × 32-in pallet board. If aisles are narrow and a straddle truck is used in the operation, then the bottom pallets are raised onto a pair of load beams. This provides straddle clearance but increases the bottom height level by a nominal 12 in; therefore, to maintain a reasonable second-level height, you decrease the pallet height. An alternative that permits additional vertical levels is to use a ladder or high-rise order-picker truck for transaction activities.

Whenever possible, building columns and fire sprinklers are placed in the flue space between the back-to-back rows. The standard pallet rack design provides FIFO

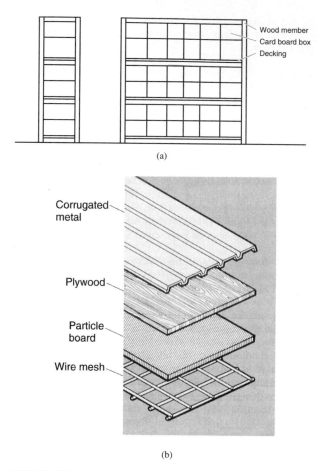

(a)

(b)

FIGURE 6.19 (*a*) Wood members and cardboard box rack arrangement; (*b*) various dock materials.

product rotation and small numbers of facings per aisle. It handles SKUs that have a high volume, heavy weight, medium cube, and a large on-hand inventory.

Disadvantages of the pallet rack design are low SKU hit density and hit concentration, two-levels height and requirement for large floor area. Advantages are good SKU access, few handlings, high inventory in pick positions, reduced reserve position requirement, and heavy and high-cube SKU handling.

Hand Stack in Pallet Rack or Wide-Span Shelving. The third storage-pick design is the hand stack of containerized or packaged product into a standard pallet rack or wide-span shelving bay opening. This concept provides the space to stack containers 20 to 36 in high per level. Wire mesh, wood, metal slats, or gravity conveyors (Fig. 6.19*b*) are placed onto the load beams to make up the SKU bottom support structure. To make smaller compartments and to separate SKUs, rack depth dividers (Fig. 6.19*c*) are installed into the hand-stack rack bay opening. When heavy product is placed on

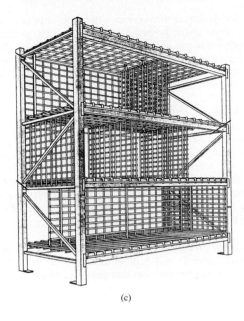

(c)

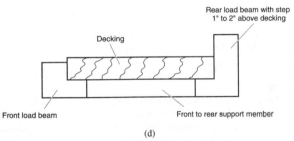

(d)

FIGURE 6.19 (*Continued*) (*c*) Various divider materials
(*Courtesy of Cargotainer.*); (*d*) high-step load beam.

the decking, then the decking is designed to support the load, or additional support cross members (front to rear) are placed between the load beams.

Tips and Insights on Handling Two- to Three-High Stacked Cartons in a Rack. When your operation calls for hand stacking of cartons of different SKUs in a rack storage bay, then the horizontal clearance, vertical clearance, and weight and height of the carton have an influence on the number of cartons per stack or bay.

The first factor is the horizontal clearance, which is the open space between two cartons or a carton and upright frame. This clearance (open space) compensates for the bulge in the carton sides. If you allow 1 to 2 in of clearance space, the employee is better able to move cartons and complete the storage transaction.

The vertical clearance is a nominal 6 in between a carton top and the rack beam. This space provides the employee with a clear space to raise the top carton of the stack and to remove the bottom carton. With the top carton raised up, the weight on the bottom carton is reduced, permitting an easy completion of a withdrawal transaction.

In a multiple-SKU-per-stack arrangement, the weight of the carton has an influence on the number of cartons in the vertical stack. If the carton is heavy (15 to 20 lb), and to maintain high employee productivity, then the carton stack should be two or three high. If the carton weight is heavier, then the carton stack is one to two cartons high. These arrangements permit your employee to raise the top carton and remove the required lower cartons.

Single-Deep Hand-Stack Cartons with Barrier Design.   The first alternative of the hand-stack rack design is the single-deep rack design that is in a 2- to 5-ft-deep rack. In the middle of the rack bay is a barrier (angle iron with the leg up, rope or pipe) that divides the rack bay length into two sections. One section faces one aisle, and the second section faces the other aisle. With one carton deep, it permits easy employee access from both aisles and handles a 24-in-long carton. When the rack depth is 25 in and has access from one aisle and there is decking on a step beam, then the load beam along the rear side has the step extend upward above the decking to act as a stop (Fig. 6.19*d*). This feature reduces material and installation cost.

Two-Deep Hand-Stack Design.   The alternative hand-stack storage-pick design is the back-to-back rack, with 3- to 4-ft-deep rack frames. With this rack depth, one or two cartons are placed into the position that provides two-deep storage-pick positions that face one aisle.

In the hand-stack rack bay, there is at least 5 to 6 in of vertical clearance between the master carton top and the bottom of the next level load beam. The horizontal clearance between two cartons is one inch. These clearances provide the sufficient clear space for the employee to efficiently perform transactions.

To use the cube space or increase the height of the hand-stack storage concept, it requires an order picker truck or ladder to access the storage positions.

With 12-in-wide cartons in a 9¼-ft-wide rack opening, the hand-stack design provides 7 facings per level; therefore, the normal two-level design has 14 facings per rack opening.

The hand-stack storage method provides LIFO product rotation, good SKU access, and a medium number of facings per rack bay. It handles medium- to heavy-weight and high-cube SKUs, SKUs with medium movement, and a medium on-hand inventory.

Disadvantages of the hand-stack method are that it increases product accessibility and handles heavy and high-cube SKUs and medium inventory in the pick position.

Versa Shelving.   The next storage-pick method is versa (multilevel) shelving that consists of 2- or 4-ft-deep upright frames, specially designed upper or lower load beams with post holes, shelf support devices, shelves, posts, and dividing material. These components divide the normal rack bay opening of 9¼ ft into smaller compartments designed as a single row with two-aisle access or back-to-back rows with one-aisle access to each rack opening.

Versa shelving upright adjustability has a nominal 1-ft-high by 1-ft-wide opening with a 1-in shelf bottom support member. Versa shelving provides 8 openings wide and 5 openings high for a total of 40 facings. If the uprights are taller, then the additional height is used for storage or to support a walkway or mezzanine. To access individual packaged or containerized product, a person rides or pushes a vehicle in the aisle.

Versa shelving provides good SKU access, medium facings per bay, and LIFO product rotation. It handles medium- to high-cube SKUs, slow-moving SKUs, medium to heavy items, and SKUs with a small on-hand inventory.

Disadvantages of versa shelving are higher capital investment and LIFO product rotation. Advantages of versa shelving are that it handles loose SKUs, provides good

SKU accessibility, handles heavy and high-cube SKUs, and has adjustable opening size and separate compartments for each SKU.

Standard Shelf Design. The next storage-pick design is shelf storage (Fig. 6.20*a*). A shelf bay consists of upright posts, shelves, and bracing. The upright post is the beaded or offset design.

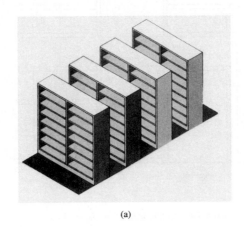

(a)

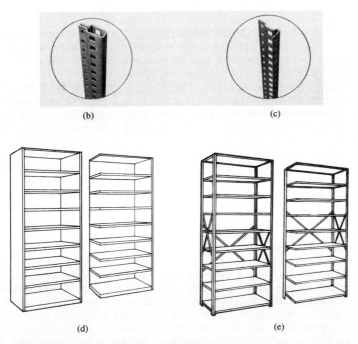

(b)   (c)

(d)   (e)

FIGURE 6.20   (*a*) Standard shelving (*Courtesy of SpaceSaver.*); (*b*) beaded post (*Courtesy of Inter Royal Corp.*); (*c*) offset post; (*d*) closed shelving (*Courtesy of Inter Royal Corp.*); (*e*) open shelving (*Courtesy of Inter Royal Corp.*).

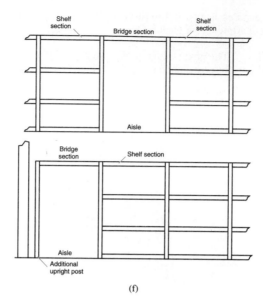

(f)

FIGURE 6.20 (*Continued*) (*f*) Bridging the aisle with shelving.

Beaded Posts. The beaded post design (Fig. 6.20*b*) permits the use of the entire shelf. When the shelf is used for bin box (container or tote) storage, then the beaded post is preferred because more of the shelf opening is available for the product. In a 3-ft-wide shelf with beaded posts, the usable shelf opening is 35½ ft.

Offset Posts. The offset shelf post (Fig. 6.20*c*) is in the shape of the letter "T." The T post design reduces the usable shelf space. With a 3-ft-wide shelf and a T post, the usable shelf opening is reduced by as much as one inch. This makes the usable shelf opening 35 in wide, due to the ½-in offset of each post, and is preferred for loose merchandise.

Closed or Open Shelving. With loose product or individual packages, the preferred shelf is the closed-type shelf (Fig. 6.20*d*) (with solid back and solid side panels). If the product is in large containers or cartons, then open-type shelving (Fig. 6.20*e*) is used with slat sides and back bracing. Standard shelf widths (down-aisle) dimensions are 3 or 4 ft, depth is 12 to 24 ft, and upright post height is 7 to 12 ft. A 1-in-high lip is extended downward and is included on the aisle side of the shelf. This permits storage position identifier label attachment. Clips or bolts and nuts connect the upright post to the various members.

At many facilities employees use the first, second, or third shelf level as a ladder to reach the higher pick positions. Heavy-duty or reinforced shelves at these lower levels reduce shelf damage. To control dust in the warehouse aisle, plates are installed to the bottom shelf lip. These dust plates extend downward to the floor. If dust or dirt collection on the shelf is a problem, then wire-mesh shelves are used on the levels.

Shelf bay openings are divided or accept containers that permit several SKU facings per bay. Shelves are designed as single rows or back-to-back rows. Both designs permit one aisle access. Building columns are designed at the end of the aisle or in between two bays. But to maintain correct alignment, a building column could fall in the shelf bay. If a mezzanine level or higher racks are designed in the area, then additional upright posts are spliced onto the heavy-duty bottom posts. This arrangement provides good space utilization, and to handle the product, an employee rides an order-picker vehicle or pushes a cart through the warehouse aisles.

Shelf storage is designed for small-cube, slow- to medium-moving SKUs, and lightweight SKUs with a small on-hand inventory. It provides good SKU accessibility, LIFO product rotation, and low facings per bay; but with divided shelves, the number of facings per bay is increased considerably. Shelf storage with three SKUs per level or six levels high provides 18 facings per bay.

Shelving to Bridge the Aisle. If your shelving storage-pick area requires additional carton or tote storage locations and you do not use mobile equipment that elevates above the top shelf, then you have an opportunity to increase your carton or tote storage positions. The increase is achieved by bridging the aisles with a shelf level (Fig. 6.20*f*).

If your shelf layout design has a middle aisle, then to bridge the aisle, the width between the one end post from each row must accommodate a shelf. The standard shelf length is 3 to 4 ft long.

If your shelf layout design has two end aisles, then to bridge these aisles, a new end (freestanding) post and a shelf must be connected to the last shelf post of the row to the new post. With the normal structural support shelf, the new end post is secured to the building wall and anchored to the floor; also, the shelf has a nut-and-bolt connection or other positive connection method to the upright post.

Disadvantages of the shelf storage concept are difficult handling of large or heavy SKUs, medium inventory in the position, shelves collect dust and dirt, and employees stand on the bottom shelf and damage it. (This can be reduced with heavy-duty shelving on the bottom.) Advantages of this concept are low capital investment, good SKU accessibility, handling of medium- to slow-moving SKUs, and product is separated and shelf storage handles loose or containerized product.

Mobile Shelving Design. The next storage-pick design is mobile shelving, which consists of standard single rows or back-to-back rows of shelves on movable bases. For access to the SKU storage position, one mobile shelf design moves the shelves into the main traffic aisle. The second mobile shelf design moves the shelves parallel to the main aisle. This movement creates a cross aisle for access to the appropriate storage-pick position. Mobile shelves are moved with an electric powered motor, mechanically operated mechanism, or manual power. When compared to the standard shelf storage-pick concept, mobile shelving reduces the number of aisles and has similar storage characteristics and design parameters, but it increases the storage density for the warehouse area.

Parallel moving mobile shelves (Fig. 6.21*a*) are designed with at least six back-to-back movable rack rows and one single-deep fixed rack row at each end of the module. The mobile shelves are electrically powered, or mechanically or manually moved to create an access aisle (Fig. 6.21*b*).

The design parameters and operational characteristics for mobile shelving are similar to those for regular shelving, except that mobile shelving is slow-moving and is best suited for very-slow-moving SKUs. Additional clearances of a nominal 1 to 2 in are designed for the movable base components.

Disadvantages of the concept are higher capital cost, slow movement to create an access aisle, and height restrictions. Advantages of the mobile shelving concept are dense storage, fewer aisles, and use for slow moving SKUs.

Various Bin Boxes. When you consider using a bin box in your small-item pick shelf or hand-stack rack design, there are three basic styles: chipboard box, corrugated box, and plastic box.

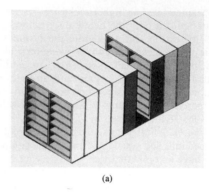

(a)

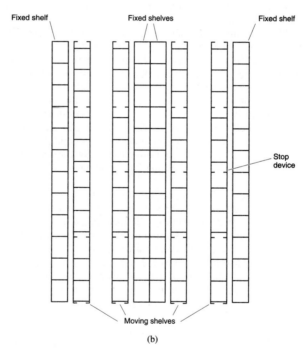

(b)

FIGURE 6.21 (*a*) Mobile shelves moving parallel to the main aisle (*Courtesy of SpaceSaver.*); (*b*) mobile shelves traveling in path along the front.

The chipboard box is a preformed box of thin pressed corrugated material that is designed to handle small, lightweight items. These boxes are available in a wide variety of sizes and have a full-height back and two sides that are full height. These sides are angled down to the front, which is a 1- to 2-in-high barrier.

The corrugated (cardboard) and plastic box is used to handle larger-cube and heavier small items. The corrugated or plastic box has two designs: the open face and the barrier at the face.

The *open-face* box (Fig. 6.22*a*) is a precut or modified standard box that has three sides, open top, open front, and a bottom. To provide strength and rigidity to the box, tape secures the bottom flap. This arrangement provides a bin box with an open face for easy and unobstructed product movement by an employee. With the open front, it requires the shelf lip or load beam to allow sufficient space for the pick position identification. The SKUs that are placed into the open-face bin box have the package characteristics to prevent it from sliding from the bin box.

The *bin box with a half- or quarter-high front barrier* (Fig. 6.22*b*) is a precut or modified standard box that has three sides, open top, and half- or quarter-high front barrier. The open-front flap is folded down and tucked under the bottom surface. To provide strength, tape secures the top and bottom flaps. This arrangement provides a bin box with a front barrier that retains SKUs which have a tendency to slide from the box and a location for the pick position identification. With low-SKU inventory in the bin box and a front barrier, the design enables the employee to remove product from the pick position.

Carton Flow Rack. The next storage-pick design is a carton (case or container) gravity flow rack that consists of frames, bracing, rollers (conveyor), guiderails (dividers), end stops, and posts. These components create individual container flow lanes. If ceiling height exists above the flow lanes, then the space is used for pallet storage or a mezzanine level. The standard case flow rack is designed with 5-ft facings (down-aisle) sections, 7-ft-high frames, and 120" and 60-, 120-, and 240-in-deep flow lanes. There are four different types of flow rack: (1) lay-back, (2) straight-back, (3) tilt-back, and (4) standard conveyor in standard pallet rack.

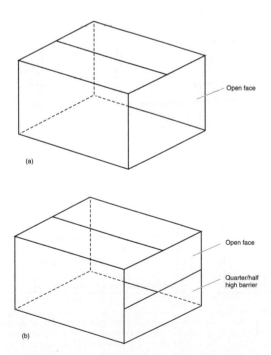

FIGURE 6.22 (*a*) Open-face bin box; (*b*) quarter- or half-high barrier bin box.

Lay-Back Design.   In the lay-back design (Fig. 6.23a) the lowest case flow lane level extends the furthest into the order-pick aisle and each subsequent flow lane level is set into the rack. This feature permits easy access to the SKUs; however, the design extends the top flow lane into the replenishment aisle. This combination of lanes makes the lay-back case flow lane width dimension greater than those in other case flow lane storage-pick systems.

Tilt-Back Design.   The tilt-back (lay-back with tilted shelves) case flow lane (Fig. 6.23b) has the lay-back frame, and at the pick aisle end of the flow lane, the flow lane is tilted (sloped) down. The tilt-back feature positions the case at an angle for easy order pick.

Straight-Back Design.   In the straight-back case flow rack (Fig. 6.23c) design the next-highest case flow lane is directly above the lowest flow lane. When compared to other case flow lane designs, it has a small width and permits excellent case handling.

Standard Conveyor in Pallet Rack.   An alternative carton flow rack arrangement is to have a skatewheel or roller conveyor section attached to the rack bay load beams.

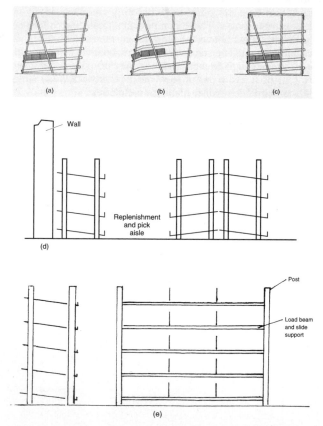

FIGURE 6.23   (a) Lay-back case flow rack (*Courtesy of Kingston-Warren Corp.*); (b) tilt-back case flow rack (*Courtesy of Kingston-Warren Corp.*); (c) straight-back case flow rack (*Courtesy of Kingston-Warren Corp.*); (d) push-back rack; (e) chute or slide rack.

(f)   (g)

FIGURE 6.23   (*Continued*) (*f*) Bins (*Courtesy of I.D. Systems Magazines.*); (*g*) storage drawers (*Courtesy of Modern Materials Handling Magazine.*).

As required, the rear load beam is elevated to provide a pitch or slope toward the pick aisle. An end stop on the conveyor stops the carton flow. A 9¼-ft centerline-to-center-line load beam contains eight boxes at 12 in width. With four levels height, the design provides 32 facings.

Case (Carton) Flow Rack Disadvantages and Advantages.   Case gravity flow rack is designed as a stand-alone system with replenishment and order-pick aisles. All cases and containers require a smooth conveyable bottom surface. (Note that most import cartons have a greater hang-up probability.) A combination of the flow rail slope (pitch) and the container weight allows gravity to move (index forward) the case through the lane. With a minimum height of four to five lanes high and width of four to five lanes, each flow rack bay provides 16 to 25 facings. A person pushes a cart, pushes a tote on a conveyor, or rides an elevating vehicle to replenish or order-pick product in the case flow rack design.

Disadvantages of the case gravity flow rack are a higher investment and hang-up of poor-quality cartons on the rails. Advantages are FIFO product rotation, large quantity of product in the lane, ease in reaching the product, and good SKU accessibility.

Push-Back Rack.   The next storage-pick design is a hybrid case flow rack, which is the push-back rack (Fig. 6.23*d*). If you have floor space, then the 5- to 8-ft-deep flow rack with end stops on both ends of the flow lane is designed as a push-back shelf. The push-back design has one aisle for both replenishment and order-pick activities. When compared to the standard flow rack, the push-back design requires a steeper slope to the rails.

The push-back design provides a LIFO product rotation and handles medium to small SKUs, small to medium inventory, medium- to slow-moving SKUs, and medium-weight to lightweight SKUs.

Disadvantages and advantages of the push-back design are the same as those of the carton flow rack except that the additional disadvantages are LIFO product rotation and limited product quantity in ready reserve.

Chute or Slide Rack. The next storage-pick design is the chute or slide (Fig. 6.23*e*), which has structural components similar to those of the standard rack designs. With the slide rack, each bay has a solid deck for loose merchandise or wire-mesh deck for containers with an end stop on the order-pick aisle side. When wire mesh is used, then the top wire strands are one piece and are run from front-to-rear (depth) of the rack. The slide is coated metal, plastic, or wood that is sloped (pitched) toward the order-pick aisle side. When compared to flow rack, the slide or chute concept has a steeper slope and raises the replenishment side of the slide. The slide concept provides a height of three to four levels for a total of 15 to 20 facings per bay.

The slide concept handles loose or containerized product, small on-hand inventory, medium to small SKUs, slow-moving SKUs, and medium-weight to lightweight SKUs. The slide rack design provides a FIFO product rotation and a medium number of SKU facings.

Disadvantages and advantages of the slide concept are the same as those of the carton flow rack except for an additional disadvantage that the slide rack has a smaller storage capacity and fewer openings.

Bin Design. The next storage-pick design is the bin (Fig. 6.23*f*), which is designed for placement in standard or mobile shelves or hooked onto a stationary or mobile board. The shelf bin arrangement consists of bins that are preformed plastic, metal, corrugated, or chipboard material. Bin length, width, and height are determined by the loose SKU's characteristics, shelf dimensions, and SKU inventory quantity. When designing this concept, you allow clearance between bin or box opening for easy SKU deposit and withdrawal by an employee's hand and SKU identification. If you do not require these two features, then the bin front has the space for the pick position identification and the other information appears on the container side. This practice reduces the order picker's productivity because the bin is partially removed to verify the SKU. But this arrangement increases the number of SKUs per shelf opening. Also, the exterior of the bin dimensions plus $\frac{1}{8}$-in clearance is used to determine the number of bin boxes that fit into a shelf opening.

Importance of the Bin Face Width. For good employee productivity and inventory control, the bin face width and height permit a person's hand to move in the bin and an identifier label to be placed on its face. Bin storage or pick concepts are used in an employee push or ride a vehicle system.

The bin storage-pick design provides a LIFO product rotation and large number of facings. With small bins, there is a possibility of 13 bins per level and 12 levels high for a total of 156 facings. The bin design handles loose or packaged product that is very-small-cube and slow- to medium-moving and handles small, lightweight on-hand inventory.

Disadvantages of this system are small inventory quantity, LIFO product rotation, difficulty in performing transaction with small bins, and additional bin cost. Advantages of the bin storage design are excellent suitability for very small items, maximum number of facings per aisle, good storage density and SKU accessibility, and mobile when attached to a dolly.

Storage Drawer Design. The next storage-pick design is the storage drawer (Fig. 6.23*g*), which consists of a large fixed wall container with many interior small fixed wall containers (drawers). These drawers are separated into numerous small compartments that have a lip for a SKU identifier. The concept provides 15 to 20 SKU positions per drawer and a height of 8 to 10 drawer levels for a total of 120 SKUs per large container. This is a very high number of facings. Drawer storage is designed for very small loose or packaged SKUs with a very small on-hand inventory. The other product characteristics are LIFO product rotation and very lightweight and slow-moving SKUs.

The drawer storage-pick arrangement is designed as a single row of drawers or back-to-back drawers with aisles of sufficient width to open drawers into. If required, the aisle width is sufficient to permit a second-order picker to pass as the drawer is open. If the storage-pick design includes an elevating vehicle, then the drawer units are stacked or placed into rack bays.

Disadvantages of the drawer design are small inventory quantity, increase in capital cost, and additional employee activity in opening drawers. Advantages of the design are that it is excellent for very-small or slow-moving SKUs, provides the greatest number of facings per aisle, has excellent storage density and SKU accessibility, improves security, and keeps dust and dirt off the SKUs.

Kit Storage-Pick Design. The next storage-pick design is the kit, which is used in a manual or automated storage and retrieval pallet load systems. The kit storage design consists of a unit-load container that has nominal 16 shelves (slots) that hold one container or a divided container. The kit pallet board with the containers provide storage-pick positions for small parts that are common components for a finished product or sale unit.

As required, the kit container is withdrawn from the storage position and placed onto a conveyor for transport to the workstation. After completion of the pick activities, the conveyor system returns the kit container to the storage aisle input station.

The kit container storage-pick system provides a LIFO product rotation and a medium number of facings in a small area. The concept handles SKUs that are small, have medium velocity movement, are medium to light in weight, and have a small on-hand inventory.

Disadvantages of kit storage are that product is stored in several locations and small inventory is in the storage position. Advantages are that this design reduces the number of trips for components for one product, facilitates filling of workstation orders, and utilizes the full pallet load storage concept.

Carousel Storage-Pick Design. The next storage-pick design is the carousel, which consists of an endless chain which has a series of bins, baskets, or trays that are attached to the chain. These containers are divided into smaller compartments. As required, the chain is turned by a motor-driven sprocket. The carousel is manually or computer-controlled to move in a forward or reverse direction. This chain movement brings the required bin to the workstation for the replenishment or order-pick activity.

The carousel provides a LIFO product rotation and high SKU facings per aisle. The carousel handles small- to very-small-cube items, slow-moving SKUs, and lightweight SKUs, and each SKU has a small on-hand inventory.

Various Horizontal Carousel Options. When you use a horizontal carousel in your storage-pick operation, you should consider several design options that increase employee productivity and space utilization: (1) a robotic tray (container) withdrawal-deposit transaction device (Fig. 6.24a) that automatically performs insert and withdrawal of trays between the carousel and a conveyor and (2) a stationary employee lift (Fig. 6.24b), which elevates your employee to a height that permits easy access to reach the elevated pick positions of a single- or double-stacked carousel.

When designing horizontal carousels, you consider employee access aisles to both sides of the carousels. In the event of a mechanical problem, this feature permits an employee to perform SKU transactions.

Also, if merchandise falls from the basket opening during rotation, then a removable barrier (netting) is attached to the opening, thus preventing the product from falling to the floor. The method of the netting attachment allows the barrier to be adjusted for inventory situation. Hooks or Velcro is the preferred attachment method.

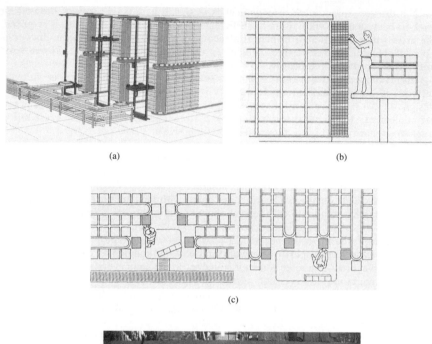

(a)

(b)

(c)

(d)

FIGURE 6.24 (*a*) Robotic tray; (*b*) employee lift; (*c*) multiple carousels (*Courtesy of White Storage and Retrieval Systems, Inc.*); (*d*) cantilever shelves (*Courtesy of Unarco Material Handling.*).

When determining the inventory capacity per bin, a reasonable bin occupancy is 50 to 66 percent of the bin space.

Disadvantages of the carousel concept are high capital investment, small inventory quantity in the position, and requirement for two shifts (one for replenishment and one for picking). Advantages are that product is delivered to the workstation, and the design requires only a small area and few aisles and provides dense storage and a high number of facings.

Horizontal Carousel Concept.   If your floor space is not a critical factor, then the horizontal carousel concept is designed for the facility. The horizontal carousel is designed with its endless chain that is driven by a top or bottom drive-and-sprocket arrangement. Most horizontal carousels are 40 ft long and 10 ft high. One carousel unit with 50 to 60 baskets holds approximately 500 to 1600 SKUs. The number of SKUs is determined by the SKU size, inventory quantity, and quantity of baskets per carousel. These arrangements (SKU positions) are one SKU per container or a divided container with at least several SKUs. The bottom of each container has a 5 to 10° pitch toward the rear. This pitch reduces product movement as the carousel is rotated to the required position.

Multiple Carousels Improve Productivity.   Multiple horizontal carousels (Fig. 6.24c) are controlled from one workstation, which increases the employee's productivity and equipment utilization because an increased number of SKU faces are available to the picker and there is a reduction in the picker's nonproductive wait time (carousel revolving time).

Vertical Carousel.   If your floor space is limited with the appropriate ceiling height, then the vertical carousel is considered for your facility. These bins rotate in a vertical direction past one or more workstations. These stations are on both floors. Most vertical carousels are 9 to 23 ft high, 3 to 5 ft deep, and 8 to 10 ft wide. If these trays are divided, then the vertical carousel unit holds approximately 500 SKUs.

Manually or Computer-Controlled Miniload.   The next storage-pick design is the manually or computer-controlled miniload, which consists of a series of upright frames, posts, support arms, and bracing. These components create the storage openings. A captive-aisle- and rail-guided storage and retrieval vehicle with an extractor and injector mechanism is used to withdraw and deposit containers between the storage positions. These containers are separated into smaller compartments that permit multiple SKUs per container or hold one SKU per container. The miniload holds approximately 600 SKUs and is 30 ft high and 40 ft long. Aisle width is determined by the container length dimension as it travels down-aisle (typical length is 48 in). The typical container is 24 in wide and 48 in long and handles a load of up to 600 lb or a 18 × 36-in tote with a weight of 250 lb. These containers hold one SKU or are multicompartment totes that contain multiple SKUs.

The miniload concept provides a LIFO product rotation and a high number of pick faces. The concept handles a medium-to-very-small-cube SKU, slow-moving SKU, and medium-weight to lightweight SKUs that have a small on-hand inventory.

Disadvantages of the concept are high capital investment and requirements for an inventory control program and a conveyorized system or vehicle to transport between pick stations and the miniload. Advantages of the miniload are like parts in one container, few aisles, requirement for only a small-square-foot area, utilization of the air space, and good security provision.

Various Tote Infeed-Outfeed Methods

In the first miniload design the container infeed and outfeed conveyors are in a U or horseshoe pattern at one end of the miniload. In the second design an infeed-outfeed conveyor network conveys the container to a remote pick station. In the third design the horseshoe infeed-outfeed stations are at both ends of the miniload system.

Cantilever Shelf or Peg Board.   When your long replacement parts are stored in a flat position and are not handled by one of the previously mentioned methods or designs, then a cantilever shelf or peg board storage-pick system handles the function. Cantilever shelving (Fig. 6.24d) consists of upright posts, top or back bracing, bases, and arms with shelves. With the shelving connected together, the storage position handles SKUs that extend beyond one shelf bay. If your merchandise is not stackable on the shelf, then a two- or three-sided container (open-ended basket) holds the merchandise in the storage-pick position. Also, metal, wire-mesh, or wood decking on the cantilever arms permits the hand stack of cartons or totes.

The peg board handles loose or packaged product that has a loop which is slipped over the peg. The peg permits the lightweight product to be stored in the full-length position and to handle a small inventory quantity.

Disadvantages of the cantilever shelf or peg board are that it requires a large area and holds a small inventory quantity. Some advantages are that this design handles long SKUs and SKUs that are stored flat.

STRUCTURAL CONSIDERATIONS FOR THE WAREHOUSE

When to Consider Installation of a Mezzanine

When your company handles or processes small items or a low volume of cartons and your operation requires additional space, then consider an additional level in your warehouse and distribution facility. Your company has two opportunities to consider an additional level: (1) during the design phase of a new facility and (2) when your company has an existing facility with sufficient clear ceiling height and is experiencing a growth in business which creates an overcrowded condition on the main floor.

The topic of this section is to look at the various alternative strategies for an additional level in your warehouse or distribution or processing facility. This additional level increases your company space to handle and process small parts or cartons.

In the design of a new or remodel of an existing warehouse or distribution facility an additional level is planned within a minimum of 15 to 16 ft of clear space above the finished floor or in a 8-ft space above an existing structural area (office). To achieve this design, the lights and sprinklers should fit within the ceiling joists or mezzanine floor support members.

Where to Locate a Mezzanine.   Reviewing the typical warehouse floor area in a 25-ft-clear-height building, we find that the required clear height exists above many of the key warehouse function areas and processing locations. Some of these areas and locations include (1) receiving, (2) shipping, (3) packaging and ticketing, (4) office, (5) picking areas (except HROS), (6) manufacturing, and (7) processing.

Reasons to Cost-Justify a Mezzanine.   Your company may consider the addition of another floor level because of the high land and building (capital) investment costs that are required for providing the additional required space on the existing floor level. The approximate cost for the construction of one floor level is $50,000 for an acre and $35 per square foot for building construction costs. The estimated economics for an additional level range from $10 to $25 per square foot of construction and installation (excluding lighting and sprinklers). From these cost factors, we conclude that the additional level is an economically attractive strategy to provide additional space.

Some other advantages of floor level addition are that it permits storage-pick-process area expansion with no additional land, increased security, increased inventory capacity within the building square-foot area, and reduced horizontal travel distances between key warehouse function locations.

Various Mezzanine Alternatives.   The various additional level alternative concepts are

- Additional constructed floor (nonmezzanine)
- Mezzanine

   Fixed mezzanine with lightweight aggregate-concrete-filled deck

   Freestanding mezzanine group of equipment supported with solid deck that is rack- (standard pallet or flow rack) or shelf-supported and equipment supported with grating that is rack-to-rack, shelf-to-shelf, or rail-to-rail

   Custom-engineered post supported with solid deck

   Construct an Additional Floor.   The first additional level alternative is to have an additional floor constructed, or have the required appurtenances incorporated during construction for a future level. This alternative is decided on in the floor or building design and construction phase of the project and has an estimated cost of $20 to $25 per square foot.

   Mezzanine.   The two basic mezzanine types are the fixed mezzanine and the free-standing mezzanine. These mezzanine designs are very flexible because they are installed during the initial construction of the building or at a later date inside an existing building. For a fast-track installation and less impact on overall warehouse operation, these mezzanine designs are considered as part of the initial building construction. This fact is especially true for the concrete-filled mezzanine.

   First, in the design and construction phase of a new building, the manager is aware of several facts. If an equipment-supported mezzanine is being installed in your new building and clear ceiling height permits a third level and is considered as a possibility for future expansion, then the manager follows these guidelines.

   With an equipment-supported mezzanine, all structural support members (posts) are designed to support the additional weight (load) and have the proper height for the splice and connection of the third-level rack and deck components without disturbing the second-level existing components and warehouse operation. In the case of adding a second level onto the first level, then the design parameters mentioned above are considered in the initial construction phase. In addition, the lights for the floor level are attached to the cross-aisle ties (support members) for the second floor. This practice permits an easy future mezzanine installation project.

   Second, if the new building does not have a mezzanine in the initial build but is considered for the future, then the anticipated finished floor for the mezzanine area has the floor construction designed with the additional footings to accommodate the future mezzanine posts.

   Third, in the initial build, make sure that there are sufficient finished floor-level emergency exits in place along the wall in accordance with code, thus reducing the need to construct fire exit tunnels.

Major Mezzanine Designs.   This section looks at each of the mezzanine designs on an individual basis. This review includes the following major mezzanine components: (1) support method, (2) decking, and (3) other criteria (handling product mix changes, handling equipment changes, and expansion capability).

When considering construction of a mezzanine, you should consider the following factors, which are generally similar among all the basic mezzanine concepts: equipment layout, movement of product between levels, movement of product on the mezzanine level, and codes and standards; therefore, each of these areas is reviewed as a general condition for all the mezzanine designs.

Concrete-Filled-Floor Mezzanine.   The first mezzanine design is concrete filled on a metal deck. This mezzanine design has upright structural members that are attached to the building columns. The decking support members are roof decking, beams, and joists. On top of these components is a concrete-poured floor. The type and depth of beams and joists and concrete depth are determined by the architect or structural engineer and is a function of the seismic area and imposed loads.

With a solid concrete floor that covers the entire area, the concrete-filled metal deck mezzanine is very flexible for handling product mix changes. It is the best mezzanine design to accommodate processing and material handling equipment changes or new technology; but expansion of the mezzanine floor is extremely difficult.

Rack (Equipment)-Supported Mezzanine.   The next mezzanine design (Fig. 6.25a) is the rack-supported mezzanine with a solid deck. When we consider a rack-supported structure, the three most frequently used rack types are (1) standard pallet rack (unit load or hand stack), (2) pallet flow rack, and (3) cantilever rack.

The rack-supported mezzanine structural members are (1) rack upright frames or posts, (2) standard load beams, and (3) specially designed floor deck support load beams.

On the floor level the upright rack frames with baseplates are spaced on relatively close centers with load beams (arms) to create an opening that holds pallet loads or hand-stacked product. The space between load beams (frame depth) is 3 to 4 ft, and the frame horizontal centers are nominal $9\frac{1}{2}$-ft centers.

Standard and special designed mezzanine support beams are attached at the top of the upright frames or posts to provide structural stability and support members (cross-aisle ties) for the metal roof deck and floor.

The solid deck on the second floor permits the area to be used for processing activities or a storage-pick position layout different from that in the floor-level arrangement. This feature permits maximum flexibility to accommodate changes in equipment or product mix. Mezzanine floor expansion is easily achieved by the attachment of additional upright frames, load beams, and deck material to the starter (existing) rack bays.

The next freestanding mezzanine design is the shelf or carton flow-rack-supported mezzanine with a solid deck. This design has the same features and components as the rack (upright frame)-supported design.

When compared to the rack-supported type, the shelf type has shelf upright posts that are on very close centers (nominal 2 × 3 ft deep and 3 to 4 ft horizontal), and the carton flow rack type has posts that are on close centers (nominal 5 × 10 ft deep and 5-ft horizontal centers).

The next equipment-supported mezzanine group is considered the double-decked type or is defined as rack-to-rack, shelf-to-shelf, or post or rail-to-post or rail.

In these equipment supported mezzanine designs the upright frame or post continues vertically upward from the finished floor to the required stacking height or structural stability height on the second floor. The concept provides a storage-pick arrangement on the floor level and the same pick position arrangement on the mezzanine level. In a building that has a 28- to 30-ft-high ceiling, it is very common with an equipment-supported mezzanine concept to have a third level.

Attached between the vertical structural members (uprights are cross-aisle ties) that support the decking and the load beams which create the rack bays for the product.

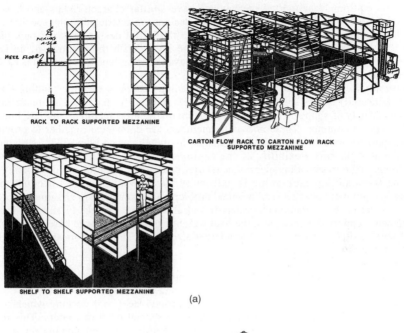

RACK TO RACK SUPPORTED MEZZANINE

CARTON FLOW RACK TO CARTON FLOW RACK SUPPORTED MEZZANINE

SHELF TO SHELF SUPPORTED MEZZANINE

(a)

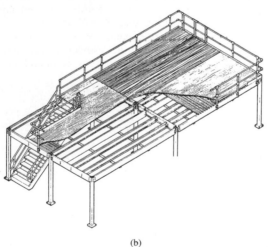

(b)

FIGURE 6.25 (*a*) Various equipment-supported mezzanines (*Courtesy of Warehouse Digest Magazine and Republic Storage Systems.*); (*b*) freestanding (custom-engineered) mezzanine (*Courtesy of Wildeck Inc.*).

For the employee walkways and aisles, the solid or grated deck material is attached to the load beams and cross-aisle ties.

Many rack-to-rack equipment-supported mezzanines have pallet loads that are placed in the rack bays by a lift truck from a replenishment aisle. The order selection activity is performed from the mezzanine walkways that are solid deck or grated-type floors.

These equipment supported mezzanines have similar characteristics and handle limited product mix changes. Typically, these designs are restricted to the specific use for the storage-pick walkway function because the initial design uses the rack pick positions both above and below the mezzanine level. With the addition of upright frames, load beams, and decking to the starter bays (existing bays), expansion is easily achieved in the area.

Freestanding Mezzanine.   The final mezzanine design is the freestanding (custom-engineered) beam- and post-supported mezzanine (Fig. 6.25*b*). This mezzanine design consists of support columns and horizontal structural members that are bolted to the support columns. These horizontal members are connected together to provide the support for long spans or heavier loads on the metal or wood deck. This structural arrangement permits flexibility of the equipment layout on the mezzanine level. Additions to the freestanding mezzanine (starter bay) are planned in any direction.

The freestanding mezzanine is extremely flexible and handles product mix changes, equipment changes, and minimal problems for expansion in any direction.

There are several common characteristics of all additional levels. First, when using equipment-supported mezzanines, the load weight can require a large baseplate on the floor level, a different-gauge steel, or a larger-sized post. These factors are considered for the floor rack and aisle layout.

Mezzanine Floor Deck Material.   Another important mezzanine component is the elevated deck support material, deck (floor) material, and design.

The most common deck support material is formed steel roof decking which is at least 16-gauge steel with 1½-in-deep ribs that are spaced on 6-in centers. This roof deck is secured to the structural members with self-drilling or self-tapping screws. If heavy equipment is anticipated on the mezzanine level, then the area is identified to the vendor, who provides additional floor support members directly under the equipment locations and a solid metal or wood floor deck under each equipment leg.

The deck (floor) surface consists of two major types: solid and nonsolid. The various deck materials include

- Solid group

  Plywood

  Plywood with a coating

  Plywood with Masonite

  Plywood with tile

  Polytexture

  Metal plate

  Solid grating

- Open group

  Grating (viz., bar or preformed deck grating, open-plank or grip-struct type)

  Perforated metal or expanded metal

The plywood deck surface is of a solid material (Fig. 6.26*a*). Disadvantages are that when there is high employee or vehicle traffic on the surface, it tends to wear quickly, and the bare plywood surface is difficult to clean. The advantage is that it is very inexpensive to install.

The plywood deck with a coated surface has the same operational characteristics as

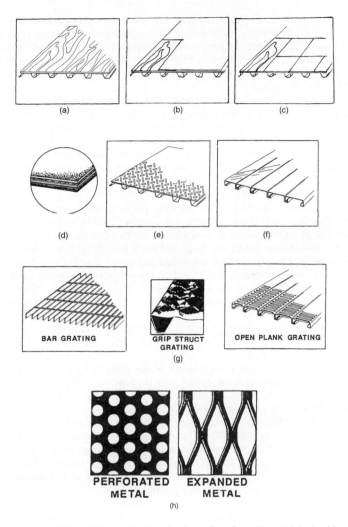

FIGURE 6.26 (*a*) Metal deck with plywood surface; (*b*) metal deck with Masonite surface; (*c*) metal deck with tile; (*d*) polytexture deck material; (*e*) metal deck with floor plate; (*f*) solid deck with plank grating; (*g*) various types of deck grating; (*h*) perforated metal and expanded metal mezzanine deck materials (*Courtesy of Plant Engineering Magazine.*).

the bare plywood deck surface. Additional advantages are that it is easier to clean and the surface reflects the light.

The plywood deck with a Masonite surface consists of a plywood deck with a Masonite overlay. The Masonite overlay is placed in the high-traffic aisles and provides a surface that resists the wear from high employee and vehicle traffic. The slight elevation change between the Masonite and plywood creates a housekeeping and possible employee injury problem (employees might trip and fall).

The plywood deck with tile surface (Fig. 6.26c) is a plywood deck with a tile cover. This solid deck surface has a long wear characteristic and an even surface for the entire area. These features permit easy housekeeping, no risk of employees tripping, and flexibility of equipment arrangement. This mezzanine requires additional installation time, even floor surface, and a mastic (adhesive) seal between the roof decking and the floor deck.

With the four previous floor deck types, the deck is fastened to the steel roof deck with flathead self-drilling or self-tapping screws that are spaced on nominal 12-in centers in both lateral and longitudinal directions of the plywood edges. The plywood deck material is a minimum of $\frac{3}{4}$ in thick, interior, APA class 1 or 2 plywood, and tongue and grooved on all edges with the C side up. Fire-treated tongue-and-grooved plywood with its high potential for uneven edges may create installation problems and result in an uneven mezzanine floor surface.

The next mezzanine deck surface is the polytexture deck material (Fig. 6.26d). This deck material has a textured high-density polyethylene overlay preattached to the plywood. This feature creates a smooth and even mezzanine surface with less installation time. The surface provides a good surface for flexible equipment arrangement. As part of the installation work, any gaps between two polytexture deck sections are filled with an epoxy filler and are smoothed even with the floor surface. The tops of the screws are coated with a coating to match the polytexture coating color.

The next solid mezzanine deck is the metal (floor) plate type (Fig. 6.26e). This deck consists of a solid metal plate that has a series of diamond shapes on its surface to improve employee safety.

The last solid deck is plank grating (Fig. 6.26f). This is a mezzanine deck that is used for walkways. For rigidity, additional cross-aisle members and the planks are crimped or welded together.

The first open-mezzanine deck concept is the grated plank type, which is the bar grating, grip-struct grating, or the open-plank deck grating (Fig. 6.26g). Each of these types provides sufficient support for an employee walkway and are used as the aisle decks in a rack-to-rack or shelf-to-shelf pick layout. When using these deck types, the concept requires additional cross-aisle structural members and for rigidity the planks are crimped or welded together. With local authority approval, a mezzanine with grating walkways may exceed the "one-third of the ground floor" criterion.

The next mezzanine deck materials are the perforated metal and expanded metal (Fig. 6.26h), which is a metal sheet that has a series of holes punched into it. This design provides a walkway between pick positions.

Methods to Protect the Open Space under the Mezzanine Pick Positions. The next mezzanine decking material is used in the pick methods to provide an employee safety surface under an open pick position bay.

In a small-item or carton pick layout, decking material under the pick positions and ready reserve positions protects an employee from falling through a vacant (empty) pick or ready reserve position. This area is covered by one of the following methods:

1. The entire mezzanine floor covering the deck area.
2. Securing the first pick position decking to the rack structural members.
3. Pick position opening decked with wire cloth, wire mesh, nylon netting, wire screen, or expanded metal. These materials provide sufficient coverage to support an employee or product that could fall into an empty rack bay.

When to Use a Closed-Deck or Open-Deck Mezzanine Floor. When you must decide between using a closed deck and an open deck on a mezzanine level, the following situations may provide some insight.

The solid deck is preferred in the following circumstances: (1) when carts or other mobile (wheeled) warehouse equipment are used on the mezzanine level, (2) when small items are handled on the mezzanine level, (3) when the equipment layout (arrangement, functions, or environmental conditions) on the mezzanine level is different from those on the floor level, (4) when the mezzanine-level activity has a high-traffic requirement, (5) when the mezzanine floor area is large, (6) when women in dresses are employees on the second level, and (7)when the local authorities permit your design to exceed the one-third area criterion with smoke detectors and sprinklers under the mezzanine deck.

The open deck is considered for the following situations: (1) when there is a short span for the walkway (aisle) requirement, (2) with a rack or shelf (equipment)-supported mezzanine, (3) when the function on the mezzanine level is the same as that on the floor level, (4) when only a few employees are working on the second level, (5) when heating and ventilation conditions are the same for the two levels, and (6) when the mezzanine exceeds the allowable area (in this case the grating deck would be preferred).

Moving Product between Levels.   The third major mezzanine component is the vertical transportation method that is used to transfer product between the two levels. The specific method is determined by the product characteristics and available building space. The various vertical transportation methods include lift truck with pallet load or cart that are transferred through a safety roll (Fig. 6.27*a*) or sliding (Fig. 6.27*b*) gate on the mezzanine level, powered belt conveyor, powered overhead trolley with hanging garments or with baskets that pass through a corral of kickplates and handrails (Fig. 6.27*c*), vertical lift or gravity-powered chutes or roller, or skatewheel conveyor. Chapter 9 reviews vertical transportation concepts in detail.

Building Codes and Standards.   The fourth major mezzanine component consists of building codes and standards, intended to improve the safety of the operation. Some of these codes and standards are as follow:

1. Kickplates and handrails should be installed on the perimeter and floor openings of the mezzanine. As the overhead trolley or conveyor breaks the handrail or kickplate perimeter, the handrail or kickplate extends inward to form a corral of at least 6 to 7 ft length. At the interior of the corral, there is a trip bar or kickplate in the product travel path but no handrail.

2. There should be sufficient employee exits and stairways.

3. When the additional level has a fire barrier penetration, the opening should have sufficient fire protection (deluge or dog house) and employee guarding.

4. The structural members should be designed for the seismic conditions and to support the loads. This includes a baseplate on the floor level to disburse the imposed load.

5. If the additional level covers more than one-third of the building square-foot area, then additional fire protection material (fireproof spray mineral fiber, cementitious dry-wall drop ceilings, or dry wall around the columns) should be added to protect the structural members. When employees work under a fiber-treated ceiling, a drop ceiling collects the fibers that periodically drop.

6. Provide sufficient fire protection, lighting, and head clearances.

7. If the mezzanine is adjacent to a mobile equipment traffic aisle, then add highway guarding to protect the structural support members.

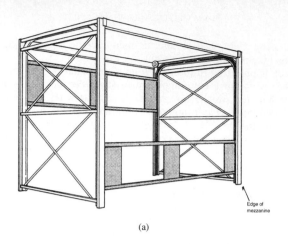

Edge of
mezzanine

(a)

(b)

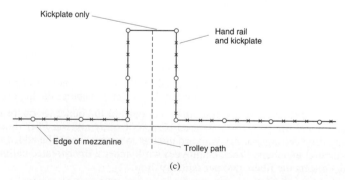

Kickplate only

Hand rail
and kickplate

Edge of mezzanine

Trolley path

(c)

FIGURE 6.27 (*a*) Mezzanine-level safety gate (*Courtesy of Roly SafetiGate, Inc.*); (*b*) sliding gate (*Courtesy of Plant Engineering Magazine.*); (*c*) corral for trolley path.

6.66

8. In warm climates or summer weather, provide sufficient air movement to assure a reasonable work environment.

A more detailed requirement is obtained from the National Fire Protection Association Code 231-C, BOCA 1987, insurance underwriter, national and local fire and building authorities, and your company security department.

Material Handling or Processing Equipment Layout. The final mezzanine component is the material handling or processing equipment layout. The major factors that determine the equipment arrangement are (1) type of additional-level structural design concept, including ceiling height and stairs; (2) weight of the equipment, product, and employees; and (3) warehouse and process functions and product flow.

Hanging-Garment Storage-Pick Design Parameters, Components, and Methods

When your distribution facility handles hanging garments, the hanging garments have two basic storage lengths: $3\frac{1}{2}$ or $3\frac{1}{4}$ ft to top of rail (TOR) for each level of short garments and $5\frac{2}{3}$ to $6\frac{1}{12}$ ft for long garments. These lengths are from TOR to the next garment or structural member. The width of most hanging garments is 27 in.

The two hanging-garment storage-pick methods are the static storage-pick method and the dynamic storage-pick method. With static hanging-garment storage-pick methods, the hanging garments are placed onto a structurally supported storage rail. In the dynamic hanging-garment storage-pick methods, the hanging garments remain on the trolley. The trolley is on a trolley rail.

Hanging-Garment Storage Rail Components. The operational design parameters for a hanging-garment storage-pick rail method is to provide the storage-pick positions for hanging garments. Components of the hanging-garment storage-rail system are (1) storage rail, storage rail support structure, and employee walkway. Standard pallet rack frames, cantilever rack posts, and tubular pipe are the support members for storage rail or tubular pipe. The pipe is attached to a single or double cantilever arm that is attached with screws or a self-locking device to the post and extends outward toward the aisle. The individual hanging garments are hung on this $1\frac{5}{16}$-in storage rail that has a support member every 6 ft. When long garments are placed into the storage area, then the storage rail is placed at 81 in height above the finished floor. When short garments are placed on the storage rails, the first storage rail is set at $3\frac{1}{4}$ to $3\frac{1}{2}$ ft above the finished floor and the second-level rail is set at $6\frac{1}{2}$ to 7 ft above the finished floor. Rubber cups on the rail ends reduce the risk of employee injury and garment damage.

Hanging-Garment Rail Capacity. The typical storage rail capacity is 150 lb/ft (linear foot) and has the capacity to hold 8 to 10 heavy garments (e.g., coats) per foot and 12 to 16 thin garments (e.g., dresses) per foot. If the storage rail arm does not permit the hanging garments to hang in front of the upright post, then the storage capacity is the internal dimension between the two upright posts. Both long and short garments require a 27-in storage depth dimension between the center of the two storage rails.

After the hanging garments are delivered by rolling racks or trolleys to the storage position, they are manually transferred from the transport vehicle to the static storage-pick rail. In the dynamic storage method, the employee activates a switch to direct the trolley onto the storage rail. The trolleys accumulate against an end stop.

Various Hanging-Garment Static Storage-Pick Methods. The basic hanging-garment static storage-pick rail method is one or two rails high above the finished floor. Variations of this are

* Floor-level storage methods

  One deep

  Two or three deep

  Raised walkway

  Rails above the pit

  Three rails high and a rolling ladder or long rod with a hook

  Rail module

  Hang rail in a rack bay

  Push-back

  Carousel

* Multilevel storage methods

  Multiple floors

  Mezzanine (tubular pipe or equipment-supported) with a floor that is regular level, raised walkway or platform, or raised above a recess (cavity)

  Order-picker truck with cantileveler or upright post rack, standard pallet rack, or tubular pipe

One-Deep with One or Two Rails High. The first floor-level storage-pick concept is used in a facility that has a low ceiling height. In this design the storage area has an aisle between two rows of storage-pick rails. The storage-pick rails are one or two high and one rail deep from the aisle (Fig. 6.28a).

Disadvantages of this method are that it requires a large area and provides low storage density. Advantages of the method include easy access to the storage positions, suitability for a low-bay building, and low capital investment.

Two-Deep with One or Two Rails High. An alternative design for a floor-level storage-pick method is the two-deep hanging-garment storage-pick design, which is similar to the single-deep hanging-garment storage design (Fig. 6.28b). The unique feature of this two-deep design is that the two-high position rails are fixed and the lower rails can be either fixed or removable rails. The structural design of the two-rail storage-pick concept requires that the cantilever arm have the design and structural support for the attachment of the two storage rails.

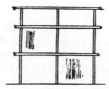 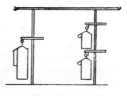

(a)

FIGURE 6.28 (*a*) One-deep-rail; one- or two-garment-high storage-pick arrangement.

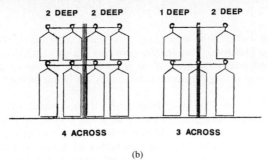

(b)

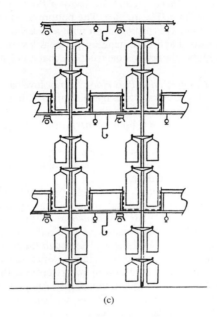

(c)

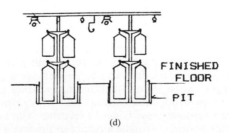

(d)

FIGURE 6.28 (*Continued*) (*b*) Two-deep-rail, one- or two-garment-high arrangement; (*c*) raised walkway; (*d*) storage-pick arrangement one rail deep and two garments high above a pit in the floor.

6.69

It is mentioned that the two-deep design is coupled with a single-deep rail to make three-deep storage positions. This feature provides design flexibility in the layout of your facility. Disadvantages of the two- or three-deep design are access to all product and inventory control of the garments. Advantages are dense storage and requirement for a smaller area.

One-Deep with One or Two Rails High with a Raised Walkway.   The next hanging-garment storage-pick design is the raised walkway (Fig. 6.28c). An elevated walkway or platform is placed in the aisle between the two storage rails. The increased employee height from the elevated walkway with a kickplate and guardrails permits the employee easy access to the second or third garment pick levels, but there is an additional cost and increased housekeeping problems.

One-Deep with Two or Three High Rails Above a Pit.   In this design the rails are raised above a pit in the floor (Fig. 6.28d). The bottom level of the hanging garments hangs into a pit in the floor. This arrangement with a kickplate and handrails on the pit perimeter makes the second and third hanging-garment levels easier for the employee to reach, but incurs higher building construction costs and increased housekeeping problems.

Rolling Ladder and Long Rod with a Hook.   In a low-bay building, a rolling ladder or long rod with a hook permits storage-pick transactions at three levels of short garments and two or three levels of long garments. With single-deep storage rails, the employee uses a rolling ladder or a hook to perform a storage-pick transaction at the higher levels.

A disadvantage of the design is a decrease in employee productivity because the employee walks or moves the ladder or hook to the requires aisle location. The advantage is an increase in the storage density.

Hanging-Garment Rail Module.   If your hanging-garment business is small-volume and you desire to use the available air space in your warehouse, then you design a hanging-garment rail module for your warehouse area. The hanging-garment rail module is a freestanding structure that consists of handrails; structural tubing, pipe, and angle iron members; and grating kickplates. The hanging-garment rail module has two or four rail levels with each rail level providing storage-pick positions for hanging garments. The tubing, pipe, and angle iron members provide the structural support for the hang rails, handrails, and grating and kickplates. An additional structural member or rope is on the far side of each handrail level to retain the pick position identification device. The grating, kickplates, and handrails provide a walkway for employees to perform transactions at the third and fourth rail levels. The warehouse finished floor provides the employees access to perform transactions at the first and second rails.

With these horizontal and vertical members and grating connected together and secured to these members, the result is a stable and sturdy freestanding rail module for hanging-garment storage-pick positions.

Hanging-Garment Bar in a Rack Bay.   When you have a low or medium hanging-garment volume that fluctuates, then you desire flexibility in your storage-pick area. The flexibility criterion provides that the storage rack opening be used for hanging-garment or hand-stack carton storage-pick positions. Since all SKU characteristics (hanging-garments and hand-stack cartons) and storage methods have different requirements, having separate areas is very expensive. The labor that is required to adapt the storage-pick position to handle a particular product's storage-pick characteristics is a tremendous expense to your operation. The storage-pick position characteristics are that the hand-stack carton requires a flat surface and hanging garments require a pipe for the hook of the hanger.

To achieve the storage-pick position design flexibility, the rack bay opening requires a solid or wire-mesh deck with two eyehole hooks that extend downward into the lower rack opening for a hang bar (Fig. 6.29a).

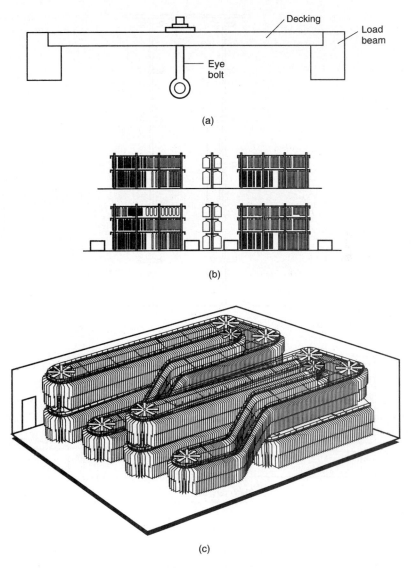

FIGURE 6.29 (*a*) Hanging-garment bar in a rack bay; (*b*) push-back hanging-garment rack (*Courtesy of Boyson, Inc.*); (*c*) hanging-garment carousel (*Courtesy of W&H Systems, Inc.*).

If wood, fiberboard, or metal is used as the deck surface, then two holes are drilled through the deck surface. The space between each hole and rack upright frame post permits one carton to be hand-stacked onto the deck. To secure the eyehole bolt straight portion onto the deck, the straight end of the bolt on the deck side has a large washer and bolt.

If wire mesh is used as the deck surface, then the straight part of the eyehole bolt is located in one of the openings of the wire mesh. The distance between the bolt and the

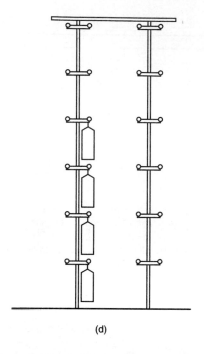

(d)

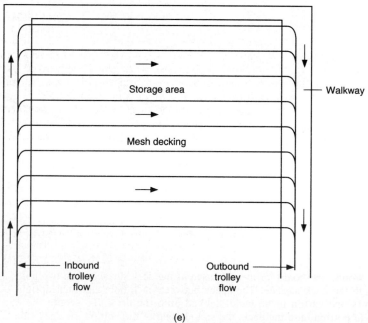

(e)

FIGURE 6.29 (*Continued*) (*d*) High-rise order-selector garment rack; (*e*) dynamic trolley storage.

upright frame allows for a hand-stacked carton. To secure the bolt, then several large washers are used to secure the bolt to the wire-mesh opening.

The bolt eyehole extends downward to a distance into the rack bay below. The hang bar (pipe) is inserted through the two bolt eyeholes and provides the hanging-garment positions. To reduce any pipe horizontal movement through the eyeholes, tape is applied to the pipe at each exterior side of the eyehole.

Hanging-Garment Push-Back Design. The hanging-garment push-back design (Fig. 6.29*b*) consists of a series of telescoping carriages that contain hooks for hanging garments. Each lane is two or three carriages deep, with the carriages facing the warehouse aisle. When the carriages are empty, they accumulate at the aisle end of the lane. As required, an employee transfers individual hanging garments onto hooks of the inner carriages. When the carriage is full, it is pushed back into the lane and is held in place by the second carriage. When the hanging-garment transfer is completed onto the second carriage, the carriage is pushed into the lane.

As the front carriage becomes empty, the second carriage is moved forward by gravity to the pick position.

A disadvantage of the push-back design is higher capital investment. Advantages of the design are that it increases storage density and number of SKUs (hanging garments) faces per aisle, and fewer aisles are required.

Hanging-Garment Carousel. The hanging garment carousel (Fig. 6.29*c*) uses the horizontal carousel design to provide the individual SKU (hanging garments) storage-pick position. The horizontal carousel is an electric powered endless chain that is top-driven, and the chain has the capability to handle individual garments on hooks. The carousel path is on both the floor level and the mezzanine level. The carousel convey-or path has an L, T, four-finger, U, E, or F shape.

As required, at the replenishment-pick station, an employee programs the carousel to rotate. The rotation brings the appropriate carousel hanging-garment position to the workstation. After completion of the transaction, the employee activates the carousel to revolve and bring the next required SKU position to the workstation.

Disadvantages of the carousel are that it requires a higher capital investment and requires controls and codes. Advantages are that it requires fewer employees; occupies less space with fewer aisles, which means increased storage density; reduces nonproductive walking time; and reduces the possibility of deposit and selection errors.

Multilevel Storage Methods. The next group of static hanging-garment storage-pick methods consists of multilevel storage arrangements. These various storage-pick arrangements are used in a building with a clear ceiling height above 15 to 20 ft. This clear space permits you to use the space for multiple levels of hanging-garment storage-pick positions.

1. *Multifloor building.* The first alternative is to use a building that has multi-structural floors. Each floor has the desired one-floor-level storage-pick arrangement. With various hanging-garment transportation methods this is a feasible alternative when the fire insurance and local codes do not permit the design of other storage concepts.

2. *Freestanding (pallet rack or tubular rail)-supported mezzanine with garments hanging in a cavity.* The second alternative is to construct a tubular or standard pallet rack (freestanding)-supported mezzanine. In this structure the rail storage arms extend outward from the tubular or rack upright members. Other arms that extend from the rack members are each level's floor surface support members. On these various mezzanine levels, the single- or two-deep storage rail design is used to store hanging garments. As another alternative, on these mezzanine levels the hanging garments are designed with a raised platform floor that has kickplates and handrails, or the rails

are above a recessed floor (cavity). If the recessed floor design is used on the mezzanine, then the cavity has a wire cargo container or solid decking installed along its sides and bottom. This material catches loose garments and employees who might fall or walk into the cavity. Some installations have solid material that is placed in the cavity such as wood or metal which serves as a fire barrier.

3. *High-rise vehicle with multilevel rails.*   The third alternative for multilevel storage is to use the manual high-rise order-picker truck (Fig. 6.29d) or another guided narrow-aisle vehicle to transport garments to the storage-pick positions. This design uses cantilever rack, standard pallet rack, or tubular pipe as the hanging garment structural support members.

Prior to implementation of the multilevel hanging-garment arrangement, your storage design and fire protection requirements are approved by your company's insurance underwriter and local authorities. These agencies could require additional fire protection equipment (sprinklers, smoke detectors, or fire barriers) in the storage-pick area.

Dynamic (Live) Hanging-Garment Storage-Pick Designs.   The various dynamic (live) hanging-garment storage-pick methods use the floor or mezzanine-level design (Fig. 6.29e). Both of these live storage concepts are similar in terms of design parameters and operational characteristics.

The live storage design consists of two walkways and trolley traffic rails, diverts, and stops. For a multilevel structure, the wire cargo container serves as a maintenance walkway and protective walkway surface between the two main traffic walkways.

An employee or powered chain pushes (pulls) a full trolley of hanging garments on the main traffic lane along the walkway to the assigned storage-pick lane. When the trolley arrives at the storage-pick lane, the divert mechanism is triggered and the trolley is diverted from the main traffic lane onto the infeed spur to the storage lane (branch lane). The storage lane at the entry end is set at a higher elevation and is sloped down at a 3°-angle pitch toward the end stop at the exit end. The trolley travel is by gravity-powered on this storage rail until it reaches the end stop, which is the pick position. Other trolleys accumulate behind this stopped trolley.

As required to remove the trolleys from the reserve lane, an employee adjusts the end stop and permits the trolleys to flow from the storage lane onto the main traffic rail.

Disadvantages of the dynamic (live) storage are higher trolley investment and inventory control. Advantages of this method are that it handles a large inventory level of one SKU and provides dense storage, and operation is easy.

SINGLE-ITEM SORTATION IS THE HEART OF A BATCHED PICK CONCEPT

The next section of this chapter on small-item handling discusses single-item (SKU) sortation. Single-item sortation is the heart (key) to the batched customer order-pick concept and ensures that your individual customer order-picked SKUs are transferred (separated) from a mixed group of customer order-picked SKUs.

The single-item sortation methods are divided into the manual or mechanized sortation groups. Each of these sortation methods requires a human-readable tag or human- and machine-readable label or tag. The human-readable concept requires a label or tag that is color-coded or has discreet alphanumeric characters. The mechanized sortation concept requires a human- and machine-readable label or tag. The mechanized sortation concepts require a product identification entry device and programmable controller (microcomputer) that controls the tracking device and sortation device. The design parameters and operational characteristics of your sortation concept are the same as your order-pick method parameters.

Various Nonhanging Single-Item Sortation Methods

The various single-item sortation methods are as follows:

- Manual sortation methods

    Pigeonhole or container—manual trolley

    Manual sort into a divided container onto a conveyor or cart

- Mechanized sortation methods

    Tilt tray

    Flap sorter

    SBIR

    Gull wing

    Bomb-bay drop

    Programmable trolley (hanging garments)

    Promech (individual hanging garment)

Manual Sortation to Pigeonhole, Trolley, or Container. The first SKU single-item sortation method is to manually sort SKUs into a pigeonhole (Fig. 6.30*a*), into a container, or onto a trolley. This method requires an employee to pick an individual SKU from a large quantity (en masse) of labeled SKUs. The employee reads the SKU label, matches the label code with the sortation location, and places the SKU into the appropriate customer location. This customer location is a pigeonhole (shelf), container, or trolley.

In the manual sortation layout, the batched customer order-picked product travels in front of the sortation-packing station. As the sortation employee or conveyor pushes and stops the cart, basket, container, or merchandise (en masse) on a belt conveyor, the employee transfers the appropriate SKU from the mixed customer group and places an individual customer's items in the appropriate customer sortation area.

In the manual trolley basket or hanging-garment sortation method the employee pushes or pulls the trolley on the rail that travels past the sortation-packing stations. With the manual method, each trolley and each sortation-packing station is identified with a manual identifier. For best results, the packing area layout has multiple packing lanes, each lane has a color-coded identifier, and the appropriate trolley has the appropriate color-coded identifier.

As the sortation employee pushes a trolley from the staging area to the packing lane, another employee matches the trolley code with the appropriate lane code. After the employee enters the packing lane, the individual SKU code is matched to the individual sortation-packing station code. After matching of the codes, the SKU is transferred to the sortation-packing station.

It is mentioned that this concept can also be used for carts or containers on a conveyor.

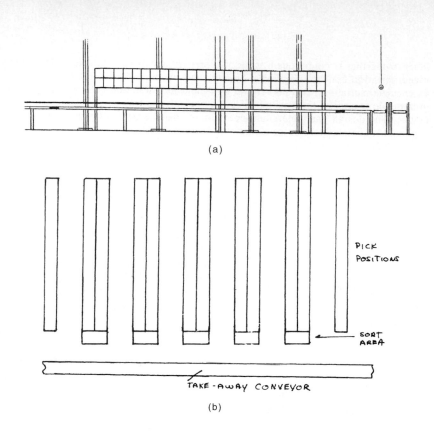

(a)

PICK
POSITIONS

SORT
AREA

← TAKE-AWAY CONVEYOR

(b)

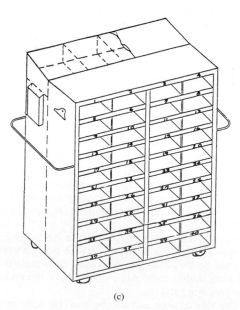

(c)

FIGURE 6.30 (*a*) SKU single-item sortation pigeonhole concept; (*b*) presort in the pick area; (*c*) sortation cart.

6.76

For best results and to achieve high employee productivity, the sortation-packing station lane must have transfer (bypass) turns and your high-volume customer pack-sort stations located on the exit side of the last sort position of the zone or lane. The customer sort-pack stations are arranged in a sequential order (low number to high number). Whenever possible, your layout has a presort area for sortation prior to the customer sortation-packing lane.

Disadvantages of this method are that it requires a large number of employees, a large floor area, and a large human-readable label and handles a low volume of merchandise with possibility of sortation errors; and with this method it is difficult to handle a large number of customers. Advantages are low capital investment, handling of all SKUs and a large volume per customer; also this system is flexible and easy to relocate, can be installed in an existing low-ceiling building, has a low impact on the product, and can be used for catalog or retail store distribution operations.

Manual Presort Product Prior to the Final Bin Sort. In the next manual sortation concept of batch-picked orders the order picker performs a presort in the pick area (Fig. 6.30*b*). In this method, the order picker picks several customer orders (batched orders). The batched orders are separated by aisle or zone. The two concepts for the presort of the batch activity are during the order-pick activity and post-order-pack activity. Both concepts require the order picker to place the SKUs into a temporary batch holding position.

Presort During the Order-Pick Activity. If sortation is performed during the order-pick activity, then the order picker pushes a cart (Fig. 6.30*c*) with separated compartments or a series of totes through the aisles. During the order-pick activity, the order picker places the merchandise onto or into the appropriate shelf or tote. When you divide your pick tote or container cavity into smaller compartments, then your tote or container has slots on the sides to hold these dividers. These dividers fit easily into the slots and have sufficient strength to remain in the slots of the divided container.

Presort Postpick Activity. In the second pick concept the order pickers pick the SKUs or merchandise en masse into the cart, tote, apron pouch, or container. At the end of the aisle, zone, or wave, the order picker transfers the individual customer order items to the appropriate end-of-aisle temporary bin sortation location. After completion of the zone or aisle, the cart, tote, or container with sorted orders is placed at the end of the aisle, pushed to another zone, or pushed to the pack area. The order picker proceeds to perform the order-pick activity for the same group of customer orders in another zone or starts a new group of customer orders.

When required to release the batch to the transport conveyor system, a consolidation employee pushes a series of containers on a consolidation conveyor or pushes the completed cart(s). Each group of customer containers or carts represent a specific batch. As the consolidation container carts pass each end of the aisle hold station, the SKUs are transferred from the customer order temporary hold position into the appropriate customer order container or cart. During the trip past these temporary hold stations as required, additional empty customer order containers are placed on the conveyor in front of the existing customer order full container. After all your customer's merchandise from the temporary hold stations is transferred to the transport containers, the appropriate batch containers are released to the transport take-away conveyor for transport, or the cart is pushed to the packing area. In the packing area, per batch, your individual customer orders are separated for customer packing.

Disadvantages of this method are that it requires a medium number of employees and a human-readable label, introduces potential sortation errors, handles a small to medium volume, requires double handling, and, if carts are used, requires a holding area. Advantages of the method are that it requires a low to medium capital investment, handles all SKUs, is flexible and easy to move, can be installed under a low ceiling of an existing building, has a low impact on the product, and handles a large number of customers and any size of order.

Mechanized Single-Item Sortation Methods.   There are 10 mechanized sortation methods. When considering implementation of one of these methods, you provide sufficient SKU accumulation prior to the sortation infeed station and between the sortation station and the packing station. The sortation station layout is in an arithmetic progression from the induction station, with even numbers on the right side and odd numbers on the left side of the sortation chain or conveyor. When designing the conveyor path in a closed or a L-shaped arrangement, an important consideration is future sortation (packing) station locations on the sortation conveyor path. These future sortation stations are located as the first sortation stations past the induction station or past the last active sortation station. Another important design criterion is that the packing station has one extra batch (third) sortation-packing sortation location beyond the first and second batch sortation locations. If the packing activity is behind schedule, then this feature permits the order-pick and sortation activities to function without interfering or mixing SKUs from two batches into one sortation location. Also, in your packing station numbering arrangement odd numbers are on the top and even numbers on the bottom in a sequential order. Other design considerations are dual-induction stations, which are suitable for sortation systems having tilt locations for shipping lanes and customer service locations.

Tilt-Tray Design.   The first mechanized single-item sortation method is tilt-tray sortation (Fig. 6.31a), which is an active-passive sortation method. This single (small)-item sortation concept consists of a series of trays that ride on a motor-driven closed-loop chain. The two types of motor-driven methods are the "bull wheel" and the "caterpillar drive." In the bull-wheel design the chain travels around the side of a large circular drive wheel that requires additional floor area and has a jerky start-up. In the caterpillar drive design there are one or several straight drive units under the chain. This design requires less floor area, has a smooth start-up, and attains higher speeds.

The trays are considered as the SKU-carrying platform and are designed to carry and tilt the largest SKU. New technology carries a large SKU on two trays. The direction of travel dimension for the tray varies from 17 to 27 in on centers with a 2-in space (gap) between trays. Heights of the tray sides are determined by the SKU characteristics, chain speed, and path (incline or decline slope). If your chain path inclines or declines by 15°, then the sides are higher to retain the SKU, which means a wider sort window. Moreover, a high-speed tilt tray requires a wider sort window.

To ensure that the chain maintains its proper speed and allows accurate tracking, the path is level and there are take-up devices to take up the slack (loose) chain. Slack chain results from weather changes and wear over time.

After the SKU is inducted (placed and product identification entered in the entry device) onto the tilt-tray system, the tray is under the control of the tracking device and programmable controller that electrically and pneumatically tilts a tray at the assigned (inducted) store sortation location. Your customer tilt location is on the right or left side of the tilt tray. Induction of the SKU onto the tilt-tray design is done manually (key station) or automatically by scanning a bar-coded label. The tray tilting action and travel speed cause the SKU to slide from the tray into the sortation location. With the present communication network, chute door designed with a flipper device(s) and mechanical technology, the sortation chute has multiple openings (Fig. 6.31b).

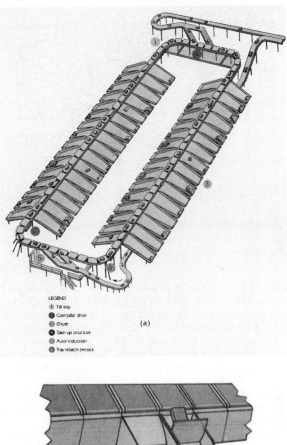

LEGEND
1 Tilt tray
2 Caterpillar drive
3 Chute
4 Take-up structure
5 Auto-induction
6 Tray relatch (recock)

(a)

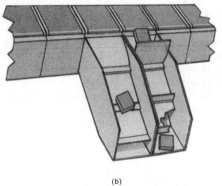

(b)

FIGURE 6.31 (*a*) Tilt-tray sortation (*Courtesy of Kosan Crisplant.*); (*b*) multiple-opening chute (*Courtesy of Litton Indl. Automation.*).

After the tray is tilted and prior to reaching the induction station, all trays pass a cleanout chute that tips all the trays. This action dumps all unsorted SKUs from the trays into the cleanout chute. The next device relatches (levels) each tray. In this position, the tray accepts another SKU.

Small-Item Divert Option. If your active small-item divert device diverts SKUs at

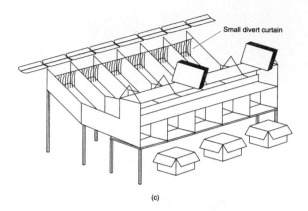

FIGURE 6.31   (*Continued*) (*c*) Small-item divert curtain.

a high speed into a chute and causes product damage, then to reduce product damage you are required to slow the SKU divert speed. With a wide variety of SKUs (sizes and weight) in your inventory, setting divert device speed at one speed to handle light-weight and heavy SKUs sometimes is very difficult. An alternative is to install a plastic strip curtain in each chute path. From the divert point, these plastic strip curtains are located at least one-third of the chute length. The curtain strips extend from a rod at the chute top downward to a length that allows for a clear travel path along the top of the chute surface. The strip arrangement slows the high SKU travel speed, thus reducing potential product damage and permitting product to flow through the chute.

With dual induction, the tilt tray handles an additional pick-sortation volume, shipping-sortation volume, or return-to-stock-plus-sortation volume.

The tilt-tray sortation system is capable of handling 65 to 200 sorts per minute, the package weight range is up to 25 to 40 lb, the trays are 17 to 27 in on centers, and the package impact in the chute is medium to rough.

Disadvantages of this concept are that it requires a human- and machine-readable label, requires a high capital investment, requires management discipline and control, and is not suitable for handling very-low-profile or round-shaped SKUs. Advantages of the concept are that it handles a high volume, assures accurate sortation, allows batched order-pick activity, handles a wide range of SKUs, sorts to the left or right, and permits dual induction and multiple sorts per location.

Novasort System.   The Novasort (Fig. 6.32) is a modular tilt-tray sortation mechanism that has a closed-loop fixed travel path. The components are (1) a train of three to four tilt trays that have a 500-lb load-carrying capacity and tilts to the left or right, (2) a programmable controller for each train or tilt trays, (3) a fixed travel path which has an electric powered bus bar, and (4) on-board sensing devices to detect objects or employees in the travel path.

The Novasort vehicle is a ceiling- or floor-supported device, travels up slight grades, and travels over 90°- and 180°-angle curves. The path directs the tilt-tray train past an induction station and delivery (sortation) stations. At the induction station, the induction employee places the product onto the tray and keys the tray tilt location into the train individual programmable controller. At the delivery station, the programma-

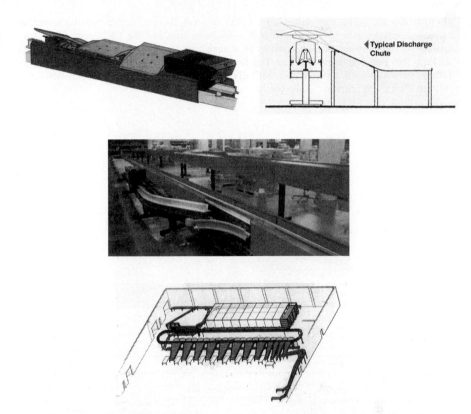

FIGURE 6.32 Novasort tilt-tray sortation device. (*Courtesy of Nova-Sort.*)

ble controller has the appropriate tray to tilt and discharge the product onto the assigned station.

With the on-board programmable controller, a train of trays is unloaded and loaded automatically or by the workstation employees. After all trays are tilted at the required stations, the train of tilt-tray carts travels on the path, with the carts accumulating prior to the induction station, or continues traveling to the maintenance spur. On this maintenance spur, the tilt-tray carriers wait for the next assignment.

As an option, the tilt tray has a fixed load-carrying surface and has the capacity to handle single-tray items, small items, and dual-tray items. As required to satisfy expansion of the area, additional track is added to the existing travel path.

Design parameters and operational characteristics of the tilt-tray train concept are similar to those of the tilt-tray sortation concept. Induction is manual or automatic with a tray that travels on a fixed closed-loop rail path between two locations. Disadvantages of the Novasort train of tilt-trays system are that it handles a medium volume, travels a fixed closed-loop path, and high capital investment. The single travel path could create queues, travel path design requiring employee and equipment access to and egress from the workstation. Advantages are transport of product as required to the workstation, ceiling- and floor-supported design, unloading and loading at the workstation, and easy expansion to the travel path (track).

Flap Sorter. The flap sorter (Fig. 6.33) is a small-item, active-passive sortation apparatus. It consists of a series of belt conveyors that are the SKU conveying surfaces. This conveying surface consists in a series of fixed surface conveyor sections and the flap belt conveyor sections. The flap sorter is designed as a unidirectional sortation conveyor with a fixed "did not" divert location or as a closed-loop conveyor system with a recirculation conveyor.

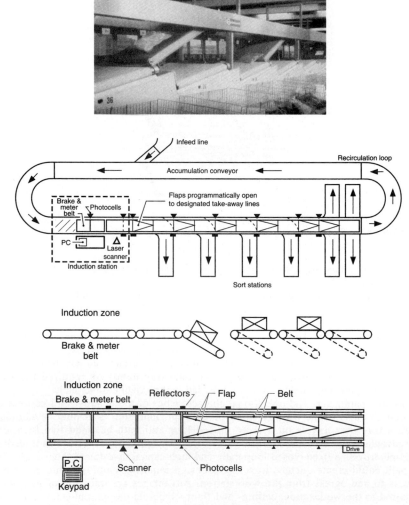

FIGURE 6.33   Flap sorter. (*Courtesy of Olympia Conveyor.*)

After the coded SKU label is read by a scanning device and is communicated to the programmable controller, the programmable controller ensures that the SKU is sorted at the appropriate discharge location. This discharge divert location is directly past and under the flap belt conveyor section.

As the SKU arrives at the discharge location by traveling over a series of fixed and flap conveyor sections, the tracking device and programmable controller device trigger the appropriate flap conveyor section to swing down. This swing-down action, with a forward travel motion and acceleration due to gravity, discharges the SKU into the appropriate customer packing station chute or infeed conveyor.

Disadvantages of the flap sorter are that it requires a high capital investment and one-way sort. Advantages include handling of batch-picked SKUs, flat and low-profile SKUs, and a high volume.

SBIR Method. The SBIR automatic sortation method (Fig. 6.34) is an active small-item sortation system that consists of a series of individual (divided) carrying surfaces (segments). These carrying surfaces are conveyor belt platforms that are mounted on wheels that are designed to travel in a closed-loop pattern. Each conveyor belt surface travels past an induction station. At the induction station, a SKU is placed onto a carrying surface. A coded label is read by a scanning device that programs a programmable controller and is assigned to divert at an assigned customer discharge location. The customer discharge location is on the right or left side of the carrying surface. Each location is designed with one-, two-, or three-deep controllable flaps (flippers) that cover a chute or customer sort location. As required, these chutes open to accept the appropriate diverted SKUs.

As the belt-driven conveyor carrying surface arrives at the assigned customer divert location, the programmable controller activates the carrying surface conveying belt to turn in the proper direction, to either the left or right of the belt. The carrying surface conveying belt turning motion and gravity cause the SKU to move from the carrying surface to the customer packing station. Arriving at the induction station, the carrying surface is ready to receive another SKU. The SBIR automatic sortation programming is achieved by an operator or by a scanning device reading a human-readable bar-code label. With a single-induction station, the concept handles 10,000 to 14,000 sorts per hour with a SKU weight of up to 44 lb.

Disadvantages and advantages of the SBIR sortation system are similar to those of the tilt-tray system, except with the additional disadvantages that with the SBIR it is difficult to handle round objects, and there is increased capital investment, and increased moving parts means increased maintenance. An additional advantage of the SBIR is that it requires less floor space and permits easy expansion.

Gull-Wing System. The gull-wing sortation system (Fig. 6.35) involves active-passive sortation. The gull-wing system consists of a series of single-item carrying surfaces that ride on a closed-loop chain. The load-carrying surface is in the shape of a bird's (gull) wing or a V-shaped carrier.

The operational characteristics, disadvantages, and advantages are similar to those of the tilt tray. After the gull-wing carrier arrives at the proper sort location, a tipper is activated that tips the carrier. This tipping action and gravity allow the single small item to slide from the carrier to the packing station. The gull-wing sortation system handles a wide variety of SKUs and performs 12,000 sorts per hour.

Bomb-Bay Drop System. The next single-item sortation system is bomb-bay drop sortation system, which is a series of platforms that are side-mounted to a closed powered loop chain. The platform is the individual SKU carrying surface. The entire batched group of customer-ordered quantity for a particular SKU is order-picked and is queued for release as a slug to the bomb-bay drop induction station. At the induction station, the computer addresses each individual SKU to a specific platform or has

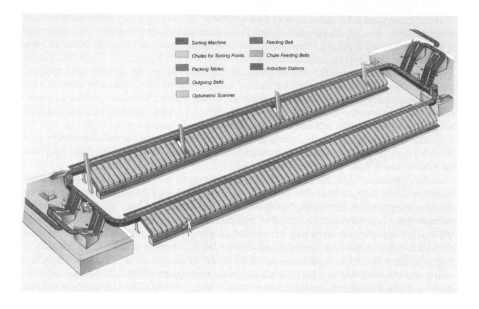

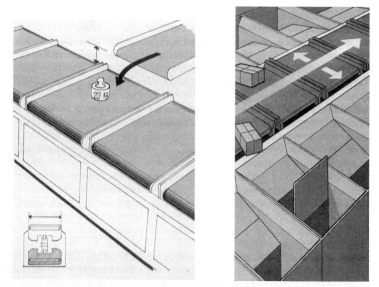

FIGURE 6.34   SBIR automatic sortation (*Courtesy of Litton Indl. Automation.*).

a bar-code label scanned by a scanning device. As required to complete the entry of the order-picked SKU, an individual platform receives an individual SKU. As the SKU passes a reading device that verifies whether the tray is occupied by a SKU, the computer addresses each appropriate platform with a SKU to a specific customer sortation station. As the platform leaves the induction station, the platform is tracked by a programmable controller. When the appropriate platform arrives at the assigned sorta-

FIGURE 6.35   Gull-wing active-passive sortation (*Courtesy of S.I. Handling Systems, Inc.*).

tion station, the platform carrying surface divides in the middle, allowing the SKU to drop into its shipping container or packing station. This platform's opening action is similar to an airplane bomb-bay door opening (Fig. 6.36).

The platform doors are relocked (closed) prior to arrival at the induction station.

Disadvantages of the bomb-bay drop are that it handles SKUs with a limited weight, requires a capital investment, and is a unidirectional sorter. Advantages are that it handles batch-picked SKUs and separates merchandise by family grouping.

Hanging-Garment Sortation Methods.   The next group of sortation methods are designed to handle hanging garments. Some of these methods handle hanging garments on trolleys or individual hanging garments.

Programmable Trolley.   The programmable trolley (Fig. 6.37a) is the device that is used to sort hanging garments on trolleys. The trolley consists of a load bar that carries 25 to 50 garments and is attached to two wheels. The wheels ride on a rail that is pulled by a powered chain. The chain has a metal downward-extending pendant

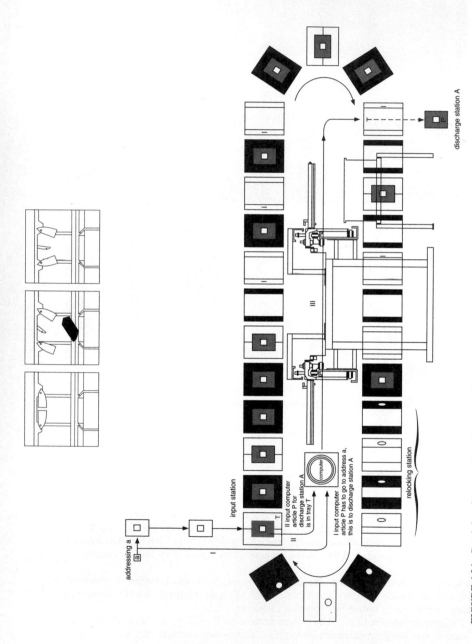

FIGURE 6.36 Bomb-bay drop sortation (*Courtesy of Bushman Co.*).

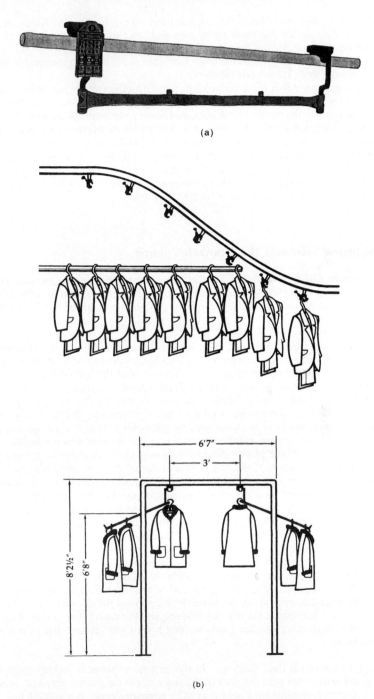

FIGURE 6.37 (a) Programmable trolley (*Courtesy of SDI.*); (b) promech hanging-garment sortation devices (*Courtesy of SDI.*).

("pulling dog") that pulls the trolley of garments over a fixed path. Along the fixed path is a reading device that reads the trolley code and triggers a divert device on the path to divert the trolley from the main travel path to a branch spur.

The three types of programmable trolley devices are (1) sliding retroreflective tabs, (2) wire prongs, and (3) bar-code labels.

Disadvantages are that there are a large quantity of garments per trolley and it is difficult to efficiently handle single SKUs. Advantages of this system are that it handles a high volume, the trolley is reusable, and some code methods are reusable.

Promech Hanging Garment Sortation.   The promech overhead hanging-garment sortation device (Fig. 6.37*b*) is an automatic individual hanging-garment sorter. This system consists of an infeed station, clasp-levered hook that is attached to a rope conveyor, a solenoid magnetically operated divert station, a slick rail, and a computer system. This system delivers the individual hanging garment to the required packing station. It handles a slug of SKUs and sorts 5000 per hour.

A Ring (Hook) Increases the Promech Volume

If your customer packing station is to receive more than one hanging garment per customer order, then a ring (hook) that holds up to six garments is used to infeed the garments. This feature increases the throughput (volume) to approximately 20,000 per hour.

Promech Manual Infeed Station.   The infeed station is a manual or automatic garment infeed station. All garment hooks are attached to the closed-loop rope conveyor system. The closed-loop conveyor system travels past the infeed station and each divert (packing) station. At the manual infeed station, the rope conveyor declines, causing the clasp hook to open for the infeed of a garment. The employee enters the hanging-garment SKU identification number into the terminal. On the display screen, a customer order quantity or the total required garments appear to the infeed operator. The first active divert station number that requires the garment appears on the screen. As directed, the employee places the garment onto the hook (ring of garments) and presses the dispatch button. If the ring device is not used in the operation, then multiple SKUs per customer location requires multiple listings and sending of the garment.

The hanging garment on the hook passes a photo-eye that activates the appropriate solenoid magnetic divert device. The divert device is the first divert station that requires the SKU. As the garment approaches the appropriate divert station, the solenoid magnetized divert station lifts the trip bar in the direct path of the levered hook wheel. As the hook wheel travels onto the elevated trip bar, and at this location, the wheel travels up the trip bar diverter, permitting the hook clasp to move upward. The action releases the garment (ring) onto the slide rail for travel (slide) to the packing station.

As the hook wheel passes to the upper end of the trip bar, it forces the trip bar to the lowered position, permitting the hook device to pass the divert device. Also, this discharge action disengages the trip bar magnet, and the divert station is in the neutral position, thus permitting another garment on a hook to travel onto the next assigned divert location.

Promech Automatic Infeed Station.   In the automatic garment infeed concept, one garment at a time is fed onto the path of the hook of the rope conveyor. The automatic infeed station is controlled by a computer. The computer controls a spindle that is driven by a motor. The control system regulates the spindle speed in sequence with the speed of the hook travel on the rope conveyor. The path of the rope conveyor elevates and declines prior to the infeed station. This decline permits the clasp hook device to open because the location and weight of the wheel forces the clasp open. When hook is open, a garment slides onto the hook.

Special sensing devices are in the spindle area to determine the existence of two hangers on the same spindle. If this situation occurs, then an alarm is sounded and the system is stopped to prevent damage. An employee removes the hanger and reactivates the system.

AUTOMATIC IDENTIFICATION METHODS

As indicated in the automatic identification section of this chapter, all sortation methods require human- or human- and machine-readable labels. These code groups are described in Chap. 13.

Label Codes

Human-Readable Codes.   The first identification is the human-readable label group, which consists of a group of alphabetic characters, numeric digits, or colored labels. These codes are used in combination with one another.

Human- or Machine-Readable Codes.   The second group is the human- or machine-readable code, which consists of the bar-code label, a retroreflective tab, or a wire prong. The bar-code label consists of a black line with white spaces, which is the background. The retroreflective tab reflects a light beam back to a reader. The wire prong extends outward and passes through a reader that triggers a divert device.

Packing Station Activity

After the SKU sortation to the packing station, the next activity is single-item packing. Your customer order packing activity is one of the last major warehouse and distribution functions in a single (small)-item warehouse and distribution facility. The packing activity assures your company that the merchandise matches your customer order and that during delivery by the transporting vehicle, the package protects the merchandise from damage or from becoming lost. Also, packing assures your company that your customer's correct delivery address is clearly stated on the package exterior.

The particular industry of your company determines how the merchandise is handled at the packing station. The two major industry groups are (1) direct customer contact (catalog or direct mail) and (2) retail store customer.

In the Catalog or Direct Mail Business the Package Appearance Means Your Business.   If you are in the direct customer contact industry, the package exterior appearance (container and label) is your customer's second physical contact with your company. The first physical contact was from your catalog appearance; therefore, a quality package appearance improves your customer's satisfaction.

The typical direct customer contact industry order size ranges from 1 to 4.75 merchandise pieces. With this order size to achieve an excellent order-pick productivity and smooth product flow, your company batch-picks customer orders that contain multiple SKUs. A batched order-pick activity requires a group of mixed customer orders (batch) that is accumulated and separated at the packing station by each individual customer order. Another packing station activity is to ensure that the picked merchandise conforms to your company standards (quality check) and is in the correct SKU quantity per the customer's order.

Operating a Pick-Pack of Singles. If you are in the catalog or direct mail business, then your company reviews the order data to determine the volume of single-item orders for specific SKUs. If you have a large quantity of single- or two-item customer orders, then you consider a pick-pack operation. In this specialized operation an employee picks the SKU en masse and packs it with an invoice in the customer shipping container. The individual customer order container is sealed, addressed, and placed onto a shipping take-away conveyor.

If your business is in the retail store distribution industry, then there exists one customer per multipiece order; therefore, your packing employee's activity is to verify that the correct merchandise conforms to your company standards and is in the correct quantity, and that the package is secured and labeled (addressed) for shipment.

Customer Order Check Methods

No matter what the industry or customer handling method, to ensure that the picked merchandise is correct and the picked quantity is correct, there are two inspection methods: manual (visual) and bar-code scanning.

Manual (Visual) Scanning. The manual (visual) scanning method requires a packing employee to make a visual comparison of the actual picked merchandise and the packing slip items. If there is a match, the packing employee prepares the package for shipment. If there is no match and there is a shortage of picked merchandise, the packing employee places the picked merchandise and packing slip together onto the "problem order" shelf. If there is no match because of an excess or surplus of picked merchandise, then the packing employee prepares the package for shipment and places the surplus picked merchandise onto the problem order shelf. This method is used in the catalog and direct mail (customer contact) industry.

Disadvantages of the manual (visual) customer order check concept are possible verification errors and no on-line tracking of the merchandise or order. Advantages are large-volume handling, easy employees training, and handling of a large quantity of customers.

Hand-Held Bar-Code Scanning. In the second customer order check method the packing employee does hand-held scanning of the bar code of each piece of picked merchandise. The packing employee scans the customer's bar-code label on the packing slip, scans each piece of picked merchandise bar-code label, and then packs the merchandise into the shipping container. The scanning device establishes a record that your customer's merchandise was handled at the packing station.

The bar-code scanning merchandise verification method is used in the retail store distribution industry.

Disadvantages of this concept are that it requires an increase in investment, greater employee training, and a bar-code label. Advantages are a high degree of accuracy, high-volume handling, and handling of a wide variety of SKUs.

Packaging Objectives

The packing station activities include the packing, sealing, and addressing of the merchandise for customer delivery and are performed at the packing station. The direct customer contact industry and retail store distribution industry use different shipping methods, but both have similar objectives to ensure that the merchandise is protected against damage or loss and that your customer's address remains on the package exterior.

Packaging Station Design Parameters. To achieve these objectives, the packing employee's workstation is designed with several factors that are taken into consideration. These factors are the type of SKU and quantity of SKUs per order, the type of order-pick system, the type of sortation system, the type of shipping system, the package in-house transport method, and other related activities such as catalog and mail stuffers.

Packing station design considerations include order-picked merchandise delivery method, work surface of the table, type of shipping container and on-hand quantity, type of packing material and on-hand quantity, package or container seal method, method of addressing packages, and type and on-hand quantity of advertising material.

Pick-Pack Is the Best Pack Method. In the most efficient packing method, the product bypasses the packing station. This situation is achieved in the catalog and direct mail customer industry for single items or two items for single customer orders. This packing activity is performed by an order picker and constitutes the pick-pack method. This method reduces the product handlings by a factor of 1 and reduces operating expenses.

If the SKU is not prepackaged, then the SKUs are picked and packed by an employee. In this activity the SKU is removed en masse from the picking area and placed adjacent to the shipping conveyor. An employee with the appropriate customer invoices, address labels, and sufficient shipping containers performs the individual picking, packing, sealing, and addressing activities. When these tasks are completed for your customer's order, then your individual customer's package is placed onto the shipping conveyor or onto a transport device. If the shipping label is hand-scanned for a manifest record, then the outbound packages go en masse to the shipping dock.

Product Accumulation Prior to a Workstation Is a Must

An important sortation system design parameter is to provide sufficient product accumulation prior to the packing station.

Various Product Accumulation Methods. The various product delivery methods are (1) manual (pallets or carts), (2) baskets or trolleys, (3) totes or containers, and (4) chutes.

If your product is manually delivered to a packing station, then the two major design parameters are that there be sufficient setdown area for two loads, to assure a constant product flow, and that there be an empty device transport take-away path or conveyor system.

Pallet Load Delivery.   If your operation uses pallet boards at the packing station, then to assure a constant product flow, you need a roller conveyor on the floor, two pallet floor positions (one infeed and one outfeed), or a unit-load turntable. These concepts accomplish the product flow design parameters.

Cart Delivery.   If your operation uses a four-wheeled cart as the product delivery method, then your packing station layout has an infeed lane and an outbound cart lane. Each lane holds at least two or three carts.

Tote or Container Delivery.   When your product is delivered in totes on a gravity-powered or on overhead trolley conveyor, then you have several alternative layouts for a packing station. If you have powered roller or skatewheel conveyor as the delivery device, then the space above the conveyor is used for KD carton storage. If the tote is pulled from the elevated conveyor system, then there is a metal slide that serves as a tote transmission path and a holding area.

In the first arrangement the transport system delivers full totes or trolleys to the packing station. The packing employee places empty totes or trolleys onto the conveyor or cart for transport from the packing station to the picking or opening area. In the second arrangement the employee stacks empty totes or trolleys onto a cart in the packing station area and has another employee remove the stack.

If your packing station layout has trolley baskets or trolleys of hanging garments delivering product, then you have two rail sections that merge together with pull-down switches prior to the packing station. The first rail section is the infeed rail, which has accumulation for at least two or three trolleys and one trolley position at the packing station. The second rail is the empty trolley or trolley basket outbound accumulation line.

Gravity Chutes.   If you have gravity chutes or slides deliver merchandise to your packing station, then the chute cubic-foot capacity is designed to handle the product quantity per batch. In the typical direct mail or catalog facility, this quantity is 50 to 60 pieces of product. You consider that the chute does not fill more than half full, which is a 50 to 66 percent utilization factor.

There are three basic chute designs for receiving product directly from the sortation system. The chute or slide is a transission section and accumulation area for the product prior to the packing station. These chute designs are (1) flat, (2) concave, and (3) spiral.

1. *Flat chute.*   The flat chute is designed with a flat bottom and solid sides from the charge end to the discharge end. The product flows in a straight path between the two systems.

2. Concave chute.   The concave chute is a slide with a concave bottom from the charge end to the discharge end. The concave design reduces product jams. Between the two locations, the product flows in a straight line.

3. *Spiral chute.*   The spiral chute is a slide with a helical bottom from the charge end to the discharge end. Between the two locations, the product travels in a direct but circular path.

There are several options and add-ons to the chute design.

1. *Step in the bottom of the chute.*   The first chute design option is a step in the bottom of the chute. The step permits product to accumulate in the chute without increasing the line pressure on the product or door. This line pressure is reduced by the new product riding on top of other lower product that is stopped by the door.

2. *Ledge at the discharge end with a lip.*   The second chute design option, at the discharge end of the chute, is an 18- to 24-in-long flat ledge with a 1-in-high lip on the

perimeter. This ledge extends outward from the chute base with the 1-in-high lip and permits product to flow from the chute and prevents it from falling onto the floor. Also, this ledge provides a surface for the packing person to make the final product-customer sortation.

3. *Door at the discharge end.*   The third chute option is to have a transport chute door or 2- to 3-in-high end stop at the discharge end. This device stops the flow product and permits product accumulation at the packing station.

4. *KD carton storage above the chutes.*   The fourth chute option is to have slots for temporary storage of KD cartons above the chute discharge end.

The final major packing station design feature is that there be a sufficient number of chutes to provide flexibility at the packing station between the order-picked batches. When there is a batch order-pick system and chute sortation method, then there are at least three chutes per packing station. The chute design can be one of the following:

Three chutes: (1) one chute for the first batch, (2) one chute for the second batch, and (3) one chute for the third batch.

Three chutes subdivided: (1) one chute separated into half for the first batch, (2) one chute separated into three-fourths for the second batch, and (3) one chute separated into five-sixths for the third batch.

This latter arrangement, which subdivides the chute into two compartments, reduces the packing station employee's final individual customer order sortation effort because of the reduction in the pieces that are handled per chute.

How to Determine the Quantity Order-Picked per Batch

The unique characteristics of most high-volume small-item sortation methods is that product are order-picked in batch groups (batches or mixed customer orders). The amount of product order-picked per batch is determined by the quantity that is handled at the packing stations.

To determine the number of order-picker packing chute length and number of packing stations, the following factors can be used as design parameters: (1) packing employee rate [(50 pieces per hour $\times$ 20)/stations = 1000], (2) order-picker rate (300 pieces per hour = 1000/300 = 3.3 order pickers), (3) average order size 2.1 pieces per order, (4) average number of orders (50/2.1) = 23.8 orders, (5) average cubic feet per piece (0.5), (6) number of packages (20 per batch), (7) number of batches 15,000/1000 = 15, (8) average cubic feet per chute (50 $\times$ 0.5) = 2.5, and (9) length of chute (2.5 $ft^3 \times 1.50$) = 4 ft^3.

Types of Catalog and Direct Mail Packing Tables

If your business is in the direct mail or catalog industry, the next important design factor is the design of the packing station work surface. There are two types: (1) the plain flat-top table and (2) a specially designed table with drawers and shelves.

Plain Flat-Top Table Method.   The plain flat-surface table (Fig. 6.38a) with a 3 $\times$ 6-ft surface, with fixed or adjustable legs, does provide the packing person with the necessary work surface to separate and package customer orders. Shelving adjacent or

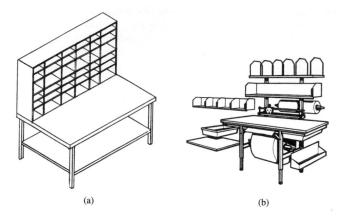

(a)   (b)

FIGURE 6.38 (*a*) Plain flat table; (*b*) specially designed table for packing (*Courtesy of Dehnco Equipment and Supplies Inc.*).

36 to 48 in behind the work surface provides additional KD carton temporary storage. This arrangement is acceptable for handling all types of product and is the preferred surface for the packing (folding or draping) of hanging garments in chipboard boxes, cardboard boxes, or paper or plastic bags.

Specially Designed Table Method.   The second packing surface is the specially designed table (Fig. 6.38*b*) that has drawers, shelves for mail stuffers, and KD carton storage space. In addition, 3 × 4-ft shelves are located behind the packing employee. The separated KD carton storage provides storage for each size of carton used in the packing operation. Additional features of the specially designed table include adjustable legs, peg backboard, or paper roll hanger under the table surface.

This specially designed table is for packing of flatwear and replacement parts.

Retail Store Distribution Packing Methods

When your distribution facility services retail stores, then you have several options for packing station design. These packing station designs are (1) plain-top (surface) or specially designed table, (2) dump, (3) hotel (rapid pack), (4) carousel, and (5) automatic fill.

Plain or Specially Designed Table Method.   Because of the merchandise volume per retail store order and the single and batch order-pick system, the plain or specially designed table is not the preferred packing table for a retail distribution facility.

Store Dump Method.   The second retail store packing work station layout is the store dump (Fig. 6.39*a*). The store dump system consists of a merchandise infeed transport system (tote conveyor, basket, or cart), packing shelves, store carton take-away system, and empty KD carton replenishment system.

After the trolley basket, tote, or cart is full of merchandise, it is pushed on the infeed conveyor and enters the packing area. The transport device has compartments

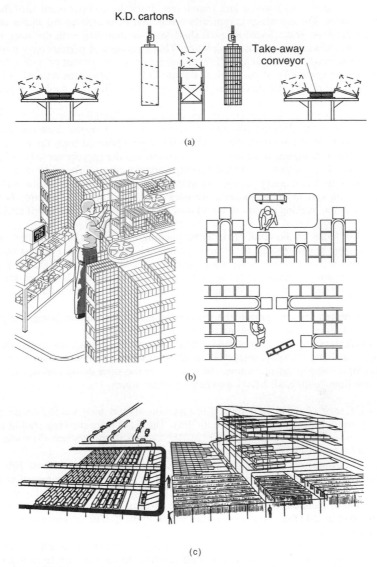

FIGURE 6.39 (*a*) Store dump; (*b*) carousel (*Courtesy of White Storage Retrieval Systems, Inc.*); (*c*) rapid pack (*Courtesy of SDI.*).

that contain specific SKUs and the associated store packing slip. These packing slips are in sequential order corresponding to the arrangement of the store packing area layout. While controlling the flow on the conveyor system, the packing person reviews each packing slip for each store's SKU amount. The transport flow is stopped at the appropriate location across from the store container. The appropriate store order quantity of merchandise is counted and transferred from the compartment into the retail store container. The container is typically a box that is located on the upper or lower shelf. At the appropriate location, each shelf level is identified with the store number and is sloped toward the packing employee. This arrangement permits easy transfer of merchandise into the box. Each shelf level has a narrow trough that provides the space to hold the store packing slips. Typically these holders are located on each level, and there are three retail store boxes per $10\frac{1}{4}$-ft space. The slip is a card or paper document.

Prior to moving the basket onto the next retail store packing location, the packing employee places the packing slip into the trough. When the retail store box is full, or at the completion of the order, the packing slips are transferred from the trough to the box. The box is temporarily or permanently sealed and the appropriate label placed on the exterior. The full box is pushed directly forward onto a take-away conveyor for transport to the manifesting station. An empty box is transferred from the KD carton replenishment system to the vacant store packing shelf location. The empty boxes are conveyed to the packing station or are placed on the floor between the full basket traffic path.

The key factor in the retail store dump packing concept is the sequential order of the retail store position along the store packing shelf or dump line. For best results, as the tote, basket, or cart enters the store packing lane, the first store location is a low-volume store and the last store location is a high-volume store. Also, there are bypass lanes for the transfer of empty baskets or full baskets through the retail packing store area.

These design features permit the packers to maximize their walking and packing activity and do not result in queuing of other packers in the area.

Disadvantages of the concept are increased possibility of packing errors, increased employee walking distance, increased capital investment, and requirement for greater management control and discipline. Advantages of the store dump concept are it handles a medium volume, all SKUs, and large-volume stores.

Multiple Carousel.   The next retail store packing station design is the single or multiple (up to four levels) carousel (Fig. 6.39b). The retail store packing station components include (1) SKU infeed system, (2) KD replenishment system, (3) packing station, and (4) take-away system.

In the carousel packing station method, as the trolley basket or tote full of one SKU travels on the infeed conveyor system, it is stopped by the accumulation device. At this location, with easy access to the merchandise, the packing employee controls the rotation of the carousel. The carousel is a series of shelves or bins that are two to four levels high. These bins level rotate on an endless top- or bottom-driven loop chain past the packing station.

Each shelf level has a human-readable or bar-coded label that discreetly identifies the store that has been assigned to the shelf location. In the shelf location is an empty box, which is the retail store packing container.

Prior to rotating the carousel, the merchandise SKU number is entered into the computer system. The computer program allows the required SKU quantity to appear in the visual display screen and permits each retail store bin to pass the packing station. As the carousel rotates, scanning devices read the shelf discreet number and indicate the required SKU quantity on the visual display screen. As each piece of merchandise is transferred to the retail store packing container, the employee presses the button to indicate a transfer. This activity reduces the visual display screen quantity by one unit. This process is repeated until the retail store quantity is zero. The packing person transfers the packing slip into the box or packing slip holder that is located on the carousel bin side. An alternative to the visual display terminal requires a packing person to use a paper document that indicates the retail store order quantity.

As required, another SKU is delivered on the infeed conveyor system and the required merchandise is transferred into the retail store container. All full store containers are transferred to the outbound shipping take-away conveyor. These boxes are temporarily or permanently sealed for retail store delivery. As required, empty boxes travel on the empty carton conveyor system.

Partially full boxes are shipped to the store or retained in the carousel bin for the next day's order-pick cycle.

Disadvantages of the carousel are that it requires a high capital investment and increased management control and discipline, and has only one packing station but multiple carousels. Advantages are that it handles a high volume and a large order quantity per store, improves employee productivity, reduces lifting activities, and reduces the chance of error.

Rapid Pack or Hotel Method. Another retail store packing station system is the rapid pack (Fig. 6.39c) or hotel system. This system consists of merchandise conveyor flow lanes, retail store packing conveyor, retail store take-away conveyor, empty container infeed system, and retail store partially full container holding area.

Each retail store has a box (container) that is placed onto a discreetly identified transport tote. This tote has a packing slip holder and is three-sided with an open side on the far side of the tote. The tote design permits easy transfer of the full shipping container from the packing conveyor to the shipping conveyor. These totes are placed onto a closed-loop accumulation conveyor system that has the tote travel past each packing station and to the holding area. The holding area is a series of accumulation conveyor lanes that hold a preassigned and predetermined number of stores (totes). The packing area accumulation conveyor is zone-controlled for each pick area.

As the tote travels to the packing conveyor, a packing station induction employee reads the human-readable label or a scanning device reads the human- and machine-readable label. The packing station consists of the container in a tote traveling on accumulation conveyor past a series of multilevel carton flow lanes. These flow lanes are separated in preassigned areas or zones for a packing persons zone (area).

In the manual system the packing employee reads the packing slip quantity and turns to the merchandise flow lane. From the merchandise flow lane, the packing person removes and places into the retail store packing container the required merchandise quantity. On completion of the transfer, the packing slip is transferred to the container. The activity is repeated for each SKU of merchandise in the packing employee's zone.

After the completion of all picks in this packing employee's zone, the retail store box is transferred to the next packing employee's zone. In this zone the packing activity is performed by an employee for the merchandise within the zone.

When the retail store box is full, the packing employee places all the packing slips inside the box. The packing box is temporarily or permanently sealed and pushed forward onto the retail box take-away conveyor for store shipment. From the empty-box conveyor system, an empty box is replaced in the appropriate retail store transport tote.

The partially full shipping boxes in the transport totes are sent to the temporary holding area or to the retail store.

Disadvantages of this method are that it requires a high capital investment, some employee nonproductive walking, and an increase in management discipline and control. Advantages are that it handles a high volume and a large number of stores, reduces the need to lift heavy loads, and reduces errors.

Retail Store Automatic-Fill Method. The automatic-fill packing method is the last retail store packing method. This concept has the following options: (1) merchandise feed system, (2) box (container) conveyor system, and (3) box (container) seal station.

There are two alternatives to the automatic-fill method, which are based on how the order-pick system releases product onto the conveyor: the remote-fill package station method and the direct-fill method.

Product Released onto Conveyor. In the remote-fill method, the automatic order-pick device has released the merchandise onto the conveyor for travel to the packing station. As the merchandise arrives at the packing station, it is transferred from the conveyor into the retail store shipping box. When the retail shipping box is full, a packing slip is placed with the merchandise, and the box is sealed and transferred to the outbound take-away conveyor. From the empty-box conveyor, an empty box is placed in the packing station fill location.

In the automatic-fill method, since there is one retail store's merchandise on the conveyor, all the merchandise is removed from the conveyor; therefore, partially full containers with packing slips are closed and released onto the take-away conveyor.

During the process on another infeed conveyor a second store's merchandise is released onto the packing station infeed conveyor. This conveyor network optimizes the packing employee's productivity.

Since the merchandise was order-picked for a single customer by an automatic computer-controlled order-picking device, the packing person is not required to count the merchandise.

Product Released into A Container. The second automatic direct-fill retail store packing method is associated with an automatic (computer-controlled) order-picking device that order-picks merchandise for one customer.

With this concept, a retail store empty box is placed onto a conveyor that is located between two series of merchandise order-pick lanes. These lanes are belt conveyors that are indexed (moved) forward toward the center belt conveyor. As the box travels on the middle conveyor, a computer-controlled system causes the appropriate merchandise lane conveyor to index forward. The forward movement permits a single item of merchandise to advance forward and fall into the box. Until the box receives a certain volume of merchandise or until all the merchandise has been order-picked, the box travels on the conveyor to the packing station. After the box or container arrives at the packing station, an empty box is placed at the beginning of the order-pick conveyor.

At the packing station, the packing employee places a packing slip (or slips) in the box, seals the box, and places the box onto a take-away conveyor. Since the merchandise was automatically picked for one customer, there is no counting at the packing station.

The disadvantages and advantages of this method are the same as those of the hotel method.

Hanging-Garment Packing Methods. In the retail store customer industry, there are several hanging-garment packing methods. Garments are (1) hung on a rolling rack, (2) sheathed in a plastic bag and hung in a trailer, (3) draped in a flat box, or (4) hung in a hanger box.

Hanging Garments Hung on a Rolling Rack. The first hanging-garment packing method is the rolling-rack method. In this arrangement the hanging garments are hooked onto a rolling rack. The rolling rack is a four-wheeled structure that has a load bar for the hanging garments. To prevent the hanging garments from falling to the floor, a cardboard sleeve is taped over the hooks.

The rolling-rack method is used in a company that has its own truck fleet, makes deliveries to the retail stores, and returns with the empty rolling racks to the facility.

Disadvantages of the rolling-rack method are potential for hangers to fall to the floor, carts secured in the delivery truck, and carts returned to the warehouse. Advantages of the rolling-rack method are that it is a reusable transport device, handles a large quantity, and can be used in either the warehouse or retail store floor area.

Hanging Garments Hung in a Plastic Bag on Rope in a Delivery Truck.   The second hanging-garment packing station method is to have the garments packaged in individual plastic bags that are hung in the delivery truck. This concept requires each garment to be placed into a plastic bag, and 5 to 10 garments are collected and transferred to the delivery truck. The garments on hangers are hooked onto rope loops that are hung from the trailer roof.

This delivery concept is used for a company that owns its truck fleet and permits the truck to make backhauls. This delivery method handles any volume.

Hanging Garments Draped in a Flat Box.   The third hanging-garment packing method is to drape the hanging garment in a flat box. In this concept the hanging garment is draped and placed into a cardboard box. After the box is sealed and addressed, it is sent to a local remote retail store or to a catalog or direct mail customer.

The disadvantage of this method is if the cartons are not reused, then this method becomes expensive. Advantages are increased security and protection to the garment and ability to ship to remote stores and catalog or direct mail customers.

Hanging Garments Hung in a Hanger Box.   The fourth hanging-garment packing method is the hanger box packing method. The hanger box concept consists of a cardboard (corrugated) box that has a wood or metal structural cross member. The member is inserted or attached to the two sides of the cardboard box. This feature adds rigidity to the box and permits garments to be hung onto it. To provide additional hanger security, the hangers are taped to the cross member or held by a metal clamp.

After the hanger box is full of garments, the carton is sealed, addressed, and loaded onto the local or remote delivery truck.

Disadvantages and advantages of the hanger box method are the same as those of the drape box, but the hanger box handles a larger volume.

What Determines Your Shipping Container.   The shipping container that is used by a distribution facility is determined by several factors: the quantity of merchandise per order, the value of the merchandise, the type of manufacturer merchandise packing, the method of shipment, and the reusability of the container.

With these factors, the distribution facility alternative shipping containers in the retail industry are the cardboard box and the plastic container.

Package Exterior Appearance Influences Customer Satisfaction.   If you are in the catalog or direct mail industry, then the type of container that is used to ship merchandise is a very important factor because it does influence your customer's opinion of their purchase. Also, your company relies on another company to transport the package to your customer.

To achieve maximum packing productivity, efficiency, product protection, and package presentation, selection of the container is based on the merchandise cubic feet and is recommended by the computer. The appropriate container is printed on the customer packing slip and becomes an instruction for the packing employee, thus improving the packing productivity.

Various Containers in the Catalog and Direct Mail Industry.   The possible containers that are used by the catalog and direct mail industry include (1) cardboard boxes of various sizes and (2) bags (corrugated or plastic).

Shipping Container Filler Material Objective.   After your shipping container, your package filler material is the next important packing material. The filler material is added to the void space inside the shipping container and provides protection to the merchandise from damage during handling and shipping. Most of the retail and catalog and direct mail distribution facilities use a filler material in the package.

Filler Material Design Factors.   The particular type of filler material used in a package is determined by a combination of several factors, such as value of the merchandise, cost of the filler material, whether filler material is recyclable, type of merchandise packaging, type of shipping container, and method of shipping.

Various Filler Materials.   The various filler materials include (1) loose (no material), (2) paper, (3) "peanuts" (Styrofoam or corn), (4) foam padding, (5) bubble sheeting, and (6) foam in place (manual or automatic).

Loose Items in a Container (Typical Retail Store Distribution).   In the retail store distribution industry or catalog agency industry, loose packing into a large cardboard box or plastic container is a very common practice. First, since the majority of the merchandise is delivered by a company truck, there is sufficient control over the transportation delivery of the merchandise and the container is returned and reused for other deliveries. Second, the majority of the merchandise that is inside the container has to some degree an outer sleeve of packaging which absorbs, to some degree, the impact of movement inside the container. Third, the cardboard or plastic container is sealed for security.

Disadvantages of the concept are increased possibility of merchandise damage, and an employee has difficulty in moving the container. Advantages are no expense on the filler and reusable container.

Paper Filler.   Another filler material that is used in both retail and catalog industry is the paper. The packing station is supplied with a supply of paper (special paper or second-hand printed paper). As required, the packing person crushes one or several sheets of paper or uses shredded paper to fill the void spaces inside the container.

Disadvantages of the paper filler method are low packer productivity; also it does not provide the best protection and requires space at the packing station. Advantages are low cost and does not require a special container or fire protection.

Peanut Filler.   Another filler material that is used to protect merchandise in the container is "peanuts" (Fig. 6.40). The two types of peanuts are the Styrofoam peanuts and processed-corn peanuts.

The Styrofoam and processed-corn peanuts are little chunks of Styrofoam and processed corn that are placed in the container void spaces. The peanuts are manually placed into the container or are automatically funneled from an overhead peanut bag into the container void spaces.

When you consider the peanut filler system, then your airduct system for new filler peanuts and vacuum-return system for spillage are the most important design features of the peanut filler system.

In some funneled applications, excess or peanut spillage is returned by a vacuum system to the overhead peanut container.

Disadvantages of the peanut filler method are that it requires an increase in investment, increased fire protection, and installation of a reserve supply. Advantages of the peanut filler method are that it is easy to handle, handles a high package volume, and improves merchandise protection.

When you consider the use of a natural corn-processed peanut filler system, you must also consider the fact that the peanuts could add weight to the package, have a tendency to stick together (especially in humid environments), increase cost, and introduce potential rodent or insect infestation problems.

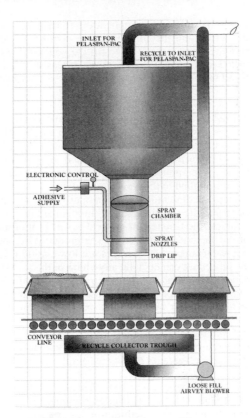

FIGURE 6.40   Automatic peanut fill to protect merchandise (*Courtesy of Quickdraft.*).

Styrofoam Sheet Filler Material.   Another filler method is Styrofoam sheets. Inside the container, as required, the Styrofoam sheets are placed on the sides of the container and thick sheets are placed on the bottom and top of the merchandise.

When compared to the peanut filler the Styrofoam sheet is more labor-intensive.

Air-Bubble Sheet Filler Material.   Another small-item package filler material is bubble sheets. These sheets are wrapped around the individual SKUs and taped to the SKUs or to the sheet to prevent unwrapping in the container. The bubble sheet consists of multiple air bubbles that are in a sheet of thin plastic which absorbs the impact.

Disadvantages and advantages of the bubble sheet are similar to those of the Styrofoam sheets.

Foam-in-Place Filler Material.   Another package filler method that is used to protect merchandise is the foam-in-place concept. The foam-in-place substance is sprayed in the container bottom. The SKU is placed onto the foam that forms around the SKU. After the foam substance hardens, additional foam is sprayed onto the top of the SKU.

The foam substance is applied either manually or automatically. The manual application is used for the low-volume operation, and the foam is applied to the top and bottom of the container. The automatic application consists of a conveyor system in which the SKUs are placed into the container. The container is conveyed under the automatic filler device that applies the foam to fill the voids inside the container.

Disadvantages of the method are that it is more expensive and difficult to use with a multiple loose packed merchandise order. Advantages are optimal protection against damage and no waste.

Cardboard Boxes Sealed with Gummed or Self-Adhesive Tape.   If you are in either the catalog or retail store distribution industry, when a cardboard box is used as the shipping container, then the package is secured with gummed or self-adhesive tape.

Manually Operated Gummed Tape.   The gummed tape is tape that contains glue on one side and has moisture applied to the surface that activates the glue. With the activated glue, the tape is applied to the shipping container.

The first gummed tape method utilizes the manually operated tape machine, which requires the operator to pull the machine handle. In this method the tape is fed through the moisture applicator and the tape length is equal to one pull of the handle.

Disadvantages of this method are low employee productivity, increased tape waste, and requirement for a water supply. Advantages are low capital investment and mobility (the gummed tape procedure can be done anywhere in the warehouse).

The second gummed tape method is the *electric tape machine* concept. The packing employee sets the desired tape length on the tape machine selector switch and presses the hand or foot operator button. This machine automatically dispenses the desired tape length.

When compared to the manually operated gummed tape method, additional disadvantages of this automatic method are increased investment and requirement for an electrical outlet. Additional advantages are higher employee productivity and less tape waste.

Manually Applied Self-Adhesive Tape.   Another tape method that is used to secure cartons is self-adhesive tape. This tape is available in many different materials and is applied to the carton by the manual, semiautomatic, or automatic tape application method.

It is also possible to manually apply self-adhesive tape with a dispenser. First, the packing employee applies the self-adhesive tape to the carton exterior, and then uses a roll of tape with a knife or uses a standard- design roll dispenser with an attached tape cutting edge.

Disadvantages of the dispenser method are waste of tape and requirement for an employee that is determined by the dispenser roll size. Advantages of the method are that it handles all package sizes and can be performed in any warehouse location.

Semiautomatic or Automatic Self-Adhesive Tape.   An alternative self-adhesive tape application method is semiautomatic taping (Fig. 6.41*a*). The semiautomatic tape system consists of several variations. First is an *automatic tape dispenser and a package conveyor (belt) system.* As the carton travels on the belt conveyor across the automatic tape dispenser, it seals the top or the top and the bottom of the carton. The automatic tape system (Fig. 6.41*b*) consists of several mechanisms that determine the carton width and height. These devices adjust the top tape dispenser height (5 to 20 in) and side grippers to fit the carton's width (6 to 20 in). Second is an *optional bottom tape dispenser.* Third is a device to *automatically close the carton flaps.* The fourth is a *package-conveying surface* that is a belt surface or a roller conveyor surface with two belt side grippers. The final device is the *top tape dispenser,* which applies the tape to the carton top flaps (8 to 24 in).

The second component of the semiautomatic tape system is the *infeed and take-away conveyor system.* The infeed system ensures that the cartons are accumulated,

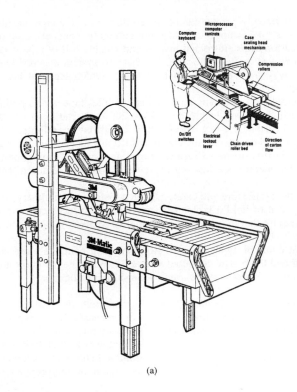

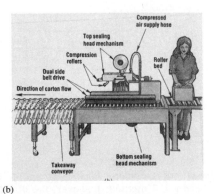

(a)

(b)

FIGURE 6.41 (*a*) Semiautomatic taper; (*b*) automatic taper (*Courtesy of Modern Materials Handling Magazine, 3M Co., and Loveshaw Corp.*).

singulated (singularized, separated) and indexed forward to the tape machine. In a special work area (this area is not required in the automatic concept), employees ensure that the tabbed carton flaps are folded inward and when a roll of tape is depleted on a machine, the employee replaces the tape. The outfeed system ensures that the cartons are transported away from the tape machine and provides accumulation prior to the next workstation.

The semiautomatic or automatic tape concept handles tape with a width that varies between 1 and 2 in and a roll length of 1000 to 2000 yd. The taper has the ability to tape random-sized cartons at a rate between 8 to 25 cartons per minute.

Disadvantages of the concept are that it is difficult to handle all small and large packages, there is downtime to replace tape, and capital investment is higher. Advantages of this method are less waste, fewer employees required, and large-volume handling.

If you are in the retail store and catalog agency distribution industry, your shipping container is required to hold a larger capacity and has a large-cubic-foot container. These features require that the shipping container have more sealing than the standard tape.

The two most common types of retail store shipping containers are the cardboard (corrugated) box and the plastic container with a lid. When heavy items are shipped to a customer, then wood containers are used as the shipping container.

Various Container Sealing Methods.   The various methods used to seal containers are (1) strapping application methods (manual or mechanical), (2) types (steel, polyester, nylon, polypropylene, or plastic), and (3) self-locking seals (metal or plastic).

If you are in a small-parts distribution facility that uses a cardboard box, wood container, or plastic container, then you provide additional shipping container strength by having the container strapped with steel, polyester, nylon, or polypropylene material.

Various Strapping Material Strengths.   The various strapping materials have the following characteristics:

Material	Strength
Steel	High
Polyester	Medium
Nylon	Fair
Polypropylene	Low

Manual and Automatic Plastic Strapping Methods.   All these concepts are applied by the manual method; however, the plastic (polypropylene) strapping material is applied with an operator (mechanized) or automated strapping device. The strapping machine layouts are similar because the machine applies a strap onto the container in the direction perpendicular to the direction in which the container travels on the conveyor. Therefore, either an employee turns the container for a second strap in the opposite direction, or the container is mechanically turned on the conveyor system prior to a second strapping station.

A strapping machine (Fig. 6.42*a*) is a device that bridges the conveyor section. Prior to the container strapping station, the containers are accumulated, singulated, and indexed forward to the strapping station. After the container is under the strapping machine bridge, the machine applies the strap to the container (Fig. 6.42*b*).

Disadvantages of this method are higher expense and downtime to replace the strapping material. Advantages of the strapping method are that it provides additional package security, is best for one-way delivery, and serves as a carrying handle.

Reusable Plastic Containers with Seals.　　If your distribution facility provides the service with your company delivery vehicles to a group of company customers and the

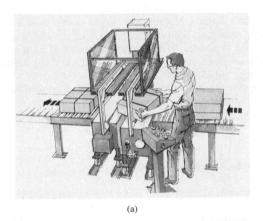

(a)

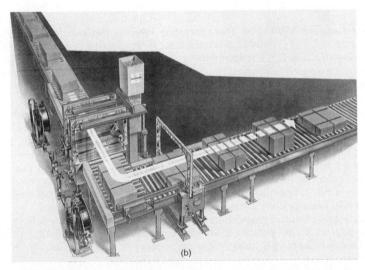

(b)

FIGURE 6.42　(*a*) Mechanized strapping machine (*Courtesy of Materials Handling Engineering Magazine.*); (*b*) double strapping machine (*Courtesy of Signode Packaging Systems.*).

vehicle returns to the distribution facility, then you use a reusable plastic container as the shipping container. To provide the required product protection and security, these containers are designed with an attached lid or a removable lid. Both lid types and the containers are designed with holes for the use of a self-locking seal. Either a plastic or a metal seal is inserted in the holes. This action secures the lid to the container top. To open the container, the seal is broken. For additional control, these seals are discreetly numbered to restrict the use of a bandit seal.

Disadvantages of the seal method are that the container is designed for a seal, seal is controlled by management, this method represents an additional expense and weight capacity is handled by an employee. Advantages are increased control and security, no downtime, and no floor space requirement.

An alternative security strategy is to use plastic strapping on the plastic container.

Corrugated or Plastic Bags.   In the catalog and direct mail industry, some of the merchandise is very small and lightweight, permitting the use of a plastic or corrugated bag as the shipping container. The various methods and components used to secure the bag are (1) self-adhesive or gummed tape, (2) staples, (3) stitching by a machine, (4) a self-locking bag, and (5) a self-sealing bag.

The bag seal method that is used in an operation is determined by the economics of the material, the volume handled, the sealing method used for the packages, material handling concept, shipping method, and appearance of the package exterior.

Labeling (Customer Address) Activity

The next packing station activity is to have the packing employee place your customer's address label on the container exterior. The labeling activity ensures that your customer "ship to" address is clearly visible and easily recognized on the container exterior.

Various Labeling Methods.   The alternative labeling methods are (1) handwritten, (2) ink-stamped, and (3) machine-printed label (self-adhesive or nonadhesive).

Handwritten Label.   The handwritten label concept is considered the basic type, where an employee handwrites your customer's "ship to" address on the container exterior surface. This concept is not preferred for use in a dynamic catalog and direct mail or retail store distribution facility.

Ink-Stamped Label.   The second package address type is the ink-stamp method, which requires the employee to type or handwrite the customer address on a special label pad. After the label pad is completed, the employee places the label pad onto a liquid ink dispenser. When pressed against a carton, the ink dispenser transfers the ink through the label pad onto the carton exterior. The ink on the carton exterior is the customer's "ship to" address. To be cost-effective, the ink-stamp concept is used for multiple cartons per one customer; therefore, it is preferred in the retail manufacturing industry that has a cardboard box or a paper label on a plastic container.

Disadvantages of this method are high expense per label, messy activity, possible unclear address on the container, low-volume handling, and requirement for a cardboard or paper label. Advantages are preprinting of all addresses and low capital investment.

Your Return Address on the Label Is Important.   The preferred retail store distribution and catalog or direct mail package address (ship to customer) concept is the preprinted "ship to" address and the distribution facility return address on the label face.

Self-Adhesive Preprinted Label.   In the retail industry, the most popular type of label is the self-adhesive preprinted label. The packing employee removes the self-adhesive label from the label roll, removes the protective backing from the label, and places the self-adhesive label onto the carton exterior. Some retail companies color-code the label printing and border. The color code indicates a delivery day of the week, store, or trailer. This procedure reduces the possibility of loading and unloading errors.

Disadvantages of the concept are capital investment for the on-site printer, color code has a limited quantity in the spectrum, print time is required, and off-site print quantity is justified. Advantages include clear and understandable print, preprinted and off-site preprinted labels, reduced handling errors, labels easily recognized, and use for a human-readable or human- and machine-readable labels.

Nonadhesive Label.   The nonadhesive label in the catalog and direct mail industry is popular. This label is a computer-printed (preprinted) label that is attached to the packing slip and has a perforated edge on the packing slip. After the package is full of merchandise and the filler material is placed inside the package, the packing employee separates the packing slip and the label. The packing slip is placed inside the package, and the employee seals the package. The nonadhesive paper label is attached to the package exterior surface. The two alternative label attachment methods are glue pot and clear self-adhesive tape.

Glue Pot.   In this method, the packing employee processes the label through a glue pot. The glue pot is a brush, warm water, and roller combination device that applies heated water-based glue to the paper label. With the glue on the label, the packing employee places it onto the package exterior.

Disadvantages of the glue pot method are that it requires a water supply, an electric outlet, and additional employee activity. Advantages are low cost per label, use on cardboard and plastic containers, low capital investment, and reduced scanning device "no read" problems.

Clear Transparent Tape.   An alternative nonadhesive label attachment method is to use clear (transparent) self-adhesive tape. The packing employee uses the same procedure to prepare the package for shipment except that the customer "ship to" address label is placed print face up on the work table surface. In the special clear transparent tape dispenser the tape side of the self-adhesive faces down and the paper label with the print faces up, permitting the print to stick to the tape. The self-adhesive tape width and length exceed the label dimensions. This overhang permits a secured and sealed label attachment to the shipping container. The employee pulls the tape, cuts it from the tape dispenser, and places it with the label onto the package exterior. The tape width is 3 to 4 in, which accommodates most companies' customer "ship to" labels.

Disadvantages of the concept are that it requires the tape dispenser to be secured to the work table, possible taping errors may destroy the print, and some scanning devices cannot read through the tape. Advantages of the method are that it provides a protective covering for the label, is easy to use, and does not require water or glue.

Preprinted Label Window on the Shipping Container Improves Productivity.   To improve packing employee productivity, scanning efficiency, and shipping and handling activity productivity, many retail store and catalog and direct mail operations have a label window printed onto the shipping container exterior. This window (area) is the preferred package location for the customer "ship to" address label.

How to Handle Shortages, Overages, and Damages at the Packing Station. In your batch-picked order-pick system with a conveyor belt or mobile equipment transport system, a customer order may have overage, shortage, damage, or wrong merchandise at the packing station. These conditions are created from an order-pick error, lost label, or merchandise that is out of stock or damaged or lost in transport or otherwise.

In the catalog and direct mail industry, when a problem for customer order occurs at the packing station, the existing merchandise and customer packing slip are reviewed and corrective action is taken by management to satisfy the customer order. To assure good packing productivity, in the packing station area there is a problem order shelf that temporarily holds the problem order merchandise and packing slip. This permits management to collect the problem orders and reshop (order-pick) the order or send the customer a shortage which becomes a back order.

Weigh and Manifest Activity

After the customer package is sealed and labeled, it is transported to the weight-manifest-shipping station, where employees verify that the package label has the correct weight, that the most economical shipping (delivery) method is used to deliver the package, and that the package was shipped from the facility.

In the catalog and retail industry, packages undergo weighing to verify that the delivery fee is correct and to prepare a manifest list indicating whether the package was shipped from the facility.

There are four basic weigh, manifest, and shipping methods: (1) manual weigh (manual scale) method, (2) determining the delivery method to be used for each package by computer-printed weight, (3) manual weigh with computer-estimated weight (verifying the correct weight on a manual scale), and (4) conveyor scale method, where the package is weighed "on the fly" (bar-code label scanned to determine package weight).

Manual Weigh. In the retail store distribution and catalog and direct mail distribution industry, with the manual weigh method, each package is weighted on a scale. The weight and package number are written onto a manifest list, or the delivery fee (stamp) is placed onto the package. The package is transferred to the appropriate mailbag or transportation conveyor.

Disadvantages of this method are low-volume handling, possibility of errors, and low employee productivity. Advantages are low capital investment and no requirement for a computer program, management control, or discipline.

Manual Weigh with Computer-Estimated Weight. In the catalog industry with the manual weigh method, the computer estimates the daily package volume weight. This weight is estimated per delivery vehicle. Each package has a two-part label. One section has the customer "ship to" address and the estimated weight on its face. This part is glued to the package exterior. The second part is a tear-off slip that contains the same information that appears on the label. After the package is sealed and labeled, it is transported to a weigh-manifest station. At this station, each package is weighed on a scale to verify that the actual weight on the scale matches the computer-estimated weight on the label. If there is no weight difference (within a specific percent), then the tear-off slip is removed from the package and placed in a file. The package is placed on the appropriate shipping system. If there is significant difference between the package actual and computer-projected weight, then the correct weight is written on the tear-off section of the label. This label section is retained, and adjustments are made to the total projected delivery vehicle weight. The package is placed onto the shipping conveyor.

Disadvantages of this method are that it requires management control and discipline, a computer program, and a medium number of employees. Advantages are low- to medium-volume handling and low capital investment.

Various Manifesting Methods.   In the catalog and direct mail industry and retail store distribution industry, the next activity is package manifesting and shipping, where each customer invoice number is listed on a shipping document. This activity ensures that your customer order (package) was processed and placed onto the delivery vehicle.

The various manifesting methods are (1) manual (handwritten) and (2) computer-printed list or labels that are hand- or fixed-position-scanned with delayed or on-line information flow.

Manualwritten Manifest List.   The basic manifest document is the manual handwritten list. An employee writes onto the manifest document the customer name, the invoice number, and the package weight. After the delivery vehicle driver verifies the package count and signs the manifest document, this document provides the company with a shipment record. The record indicates that the packages were shipped and are the responsibility of the delivery company.

Disadvantages of this method are low-volume handling, transportation errors, and handling of a limited number of delivery methods. Advantages are low capital investment and easy method implementation.

Two-Part Label Manifest.   In this method a two-part printed label is attached to the package exterior. Each section of the label contains the customer address, invoice number, and package weight. While placing the package onto the delivery vehicle, the driver marks the tear-off slip. This slip serves as the delivery record.

Disadvantages of this method are that it handles a low volume and a limited number of delivery methods. Advantages are error-free transportation and low to medium capital investment.

Hand-Held Bar-Code Scanning Manifest.   In this method each package has a discreet bar-code label and each manifest station employee has a hand-held scanner.

The manifest employee is required to direct the hand-held scanning device light beam to read the package label. In the catalog and direct mail industry, after the label is read by the scanning device, the package is transferred to the appropriate shipping conveyor. The scanning device is on-line to a computer system or held in the hand-held scanning device for delayed computer entry. In the retail industry, the hand-held scanning device is portable and programmed by the employee for each dock location to read a specific customer package label. This feature is a check system to ensure that the package is loaded onto the correct delivery vehicle. In this operation, the scanning is done using the delayed or on-line computer entry method. For good employee efficiency and reduced equipment damage, the employee has a scanner holder that is attached to the waist belt and is similar to a pistol holder.

Disadvantages of this method are higher capital investment and need for a bar-code label. Advantages are a high-volume handling, accurate records, and handling of non-conveyables.

In the catalog and direct mail industry, the hand-held scanning manifest system is designed with the package sortation feature.

If the manifest system serves as the infeed system for a sortation system, then the packages are placed on a conveyor for package transportation to the sortation system.

If the manifest system is the infeed system for a multiple direct load system, then the manifest station is designed with multiple sortation locations. Each device or conveyor transports packages to the appropriate delivery vehicle (United Parcel Service, U.S. Postal Service, etc.). The manifest station has two, three, four, or five separate take-away conveyors.

Conveyorized Weigh-on-the-Fly Manifest. The last manifesting method is the most sophisticated manifesting system. This manifesting system is part of a conveyorized system in which each package is scanned by a fixed-position scanning device and weighed on a scale. If the bar code does not specify the shipment method, then the computer determines the least expensive method and instructs the package conveyor sortation system to divert the package to the appropriate shipping lane. In this system all packages are singulated and in the correct position for bar-code label reading by the scanning device. The scanning device sends the information directly to the computer or is held in a backup memory for later computer entry. The package is diverted (sorted) to the appropriate shipping door.

Disadvantages of this method are high capital investment and requirements for a sortation system and a bar-code label. Advantages are that it handles a high volume, requires no employees, provides accurate information, and provides on-line information.

Several factors should be reviewed prior to implementation of bar-code scanning manifest. These factors are for your delivery company's manifest.

Do you require a manifest for the workday or for the truck that is at your facility?

Do you require a package manifest computer entry?

Do you require a backup memory?

Does your delivery company accept this type of manifest?

Does your plastic package cause scanner problems?

Loading and Shipping Activity

The next warehouse activity is the package loading and shipping function. This activity ensures the company that the customer's package is placed onto the delivery vehicle.

In the retail industry, this activity is loading the merchandise onto the company's delivery truck, common carrier, or contracted delivery truck. This merchandise is unitized or directly loaded onto the truck floor. Some retail companies place early selected merchandise in a temporary holding area and then load the packages when required to meet the delivery schedule.

In the catalog and direct mail industry, the packages are unitized on the dock or loaded directly onto the delivery vehicle. The type of delivery method determines the loading method. The various catalog and direct mail delivery methods are (1) USPS (U.S. Postal Service); when using this method be sure of the dimensions and door location of the Bulk Mail Carrier (BMC), (2) United Parcel Service (when using this method, have extra vehicles available and determine if the trailer has a fixed-gravity conveyor in the truck), (3) Air Cargo (when using this method, be sure of the dimensions and door locations of the container), and (4) overnight or next day, foreign, and insured delivery (these methods require special manifesting and loading procedures).

Customer Returns

The next single-item (catalog or direct mail) distribution industry activity is the handling of customer returns, out-of-season product, and retail store transfer product.

The customer return activity is a warehouse activity in all industries. It is most evident in the catalog and direct mail industry with a volume that can range from 5 to 38 percent of the volume shipping to customers.

Reasons for Returns. The reason for the return varies; the customer may have received the wrong merchandise (pick error or customer order error), the package may have been received late, there may have been an overstock situation, the merchandise may have been damaged, the customer may have ordered all the colors, the merchandise may not have matched the catalog picture, or the merchandise may have been out of date.

Customer Return Activity Objective. Customer return activity assures you that your customer's return order was received at your facility. Also, after opening the return package, you can verify that the return merchandise agrees with the packing slip and that the merchandise follows one of these routes: (1) entered into the inventory and physically transferred into the pick position (since the inventory adjustment is one or two pieces, then the pick position is the location for the return because it is not a unit of storage), (2) sent to an outlet store, (3) donated to charity, or (4) disposed of in the trash.

Steps to Improve Customer Return Activity. Several factors improve the returns activity. These are to ensure that the outbound SKU has accurate pick label identification. On your return slip request that the pick label be attached in the proper area or enclosed in the package; as part of your return slip have a return label attached to it or have a bar-code label as part of the merchandise tag. If the original order was for one item, then it is preprinted on the label with an identifier indicating the type or merchandise (such as "H" for hanging garment). If the return has multiple items, then have the customer write the number on the return label, keep your return document simple, and approve your customer's credit for the return.

Also in the returns area and as part of the KD carton storage area, the peak return cartons are allocated to the storage positions. This storage is in palletized or secured loads or in nestable or collapsible containers.

How a Fixed SKU Pick Position Improves Return Productivity. During the life cycle of a SKU (merchandise or product item), a fixed SKU pick position theory has one pick position for the SKU. This system is flexible, allowing a SKU, as it declines in popularity (fewer picks) to be moved to a pick position that optimizes order-picker productivity. However, this relocation (movement from zone A to zone B and from zone B to zone C) occurs twice or three times in the life cycle of a product at your distribution facility. In the catalog and direct mail industry, this SKU life cycle occurs with the release of a new catalog.

As part of the customer return process activity, the SKU is returned to the SKU pick position. Your return employee identifies the SKU pick position and attaches it to the product. With the one (fixed) pick position for a SKU and if the return SKU does not have pick label attached to it, your return employee looks up the SKU's pick position in the inventory file to indicate the current pick position. If the return SKU does have a pick label, the SKU is automatically identified by the label. This fixed pick position on the label means an improved return processing activity.

In addition, when the merchandise is physically returned to the pick position, the employee ensures that the return SKU matches the SKUs that are in the pick position.

Various Return Opening, "Verifying the Merchandise Quality," Sorting, and Other Returns Methods.   There are four return open, verify, sort, and return-to-inventory methods: (1) mobile cart, (2) double- or triple-stacked conveyors with slides to individual workstations, (3) package conveyor over a trash conveyor with sort bins on both sides, and (4) a transporter conveyor.

Mobile Cart.   With the mobile cart (Fig. 6.43*a*) at the receiving dock, the return packages are placed into four-wheeled carts. These carts have four sides of wire-mesh or solid material. One side is removable for easy access to the cartons. As required, an employee moves a cart adjacent to the opening station. At this location, the opening employee removes a carton, opens the package, places the trash onto a conveyor belt or into a container, verifies the return, "preps" (places the merchandise in a plastic bag) the merchandise, and sorts the merchandise into a return container or bin. A conveyor under the trash conveyor transports the preliminary sort containers or bins to the final sort station.

Disadvantages of the mobile cart method are that it handles a low volume, requires an employee to move the cart, and requires a larger square-foot building area, and package flow is not constant. Advantages are that the cart can be used in the receiving and opening activities, capital investment is low, and the procedures can be relocated (moved to another location in the warehouse).

Double- or Triple-Stacked Conveyor with Slides.   This arrangement consists of two or three conveyors and a series of slides that flow to each workstation (Fig. 6.43*b*). In the double-stacked method, the upper-level conveyor is for return packages and the lower is for trash. With the triple-stacked conveyor, the opening area is on a platform, and the upper level of the conveyor is for trash, the middle level is for return packages, and the lower level is for container or tote transport to the final sort station. The gravity slide permits the opening employees to easily remove and transfer return packages to the work table.

After opening the package, the employee places the trash into the trash conveyor, then processes the paper document and the merchandise. The prepped merchandise undergoes preliminary sorting into containers or totes.

Disadvantages of this method are investment in conveyor and platform and fixed-square-foot building area. Advantages include constant package flow to the workstation, less effort to move packages, high-volume handling, and less nonproductive walking time.

Package Conveyor over a Trash Conveyor with Sort Bins on Both Sides.   The return package infeed conveyor over a trash conveyor with sort bins on both sides arrangement (Fig. 6.43*c*) concept consists of a package conveyor over a trash conveyor, a series of sort bins that flow to a workstation, and preliminary sort containers or bins.

As the return packages travel on the infeed conveyor, an employee removes and opens a package, verifies the customer name on the package and packing slip, and transfers the package and packing slip into a bin. The trash is placed into the trash conveyor. The sort bins are separated into four different types of returns (e.g., flatwear, accessories, single-item returns, multiple-item returns) that direct the flow of merchandise to the workstation. All hanging-garment returns are placed onto a hanger that is placed onto a slide rail for transport to the workstation.

If your customer name or address is not clear on the packing slip or there are other

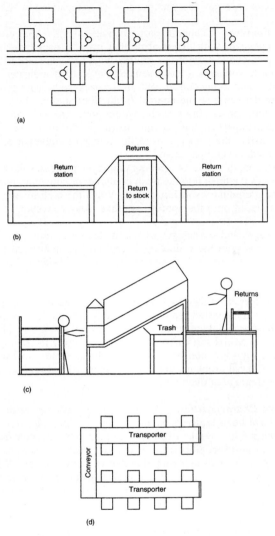

FIGURE 6.43 (*a*) Mobile cart; (*b*) stacked conveyor with slides; (*c*) conveyor with sort bins; (*d*) transporter conveyor.

problems, then the customer return merchandise and packing slip are placed in a tote and transferred to the problem return station.

At the processing station, the employee processes the return merchandise and paper documentation and the employee transfers the merchandise to the preliminary sort container or tote. As required, the sort container or totes are transferred to the final sort area.

Disadvantages of this method are similar to those of the other conveyor method.

The additional advantages are that the return work is separated by returns and to a different area.

Transporter Conveyor.   The transporter conveyor (Fig. 6.43*d*) consists of an over- and under-belt conveyor with divert arms that transfer packages to the appropriate workstation. The unique feature of this double-level belt conveyor is that it consists of only one belt. Each workstation has a sensing device that indicates an empty condition. This signal is sent to the control station for package transfer to the station. After the packages are delivered to the control (induction) station, the control panel indicates an open returns processing station. An employee places a package onto the top belt conveyor that transports the package to the workstation. At the workstation, a divert device transfers the package from the transport conveyor to the workstation slide. The slide holds several packages.

At the workstation, the package-opening employee processes the return (paperwork and product), places the trash into the trash container or conveyor, and places the merchandise into the preliminary sort bin or tote. When the preliminary sort container or tote is full, it is placed onto the appropriate take-away conveyor for transport to the final sort area.

The disadvantages and advantages of this method are similar to those of the double-conveyor method with the following exceptions: an additional investment in the conveyor controls and method designed for almost any product flow pattern and any building layout.

Other Return Process Area Design Features.   In all these methods, the preliminary sort containers are transported manually or by trolley to the opening workstation. Then all difficult or problem return packages are placed into a special colored container and placed onto the final sort transport conveyor. During its travel, this container is diverted from the transport conveyor to the problem return workstation for special attention. Some of these problems are poor or unclear handwriting, old merchandise, and merchandise shortage or overage (surplus).

Customer Return Characteristics.   When handling customer returns, the following characteristics have been realized by the writer: (1) the majority of the customer returns occur within 4 to 5 weeks from the shipping date; (2) after the merchandise is returned to the pick position, 80 percent of the SKU pick transactions occur within the first 2 weeks; (3) when small items are handled in a bin box, then the bin box has an open front for easy employee deposit and pick transfer of the merchandise; (4) with the above statistics, your total number of return activities (packing and shipping) are kept at a minimum; (5) when locating your customer returns area for maximum employee productivity and cube utilization, the pick position area is close to the receiving area and has multiple levels with dividers (it is not located near the shipping area or take-away conveyor); (6) there are three types of pick positions in the returns pick area (one for small-sized SKUs, one for medium-sized SKUs, and one for large-sized SKUs); and (7) for accurate inventory control, use the bar-code scanning method with the pick position bar code on the pick position.

How to Handle Your Customer Returns to the Pick Area.   If your warehouse operation is required to handle customer returns, then the activity is similar (or reverse) to order picking of slow-moving SKUs.

Customer Return Placed in the Active Pick Position.   When the customer returns are placed in the active pick position, an employee must look up the active pick position for the SKU and manually replenish the single item or carton to the SKU pick position. This activity is considered a reverse order-pick activity that handles SKUs with a low hit density and hit concentration, the returns employee has a great travel distance between two SKU pick (return) positions, there is a random sequence of return SKUs in the tote or on the pallet load, and the position can be located in any warehouse aisle.

Some disadvantages of this method are that it is a labor-intensive, expensive, and slow process, and in a floating-slot inventory method it is difficult to obtain the active pick position. Advantages of the return method are that the SKU inventory is in one pick location and it is possible to have the oldest SKU inventory pick first for the next customer order.

Customer Returns Handled in a Special Warehouse Zone. An alternative customer returns handling method is to lay out a special zone in the warehouse to handle your customer returns (Fig. 6.44). In this special zone you establish SKU pick position sizes that are designed to accept three different sizes (one for small SKUs, one for medium-sized SKUs, and one for large SKUs). The number of each SKU pick position is sufficient to accommodate a predetermined number of weeks of customer returns which is based on your past return experience.

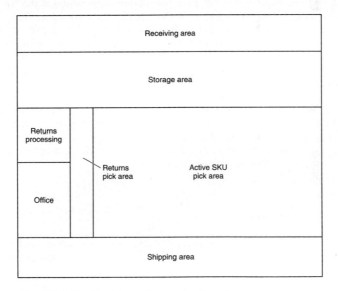

FIGURE 6.44 Special zone for merchandise returns.

As your return employee travels through the special return aisle or zone, at the first vacant pick (return) position, the employee transfers the SKU from the vehicle load-carrying surface to the return pick position. After the transfer of the SKU to the pick position, the employee writes the SKU pick position identification number and the SKU identification number on the return transfer form. Al alternative information transfer is the bar-code scanning method, where your employee hand-scans the SKU identification bar code and the pick position bar code. With both methods, the SKU and pick position identification numbers are entered into the computer inventory files.

To maintain a good inventory turn and good warehouse space utilization, this return inventory in these temporary pick positions is held for a 4- to 8-week period and then is withdrawn en masse. This SKU return inventory is sent to the discount (outlet) store, donated to charity, or placed in the trash.

Various Inventory Control Methods for Returns.   After the inventory update, you establish the return inventory priority. The first alternative is to have the return inventory handled as a new receipt of product on a last-in first-out (LIFO) rotation. With this method, when the existing inventory in the (normal) present pick position is depleted by your customer orders, then the return pick position becomes the next active pick position for your next customer order. This is a floating-slot inventory arrangement for this special return area.

Disadvantages of this method are SKU inventory in two warehouse locations and oldest SKU withdrawn last. Advantages of the method are that, for your regular merchandise order-picking activity, it maintains your order-picker productivity rate and has high return-to-stock employee productivity.

In an alternate method the return inventory is handled as the oldest inventory in the warehouse on the first-in last-out (FILO) product rotation. In this arrangement the computer treats the return SKU inventory with the highest priority. With the next customer order, the computer prints the pick instruction document for the SKU in the return pick position. When the inventory is depleted by the customer orders, the computer transfers the next pick instruction to the normal SKU pick position.

Disadvantages are low employee pick productivity during peak activity because of increased travel time through the returns area and SKU inventory in two locations. Advantages of the method are that at the end of the period, it reduces the quantity of returns inventory and removes the oldest SKU from the inventory.

Final Return Sort Area.   The final sort area is the location that consists of a series of bins or a mechanized sortation system (tilt-tray, bomb-bay drop, flap sorter, SBIR, or gull-wing) that separates the return merchandise according to a specific warehouse zone. After this sortation, the return merchandise is sent to the appropriate warehouse zone, where it is placed into the pick position.

Out-of-Season Merchandise

The out-of-season product handling activity is a temporary warehouse holding function for retail store merchandise that did not sell (or did not sell in the catalog) and has been returned to the warehouse for storage and management's decision on how to handle the merchandise. Until top management decision is made on the disposition of the merchandise, this merchandise is held in the storage area. In the storage are, for best operational results, the merchandise is placed in the most remote dense storage positions.

Retail Store Transfers

Retail store transfers consists of an overstock product from one retail store that is transferred to another retail store that has high sales of the product. Store transfers with the proper paperwork are handled as across-the-dock merchandise at the warehouse.

For an accurate and efficient store transfer system, the retail distribution company requires that each carton be identified with the appropriate store destination label and that the proper paperwork accompany the transfer.

Trash Handling Activity

In the small-item (broken-case) or garment-in-the-box industry, the nature of this warehouse and distribution business creates a large volume of trash. Your business receives large quantities of merchandise (product) in pallet loads or cartons and serves customers with smallest quantities of merchandise. The merchandise flow and operation create trash such as vendor cartons, interior packaging material, inner packs, plastic wrapping and strapping, wood, metal, and paper.

What Activities Create Trash. The trash handling activity assures you that the trash is removed in a timely manner from the work area. The work area is one of the following ares: receiving and packaging area, opening-prior-to-sort area, and count-and-pick area.

Trash Handling Design Parameters. The type of trash handling method that is used in your operation is based on the location of the work area and the trash disposal location: the type and size (length, width, height, and weight) of the trash that is created from each activity, the volume of the trash created from the work activity (average and peak), the available clear space for the various trash (material) handling concepts, available capital investment, and whether you separate the trash for a recycle program.

Various Trash Handling Methods. The various methods of removing trash from the work area are divided into two general groups: manual and mechanized. These various methods are (1) manual methods—containers and (2) mechanized methods—belt conveyor or overhead trolley conveyor.

Trash Container Methods. The trash container is the most basic and simple trash handling device. The components are trash barrels, baskets, or self-dumping hoppers that are placed adjacent to (within easy employee reach of) the workstation. As required, the employee deposits the trash into the container, and another employee with material handling equipment transfers the full container to the trash disposal location. These containers are periodically washed, or disposable plastic container liners are used to line the interior of the container. In the food industry or other industries that require a high level of sanitation, this is a common practice.

Disadvantages of the trash container method are increased number of trash handlings, requirement for another employee to handle the trash, transportation and cleaning expense, and mixed trash types in one container. Advantages are no or low capital investment and flexibility (procedure can be done anywhere in the warehouse).

Mechanized Trash Handling Methods. These include trash (material) handling methods in which the trash is mechanically moved from the workstation to the trash disposal location. At the trash disposal location, the trash is either mechanically or manually transferred into the disposal system.

The *belt conveyor* is a very common trash handling apparatus that handles a mix of trash. The motor-driven belt conveyor consists of a rubber or fabric belt that travels a fixed path past the workstation. The conveyor path is elevated or below the work surface and has side guards on both sides of the conveyor. The far side guard farthest from the employee or in all transport locations is 24 to 36 in high. In front of the workstation, the sideguards and E (emergency)-stop pull cords are low. On long runs, water from the first conveyor falls (dumps) onto a second conveyor which has jam controls at this location. At the workstation, the employee places the trash onto the belt conveyor surface. The belt conveyor transports the trash from the work area to the trash disposal location. At the trash disposal location, the belt conveyor surface returns over the pulley, thus permitting the trash to fall into the trash disposal system hopper. The hopper has jam controls and photo-eyes that control the trash disposal systems operation or cycle of the trash compactor ram. Disadvantages of the belt conveyor trash handling system are that it mixes trash types, requires an investment, and does not handle large trash items; and since it is a fixed-path system, it is designed into the work area layout (is not flexible or mobile). Advantages are that it handles a high volume, eliminates the need for employee handling, and eliminates trash from your employee workstation area.

The *overhead trolley trash handling system* is another mechanized trash handling method, consisting of a closed-loop powered chain. As the chain path travels past the workstation, there are E-stop pull cords. The chain has a trash item carrying device on 5- to 6-ft centers, and these extend downward to an elevation that permits an employee to place or attach the trash to the carrier, hook, or clasp. At a rate of 10 to 12 carriers per minute, the carriers transport the trash from the workstation to the trash disposal system hopper or onto a belt conveyor, or an employee removes the trash from the transport device. To provide an employee path across the conveyor, lift gates are located at a location where the overhead trolley provides a $7\frac{1}{2}$-ft clear path between the bottom of the carrier and the floor. Disadvantages of the overhead trolley are that it handles cartons only, requires a capital investment, is designed into the layout, and is a fixed-path system. Advantages are lower investment (than for the belt conveyor), fewer employee handlings, and elimination of trash from the work area.

With the overhead trolley chain method, there are three devices to transport trash: the *flat carrier,* the *hook carrier* (single or double), and the *clasp carrier.*

The *flat carton carrier* has a carrying surface of wire mesh that has a slight slope to the rear. At the rear, there are two metal strands as backstops and a main connecting device to the chain pendant. The flat carrier is used to replenish cartons to the work area. One or two empty cartons or a carton with trash is placed onto the carrier. At the disposal station, the trash is manually or automatically transferred to the disposal location. A disadvantages of the flat carrier is that it requires a wider path. An advantage is that it requires less vertical clearance because the carton is carried in a low profile.

The *hook carton carrier* has a single or double hook extending downward from the overhead trolley chain pendant. The end of the hook has an upward pitch to ensure that the carton is retained on the hook. At the trash disposal station, the carton is manually or automatically transferred to the disposal system. Disadvantages of the hook carrier are that it handles cartons only and requires a high, clear ceiling. An advantages is that it requires a narrow transport path.

The *clasp carrier* is a single-carrying device consisting of two components: a flat metal strip and a sawtooth clasp which faces the metal strip. After an employee inserts a carton flap between the two components, the carton is secured and hangs downward. As the clasp arrives at the trash disposal station, a trigger unlocks the clasp and the carton falls into the trash disposal system. Disadvantages of this device are that it handles cartons only and requires a higher ceiling. An advantage is that it requires a narrow path.

CHAPTER 7

CARTON (FULL-CASE) HANDLING IN WAREHOUSE AND DISTRIBUTION OPERATIONS

INTRODUCTION

The objective of this chapter is to identify and evaluate carton handling equipment applications, methods, and technologies that make carton (case) warehouse and distribution operations more efficient and effective. These factors improve profits and service to a company's customers.

This chapter reviews the key warehouse and distribution functions of carton receiving, storage, order pick, sortation, and shipping. Included in the chapter are reviews of the various manual, mechanized, and automatic storage, pick, and sortation methods. This includes an analysis of each order-pick and sortation method operational characteristics, disadvantages, advantages, and design factors.

PURPOSE OF A CARTON WAREHOUSE OR DISTRIBUTION OPERATION

The purpose of a carton (case) warehouse and distribution facility is to ensure that the right SKU is in inventory, available at the appropriate time, in the correct condition, order-picked on schedule, in the correct quantity and delivered to the required location to satisfy your customer order.

VARIOUS ACTIVITIES

To obtain an efficient, accurate, and profitable carton warehouse, you arrange the rack layout and carton handling equipment, establish procedures, and organize the employees to optimize the following activities:

- Pre-order-pick activities

  Unloading
  Identification
  Deposit into storage
  Receiving and checking
  Internal transport

- Order-pick activities

  Identification
  Sortation

- Post-order-pick activities

  Storage
  Internal transport
  Loading and shipping
  Replenishment
  Manifesting
  Customer returns and out-of-season and store transfers

Unloading Activity

The first key warehouse activity is the unloading of the delivery vehicle such as a common-carrier truck, vendor truck, express delivery, railroad car, container, or company truck (backhaul).

Typically, the outside (nonemployee) truck driver unloads the common carrier or vendor delivery truck. A company employee is responsible for unloading the railroad car, container, and backhaul delivery vehicles.

Receiving Activity

The second key warehouse activity is the receiving activity to assure that the vendor actual delivery quantity matches your company's purchase order quantity.

When your company receiving department verifies the vendor delivery quantity, either of two methods is used by your receiving operation: the manual or mechanized operations, which were reviewed in Chap. 4.

Order-Pick Activity

Order-Picker Instruction Methods. The fundamental component of any carton order-pick method is the order-picker instruction, which directs the order picker to the SKU pick position and indicates the case quantity that has been ordered by your customer or workstation.

The order-picker instruction method is matched with the proper order-picker method. These order-picker instruction methods are considered for implementation in a new or existing carton order-pick facility. This section of the chapter lists the order-picker instruction concepts. In Chap. 5, each order-pick instruction concept was reviewed in more detail.

The various order-picker instruction concepts are

- Printed paper or card document, manual or machine (computer)-printed
- Paper labels, computer-printed
- Paperless
- Voice-directed

Customer Order Handling. Another major factor of a carton order-pick method is how your customer orders are handled in the pick area. This factor determines the number of order pickers in the warehouse aisle. The various order handling methods are (1) no method; (2) single customer order, single order picker; (3) single customer order, multiple order pickers; (4) zone (variable or fixed); (5) batched (grouped customers); and (6) automatic sequential order pick.

A more detailed analysis of each customer order handling concept is presented in Chap. 5.

How to Route the Order Picker through the Aisles. The next factor of a carton order-picker system is how the order picker is routed through the warehouse facility to the appropriate pick positions. This factor determines the amount of travel time that is required between two pick positions. The various order-picker routing concepts are reviewed in Chap. 5. A list of the concepts is as follows:

- Nonrouting
- Sequential routing

  Single-side order-picker (one or two order pickers per aisle)

  Loop routing pattern

  Horseshoe or U routing pattern

  Z routing pattern

  Multilevel [HROS (high-rise order-selector)] pattern (one- or two-way aisle; four or six levels)

  Zone (fixed or variable)

SKU Location System Affects Picker Productivity. The next factor that influences carton order-picker productivity is the type of SKU locator system used in the warehouse. The inventory (SKU) location determines the SKU pick position in the warehouse aisle. The two alternative SKU location concepts are the fixed-slot or floating-slot concept. The fixed-slot concept requires fewer aisles and an increase in potential stockouts and improves the SKU hit concentration and density. The floating-slot concept requires 20 to 25 percent additional pick positions (floor area), reduces the stockout potential, and lowers the SKU hit concentration and density (order-picker productivity).

These location concepts are reviewed in detail in Chap. 12.

Definition of a Carton Order-Pick Concept. A carton (case) order pick concept is defined as the combination of rack or shelving layout, material handling equipment, and labor group that reads your order-picker instruction form and physically transfers the cartons from the pick position to the vehicle load-carrying surface or conveyor. These cartons are sorted, assembled in a shipping staging area, and loaded onto your customer's delivery vehicle.

Why Order-Pick Activity Is Crucial to Successful Operation. There are several reasons why the carton order-pick concept is considered as the key to a successful carton handling operation.

First, a carton distribution facility objective is to withdraw from inventory, on schedule, cartons of product that are on a customer's order instruction form. When completed, this activity maintains a satisfied customer group, allows the manufacturing department to maintain its production schedule, or permits the retail store to have product on hand for retail sales that satisfies the final customer.

Second, a review of any distribution facility annual expense budget indicates that the order-pick labor budget line item has the highest budget or dollar expense value and correspondingly in the distribution facility, the order-pick activity has the greatest number of employees.

ORDER-PICK METHODS

The purpose of this section is to discuss the three basic types of order-pick methods for the picking of cartons and help you determine which is best for your facility. The three basic types are described here:

1. *Manual order-pick methods.* In these methods an employee walks or rides on a vehicle to the SKU pick position, removes the appropriate carton(s) from the pick position, and transfers the carton(s) on a vehicle load-carrying surface for transport to the workstation, dock staging area, or customer delivery truck.

2. *Mechanized carton order-pick methods.* The main characteristics of these methods are that the carton(s) is (are) manually removed from the pick position and then transported by conveyor from the pick position to the dock staging area, to the manufacturing workstation, or to the customer delivery truck.

3. *Automatic carton order-pick methods.* The unique characteristic of this group is that the entire carton order-pick activity is controlled by a computer. According to the SKUs on your customer's orders, the computer network automatically or mechanically releases the appropriate cartons from the pick positions onto a take-away conveyor. The conveyor network transports the cartons from the pick area to the dock staging area, to a workstation, or onto the customer delivery truck.

If these basic carton order-pick methods are to make your operation highly productive and efficient, your distribution facility should be designed with an understanding of the proposed carton order storage-pick concept, the inbound and outbound product flow patterns, and the product replenishment requirements.

The basic carton order-pick methods available for design and implementation in a new or existing distribution facility are manual methods. For best operational results, the mechanized and automated order-pick methods are designed for implementation in a new distribution facility. These facilities are designed and constructed to house the carton order-pick area, replenishment area, receiving and shipping dock areas, the inventory in the storage area, and the support facilities.

Manual Order-Pick Methods

Manual carton order-pick methods are considered the simplest of the various order-pick methods. An employee walks or rides a manually or remotely controlled mobile order-pick vehicle. These vehicles travel through 150- to 300-ft-long aisles that are in between pick positions. These pick positions are normally pallet rack rows that are one or two levels high.

In a manual carton order-pick concept all inbound pallet loads are deposited into pallet reserve positions. As required, replenishment pallet loads are let down (transferred) from the reserve position to the pick position. The aisle clear width is designed for the pallet load handling vehicle right-angle turn requirement and to permit two-way order-picker vehicle traffic.

The manual carton order-pick method with one- or two-high rack or floor pick positions is designed for a manual pushcart, manually or remotely controlled tugger, or manually or remotely controlled single- or double-pallet truck. The remote-controlled vehicles are considered the most sophisticated and provide the best opportunity for excellent employee productivity. If your customer delivery method uses a four-wheeled cart, then an employee pushes a cart or uses a tugger to tow the carts through the warehouse pick aisles. The tugger concept is the most productive order-pick vehicle for carts.

In a three- or higher-level (multilevel) order-pick concept, the high-rise order-pick vehicle (HROS) is preferred for this method.

Floor-Stack Position.   Floor stack position is a basic factor for a pick area. The unitized loads of the same SKU are set on the warehouse floor. A floor-stack pick position is on the floor and is adjacent to the aisle. Behind this pick position are reserve pallet loads that are stacked on top of one another.

Drive-in or Drive-through Rack.   The design for this method is similar to that of the floor-stack method. The drive-in or drive-through methods handle unit loads of the same SKU that cannot support another unit load. In this rack concept, the first floor pallet position adjacent to the aisle is the pick position. The remaining pallet positions in the drive rack lane are the ready reserve positions in the rack or on the floor.

Pallet Flow or Push-Back Rack.   In these arrangements there is only one SKU per lane. The unique feature is that the setup is only one pick position high and the ready reserve pallets flow automatically to the pick position.

Standard Pallet Rack.   In this arrangement the first and second rack levels serve as the pick positions. Each pallet position constitutes one SKU. The remaining rack levels (positions) are pallet load ready reserve positions.

Pick Modules.   A typical manual order-pick system pick module of racks and aisles fits into a 40-ft-wide bay. This module has four rack rows (each holding a 4-ft-deep pallet), two 13-ft-wide aisles, and three flue spaces 6 in wide. An HROS pick module best fits into a 42-ft-wide bay. This module consists of six rack rows 4 ft wide each, three very narrow aisles of $5\frac{1}{3}$ ft width, and four flue spaces of 6 in width.

Manual Order-Pick Operation.   The operation of a manual carton order-pick method requires the following activities. The order picker pushes a floor truck or rides on a vehicle through the warehouse aisles to the pick positions that appear on the order-picker instruction form (document or labels). The instruction form has the required pick position numbers in a numerical sequence that directs order picker travel in an organized progressive path through the warehouse aisles to each required pick position.

At the required pick position, the order-picker instruction form directs the employee to transfer the required SKU (carton) from the pick position onto the mobile vehicle carrying platform.

With the remote controlled vehicle, the order picker walks in the aisle between the vehicle and pick positions. From this aisle location, the order picker has the ability to control the forward horizontal movement of the vehicle to stop at each required pick position that is indicated on the order-selector instruction form. When compared to the other carton order-pick methods and vehicles, this remote-control feature allows high employee productivity because of the decrease of the employee walking distance.

When the order picker completes the required picks or builds an outbound pallet load to a predetermined height, the order picker transports the pallet load to a preassigned outbound dock staging area. In this area, the outbound pallet load is deposited in the dock staging area. With the completion of your customer order, these pallets are loaded onto your customer delivery vehicle. The order picker travels to the order-picker dispatch station and receives a new set of order-pick instructions. These order-pick instructions are for the same customer order or for a new customer order.

Putaway and Replenishment Activities. The pallet load putaway and replenishment activities in the manual carton distribution facility that uses a fixed pick position concept has one or two alternative variations.

Manual Putaway and Replenishment. With the manual method, the lift truck operator decides to remove an empty or partially depleted pallet load from a pick position and transfers a new full pallet load from a reserve position to the SKU fixed pick position. With this method, the inbound pallet loads are put away to pallet reserve positions which are random pallet reserve positions. These positions are typically above the pick positions and are determined by the lift truck operator.

Computer-Controlled Document Method. In this method the lift truck operator has a computer-printed document that indicates a specific pallet load in a ready reserve position that is scheduled for transfer to the depleted pick position. The inbound load is assigned by computer to the ready reserve position, and the computer determines the time that the pick position is depleted of inventory and requires a new pallet load.

Disadvantages and Advantages of the Manual Order-Pick Method

Disadvantages. The manual carton order-pick method has many disadvantages: (1) it requires the greatest number of order pickers (labor expense) and associated order-pick vehicles; (2) the lowest productivity because the order picker picks one customer order per trip; (3) the best order-selector productivity averages 125 to 150 cartons per hour (with a small order, order-picker productivity probably declines to 75 to 100 cartons per hour); (4) with two-high pick positions, the rack layout for the required pick positions requires a large-square-foot area; (5) this method requires the greatest number of employees and amenities at the facility, which means a large-square-foot building, land (site), and parking area; (6) the method has the greatest potential for order-pick errors and product and equipment damage; and (7) implementation of this method requires the largest number of vehicles, which require a large battery-charging area.

Advantages. Advantages of the major manual order-pick methods are many: (1) the method is the easiest for management to implement and control, (2) the method handles all SKU types with any dimension, weight, or packaging characteristics; (3) the method has the lowest capital investment, (4) the method is flexible and permits a paper document or self-adhesive labels as order pick instruction, (5) the method is dynamic and facilitates adjustment of labor force and equipment for volume fluctuations; (6) the method is easily transferred into an existing facility or relocated into another facility; and (7) family grouping of product is designed into the storage-pick areas.

In a comparative carton order-pick method economic and operational feasibility study, the manual carton order-pick method was considered as the conventional or basic method.

The various manual order-pick methods and vehicles include

- Employee carry
- Platform truck or dolly
- Manual pallet truck or jack
- Electric pallet truck or jack
- Remote-controlled vehicles
- Two-wheeled hand truck
- Burden (employee) carriers
- Lift trucks and attachments
- Manual tractor or tugger
- HROS (multilevel)

Employee Carry and Two-Wheeled Truck. The employee and the two-wheeled hand truck (Fig. 7.1*a*) are the first and second order-pick vehicles, respectively. The truck is a nonpowered and nonride vehicle. The two-wheeled truck is designed with an H frame that has two handles on the upper end, with horizontal bracing between the frame legs, with two wheels on the axle at the H-frame lower end, and with a load-carrying surface that has a lip or "nose" ahead of the wheels.

The two-wheeled hand truck is used for a SKU that weighs up to 2,000 lb or the number of cartons limited to the frame or lip space. To operate the truck, an employee picks up or places the cartons onto the load-carrying surface, balances the truck, and pushes the two-wheeled hand truck from the SKU pick position to the shipping location. The two-wheeled hand truck is used in a floor or two-high rack position system.

Disadvantages of the two-wheeled hand truck are that during the transport activity, the employee must balance the truck, and must also push the load, which causes fatigue. Also, when the truck is used to transport the SKUs a short distance, the load-carrying surface transports a limited square-foot load (number of cartons), and additional strapping is added to secure the load to the truck.

Advantages of the two-wheeled hand truck are low capital investment, it requires little employee training and very little maintenance requirement, no requirement for replacement fuel or a battery-charging location, and empty pallet board is not required for the activity.

Dolly and Four-Wheel Platform Truck. The second and third manually controlled order-pick vehicles are considered as one group. This group consists of a dolly (Fig. 7.1*b*) and a four-wheeled platform truck (Fig. 7.1*c*). Both are nonpowered and nonride vehicles. These vehicles are designed with three or four wheels (at least two are steering wheels or casters) and a solid or slat load-carrying surface. The dolly does not have push handle, whereas the platform truck does. To move the dolly, an employee

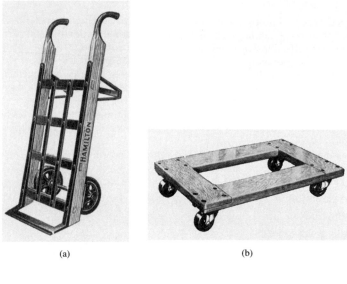

(a)   (b)

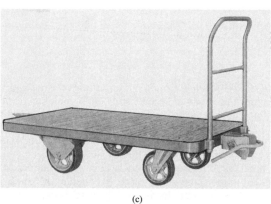

(c)

(d)

FIGURE 7.1 (*a*) Two-wheeled hand truck (*Courtesy of Hamilton Caster and Mfg. Co.*); (*b*) dolly (*Courtesy of Hamilton Caster and Mfg. Co.*); (*c*) four-wheeled platform truck (*Courtesy of Faultless Nutting Industries.*); (*d*) burden (personnel) carrier (*Courtesy of Cushman.*).

pushes against the case exterior surface. This feature, and the small wheels which have swivel casters, allow the dolly to handle lightweight and rigid cartons. This characteristic makes the dolly difficult to steer. The platform truck has large wheels (the two rear wheels are rigid and the two front ones are swivel) and a push bar that handles a wide variety of loads of up to 4000 lb. This feature makes the platform truck easy to steer.

To operate these vehicles, the cartons are placed by an order picker onto the vehicle load-carrying surface. The order picker pushes against the cartons on the dolly or the platform truck handle. This force moves the truck between two pick locations. These vehicles are used in a floor stack or two-high rack pick position method.

Disadvantages of the dolly and platform trucks are the same as those of the two-wheeled hand truck. An additional disadvantage is that unloading of the vehicle requires another labor activity, which lowers the total facility productivity. An additional disadvantage of the dolly is that without a push bar or handle, only a limited number and type of cartons can be stacked onto the dolly load-carrying surface.

When compared to the two-wheeled hand truck, these vehicles have several additional advantages. First, there is less employee fatigue, and an employee can travel longer distances. Second, the platform truck handles larger and heavier loads.

Burden (Personnel) Carrier. The fourth manual order-pick vehicle is a burden (personnel) carrier (Fig. 7.1*d*), which is a LP gas-, gasoline-, or electrically powered rider vehicle. It has three or four wheels (at least one is for steering), a steering device, an operator's position, forward and reverse controls, and a solid deck load-carrying surface.

In comparison to the other three manually controlled and nonpowered vehicles, the burden carrier is used to transport the order picker and a load over a greater distance, with no fatigue and in less travel time. However, the burden carrier power and load-carrying surface restrict its ability to transport loads more than 500 lb, and its carrying surface is designed for a small quantity of cartons.

While traveling through the rack-and-aisle layout, the operator stops the vehicle at the required SKU location, places the cartons onto the load-carrying surface, activates the power source, and steers the vehicle to the next required location. The burden carrier is used in a floor or two-high rack pick position system.

Disadvantages of the burden carrier are limited weight-carrying capacity, ability to transport only a limited number of cartons, high equipment cost per carton throughput, employee handling required, increased maintenance costs, floor space required to store fuel or recharge batteries, and low total facility productivity due to additional labor required in the staging area to transfer the cases from the order pick vehicle to the shipping vehicle or device.

Advantages of the burden carrier are that it transports the order picker and cartons a greater travel distance and is able to travel the required travel distance in a shorter time period, and the employee's physical effort is reduced to transport the cartons, which means less employee fatigue.

Hand-Operated Nonpowered Pallet Truck. The fifth manual order-pick vehicle is the hand-operated nonpowered walkie pallet truck (jack) (Fig. 7.2*a*). The hand-operated pallet truck is designed with an operator's handle that activates the hydraulic elevating mechanism, hand- or foot-operated hydraulic pressure-release lever, one or two steering wheels, and two load-carrying wheels under each fork. To prevent cartons from falling from the pallet board, a backrest is attached to the vehicle. The hand-operated pallet truck handles loads up to 3000 lb. For best results the cartons are placed onto a pallet board or skid.

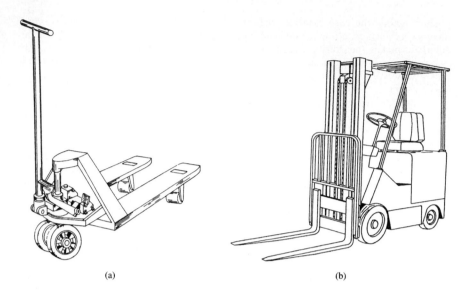

(a)   (b)

FIGURE 7.2   (*a*) Manually operated pallet truck (*Courtesy of Plant Engineering Magazine.*); (*b*) powered lift truck (*Courtesy of Plant Engineering Magazine.*).

The vehicle requires an order picker to pick up an empty pallet board, to set the hydraulic lever in the "elevate" position, and to elevate the pallet board. The order picker pulls or pushes the vehicle through the warehouse and assembles cartons onto the pallet board to a predetermined height or until all lines are completed on the order list. With completion of the order-picker task, the pallet board with the outbound cartons is deposited in the assigned area. To deposit the load, the operator sets the hydraulic lever in the "to lower" position. The hand-operated nonpowered walkie pallet truck is used in a floor or two-high rack pick position system.

Disadvantages of the hand-operated pallet truck are that it is a slow means of moving an order through the facility aisles, requires an empty pallet board, and tape or string is used to secure the cartons as a unit load. Typically, the string becomes wrapped around the vehicle wheel and operator becomes fatigued from pushing or pulling the vehicle through the warehouse aisles.

When compared to the other manually powered order-picker trucks, advantages of the hand-operated pallet truck are the equipment economics per throughput, higher number of cartons handled, for the nonpowered group, and very little employee training required. Also with a pallet board, the vehicle handles a wide variety of item shapes (SKUs), the vehicle requires a low capital investment and very little maintenance, and if all orders are on a pallet board, then there is an increase in loading productivity.

Powered Lift Truck.   The sixth manual order-picker vehicle is the rider-powered lift truck (Fig. 7.2*b*), which is either LP gas-, gasoline-, diesel-, or electric battery-powered. Generally, an indoor lift truck is electrically powered or in some instances LP gas-powered. It has an operator's platform that has the truck's forward-reverse movement controls and fork elevate-lower controls. Other design features include a set of vertically moving forks with a backrest that is attached to a telescopic mast, and three or four travel wheels, at least one of which is a steering wheel.

After an empty pallet board is picked up, the lift truck has the capacity to handle a customer's order that weighs 2000 to 4000 lb. Typically, an electric powered truck has a 12- or 24-V battery.

When the lift truck is used as an order-picker vehicle, the operator picks up an empty pallet board, maintains the forks at an elevated level, drives the vehicle down-aisle to the SKU location, stops the vehicle, leaves the operator's platform, places the required cartons onto the empty pallet board, returns to the operator's position, and travels down-aisle to the next position. After completing all the required order-picker tasks, the operator travels to the assigned dock area and deposits the customer order in the assigned area. With the appropriate aisle width, the lift truck is used in a floor or two-high rack pick position.

Of all the available vehicles, the lift truck is the only vehicle that accommodates a basiload (Fig. 7.2c), clamp (Fig. 7.2d), or pipe device. These devices allow the lift truck without forks to handle a layer of cartons, large cartons, or household appliances such as a washer, dryer, refrigerator, or roll of carpet.

Disadvantages of the lift truck as an order-picker vehicle method are that the vehicle is not designed to perform the carton order-pick task unless it is used to handle a layer of cartons, appliances, or a roll of carpet. Further disadvantages are low employee productivity when picking cartons because the operator has increased nonproductive time of moving to and from the operator's platform and vehicle operation requires employee training; also, the vehicle requires a high capital outlay and higher maintenance costs and a replacement fuel storage and battery-charging area.

Advantages of the lift truck as a carton order-picker vehicle are that when all other mobile equipment is not available, it is a vehicle of last resort; the backrest reduces the cartons from falling or becoming damaged; with the outbound unit load on a pallet board, the loading productivity is increased at the facility; the vehicle is used for storage, transport, loading, and unloading activities; and with an attachment, it handles a unit that an employee cannot reasonably handle.

Electric Powered Rider or Walkie Pallet Truck. The seventh order-picker vehicle is the electric powered end rider (Fig. 7.3a) or walkie single-pallet jack (truck) (Fig. 7.3b) that has one drive (steering) wheel and two load-carrying wheels. These load-carrying wheels are under the forks. The rider truck has an electric battery. An operator's platform is located in the middle or at the front end of the vehicle and has the complete controls for the vehicle and a horn. At the rear is a backrest with a set of forks that elevate and carry the unit load. The truck requires a pallet board or skid and has the capacity to carry a 2000- to 4000-lb load. Some trucks do not have an operator's platform; thereby, the operator walks and controls the vehicle with the vehicle handle controls. Typically, these vehicles have 12- or 24-V battery power source.

These trucks have the same operational order-picker procedures as the hand pallet truck except that the driver mechanically controls the horizontal and vertical movement of the truck or forks. Also, with the rider model, the employee rides between warehouse locations. These trucks are used in a floor or two-high rack pick position concept.

Disadvantages of the electric powered end-rider single-pallet truck are limited transport (generally one unit load), higher economic investment and maintenance costs (pallet truck), requirement for a battery-charging area, and additional employee training required. Advantages are many and make the vehicle one of the best case order-pick vehicles. These advantages include improved productivity because the truck is powered, thus reducing employee fatigue; the truck can be dispatched to travel greater distances and can transport a heavier load; a backrest is provided, which allows a higher unit-load which reduces cartons from falling to the floor or becoming damaged;

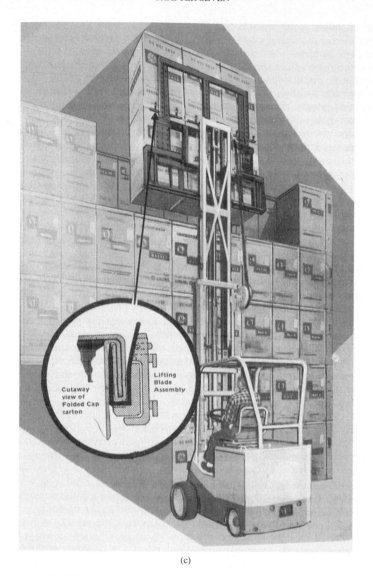

(c)

FIGURE 7.2   (*Continued*) (*c*) Basiload (*Courtesy of Basiloid Products Corp.*);
(*d*) clamp (*Courtesy of Cascade Corp.*).

forward and reverse controls on the vehicle handle allow the order picker to control the vehicle while walking or riding down-aisle; with the outbound order on a pallet board, loading productivity is improved; and the truck can transport a unit load up a grade and can be used as an unloading, transporting, or unloading-loading vehicle.

Manually Operated Electric Powered Rider Double-Pallet Truck.   The eighth order-pick vehicle is the manually operated electric powered rider double-pallet truck

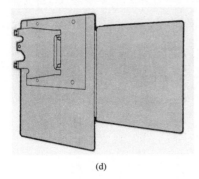

(d)

FIGURE 7.2   (*Continued*)

(jack). This truck has two designs. The first is an end-rider type (Fig. 7.3*c*) and the second is a midcontrol and remote-control type (Fig. 7.3*d*). The design and operational characteristics are similar to those of the manually operated electric powered single-pallet truck. The difference between these two vehicle types is that the double-pallet truck has the capacity to carry two unit loads or 6000 lb. These unit loads are located behind the operator's platform. The pallet truck power source is a 12- or 24-V battery.

Disadvantages and advantages of the double-pallet truck are the same as those of the powered single-pallet truck. Additional disadvantages are that the vehicle requires a wider intersecting (turning) aisle, travels up a lower grade, and requires additional employee training. An additional advantage is that the vehicle is more productive because it transports two unit loads, making it one of the best manual case order-pick vehicles. It is used in a floor-stack or two-high rack pick position system.

Manually Operated, Powered Midcontrol Rider Pallet Truck with a Step Platform. The next vehicle is a manually controlled electric powered midcontrol rider double-pallet truck (jack) with a step platform. The vehicle design and operating characteristics are the same as those of the double-pallet truck.

One additional advantage is that the step platform allows the order picker to step onto an elevated step to reach a case from the rear of the second-level pick position. This vehicle is designed for a floor or two-high rack pick system.

Empty Pallets Stacked and Dispensed at the Order Control Desk.   It has been the author's experience as a warehouse manager of a manually operated electric powered jack operation that a stack of empty pallet boards at the order control desk improves order-picker productivity. When you have available pallet boards in the area, you prevent the order picker from making nonproductive trips to search for empty pallet boards, damage to rack legs from the pallet jack pickup process, and product damage by the employee removing a partial depleted pallet board onto the floor.

Your two options are at the order control desk: a stack of empty pallet boards that are placed by a lift truck or pallets released by a pallet dispensing machine.

The stack of empty pallets that are placed by a lift truck is very simple to operate and control. The empty pallet boards are stacked in an assigned area and as required, the order picker removes a pallet board from the stack.

Disadvantages of the method are that employees lift the pallet board and high stacks can fall, which could potentially damage the product. Advantages of the concept include no capital cost and no maintenance expense.

An alternative method is the pallet dispenser machine (Fig. 7.4); in this method a lift truck places a stack of empty pallet boards into the top of the machine. When the pallet dispenser is full, the safety gate secures the pallet stack in the tower. As the order picker removes a pallet board from the floor-level pallet take-away position, it becomes empty. The pallet dispenser tower has a sensing device that controls a motor-driven device to automatically lower an empty pallet board onto the floor position. The pallet dispenser is one or two pallet positions deep.

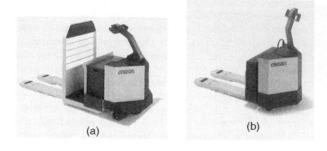

(a)

(b)

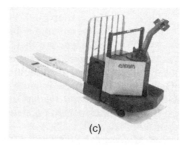

(c)

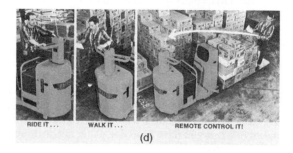

RIDE IT... WALK IT... REMOTE CONTROL IT!

(d)

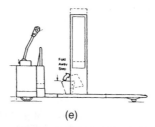

(e)

FIGURE 7.3 (*a*) Rider pallet truck; (*b*) walkie pallet truck (*Courtesy of Crown Equipment Corp.*); (*c*) double pallet truck with end rider (control); (*d*) double pallet truck with midcontrol and remote control (*Courtesy of Barrett Indl. Trucks, Inc.*); (*e*) double pallet truck with a platform step (*Courtesy of Raymond Corp.*).

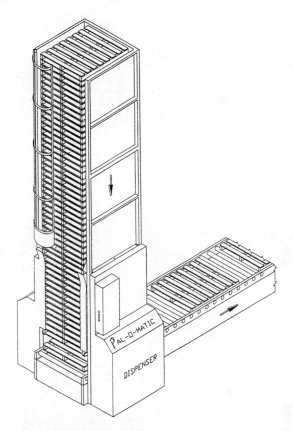

FIGURE 7.4　Pallet dispenser. (*Courtesy of Pal-O-Matic Corp.*)

Disadvantages of the pallet dispenser tower are increased investment and additional lift truck operator time to travel and deposit the pallet boards into the tower. Advantages are increased management control, reduction in nonproductive order-picker empty pallet search time, and reduced rack and product damage.

Tugger and Cart Train.　The next vehicle is a manually operated powered rider tugger (tractor) (Fig. 7.5*a*). The vehicle power source is either LP gas, gasoline, or electric battery. The tugger is designed with an operator's platform equipped with forward and reverse controls, a steering device, three or four wheels with at least one a steering wheel, and a towing hitch or coupler. Generally, the tugger capacity is a 10,000-lb rolling load or a 750-lb normal draw bar pull for a train of carts. Typically, a train consists of four to five carts, each weighing 2000 lb. The vehicle is used in a floor or two-high rack pick position system.

To operate the tugger, the operator attaches the carts to the vehicle hitch, climbs on board the operator's platform, travels down-aisle, stops at the required SKU location, walks to the rack position, transfers the carton to the cart load-carrying surface, returns to the operator's platform, and travels down-aisle to the next pick position.

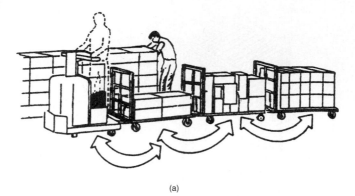

(a)

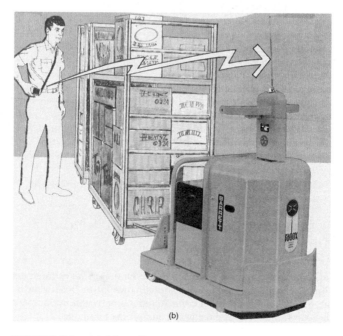

(b)

FIGURE 7.5 (*a*) Manually controlled tugger (*Courtesy of Sims Consulting Corp.*); (*b*) remote-controlled tugger (*Courtesy of Barrett Indl. Trucks, Inc.*).

Disadvantages of the tugger are that it can be used only to tug (pull) carts, requires LP gas, and requires a gasoline and battery-charging area, and the cart train requires a wide-turning (intersecting) aisle. Advantages of the tugger are that if the delivery system is a cart system, then the tugger provides the best order-picker and loader productivity and requires little employee training and the total facility productivity improves with less handling of outbound cartons at the delivery dock because the cartons are on the shipping cart.

Remote-Controlled Electric Pallet Jack and Tugger.   The next group of vehicles in the remote-controlled electric powered walkie-rider vehicle group (Fig. 7.5*b*) consists of the remote-controlled electric powered walkie-rider single- or double-pallet truck and tugger. These vehicles are the most sophisticated vehicles and provide the best case order-picker productivity in a floor or two-high rack pick position system.

A remote-controlled vehicle is a radio-controlled vehicle whose forward and directional movements are controlled by an employee with a transmitter. The transmitter sends a signal which is received by the vehicle receiver, which causes the vehicle to respond to the signal. While controlling the forward movement and steering of the vehicle, this communication network and device allows the order picker to walk in the aisle between the cart load-carrying surface and the rack (floor) positions.

Operation of the vehicle is as follows. The operator-controlled vehicle stops at the required aisle rack position, the operator places the cartons onto the vehicle load-carrying surface, and the operator walks beside the vehicle as it travels down-aisle to the next required SKU pick position.

Additional disadvantages of the remote-controlled vehicles are several: (1) when compared to the other vehicles, the capital investment cost is the highest; (2) when there are more than four vehicles per facility, then there is an additional cost for the remote-control system; and (3) the remote-control feature requires additional maintenance and employee training, and in some buildings, the extensive metal structure makes it difficult to transmit and receive signals.

The additional advantage of the remote-controlled vehicle group is a very high level of order-picker productivity as an employee's time to walk between the vehicle operator's platform and rack position is reduced.

Electric Powered High-Rise Order-Picker Truck.   The next order-picker vehicle type is the electric powered but manually controlled rider (high-rise or multilevel) order-picker. This vehicle type consists of the platform, spiral chute, side-loading, counterbalanced, and straddle order-picker and hybrid trucks. These vehicles operate best and provide excellent order-picker productivity in a guided aisle that is between at least four vertical levels of rack positions. When these vehicles are used in a floor or nonguided rack system, then the vehicle provides low employee productivity.

The most common trucks in this group are the counterbalanced and straddle order-picker trucks. These trucks have an operator's platform with steering, elevating, declining, forward, and reverse controls and an operator's harness and overhead guard that elevates and lowers with the load-carrying surface. These vehicles have three or four wheels (one of which is a drive and steering wheel) and a pallet or cart securing device.

Counterbalanced Truck.   The counterbalanced truck (Fig. 7.6*a*) has a set of load-carrying forks. The weight of the vehicle is forward (ahead) of the operator's platform and offsets the unit-load weight. This feature allows the counterbalanced truck to handle a unit load of 48 in length (typically a pallet board) and 3000 lb at a height of 10 to 20 ft.

Straddle Truck.   Operation of the straddle order-picker truck (Fig. 7.6*b*) is similar to that of the counterbalanced truck. The straddle truck is designed with two straddles that extend behind the operator's platform. With an elevated unit load, two straddles stabilize or support a unit load of 84 in length (typically a cart) and 3000 lb weight to a height of 20 to 30 ft.

Platform Truck.   The platform truck (Fig. 7.6*c*) is the next high-rise order-picker truck. Its operational characteristics are similar to those of the counterbalanced truck.

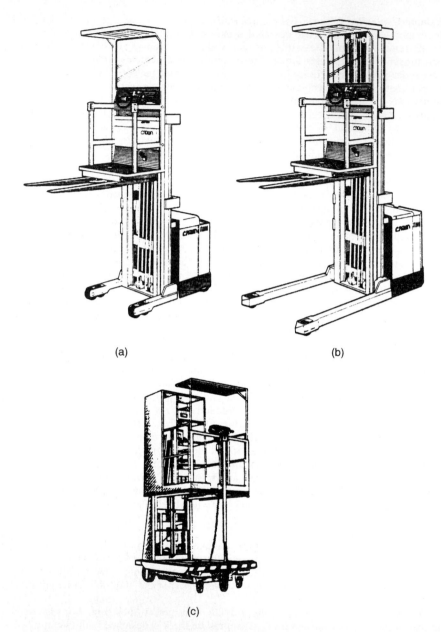

(a)

(b)

(c)

FIGURE 7.6   (*a*) High-rise order-selector (HROS) counterbalanced truck (*Courtesy of Crown Equipment Corp.*); (*b*) HROS straddle truck (*Courtesy of Crown Equipment Corp.*); (*c*) platform truck (*Courtesy of West Bend Equipment Co.*); (*d*) spiral chute (decombe) truck (*Courtesy of Rapistan Demag Corp.*).

(d)

FIGURE 7.6   (*Continued*)

The platform truck is designed to lift the operator and platform to a 20 ft height. This platform is designed to handle small cases, bins of small items, or single unit loads that weigh under 500 lb.

Decombe (Spiral Chute) Truck. The spiral chute (decombe) truck (Fig. 7.6*d*) is the next high-rise order-picker truck. This truck has an operator's platform with the elevating controls and steering mechanism that elevates the operator to 20 ft. The spiral chute truck does not have a load-carrying surface. Ahead (forward) of the operator's platform is a gravity spiral chute that extends the entire height of the rack positions. The chute and the movable platform allow the order picker to place required cases onto any part of the chute. Once in the chute, gravity and side guardrails direct the case flow to the floor level take-away conveyor. This feature limits the ability of the machine to handle sealed conveyable cartons, and it is difficult to access pick positions from both sides of the aisle.

Side-Loading Truck. The side-loading truck is the next high-rise order-picker truck. This truck has an operator's platform with controls for vehicle vertical and horizontal movement. As the truck travels down-aisle, the unit-load-carrying surface or forks face the rack position side of the aisle. With this feature, as the truck travels down-aisle, the operator and load carrying device face one side of the aisle pick positions, allowing one-side (unilateral) order-picking per down-aisle trip. This particular truck is used to order-pick long and light- to medium-weight SKUs such as pipe or lumber.

AS/RS (Automated Storage and Retrieval System) and Employee-up Turret Trucks. The operator-controlled AS/RS and employee-up turret truck (swing reach or turn-a-load) are high-rise order-picker trucks. In these trucks the operator and control platform rise up with a load-carrying platform. As the truck travels horizontally and vertically down the aisle to the required pick position, the operator removes the carton from the pick position and places it onto the load-carrying platform. As required, the activity is repeated for the number of cartons on the customer order or until the load reaches a specific height.

Typically, these trucks have a 36- or 72-V battery and an employee activated brake system. Some models have direct electric power.

Disadvantages of the high-rise order vehicle are one truck per aisle, and for optimum employee productivity the truck has a guidance system that is added to the truck high capital investment, the high-rise truck requires a multilevel rack structure, the operator requires additional training, the operator is required to wear a safety harness, and unless the pallet load replenishment is made from the same aisle, the SKU replenishment is made from a separate aisle.

Advantages of the high-rise order-picker trucks are many, especially the high employee productivity and improved space utilization achieved with these trucks. These machines allow the order picker to order-pick cartons from all vertical rack pick positions. This feature provides high SKU hit concentration and hit density and good space utilization when picking small or slow-moving SKUs, which means high employee productivity; also the rail guidance system reduces mobile equipment and rack damage and allows the vehicle to travel down-aisle at high speeds, and when compared to the other order-picker vehicles, these vehicles minimize the building footprint area to house the required pick positions.

How to Pick with the Order-Picker Truck. The operator elevates the operator's platform to the required rack level and drives down-aisle to the required pick position, places the required cartons onto the pallet board or cart, and travels down-aisle to the next required pick position. On completion of the order or when the unit load reaches a predetermined height, the order-picker deposits the outbound unit load at the end of (outside) the aisle in an assigned location.

Picking Cage.   A high-rise pallet picking cage helps reduce the employee fear of high levels and product damage. The pallet cage is a three-sided, bottomed, wood, metal, or plastic structure that holds a pallet board and has a bottom support device with a middle stringer. The open side permits the order picker to perform order-pick transactions. Elevation heights of the sides are predetermined by the cube of the order-pick computer program and prevents cartons from falling to the floor. The bottom support device permits a pallet board to set in the cavity and a high-rise vehicle securing device to clamp onto the middle stringer.

Mechanized Order-Pick Methods

The next carton order-pick methods group is the mechanized, or "product to order selector," group. These mechanized order-pick methods (SKUs are transported by conveyor or cart to and/or away from the pick position) are (1) walk pick-to-belt (standard rack or flow rack), (2) ride (pick car or decombe), (3) cartrac, (4) single or multiple carousel units, and (5) miniload.

The basic feature of the mechanized group is that your customer order-picked cartons are transported from the pick position by conveyor or cart to the staging or shipping area.

Walk Pick-to-Belt Method.   The walk pick-to-belt method (Fig. 7.7a) is considered the basic mechanized method. It is designed with a belt take-away conveyor and with human aisles on both sides of the conveyor. The conveyor and aisles are designed between two rack rows or pick positions. On the other side of the rack pick positions or pick module is the lift truck replenishment aisle.

The conveyor, racks, and aisles are designed on mezzanine levels that permit utilization of the entire distribution facility cubic-foot space. A series of stairways allow the order pickers to walk and to pick cartons from all pick positions on the floor and mezzanine levels.

The typical pick aisle length is 150 to 200 ft. The rack positions are 5- to 20-ft-long carton flow lanes, one or two standard rack positions high, or two- or three-deep pallet flow rack positions. The number of rack pick positions per aisle is determined by the carton or pallet load dimensions and required clearances. At predetermined intervals, empty rack positions are designed to hold depleted pallet boards. Slow-moving SKUs are placed in hand-stack rack, slide rack, or carton flow rack.

All lift truck inbound load putaway and pick position replenishments are performed from a separate replenishment aisle. In a standard rack and mezzanine facility, these reserve positions are in the additional rack positions that are located in the reserve racks between the pick modules. In a two- or three-deep pallet load flow rack design, the pallet load reserve positions are directly behind the pick positions. If additional pallet load reserve positions are required for SKUs, then these reserve positions are located in rack positions between two pick modules.

The typical standard rack walk pick-to-belt module is 42 ft wide and consists of two 13-ft-wide aisles, one conveyor, two rack rows, and two employee pick aisles. The three-deep flow rack module is 52 ft wide and has two (2) three-deep flow racks, one conveyor, and two employee pick aisles. The two-deep flow rack module is 30 ft wide

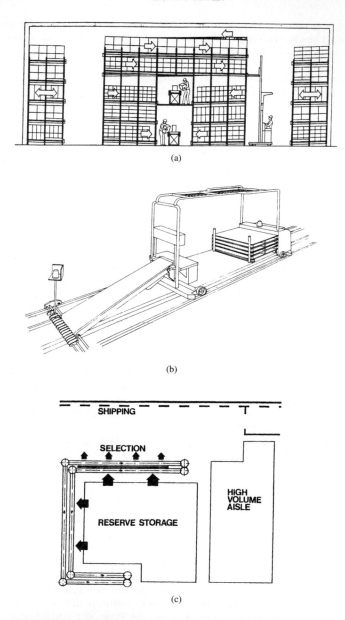

FIGURE 7.7 (*a*) Walk pick to belt (*Courtesy of Webb-Triax Co.*); (*b*) pick car to belt (*Courtesy of Rapistan Demag Corp.*); (*c*) S.I. Cartrac (*Courtesy of S.I. Handling Systems, Inc.*).

and consists of two replenishment order-picker truck aisles or conveyor aisles, two each of 5-ft-deep carton flow racks, one pick conveyor, and two each employee pick aisles.

The operation of the walk pick-to-belt concept requires, from the order control desk, the order picker to obtain a series of human- or human- and machine-readable self-adhesive labels. These labels are printed in a sequence that corresponds to the pick position number sequence. This sequence directs the order picker down one side of the pick module aisle and then down the other side or from the mezzanine level to the lower level. With crossovers, stiles, or incline belt conveyors, the picker is able to cross through the conveyor path and pick from both sides of the module. An alternative is to have two order pickers pick in the same module. Self-adhesive labels allow the order picker to pick several of your customer orders in one trip through the aisle. This batched (group of customer orders) pick method improves the SKU hit density and concentration, thus improving the order picker's productivity.

After receiving the human- and machine-readable labels, the order picker walks to the appropriate pick module pick position. At the pick position, the picker removes a carton from the rack position, places the label onto the carton with the human- and machine-readable characters facing in the proper direction (up, side, or front), and places the labeled carton onto the belt conveyor. If the next label requires the same SKU, then the picker repeats the order-pick activity. If the next label requires a new SKU, the picker walks to the required pick position and performs the order-pick activity. As required, the order picker removes all depleted pallet boards from the SKU pick positions and transfers the pallet board to the assigned empty pallet board position. A 2- to 3-in-wide metal flat surface full length along the rack front facilitates the transfer of empty pallet boards.

On all elevated-level pick positions and return pallet positions for improved employee safety, the bottoms of the pallet flow lanes or standard rack positions are decked to support an employee. An alternative to the solid deck is nylon or wire netting.

The order-picked and labeled cartons travel on the pick-to-belt conveyor to merge with the main transport conveyor. As the conveyor flow control network allows the cartons to travel from the pick conveyor onto the transport conveyor, the carton arrives at the metering belt. This belt creates a gap (open space) between two cartons. From the metering belt, the carton travels to the encode station. At the encode station, by key entry or an automatic scanning device, the carton shipping lane is programmed into the programmable controller and computer memory for manifest and shipping document preparation. After the encode station, the carton travels on the sortation conveyor. As it travels on the sortation conveyor, the carton is tracked by the tracking device to the appropriate divert station. Arriving at the divert location, the tracking device signals the programmable controller to activate the appropriate divert device to divert the carton from the sortation conveyor onto the loading lane. At the end of the loading lane, the carton is unitized onto a shipping device (pallet board or cart) or continues on an extendible conveyor and is placed onto the delivery truck floor.

In the walk pick-to-belt system, the lift truck operator removes the employee board stack from the assigned empty pallet board position.

Disadvantages of the walk pick-to-belt order-pick method are (1) it requires a human- or human- and machine-readable label; (2) it requires more exact carton volume, SKU quantity, and SKU characteristic information; (3) it requires on-time orders and increased management control along with order-picker and maintenance employee training; (4) some of the SKUs are nonconveyable SKUs and are handled off-system and are matched with the conveyable SKUs on your customer's delivery truck; and (5) when compared to the manual method, the capital investment for this method is higher for the walk pick-to-belt concept.

Advantages of the walk pick-to-belt order-pick system are (1) it handles single or batched (grouped) customer orders; (2) with the capability to add order pickers to the pick modules, it minimizes throughput volume fluctuations or downtime effects; (3) the order-picker productivity rate, on the average, is 300 to 500 cartons per hour; (4) this method provides an accurate and automatic carton sortation and shipping manifest document; (5) when compared to the manual method, the walk pick-to-belt method requires fewer employees to handle the volume; (6) with multiple levels of pick positions and two- or three-deep flow rack, the square building is medium-sized; (7) with two- or three-deep flow racks, there are fewer product handlings, which means less product and equipment damage; and (8) with additional controls and accumulation conveyors, family grouping is designed into the system.

Pick Car. The next mechanized carton order-pick method is the pick car-to-belt (Fig. 7.7*b*). This system is designed with a pick car and its carton take-away conveyor, which travels in an aisle between the two rack rows of pick positions. These pick positions are two or three levels high. The pick car aisle, standard pallet racks or flow rack pick positions, and two wide lift truck replenishment aisles require a 42-ft module.

The pick car rack layout is designed with a human pick-to-belt mezzanine level(s) located above the pick car module. This feature allows utilization of the entire available distribution facility cubic-foot space. The aisle length and number of aisles are determined by the design-year SKU throughput volume, SKU inventory quantity, and SKU characteristics. Typically, a pick car aisle is 150 to 300 ft in length and the rack row pick positions are one, two, or three levels high, corresponding to the pallet load height. Slide or carton flow rack provides additional pick positions. The self-propelled pick car requires floor rails, a travel aisle of approximately 4 ft 10 in width and an overhead clearance to elevate or lower the order picker to the required pick position. The order-picker cab is furnished with manual controls and an employee pallet storage magazine. Automatic controls, videodisplay heater, and label printer are options. The pick car has a pivoting carton take-away belt conveyor that travels through the pick car and declines to the floor.

The lift truck replenishment aisle is a separate aisle and is the same as in the walk pick-to-belt layout.

Pick car operation requires the order control desk to issue the order picker the required series of pick labels. These labels are the human- and machine-readable label type. After the order picker is on board the pick car, the controls are set for a trip to the first label pick position number. At the pick position, the order picker removes the carton from the rack pick position, places the label onto the carton, and places the labeled carton onto the belt take-away conveyor with the label facing front, to the side, or on the top. Recent models have semiautomatic or fully automatic controls and mechanisms to provide the identification of the pick location, label generation, and application to the cartons.

The labeled case flows on the take-away conveyor that has the same route as that described for the walk pick-to-belt system. The next label directs the order picker to withdraw another carton from the pick position or to travel down-aisle to the next pick position.

As required, the order picker removes all depleted pallet boards from the depleted SKU pick positions and places the pallet boards onto the pick car empty pallet position. Periodically, the empty pallet stack is transferred to an assigned empty pallet position.

Many disadvantages are associated with the pick car method: (1) it requires a human-readable label; (2) the pick car horizontal travel is restricted to one aisle, and vertical travel is restricted to the height of three to five pick positions; (3) when compared to the walk pick-to-belt method, investment for the pick car is higher; (4) because one person rides in the pick car, it has a limited ability to adapt to volume fluctuations; (5) the method requires employee and management training and controls; and (6) some SKUs are inherently nonconveyable.

The pick car method has many advantages: (1) the order picker picks from three pick positions high; (2) there is higher order-picker employee productivity because of reduced walking time; (3) there is a reduction in order-selection errors and damage to product and equipment; (4) the method allows automatic carton sortation, manifest, and shipping document preparation; (5) the pick car is heated, allowing employees to work for longer periods in refrigerated areas; and (6) with additional controls and an accumulation conveyor, family grouping is designed into the concept.

In a carton order-pick method economic and operational feasibility study, the pick car was considered as a mechanized alternative to the manual method.

S.I. Cartrac System. The next mechanized carton order-pick method is the S.I. Cartrac (Fig. 7.7c). The S.I. Cartrac system is designed with a series of carts with load-carrying surfaces that travel on a fixed-path track. The fixed cart path is past the pallet load replenishment area and past the carton order-pick area. The system is designed to make 90° (right-angle) or 180° (curve) turns. At the replenishment and pick stations, replenishment and order pickers stop the cart travel for replenishment or pick transactions.

The S.I. Cartrac layout is determined by the order-pick area, replenishment area, and turn requirements. The typical cart width is 54 × 50 in with a travel speed of 180 to 300 ft/min and an employee-activated car stop device.

The pallet load reserve area is designed to have immediate access from the replenishment area to the S.I. Cartrac area. The order-pick area is determined by the number of required order-pick stations (positions) and outbound staging area. There is a direct path between the pick area and the ship area.

Operation of the S.I. Cartrac system requires the order picker to obtain pick instructions [tickets or an order recap (recapitulation) form] from the order control desk. With these instructions, the order picker walks to the assigned order-pick station. After arriving at the order-pick station, the employee places empty pallet boards or carts into the required positions or turns on the take-away conveyor.

The S.I. Cartrac system moves the carts with a pallet load of a SKU from the replenishment area to the order-pick area. As the cart travels on its path past the order-pick area, all order pickers compare their order-picker instruction SKU numbers and quantities to the S.I. Cartrac cart SKU number. If the two SKU numbers are a match, then the order picker stops the cart and removes the required number of cartons for your customer's orders. The order picker labels the cartons or indicates on a list whether your customer's SKU was picked and transferred to one of the outbound shipping devices. On completion of the order-pick activity, the order picker allows the empty or partially full cart to continue its travel on the track to the next order-pick station. The order picker repeats this activity for each SKU. When the outbound shipping device(s) reaches (reach) a predetermined height or your customer's order is completed, it is transferred to the outbound staging area.

In the replenishment area of the S.I. Cartrac system, the replenishment lift truck transfers the required unit load from the reserve rack position to the SKU-depleted pick cart. The pick cart travel is stopped, empty pallet is removed, the new pallet load is placed onto the cart carrying surface, and the cart continues its travel on the track. The SKU pallet load replenishment is controlled by a computerized let-down program or the lift truck driver's decision.

Many disadvantages are associated with the S.I. Cartrac: (1) the cart circulation and cycle time required to make one complete trip around the track restricts its flexibility; (2) uniform pallet loads are on the cart; (3) with the daily start-up of the system, the carts are queued at the first order-pick station; (4) the very-fast-moving SKUs require several carts or are handled off-system; (5) when compared to the manual system, the S.I. Cartrac requires a higher capital investment; and (6) the system is designed for a fixed throughput volume, and additional volume requires overtime.

Advantages are that (1) the S.I. Cartrac system reduces the number of order pickers and mobile vehicles that are required to handle the business volume, (2) the system handles a wide variety of SKU types, (3) the system handles single or batched orders, (4) the method requires less battery charging area and thus has lower maintenance cost, (5) there is less product or rack damage, (6) the order picker is more productive because there is no walking, and (7) the cart with a shelf carries slow-moving SKU, thus increasing the order picker's productivity.

In a carton order-pick system economic and operational feasibility study, the S.I. Cartrac was considered to be a mechanized alternative concept to the manual concept.

Carousel System. The next mechanized carton order-pick method is the carousel system (Fig. 7.7*d*). A carousel system consists of a series of shelves that contain very-slow-moving cartons. These shelves travel in a forward or reverse direction on a powered chain that travels past one pick station and replenishment location. The powered chain is a top- or bottom-driven chain.

When required to make a pick or replenishment transaction from the carousel shelf, the employee activates the carousel to rotate until the required shelf arrives at the pick station. When the shelf arrives at the pick position, the order picker removes the required quantity of cartons from the shelf.

If the order picker is picking for an individual customer order, the cartons are placed onto your customer's shipping device. If the order-pick activity is a batched customer order-pick arrangement, then the cartons are placed onto a take-away conveyor system or onto the appropriate pallet board.

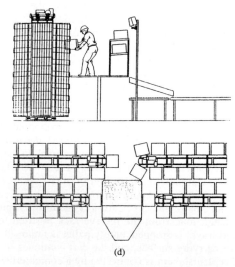

(d)

FIGURE 7.7 (*Continued*) (*d*) Carousel for mechanized carton order pick. (*Courtesy of Raymond Corp.*)

The carousel design is based on the "bring the stock to the order picker" concept, which reduces your order picker's unproductive walking time. The carousel is computer-controlled to rotate forward and backward and for improved carousel rotating time. If you have multiple (two or four) computer-controlled carousels, the order picker selects from all four carousel units. For slow-moving SKUs, this arrangement increases your order picker's productivity. The carousel method provides the order-picker instructions as a paper document, in label form, or as a paperless picking method.

Ministacker.   The ministacker is considered as a mechanized carton order-pick system. Via a captive aisle vehicle, required cartons are transported to and from the end-of-aisle input/output stations. These stations consist of conveyor systems that bring the cartons (slow-moving and small SKUs) to the order-pick station. The order-pick stations are located at the immediate front of the aisle or in a remote location. At these pick stations, the order picker removes the required cartons from the container or carton for an individual customer or a batched group of customers. The carton is identified and placed onto a take-away shipping device or conveyor. The method permits the use of a paper document, label, or paperless picking as the order-pick instruction.

The ministacker is controlled by an operator at the end of the aisle control station or by a computerized automatic stacker. The ministacker has limited applications for carton handling operations because of its small inventory capacity, long cycle time, and high capital investment.

Automatic Order-Pick Methods

Two automatic order-pick methods are described in this section:

- Ordermatic (single-, dual-, three-, or four-quadrant)

  Depalletize station and conveyor system

  Lane loader

- Vertique carton order pick

The Ordermatic and the vertique carton pick methods are computer-controlled systems that have three sections: the SKU reserve and replenishment section, the order-pick section, and the outbound take-away conveyor and unitizing section.

Ordermatic System.   The Ordermatic (Fig. 7.8a) is designed as one-, dual-, three-, or quad (quadrant) system. The one-quad design consists of the following components: pallet load reserve positions, depalletizing station, replenishment station (system), five-level-high carton order selection flow lanes, take-away conveyor network, outbound unitizing station, and computer-control office.

The typical two-, three-, and four-quad Ordermatic systems are similar to the one-quad. The two-quad system is designed as a stacked Ordermatic with vertical or side-by-side (horizontal) stacking. The three-quad is a very rare design but is a double-stacked Ordermatic and a stand-alone quad. The four-quad has two double-stacked Ordermatics.

The typical five-carton-high SKU flow lane is 100 in wide and represents one bay. One bay, on the average, has the space to handle 20 SKUs. A four-quad system with 4800 lanes handles, over a two-shift operation, 4400 SKUs that have an 80,000-carton

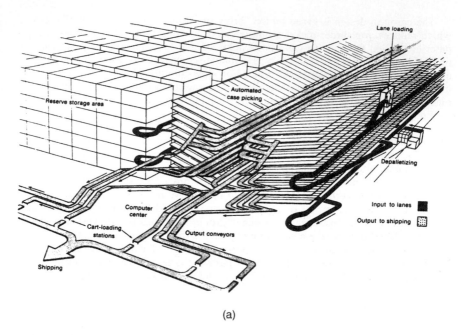

(a)

FIGURE 7.8   (*a*) Ordermatic; (*b*) depalletizer; (*c*) lane loader. (*Courtesy of S.I. Handling Systems, Inc.*)

throughput volume. The 80,000-carton throughput volume is reduced by a utilization factor of 15 to 20 percent. An average four-quadrant manual replenishment Ordermatic module (cross section) is $77\frac{1}{2}$ ft plus a lift truck aisle. An Ordermatic with a mechanized replenishment lane loader has a module cross section of $87\frac{1}{2}$ft.

Ordermatic operation requires that your customer's order be entered on schedule into the order processing computer. Each customer-ordered SKU is sent to each quadrant controller that controls the quadrant SKU flow lane level. The controller starts at the quadrant outbound side (shipping side) and travels to the quadrant opposite end. During its travel, by a computer-controlled picking impulse, the Ordermatic releases your customer-ordered single carton from each of the five lanes. The appropriate single carton slides from the pick lane onto each individual level take-away conveyor and travels to the outbound conveyor network. If additional cartons are required on your customer order, then the SKU pick position receives additional pick impulses.

The carton travels on the appropriate quadrant-level take-away conveyor to the merge location for transport to the unitizing or delivery vehicle loading station. At these merge points, the outbound carton travel is controlled by sensors.

The SKU replenishment in the Ordermatic requires that a depleted SKU flow lane be replenished prior to a stockout occurrence. This situation requires that the reserve pallet load be transferred on schedule from the reserve area to the depalletizing station. At the depalletizing station, the required cartons are removed from the pallet load and placed into the appropriate SKU flow lane replenishment side. The partially depleted pallet load is returned via lift truck to the assigned reserve pallet load position. The lift truck transfers another required SKU pallet load to the depalletizing station.

Because the Ordermatic has a "pick or no pick" situation, the SKU replenishment system is computer-controlled and is on schedule. A failure in the replenishment activity means that stockouts occur at the picking lane. There are several replenishment methods for the Ordermatic: pallet position directly behind the Ordermatic, depalletize onto a conveyor network, and mechanical lane loader.

Pallet Position Directly behind the Ordermatic (Manual Method).   In this design the replenishment pallet load is transferred to the rack row pallet position that is located directly behind the Ordermatic. A walkway is between the Ordermatic and the rack row. A replenishment employee in the walkway or on the mezzanine floor manually transfers the required cartons from the pallet load to the SKU flow lane. After the required cartons are transferred to the flow lane, a lift truck returns the partially depleted pallet load to the reserve rack area.

Depalletize Station and Conveyor System (Manual and Conveyor Method).   In this design the pallet load is transferred to a floor-level depalletizing station (Fig. 7.8*b*). At the depalletizing station, employees transfer the required cartons from the pallet load onto the replenishment conveyor network. The cartons travel on the replen-

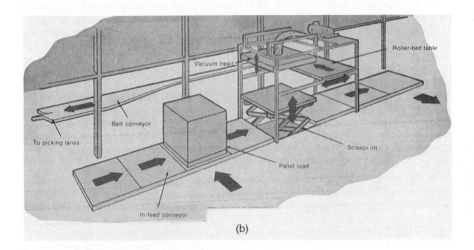

(b)

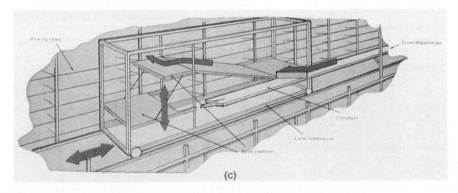

(c)

FIGURE 7.8   (*Continued*)

ishment conveyor to the assigned SKU replenishment lane. At the SKU lane, an employee or mechanical device (lane loader) places the cartons into the SKU replenishment lane. After the replenishment activity, a lift truck returns the partially depleted pallet load to the reserve area.

Lane Loader (Mechanical Method). This is a floor-level mechanical device (Fig. 7.8c) which removes a layer of cartons from a pallet load and transfers the cartons onto a movable infeed carton station. At this station, employees feed the cartons onto the replenishment conveyor system.

Disadvantages of the Ordermatic are that (1) it handles only conveyable SKUs; (2) fast- and slow-moving and nonconveyable SKUs are handled off-system; (3) there are multiple SKU flow lanes; (4) the system handles only a fixed carton throughput volume; (5) each customer order must be entered on schedule into the computer because downtime is difficult to compensate; (6) the method requires a computer replenishment system and tight management controls; (7) only one pick passes across the flow lane, and thus the conveyor requires a specific length of accumulation conveyor on-time outbound unitizing and direct loading activities; (8) stockouts are possible because the flow lane holds 10 to 20 cartons, and during order-pick periods, SKU replenishment to all flow lanes (all SKU lanes in a full condition) is very difficult to maintain; (9) package price-marking is difficult; and (10) capital investment is high.

Advantages are a computer-controlled system that minimizes order-pick errors, has the fewest order pickers (order-pick vs. labor expense) to handle the throughput volume, and selects SKUs by family group. Also, all your customer orders are picked per the order entry or delivery schedule, and operation within a freezer environment is possible.

Vertique Carton Order-Pick System. The vertique is an automatic carton order-pick system (Fig. 7.9) that consists of replenishment, SKU pick towers, shipping conveyors, and computer controls and network. This method involves automatic single or batched carton order-picking. During the pick activity, the SKU towers are replenished by the replenishment conveyor system.

In the operation of the vertique carton order-pick method SKU pallet loads are delivered to a depalletizing station. At the depalletizing station, the cartons are manually or mechanically transferred from the unit load to the replenishment conveyor. The carton identification number is entered in a computer-controlled or tracking device that diverts the carton onto the appropriate charge end (highest elevation) of the pick tower.

The pick tower consists of a series of flipper (pivoting) trays that raise and lower to accept and discharge cartons. As the carton enters the pick tower charge end, the carton is received in angled position and is stored in the flat position. In the flat position, the tray levels are the ready reserve position.

As the computer-controlled pick impulse releases a carton from the pick tower bottom level, the next pick tower level carton tray pivots, allowing the cartons to index (move forward) from the ready reserve level to the next-lower ready reserve level or pick position. As required, a second computer impulse releases another carton from the same SKU tower or travels to the next required SKU pick tower on the customer's order. These cartons travel on the shipping conveyor from the pick area to the sortation area or shipping area.

The pick towers are designed with heights of 10 to 80 ft, the flippers (pivot devices) handle cartons that weigh 10 to 100 lb, and the pick lane is a nominal 12 in wider than the carton.

The disadvantages of the vertique concept are high capital investment, requirement for computer hardware and software, handling of only conveyable cartons and a fixed throughput volume, requirement that the delivery trucks be at the docks, and requirement for carton accumulation prior to unitizing stations.

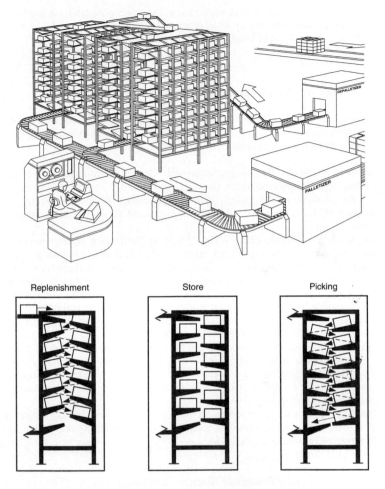

FIGURE 7.9 Vertique carton order-pick system. (*Courtesy of Vertique.*)

Advantages are accurate and on-schedule order picks, coordinated picks to the shipping schedule, first-in first-out (FIFO) product rotation, requirement for a small-square-foot facility and few employees, and family group picks.

Ordermatic and Vertique Ranking. In a carton order-pick concept economic feasibility and operational study, the Ordermatic and vertique methods are considered as automatic counterparts to the manual concepts.

CARTON STORAGE ARRANGEMENTS AND WHEN TO USE THEM

The objective of this section of the chapter is to develop an understanding of the various carton storage and pick position concepts with their alternative designs and operating characteristics. The section looks at each storage-pick position method ability to satisfy your distribution facility objectives of SKU accessibility, ability to handle the throughput volume, type of product rotation, and storage-pick density.

General Rule of Thumb for Carton Storage-Pick Positions

The general criteria that determine which carton storage-pick method to use are based on SKU volume and size. Guidelines for carton storage-pick position are given in Fig. 7.10. Various storage-pick methods are described below.

Floor-Stack Method.   The first storage-pick method is floor-stack or block storage, which has two alternative designs: the 90°-angle and 45°-angle stack arrangements.
    90°-Angle Stack to the Aisle.   The most common floor-stack design is the 90°-angle stack (Fig. 7.11*a*), which provides the greatest number of SKU (pick) openings per aisle.
    45°-Angle Stack to the Aisle.   The 45°-angle floor-stack arrangement (Fig. 7.11*b*) allows for a narrow right-angle stack turning aisle but fewer positions per aisle.
    Other Floor-Stack Features.   In a carton distribution facility, the floor-stack design has slip sheet or pallet unit loads that are two to six unit loads deep per storage lane. A storage lane is a single row or several back-to-back rows. At the aisle position, the pick position is one-high, permitting an order selector to handle cases. To utilize the cube (air space), the pallet loads that are deeper in the lane are the ready reserve pallets and are stacked two to four unit loads high.
    This stacking feature requires that the SKU in the storage lane aisle position serve as the SKU for the entire storage lane, and also that the pallet loads or cartons be capable of self-supporting the stacked weight. In some operations individual cartons are placed directly on the floor by a lift truck with a clamp device or basiload device or are hand-stacked by employees.
    The floor-stack pick arrangement provides high storage density; is used for large-cube, fast-moving SKUs; provides a LIFO product rotation; requires a low capital investment; and provides a good SKU hit density and concentration for a manual carton order-pick method.

Tier Racks or Stacking Frames or Nestable Stacking Frames.   The next carton storage-pick position design is the tier rack, stacking frame, or nestable stacking frame. When the SKU that is placed into floor storage is crushable or is not self-supporting,

Carton movement	Carton size	Type of storage-pick position
0–3	Small to medium	5-ft flow rack or shelf
0–3	Medium to large	Hand-stack
4–5	Small to medium and large	Hand-stack or 10-ft flow rack
6–10	Small to medium and long	Hand-stack or 20-ft flow rack
11–40	All sizes	Standard rack
41–80	All sizes	Standard rack
≥81	All sizes	Pallet flow, floor stack

FIGURE 7.10   Carton storage-pick position arrangements.

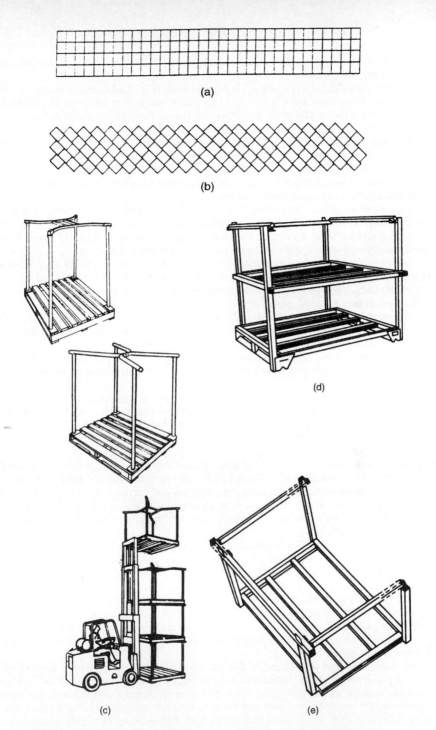

FIGURE 7.11 (*a*) The 90°-angle floor-stack arrangement; (*b*) 45°-angle floor stack; (*c*) tier rack (*Courtesy of Tier Rack Corp.*); (*d*) stacking frame rack (*Courtesy of Flexible Material Handling.*); (*e*) nestable stacking frame rack (*Courtesy of Flexible Material Handling.*).

7.33

one of these designs is used to make stackable and uniform unit loads that optimize the cube (air) storage area.

The tier rack (Fig. 7.11c) or stacking frame rack (Fig. 7.11d) has four legs at the four corners of its base. Each is attached to a pallet board. The four legs are connected (intersect) in the middle at the top. Nails are driven through the tier rack baseplate. This makes a more secure attachment to the wood pallet board.

Stacking frames have four legs that are attached and extend straight upward from the platform. At the top, there are four cross members that tie the four legs together. To prevent stacking frames from sliding on the forks, fork sleeves or stir-ups are added to the bottom. Rubber bands, stretch wrapping or netting reduces the product from falling off the stacking frame surface.

The nestable stacking frame (Fig. 7.11e) has three structural members that connect the four legs at the top. This feature creates an open side, and when the stacking frame is not being used, it permits the stacking (nesting) of up to 8 to 10 frames. This feature reduces storage space and future employee assembly time.

All three devices have a structural member base or a solid base. If they have a structural member base, then, as required, a wood or wire mesh base is used on the top of the base. This flat base permits cartons to be stacked without developing dents. In the pick area, a lift truck places one stacking frame on top of another.

The next stacking frame feature allows two-high pick positions. This stacking frame has 30-in-high vertical structural members which permit easy access to the second pick level; but this also increases the lift truck handlings.

The operational and design characteristics are the same as those of the floor-stack method. Your local fire codes are reviewed prior to implementation of the system in your distribution facility.

Disadvantages are that there is an additional capital investment. Second, when compared to standard rack transactions in the elevated position, lift truck productivity is lower. Third, if the tier rack or frames are not the nestable type, then storage space and assembly labor expense are required.

Standard Pallet Rack.   The next storage-pick method is the standard pallet rack design (Fig. 7.12), which is widely used in the carton distribution industry. The standard rack is designed for pallet loads or hand-stacked cartons. With the pallet load method, the rack opening has one, two, or three pallet loads. These rack bays consist of upright frames and a pair of load beams. The rack bays are designed as single-deep rows or back-to-back rows.

The rack structural bay is designed to have the first pallet load on the floor and the second pallet load approximately 44 to 48 in above the floor. In many distribution facilities, front-to-rear members (cross members) are used in the rack bays above the pick positions to improve product protection and employee safety.

The rack bays are considered the SKU pick positions. The SKU pick positions (two levels high) are designed for four pallets with the 40-in open side facing the aisle, four pallets with the 48-in stringer side facing the aisle, or with six 32-in-wide pallets facing the aisle. In most applications, there is at least 4 to 5 in between the product and rack members.

Disadvantages and advantages of the various pallet load pick position method are (1) two pallets wide (40-in dimension with stringer facing the aisle)—a disadvantage for some cartons, as it requires the order picker to reach deep into the pick position, but advantages are additional pick positions in the aisle and easy lift truck pallet handling; (2) two pallets wide (48-in dimension with fork opening facing the aisle)—disadvantages are fewer pick positions per aisle and requirement for good lift truck skills; an advantage is that it is easy for the order picker to reach cartons in the pick

Beam Span (Pallets without Overhang)

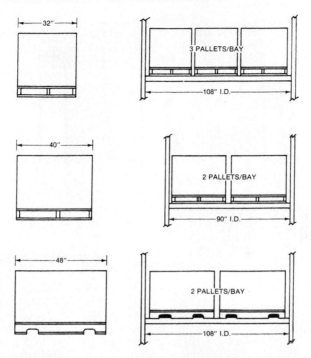

FIGURE 7.12   Various pallet load arrangements in a rack bay.

position; and (3) three pallets wide (32-in dimension with fork opening facing the aisle)—disadvantages that this is not a standard pallet board and holds a smaller carton quantity; advantages that it provides maximum pick positions per aisle and it is easy for a lift truck to move.

The rack bays above the pick positions are considered the preferred pallet load ready reserve positions for surplus SKUs. The standard rack concept provides good hit density and hit concentration, excellent SKU accessibility, FIFO product rotation, and only a medium capital investment.

Gravity Pallet Flow Rack.   The next carton storage-pick method is pallet gravity flow or flow through rack (Fig. 7.13*a*). The gravity pallet flow rack is designed as a stand-alone concept that has a minimum of two pallet loads and a maximum of 20 pallet loads deep and three to four unit loads high. There is one aisle for SKU deposit (entry) and a second aisle for SKU order pick (exit).

Gravity pallet flow racks consist of upright frames, upright posts, bracing, arms, brakes, end stops, pallet load separaters, and a skatewheel conveyor.

Air Flow Rack.   An alternative to the gravity flow rack is the air flow rack (Fig. 7.13*b*). Air pallet flow racks are designed with pallet load support members that have a series of tiny air holes in the entire length of the two rail members. As an air compressor pumps air into the pallet load support structure, the pallet load is indexed

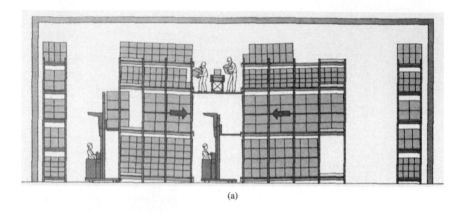

(a)

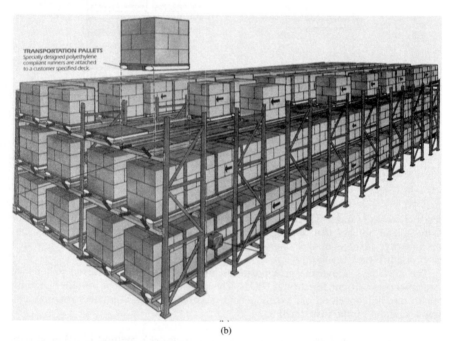

(b)

FIGURE 7.13 (a) Pallet gravity flow rack (*Courtesy of Webb-Triax.*); (b) air flow rack (*Courtesy of Airsail Enterprises.*); (c) push-back rack (*Courtesy of Webb-Triax.*); (d) pick tunnel for carton pick operation (*Courtesy of Webb-Triax.*).

through the racks. Air flow racks are best as the storage rack design for high-volume SKUs in the mechanized or automated carton handling operation.

Flow Rack Characteristics.   Lift trucks make deposits of pallet loads at the entry end of the flow rack lanes, and order pickers remove the individual cartons from the pallet load at the exit end. When the pallet board is empty, the order picker removes it from the rack opening to allow the next pallet load to index forward to the pick position.

The index movement is created by gravity force and the pallet board weight. When compared to the floor-stack concept, the pallet flow concept has a wider facing in the aisle, which means fewer pick positions per aisle.

The pallet flow concept provides high storage density, FIFO product rotation, handles a high volume, and has the fewest lift truck movements. The latter means lower product or equipment damage and high employee productivity. In a mechanized pick-to-belt system, the two- or three-deep pallet gravity flow rack that is two levels high is the preferred design in the pick area. This design provides fair SKU hit density and hit concentration. For a manual operation, the pallet gravity flow or air flow rack is designed for SKUs with a great number of pallet loads. In any type of carton order-pick method, pallet flow rack with one SKU per lane is an excellent storage-pick feature for high-volume SKUs.

Push-Back Rack.   The push-back rack is the next storage-pick design, three to four unit loads deep per lane and three to four unit loads high. There are two basic push-back rack designs: the standard pallet gravity flow rack with end stops on both ends of the flow lanes (Fig. 7.13*c*) and racks with nesting carriages that ride on two rails. Both push-back rack concepts provide one- or two-high pick positions with ready reserve pallet positions behind the pick positions and one SKU per lane.

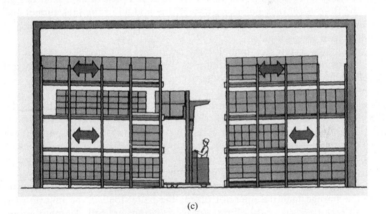

(c)

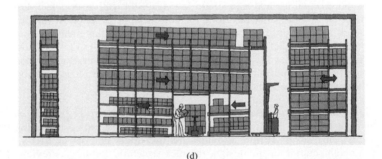

(d)

FIGURE 7.13   (*Continued*)

The push-back rack concept requires a lift truck operator to set a new pallet load in the first rack position of the lane. As required, with additional unit loads, to make a deposit in the push-back rack, the lift truck operator sets the new unit load at the aisle position. Until space is created for the new unit load, the lift truck operator pushes against the existing unit load. As the space is created in the lane, the lift truck operator deposits the load onto the flow lane. As required to make withdrawals, the unit load in the aisle position of the flow lane is removed from the pick position, and the next unit load is indexed forward by gravity force to the aisle position.

The push-back rack is a stand-alone rack that is placed along a building wall or has back-to-back rows. With this design, the product rotation is LIFO, with one SKU per lane. This system handles a high volume, provides a low number of facings per aisle, but provides high storage density.

Pick Tunnel.   This unique arrangement of push-back and gravity flow rack is that you can design a pick tunnel (Fig. 7.13d) for a carton pick operation. The pick tunnel system consists of gravity flow rack for the pick positions and push-back rack for the reserve positions. The pick tunnel is created in the middle of the floor under the two interior push-back racks. The pick faces are the gravity flow rack openings. This creates an aisle for a pick-to-belt operation or a pick-to-pallet truck operation with a large number of ready reserve positions. When you design a pick tunnel, your design specifies that the floor level rack positions flow to the middle of the pick tunnel aisle, with protection of the pick tunnel upright frame posts, and additional structural support members to span the pick tunnel.

A disadvantage of the pick tunnel is the additional investment in structural support members. Advantages include maximum use of air space, increased storage density, improved lift truck replenishment productivity, and good product rotation.

Drive-in or Drive-through Rack Design.   The next storage-pick method is the drive-in (Fig. 7.14a) or drive-through rack (Fig. 7.14b). A drive-in or drive-through rack consists of upright frames, upright posts, support arms, guiderails, support rails, and required side and top bracing members that form storage lanes. Drive-in racks have an end stop as an additional component, and at the entrance to each storage position, drive-through racks have additional top bracing members.

With the drive-in rack, the lift operator enters and exits the storage lane from one aisle. In the storage lane the operator places the pallet load in the innermost position and fills all rack positions until all the storage positions are full. To maintain this good inventory control, this feature requires one SKU per storage lane. The drive-in rack is best used in a manual method as a single row against a wall or in a back-to-back row arrangement to provide pallet positions for nonstackable product. In the design for a drive-in rack in a carton distribution facility, the storage lane should not exceed a depth of two to four pallet loads per storage lane. The drive-in rack provides a LIFO product rotation, provides fair SKU hit density and hit concentration, handles a medium volume, and is a good storage method for the storage area in any facility that handles high-volume nonstackable SKUs.

Two-High Unit Loads on the Floor.   If your unit loads are of the slip sheet type, then the floor unit-load position is double-stacked with one pallet supporting two slip sheet loads. This design feature provides the maximum number of SKUs per aisle. Prior to implementation of the design, the rack structural stability is reviewed by the rack manufacturer. Also, post protectors in front of the upright posts reduce equipment damage.

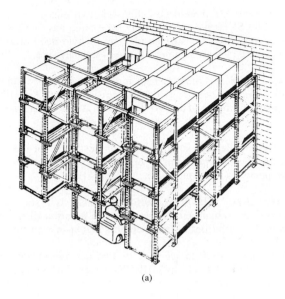

(a)

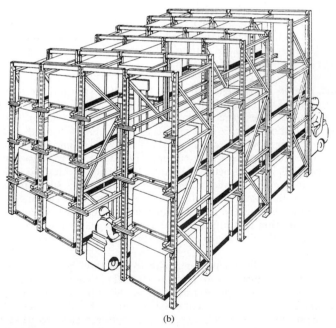

(b)

FIGURE 7.14   (a) Drive-in rack. (b) drive-through rack. (*Courtesy of Unarco Material Handling.*)

A drive-through rack in a three-deep pallet with two pick levels is designed with one aisle that has two SKUs per side, and the other side has one SKU per side pick position. This design divides the storage rack into a two-deep arrangement on one aisle side and a standard rack arrangement on the other aisle side.

Two-Deep Rack. The next storage design is the two-deep or double-deep storage system. This rack design is used in any type of carton order-pick facility for medium-to fast-moving product storage. The two-deep rack is not considered for pick positions.

Mobile Rack. The mobile rack storage design is best used for the remote (backup) reserve positions in a manual, mechanized, or automated carton order-pick system that has ready reserve positions in the pick area.

Bridge Across the Aisle. The bridge across the aisle is a standard rack storage design that increases the remote reserve positions in a carton distribution facility. It has pairs of load beams with front-to-rear members that span a traffic aisle.

HROS Rack. The next carton storage-pick method is a hybrid standard rack system, using the high-rise order-pick (-selector) (HROS) design. This system is designed with a rack structure with storage-pick positions from the floor level to at least 8 ft height. The product positions are serviced by an AS/RS, MS/RS, high-rise order picker, or man-up turret truck. All these vehicles require an in-aisle guidance system and are capable of both vertical and horizontal in-aisle travel. The AS/RS, MS/RS, and turret trucks have the ability to handle unit-load deposit transactions. With an order-picker truck, the operator hand-stacks cartons into the warehouse position, or the layout design has a separate replenishment aisle.

Hand-Stack Carton Alternative Designs. The HROS storage-pick position alternative designs are (1) the single-deep pallet rack and (2) the two- or three-pallet-wide rack. Alternative 1 is designed for one-deep hand-stack cartons. In this design the cartons are stacked on the rack decking full depth, or the rack depth is divided by a barrier. The decking is wire-mesh, plywood, fiberboard, metal slats, or conveyor. As required, front-to-rear members provide support for the hand-stack cartons. At least 1 in of open space is allowed between cartons. A barrier divides the rack depth into two pick positions with each rack position facing one aisle. When the rack depth is one carton deep, then the rear load beam has a high step. The high step extends upward above the decking and serves as the carton stop. This concept increases the SKU hit concentration and hit density. Disadvantages of the concept are that it requires double handling of the product and the pick position has a small inventory capacity.

Alternative 2 has one rack opening with two or three pallets per opening. The opening faces the aisle that is used by the order-picker vehicle. If your facility uses a straddle truck, then the order-picker truck and straddle truck operate in the same aisle. If your facility uses a turret truck and order-picker truck with the same wire or rack guidance system and rack-to-rack clearance, then both vehicles use the same aisle. An alternative single-deep pallet design is to provide a separate replenishment aisle for wide-aisle vehicles and another aisle for order-picker trucks.

With the pallet load HROS, this design provides medium SKU hit density and concentration and a medium number of SKU facings per aisle. The advantage of this design is that it requires fewer replenishment transactions.

The other storage-pick designs that are used in HROS for slow-moving SKUs are the hand-stack and flow rack designs.

Carton Gravity Flow Rack. The case (carton) gravity flow rack is the next single carton order-selection design which is used for slow-moving SKUs. The case gravity flow rack has three alternate designs: tilt-back, slant-back, straight-back, and standard rack with conveyor.

Straight-Back Carton Flow Rack. In the straight-back frame design each flow lane is located directly above the lower flow lane. It is considered the best flow rack method for handling cartons and is implemented as a single-level, multilevel, or mezzanine operation. In a multilevel operation, the mezzanine floor is supported by floor-level case flow rack upright support members.

The slant-back and straight-back gravity flow lane designs are used for carton pick positions but are preferred for single-item pick positions.

The carton (case) gravity flow rack consists of straight-back upright frames, support members, bracing members, lane dividers, end stops, and small skatewheel conveyor sections. The conveyor rails are pitched (sloped) from the replenishment aisle side to the order-pick side. This feature allows gravity to move the cartons from the entry side to the exit side. With the normal 24 in long $\times$ 10 in wide $\times$ 11 in high carton, the 10-ft-deep case gravity flow lane provides 20 to 25 pick faces per bay. The rack bay is a 5-ft-wide bay opening and five cartons deep.

With the case flow rack concept, cartons of slow-moving SKUs are replenished from the replenishment aisle at the rear of the flow rack. Per the customer order-pick instruction and from an aisle at the front of the flow rack, order pickers remove the appropriate carton from the pick position. The carton is placed onto a take-away conveyor or onto a vehicle load-carrying surface.

The flow rack carton design does require ready reserve storage and a setdown spot for flow lane replenishment. In an alternative replenishment system the replenishment cartons are conveyed to the pick area.

Disadvantages of the carton flow rack design are high capital investment and the fact that flow lanes hold a small inventory quantity. Advantages are excellent SKU hit concentration and hit density, good order-picker productivity, FIFO product rotation, excellent SKU access, and good storage density.

Standard Shelf Design. The next storage-pick design is the standard shelf, which consists of upright posts, shelving, bracing, and connecting components. Shelves are designed as single or back-to-back rows. When shelves are used to handle very-slow-moving SKUs that are small to medium in size, then closed shelving is used as the shelving design. When the shelves handle medium to large cartons or small cartons in containers, then the open shelving design is in the operation.

The shelving design has very limited application in a carton order-pick facility because it has a low pick position inventory capacity.

Cantilever Rack. The next carton storage-pick position design is the cantilever rack, which consists of upright posts, support arms and legs, decking, and bracing. The cantilever rack is designed for storage-pick positions for long SKUs such as pants or trousers, furniture, and other SKUs. The rack concept is designed with single- or double-arm rows.

Side-loading lift trucks are used to make long load transactions to the pick positions. When the cantilever rack is used for carton SKUs, the hand-stack design on wire-mesh or wood decking is in use.

PALLET, CARTON, AND STORAGE-PICK POSITION IDENTIFICATION

Pallet and carton load identification provides the warehouse employee with the means to match the product storage transaction instructions to the assigned warehouse location and to identify the warehouse storage-pick position with the order-pick instruction. The various pallet and carton load identification methods are (1) no method, (2) manual printed tags and labels (plain or colored), (3) computer-printed tags and labels (human-readable or human- and machine-readable), and (4) self-adhesive or nonadhesive.

These designs are reviewed in more detail in Chap. 5.

The identification of the SKU carton storage-pick position is another important factor that enhances employee productivity. The identification permits the order picker to match the order pick instructions to the warehouse pick position. The various concepts are (1) no method, (2) manually printed labels/tape onto the rack, (3) preprinted self-adhesive labels and tags, and (4) digital display.

Putaway (Deposit) and Replenishment Transaction Activity

The product pallet load and carton putaway and replenishment methods and systems assure that the pallet or carton deposit and replenishment transactions are accurate and completed on time. The various methods and systems are (1) method or system (manual, computer-controlled, and random) and (2) quantity (pallet load, layer of a pallet load or carton).

This activity is described in more detail in Chap. 12.

SORTATION

Batched Customer Order Carton Sortation

A carton sortation method is a basic requirement for a batched (grouped customers) carton order-pick system. The carton sortation method assures you that a group of your mixed customer order-picked cartons (batch-picked) are separated into your appropriate customer temporary holding or shipping area.

The various carton sortation methods are manual, active, passive, and a combination of active and passive.

When the order-selection activity is the batch-picked mode (in which your customer's specific ordered cartons are withdrawn from the pick positions in conjunction with your other warehouse functions that handle the customer-ordered cartons), then the carton sortation concept ensures that all your customer cartons are separated from the mixed (batched) customer cartons for delivery to the appropriate dock area or workstation.

One Customer, Multiple-Order-Picker Manual Sortation. In a single-customer multiple-order-picker order-handling system and in a manual carton order-selection system, to some degree, the order pickers sort customer orders while depositing their individual portions (pallets) of your customer's order in the assigned outbound staging area.

Temporary Storage Sortation. In a batched carton order-pick method, the outbound carton sortation has two alternative methods that satisfy the sortation objective: temporary storage sortation and shipping sortation. In a temporary storage sortation, the sorted cartons are placed into a temporary holding area for later shipment to your customer.

The temporary storage sortation method is designed to hold your customer order-picked cartons that are staged prior to the loading of your customer delivery truck. When your carton distribution facility handles a large volume of across-the-dock (flowthrough) product or order picks in advance of your customer's delivery schedule, then the temporary storage sortation method is a viable alternative. This method is very similar to the normal pick-to-belt carton order-pick method except that the cartons are removed from the conveyor and placed into the temporary (rack) storage position.

In this method, the across-the-dock customer order cartons are unloaded from the vendor delivery truck. The initial distribution of your customer cartons are labeled, conveyed, and sorted to a particular customer-assigned aisle. In this aisle, an employee transfers the cartons from the conveyor into the appropriate temporary rack position. The rack position is a standard rack or pallet flow rack. These concepts are outlined in Chap. 4.

Other cartons from the standard storage-pick area are sent to your customer's temporary holding area. These are the early order-picked cartons of your customer order. These cartons are picked from the on-hand inventory prior to the delivery schedule and are transported to the temporary holding area for palletizing into the appropriate rack position.

After the pallet load attains a predetermined height, the full pallet load is allowed to flow on the pallet flow rail, or your employee hand-stacks in another standard rack bay. With these methods, additional cartons on your customer's order are placed in other pallet positions or are accumulated in the flow rack.

When your customer's order is required at the shipping dock, per your loading schedule an outbound depalletizing employee removes the prelabeled cartons from the pallet loads in the temporary holding area. These pallet loads are at the exit end of the pallet flow rack or in standard rack position. The employee places the labeled cartons onto the outbound conveyor. The outbound conveyor transports these cartons from the temporary holding area to the shipping sortation area for sortation to your customer's shipping door and loading onto the delivery vehicle.

Disadvantages of this method are additional building and equipment investment requirement, double handling, and palletizing errors. Advantages are handling of a large number of customers, reduction in on-hand inventory, increased inventory turns, and removal of peaks and valleys in your shipping activity.

Shipping Sortation. In shipping sortation (which is more common than temporary storage carton sortation), your customer order-picked cartons are sorted for placement (unitizing) onto your customer's delivery device (cart or pallet board) or directly (fluid) loaded onto your customer's delivery vehicle.

In shipping sortation your batch-picked customer cartons are sorted for placement onto the appropriate customer's delivery device (cart or pallet board). Then the order picker labels each batch-picked carton. As the labeled carton leaves the picking area, it travels to the encoding station. In this area, the carton label sortation number is entered into the programmable controller. This information activates the appropriate divert device at the required time to divert the carton from the sortation conveyor to the assigned shipping lane.

On the shipping conveyor, the carton travels to the off-loading station. At this particular location, there are two alternative methods of handling the customer's cartons: (1) direct (fluid) loading of cartons onto your customer's delivery truck or (2) unitizing (palletizing) your customer cartons onto a cart or pallet. The cart or pallet is staged on the dock and at the appropriate time loaded onto your customer's delivery truck.

Direct Loading. In direct (fluid) loading your customer's outbound cartons travel on a gravity conveyor from the sortation conveyor onto an extendible conveyor into your customer's delivery truck. In the truck, the cartons are manually removed from the conveyor and stacked onto the delivery truck floor.

The extendible conveyor is a powered belt, a powered roller, a nestable skatewheel conveyor, or a gravity nestable skatewheel or gravity roller conveyor.

Disadvantages of direct loading are additional conveyor investment, with requirement for a recirculation section and increased need for management control and discipline for customer order entry and delivery trucks at the dock position. Advantages are less dock or holding area requirement, no double handling, and reduced loading errors.

Shipping Carton Unitizing. In the shipping carton unitizing method the cartons travel on accumulation conveyors from the sortation area to the customer's unitizing station (Fig. 7.15). At this station, the customer's cartons are stacked onto carts or pallets. After each shipping device is unitized, the completed unit load is transferred to the outbound staging area or is placed onto the customer's delivery truck.

Disadvantages of this method are that it requires a unitizing area plus dock staging area, management control and discipline, a unitizing and dock area for utilization of the air space, and a recirculation conveyor. Advantages are that it permits easy loading of unitized loads, allows the customer's unit loads to be wrapped in a protective wrapping or netting material, allows the customer's cartons to be handled without the delivery truck at the dock, and is preferred for a customer with the cart delivery method.

Carton Sortation Design Factors and Parameters

Each batch-picked carton sortation method requires several important design characteristics: (1) a customer identification label that is at least human- or human- and machine-readable, (2) a conveyor system with the ability to accumulate cartons prior to the induction and unitizing stations, and (3) capacity to handle the number of cartons and customer volume of the batch (wave).

Sortation Is the Heart of the Batched Order-Pick Method. If your distribution facility handles a large volume of cartons for a larger number of customers, then to handle the volume, the operation uses the batched order-pick method. Carton sortation is the heart of a batched order-pick method.

Prior to the design and implementation of a carton sortation concept, you determine carton size (minimum, maximum, and average length, width, and height), carton weight (minimum, maximum, and average), exterior surface of the carton (top and bottom), average and peak number of cartons per day or week, number of hours for each workday, type and size of label, bar coding and how the label is read, required gap between cartons, crushability and fragility of the carton, ensuring that all cartons are sealed, and providing necessary space for accumulation and activity stations.

It is mentioned that the implementation of a conveyorized sortation method with-

(a)

(b)

FIGURE 7.15 (a) Unitizing stations; (b) direct load. (*Courtesy of S.I. Handling Systems, Inc.*)

out sufficient carton accumulation creates an out-of-balance situation between the pick and sortation functions which does not permit the order pickers and loading employees to achieve the desired productivity rates.

After these design parameters have been determined for the carton sortation method, then you design a building to house the carton sortation system or design a carton sortation system to fit within an existing building.

After a group of your customer orders (batched group) cartons are labeled and placed onto a conveyor system, there are two basic carton sortation methods: manual and mechanical sortation.

Manual Sortation.   Manual sortation is considered more basic and simple than mechanical sortation because it requires a human-readable label. The label is placed on the carton in a location that is easy to read (front, side, or top) by the sortation employee. In this manner, the batch-picked cartons are transported to the manual sortation area which is located in the shipping area. The sortation employee reads the carton human-readable label and matches the label identification number to a customer's unitizing station number. When there is a match, the carton is transferred from the sortation conveyor onto the appropriate customer shipping device.

Few and large characters and digits means better productivity. In manual sortation, the sortation employee handles 10 to 15 cartons per minute with little damage or impact on the carton contents. The maximum carton weight is 40 to 50 lb, and the average weight is 20 to 30 lb. To assure efficient product handling, you maintain a 3- to 6-in gap between two cartons, and the carton accumulation conveyor should have no (zero) pressure.

There are four basic conveyor designs for manual sortation: one conveyor, double-stacked conveyors, recirculation loop, and apron.

One-Conveyor Design.   In this manual sortation conveyor design the sortation accumulation conveyor lane is in front of your customer unitizing stations. The zero-pressure accumulation conveyor has a 1-in-high one-side (far side from the unitizing station) guardrail and an end stop. This feature allows the sortation employee to move a "did not sort" carton in a reverse direction on top of the conveyor to the appropriate sortation location. Then, when an employee stops a carton on the conveyor, there is no forward pressure on the carton.

Disadvantages of this design are sortation errors, requirement for a large and clear label, handling of only a low volume and few customers, increased effort to return "did not sort" cartons, large number of employees required, and increased potential for employee injury. Advantages are low impact on the carton, low capital investment, method simple to operate and control, and sorts to both sides (bilateral sortation).

Double-Stacked Zero-Pressure Conveyor.   In the double-stacked zero-pressure carton accumulation conveyor, each level has a 1-in-high far-side guardrail and a fixed end stop. The lower conveyor level is for sortation, and the upper (elevated) level is for the "did not sort" cartons, and the two conveyor levels move the cartons in opposite directions. If the sortation employee at the first sort station did not sort all the cartons, then the employee at the next sortation station removes the "did not sort" cartons from the lower conveyor level and places them on the upper (elevated) conveyor level. The elevated conveyor direction of travel returns these unsorted cartons from station 2 to station 1.

Disadvantages of this system are that it requires the greatest number of employees, possible sortation errors, handles few customers and a low volume, requires a human-readable label, and requires an employee to lift the carton. Advantages of the design are low impact on the carton, medium capital investment, simple operation, easy to control, and less "did not sort" carton transfer effort.

Recirculation Loop Conveyor Methods.   In the recirculation loop conveyor design, all cartons travel past all sortation stations and the "did not sort" cartons reenter the sortation conveyor prior to the first sortation station. The recirculation loop conveyor has two alternative designs.

1. *Powered 180° curve.*   In this recirculation loop conveyor concept two powered 180° curves and a merge conveyor automatically recirculate the "did not sort" cartons. The entire conveyor system has one far-side guardrail and at the merge junction, a photo-eye with controls. These devices control the flow of the cartons from the picking area to the sortation area and the flow of cartons on the "did not sort" recirculation

loop to the sortation conveyor. The infeed conveyor, which transports cartons from the pick conveyor, is located between the two straight conveyor sections. Disadvantages of this design are the same as those of the double-stack design. Additional advantages are that employees do not lift the "did not sort" cartons, there is increased capital investment, and a higher volume is handled.

2. *Two straight conveyors with end stops and gap plate.*   In this design two straight-line, side-by-side conveyors with end stops, move cartons in opposite directions. Between the two conveyors is a gap plate that is slightly higher in the middle than the conveyor rollers. The gap plate is a solid piece of metal with a bowed middle. It serves as a far-side guardrail and provides a bridge between the two conveyors for easy transfer of cartons between the two conveyor sections. If a "did not sort" carton appears, the employee at the next sortation station gently pushes the carton across the gap plate onto the second conveyor. The second conveyor moves the carton to a location prior to the first sortation station. At this location, the sortation employee at the second sortation conveyor stops the inflow of cartons from the picking area and pushes the "did not sort" carton across the gap plate, thus reintroducing the carton to the first sortation station.

When compared to the other manual recirculation sortation methods, the two side-by-side conveyor system (item 2, above) has a lower capital investment and handles a lower volume.

Circular Apron Conveyor.   The final manual sortation conveyor method described here is the circular apron conveyor. The circular conveyor and photo-eye control network allows pick area cartons and the "did not sort" cartons to automatically pass and recirculate in front of the first sortation station.

The concept has the same advantages and disadvantages as the recirculation loop conveyor system, except that there is an additional investment in the apron conveyor.

Mechanical Carton Sortation

A mechanical carton sortation system consists of an order-pick area conveyor; transportation conveyor; induction (infeed) station area with brake, meter, and induction conveyors; "no read" conveyor line; sortation conveyor line; shipping lane(s); and cleanout and recirculation line.

When you consider implementation of a carton sortation conveyor system, you ensure that there is adequate electrical power in the required locations and that there is a drain for the air compressor.

The order-pick area conveyor is the transport conveyor for carton travel from the pick area to the transport conveyor.

Transport Conveyor.   The transport conveyor section ensures that there is sufficient carton accumulation and that the cartons travel from the pick area to the induction station area. Prior to the induction station, all cartons are singulated (lined up in a single file) and a gap is created between two cartons by brake and metering belt conveyors. This gap (open space between two cartons) ensures that your induction employee or the scanning device reads (enters) the carton label and that the divert device has sufficient space to complete the divert.

Induction Conveyors.   The induction (infeed) station is the location where your customer's carton identification number is entered into the sortation conveyor programmable controller. The induction area design has sufficient conveyor accumulation to provide a constant flow of cartons to the induction area.

There are three types of induction conveyors: manual (keypad), semiautomated (hand-held scanner), and automated (fixed-position scanning device).

Manual Induction.   In manual induction, a carton has a large human-readable label on its top or side surface. The keypad is connected to the programmable controller and tracking device.

An induction employee reads each carton human-readable label address code and keys the address code into the keyboard. The keyboard transmits the address code to the programmable controller and tracking device.

Most keypads have 10 digit (0 to 9 keys) buttons with a repeat button, scan on-off button, error signal, cancel key, five-digit LED (light-emitting diode) display, emergency stop button, and a send key.

To achieve high manual induction productivity, the address code must be large and contain the minimum number of digits.

Disadvantages of the manual induction method are increased need for employee training, low-volume handling, potential induction errors, requirement for the greatest number of employees, and all labels facing one direction. Advantages are low capital investment and fact that this method serves as a backup system for the other induction methods.

Semiautomatic (Hand-Held Scanning) Induction.   In semiautomatic induction the carton has a human- and machine-readable label on its top or side. As the carton arrives at the induction station, the brake belt and metering belt create an open (gap) space between two cartons. At the induction station, an employee takes a hand-held scanning device (wand or pen) and scans the carton label. After scanning the label, the carton passes the induction photo-eye and is released onto the sortation conveyor surface under the control of the programmable controller and tracking device. At the appropriate divert location, these devices trigger the appropriate divert device to complete the sortation activity. The label information is sent to the manifesting system.

Disadvantages of this method are additional investment and employee training. Advantages are reduced induction errors, and the pen handles a medium volume and the wand handles a high volume.

Automatic Scanning Induction.   In automatic induction a bar-code label appears on the front, side, top, or bottom of the carton.

After the carton achieves the proper spacing on the conveyor system, it travels to the scanning device. The carton label is read by a fixed-position fixed-beam scanning device or a fixed-position moving-beam scanning device. As the carton passes the induction photo-eye, it transfers onto the sortation conveyor. The label information is transmitted to the conveyor programmable controller, tracking device, and manifest system to assure an accurate and complete sortation and manifest document. This automated induction concept is used on a belt conveyor; roller conveyor; or tilt-tray, gull-wing, or tilt-slat conveyor.

Disadvantages of the automatic scanning induction method are additional capital investment and requirement for a human- and machine-readable label. Advantages are reduced induction errors, requirement for the fewest employees, and high-volume handling.

"No Read" Conveyor.   In all sortation systems, and especially in an automatic induction scanning system, a second important activity is the reintroduction of "no read" cartons to the induction conveyor. After the automatic label-scanning device station, a "no read" conveyor spur is located on the sortation conveyor. The "no read" spur is 12 to 15 ft from the scanner and is the first divert location past the scanning device. It is designed to receive all labeled or unlabeled cartons that were not read by the scanning device. The "no read" conveyor returns the cartons to the induction station. At the

induction station, the carton is read a second time or manually or semiautomatically inducted onto the sortation conveyor.

After the induction process, the carton passes the induction eye and is transferred to the sortation conveyor surface, and the programmable controller and tracking device ensures that the carton arrives at the divert location that corresponds to the address code. At the divert location, a device transfers the carton from the sortation conveyor onto the customer staging area.

Various Sortation Surfaces and Conveyor Paths

A major component of any carton sortation system is the carton sortation conveyor surface. The various conveyor surfaces that are common in the warehouse and distribution industry are roller, smooth-top belt, slat tray, tilt tray, SBIR, and gull wing.

The conveyor surface provides the carton travel path from the induction station to all sortation locations. The sortation conveyor has two basic alternative designs: the single straight line and the endless loop.

Single-Straight-Line Sortation Conveyor Path. In this design the sortation conveyor path is a straight line from the induction station to the last sortation station. After the last sortation station, the sortation conveyor ends. The alternative straight-line designs are (1) L-shape or straight-line layout and horseshoe or U-shape layout.

The unique characteristic of the straight-line conveyor sortation system is that there is no automatic recirculation of the "did not sort" cartons. In these designs, all the "did not sort" cartons that are created by a full sortation lane or a divert device malfunction are manually returned to the induction station or manually transferred to the unitizing station. These "did not sort" cartons are handled in either of two ways: (1) cartons are diverted at a special divert location or (2) after the last divert station, the sortation conveyor lowers to the floor on a "runout" conveyor. At the end of this runout lane, an employee unitizes the cartons onto a cart or pallet board. These cartons are manually transported to the induction station or to the proper customer unitizing station.

Disadvantages of this design are that it requires additional employees and handles a low volume. The advantage is low capital investment.

Endless-Loop Sortation Conveyor Path. In the endless-loop conveyor design, the sortation conveyor path starts at the lead (charge) end of the induction station and ends at the charge end (at the same location) of the induction station.

The endless-loop conveyor has four designs: L-shape, U-shape, O-shape (elliptical) and rectangular.

In the endless-loop sortation design all "did not sort" cartons are automatically recirculated via a conveyor to the induction station for rescanning by the scanning device.

A disadvantage of the endless-loop design is increased capital conveyor investment. Advantages of the endless loop are that it requires fewest employees, handles a high volume, and provides automatic carton recirculation.

Mechanical Divert Component

The next important carton sortation component is the mechanical divert. Divert devices provide the means to transfer your customer addressed carton from the sortation conveyor onto the customer's unitizing or shipping lane. These divert devices pull, push, tip, slide, or pass the carton from the sortation conveyor onto the divert lane.

The design of the sortation conveyor divert lanes requires an arithmetic progression from the induction station to the last sortation station. The stations are in an arithmetic progression with all odd number on the left and all even numbers on the right.

There are three basic mechanical carton divert designs: active sorter, active-passive sorter, and passive sorter.

The active sorter design has a powered induction (infeed) station, powered conveyor surface, and a powered mechanical device that pushes or pulls the cartons from the sortation conveyor to a sortation lane.

The active-passive sorter design has a manual induction (infeed) station, and a powered conveyor surface and tips or plows the carton from the sortation conveyor onto the sortation lane.

The passive sorter design has a manual induction (infeed) station and a powered conveyor, and gravity force removes the carton from the sortation conveyor to the divert lane.

Solid Metal Deflector.   The solid metal deflector, a passive sorter device, is the least expensive of the divert designs and is considered the basic method for diverting cartons from a conveyor surface. This basic deflector is a solid metal pneumatically or hydraulically actuated bar (arm). As required, the deflector is extended across the conveyor surface and blocks the travel path of the carton. The power of the conveyor carton forward movement and the deflector angle guides the carton from the smooth belt or roller conveyor surface onto the divert lane. When not required to divert a carton, the deflector is retracted to the sortation conveyor side.

The deflector divert design does not change the carton direction of travel and handles a low volume with a rate of 20 to 40 cartons per minute. To achieve the high sort rate, it handles a slug of cartons for one divert location. The carton weight is 10 to 50 lb, a 3- to 5-ft distance is required between divert locations, and the carton impact is medium to rough.

Pusher Diverter.   The pusher diverter, an active sorter device, has two alternative designs: (1) the side-mounted pusher-puller and (2) the overhead pusher (paddle). These two designs differ with respect to attachment to the conveyor and pusher movement across the conveyor surface.

The pusher divert system requires a pusher at each divert location. After the carton is inducted onto the sortation conveyor and as it arrives at the divert location, the pusher (divert) device is activated by the programmable controller impulse from the tracking device. As the divert device pushes its blade across the sortation conveyor path, the carton is pushed onto the divert or shipping lane. The shipping conveyor and the sortation conveyor move the cartons in opposite directions. The average number of pusher sorts per minute is 25 to 30, the maximum handling weight is 125 lb, the nominal spacing between divert locations is maximum carton length plus 6 to 12 in, and impact on the carton is medium.

The puller divert device sort rate is 35 to 40 sorts per minute, handling capacity is 10 to 100 lb, and impact on the carton is medium to rough. The mechanical functions of the puller sortation system are similar to those of the pusher system except that the carton is pulled across the conveyor onto the divert location.

Overhead-Mounted Pusher Diverter.   The overhead-mounted pusher diverter handles approximately 30 diverts per minute. The overhead pusher diverter with divert locations on both sides of the sortation conveyor pushes a carton from the sortation conveyor on the outgoing stroke and pushes another carton on the return stroke. This feature allows the overhead method to increase the throughput volume to 35 to 40 sorts per minute, but requires increased control of carton infeed.

Powered Belt Diverter.   The powered vertical belt diverter is an active sortation divert system. The divert concept is a series of arms that are located along the side of the sortation conveyor. The sortation conveyor is roller conveyor. Each divert arm has a moving belt. After the carton is inducted onto the sortation conveyor and prior to its arrival at the divert location, the divert arm swings across the sortation conveyor path and the divert arm belt starts to move. This divert arm position and belt movement quickly grabs the carton by the belt and moves it off the sortation conveyor and onto the shipping lane. On the shipping lane, the carton direction of travel is the same as on the sortation conveyor. The typical carton weight range is 1 to 50 lb, divert locations are 3 to 5 ft apart, and impact on cartons is medium to rough.

The powered vertical belt diverter handles 20 to 30 cartons per minute. When there is a slug of cartons for one divert location, then the divert rate is increased to 30 to 40 cartons per minute because the reposition of the arm is not required for the next carton divert.

Plow Diverter.   The plow diverter is an active-passive sortation device consisting of a series of curved metal arms (plates) that swing across the sortation conveyor path. The sortation conveyor is a roller conveyor. The plow diverter is designed with one end of the plow attached to the far side of the sortation conveyor and the other end moving to the near side of the sortation conveyor. The position of this plow is across the sortation conveyor path that is in the carton direction of travel. The plow metal curved arm position and the carton forward movement on the roller conveyor surface cause the carton to slowly follow the plow curve. The curve directs the carton from the sortation conveyor onto the shipping lane or take-away spur.

The plow diverter is best used on a sortation system that sorts (handles) a slug of cartons that are diverted to one shipping lane. The design parameter is 2 to 5 ft between the divert locations. The plow diverter does not change the carton direction of travel. With a relative slow plow movement, the plow diverter handles 15 to 20 cartons per minute. The cartons have a weight range of 1 to 50 lb and the impact is medium to gentle.

SBIR.   The SBIR automatic sortation system is an active and very sophisticated carton sortation system that consists of a series of individual (divided) carton carrying surfaces (segments). These carrying surfaces are individual conveyor belt surfaces that are mounted on wheels and are designed to travel in a closed-loop system. Each conveyor belt surface travels past an induction station. At the induction station, a carton is placed onto a carrying surface. A coded label is read by a scanning device that programs a programmable controller and assigns the divert (discharge) location. This divert location is on the right or left side of the load-carrying surface. As the belt-driven conveyor carrying surface arrives at the divert location, the carrying surface conveyor belt turns in the proper direction to discharge the carton.

The carrying-surface conveyor belt turning motion moves the carton from the sortation conveyor to the divert location. The carton orientation or direction of travel varies between the sortation conveyor and the divert location. The SBIR divert system has a gentle or no impact on the carton and handles a carton weighing up to 50 lb. The divert device performs a two-way sort but has a higher capital investment.

Tilt Tray.   The tilt-tray sortation concept is an active-passive divert method. This carton sortation system consists of a series of trays that constitute the carton carrying surfaces (platforms) and that ride on a motor-driven closed-loop chain. There are two types of motor-driven chain systems: the bull wheel and the caterpillar.

After the carton is inducted (placed) onto a tray, the tray is under the control of a programmable controller to electrically or pneumatically tilt at the assigned sortation location. New technology permits one long carton to ride on two trays. This tilt location is on the right or left side of the chain. For best results, odd-number divert locations are on the left and even-number locations are on the right of the chain, with the lowest-numbered location as the first one past the induction station.

The tray tilting action and the travel speed cause the carton to slide from the tray onto the shipping lane. After the tray is tilted and prior to the induction station, all the trays pass a cleanout lane that tips all the trays to tilt all the unsorted cartons. These cartons are reintroduced to the sortation system. After this station a device relatches (levels) all the trays and permits the tray to receive another carton.

Most carton tilt trays are set on 27-in centers, and the divert locations are on 3- to 4-ft centers. The tray and divert centers vary according to carton size and speed of the tilt-tray chain. The nominal conveyor tilt-tray rates are 180 to 205 trays per minute, and handling capacity is 25 lb maximum per carton. The carton orientation (direction of travel) is either changed or not changed on the shipping conveyor, and impact on the cartons is medium.

Novasort. The Novasort is a modular tilt-tray sortation device that has a closed-loop, fixed travel path. The components are a train of three to four tilt trays, each with a 500-lb load-carrying capacity and tilts to the left or right, a programmable controller on each train of tilt trays, and a fixed travel path which has an electric powered bus bar and on-board sensing devices to detect an object or employee in the travel path.

The Novasort vehicle travel path is ceiling- or floor-supported, travels up slight grades, and travels over 90° and 180° curves. The path directs the tilt-tray train past an induction station and delivery station. At the induction station, the induction employee places the carton onto the tray and assigns the tray tilt location into the programmable controller. At the delivery station, the programmable controller provides the appropriate tray to tilt and discharge the product onto the assigned station.

With the on-board programmable controller, if required at the workstation a train of trays is unloaded or loaded by the workstation employee. After all trays are tilted at the required stations, the train of tilt-tray carts travels on the path and accumulates carts prior to the induction station or continues traveling to the maintenance spur. On the maintenance spur, the tilt-tray carriers wait for the next assignment.

As an option, the tilt tray has a fixed load-carrying surface and has the capacity to handle single-tray items, small items, and dual-tray items. As required to satisfy expansion of the system, additional track is added to the existing travel path.

The design parameters and operational characteristics of the Novasort system are similar to those of the tilt-tray sortation system. Induction is manual or automatic with a tray that travels on a fixed, closed-loop rail path between two locations. Disadvantages of the Novasort train of tilt trays are that it handles only a medium volume and has a fixed, closed-loop path; the travel path design requires employees and equipment to have an access-egress path from the workstation; capital investment is high; and single travel could create queuing. Advantages are that the product is transported as required to the workstation, the system is ceiling- and floor-supported, unloading and loading is done at the workstation, and the travel path (track) is easily expanded.

Tilt Slat. The next carton sortation method is tilt-slat sortation (Fig. 7.16a), which is considered an active-passive sortation system. This system consists of a series of slats

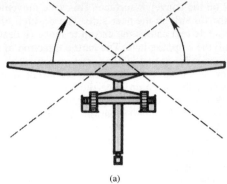

(a)

FIGURE 7.16　(a) Tilt-slat sortation. (*Courtesy of Acco Systems.*)

that ride on a motor-driven closed-loop chain. A predetermined number of slats serve as the carton carrying surfaces.

After the carton is inducted and placed onto the tilt-slat conveyor system, the carton travels to the assigned tilt location. The tilt location is on either the left or right side of the conveyor path when the carton is tilted onto the shipping lane. The carton travel on the shipping lane is different from that on the carton orientation (travel) on the sortation conveyor.

Operational characteristics of the tilt-tray and tilt-slat sortation systems are similar. An advantage of the slat method is that it handles heavier and larger-cube cartons.

Gull-Wing.　Gull-wing sortation is an active-passive carton sortation method. The gull-wing system consists of a series of carton carrying surfaces that ride on a closed-loop chain. The carton carrying surfaces are in the shape of a bird's (gull's) wing or a V shape.

The operational characteristics, disadvantages, and advantages of the gull-wing design are similar to those of the tilt-tray sortation system. The gull-wing sortation system handles a wide variety of product types and a lower volume and changes the carton direction of travel.

After the carton is placed onto the gull-wing carrier, it is inducted onto the conveyor system. When the gull-wing carrier arrives at the assigned sort location, a tipper is activated that tips the carrier. The tipping action and the conveyor speed permit the carton to slide from the carrier onto the shipping lane. Carton orientation or direction of travel on the shipping conveyor lane is different from that on the sortation conveyor. Impact on the product is gentle to medium.

Sliding Shoe. The "sliding shoe" is an active-passive carton sortation system consisting of a roller conveyor surface with sliding shoes. The roller conveyor allows the shoes to slide in the open space that is between the conveying surface. The sliding shoe has a single-side divert design (Fig. 7.16*b*) or a dual-side divert design (Fig. 7.16*c*). These shoes are located on the opposite side and prior to each divert location. This feature requires the sortation conveyor width to accommodate the widest carton plus the sliding shoe(s) dimension.

After the induction station and as the carton travels on the sortation conveyor and prior to its arrival at the assigned sort location, these sliding shoes start to move across the direction travel on the conveyor surface. This shoe movement across the sortation conveyor is from the far side to the near side. The position of shoes on the sortation conveyor forms an angle that causes the carton to move (index) forward from the sortation conveyor onto the shipping lane. The carton direction of travel or orientation on the shipping lane is the same as that on the sortation conveyor.

After the appropriate carton is diverted from the sortation conveyor, and if the next carton is not diverted at this location, then the sliding shoes return to the far side of the sortation conveyor. If the next carton is required at the same divert location, then the sliding shoes remain in the divert position. This feature requires at least 2 to 4 ft between divert locations.

The sliding-shoe carton divert system handles 130 to 150 sorts per minute or a conveyor speed of at least 400 ft/min with a gentle impact on the carton.

Pop-up Diverter. The pop-up diverter design group is an active sorter group that consists of the pop-up wheel and the pop-up roller. These sortation designs have similar characteristics and are installed slightly below the live roller or chain-driven roller conveyor surface. Another similarity is that these sortation methods handle the bottom of the carton.

Pop-up Wheel Diverter. The pop-up wheel diverter (Fig. 7.16*d*) consists of one wheel or two strings of wheels that are located on the sortation conveyor at the same level as the sortation conveyor rollers. At the beginning of the sortation location, these pop-up wheels are angled (skewed) to the left or right of the conveyor path. The direction of the skew is the direction of the sort location. As the carton leaves the induction station and arrives at the sort location, the programmable controller activates the pop-up wheels to rise above the sortation conveying surface. This elevation and the wheel angle motion grabs the bottom of the carton and directs the carton onto the shipping lane.

After the carton sortation, the pop-up wheels return to the lowered position that is below the sortation conveyor surface. This position allows other cartons to travel onto the next appropriate sort location.

The pop-up wheel carton sortation system handles 65 to 150 sorts per minute with a gentle impact on the carton, divert location spacing is 4 to 5 ft, and handling capacity is a carton weight of 30 lb. The pop-up roller makes 15 to 20 sorts per minute with divert locations on close centers and handles a carton weight of 200 lb.

Both the pop-up wheel and the pop-up roller carton sortation systems maintain the carton orientation or direction of travel on the shipping lane.

(b)

(c)

FIGURE 7.16 (*Continued*) (*b*) Sliding shoe sortation with single-side divert design (*Courtesy of Mathews.*); (*c*) sliding shoe sortation with dual-side divert design (*Courtesy of The Bushman Co.*).

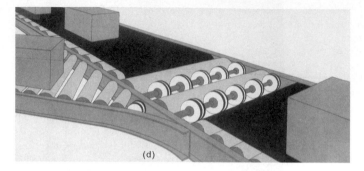

(d)

FIGURE 7.16 (*Continued*) (*d*) Pop-up wheel on the sortation conveyor. (*Courtesy of Mathews.*)

Pop-up Chain.   The pop-up chain is an active carton divert system that transfers a carton from a roller conveyor surface onto a right-angle shipping conveyor. This divert from the sortation conveyor does change the carton direction of travel or orientation on the shipping lane.

After the carton is inducted on the conveyor surface and the carton arrives at the divert location, two strands of chain that are sandwiched between two rollers are raised above the conveying surface to engage the carton bottom. With the carton bottom under the chain, the carton is above the conveying surface and is transferred from the sortation conveyor surface to the shipping lane.

An alternate to the pop-up chain is to have a chain with a blade (bar). As the divert is required, the chain turns and the blade (bar) engages the side of the carton, moving the carton onto the shipping conveyor.

The pop-up chain sortation system handles approximately 10 to 20 sorts per minute and a carton that weighs up to 75 lb. The pop-up chain divert location is on 4- to 5-ft centers. There is a gentle impact on the carton.

Rotating Paddle.   The rotating paddle pusher is an active sortation device whose operational characteristics are similar to those of the pusher divert system. The rotating paddle is a sortation component with a spherical shape, three to four paddles, and four flat surfaces. When not activated, the paddle device does not extend into the carton sortation conveyor travel path.

After the carton is inducted onto the system and arrives at the assigned sortation location, the paddle rotates so that one of the paddles comes in contact with the carton side. This contact, and the carton forward travel motion, forces the carton from the conveying surface and onto the shipping conveyor. In this system, cartons travel in different directions on the shipping conveyor and the sortation conveyor.

The rotating paddle handles a carton weight of 1 to 75 lb with a medium to rough impact on the carton and performs 50 to 70 sorts per minute. The divert locations are on 9-ft centers.

Shipping Lane Conveyor

The next carton sortation system section is the shipping lane, which is the carton travel surface between the divert location and the next workstation. Photo-eyes are located on each shipping lane, and when blocked by an accumulated carton, they deactivate the divert device, causing additional assigned cartons to recirculate on the conveyor sortation system.

The various types of shipping lane surfaces are gravity metal, plastic-coated wood slide (chute), gravity skatewheel or roller conveyor, and powered live, zero- or low-pressure roller conveyor.

Gravity Slide. The gravity slide is basically galvanized sheet metal, plastic sheet, or coated wood that has a flat or concave design. The slide (chute) relies on gravity to convey the carton from the sortation conveyor surface to the next workstation.

To achieve carton accumulation on a slide, a step is added in the slide bottom that permits cartons to ride on the surface of other cartons. The slide interfaces with an active or passive sortation device and provides limited accumulation. The slide handles a wide variety of carton sizes and shapes; but because of uncontrolled line pressure, the product is not crushable or fragile. The slide concept conveys product a short distance. To reduce carton jams in a chute, the preferred design is the concave type. Most slide designs do not change the carton direction of travel.

Gravity Conveyor. The gravity skatewheel or roller conveyor shipping lane is a concept that requires the carton sortation conveyor higher than the discharge location. For best results, the gravity conveyor should be an extension of an active sortation device and the carton should be the same direction of travel on the shipping lane as on the sortation conveyor. To reduce hang-ups and product damage, the carton sizes and shapes should be standard and the carton should not be fragile or crushable.

If the carton travel speed is excessive on the shipping lane, then to slow carton travel speed the gravity conveyor is tilted to one side, causing the carton to rub against the guardrail, or the skatewheel or roller turning ability is restricted to slow the carton travel speed. To reduce carton line pressure, brake devices are added to the conveyor path at specific intervals.

Powered Roller Conveyor. The powered roller conveyor does not require that the sortation conveyor be higher than the shipping conveyor. For best results, the powered roller conveyor should interface with an active sortation device.

Cleanout or Recirculation Conveyor. The cleanout or recirculation conveyor is the section of conveyor that transports all "did not sort" cartons back to the induction station.

Mimic Display and Control Panels. The last component of a carton sortation system is the mimic display and control panels. These panels indicate to your manager the actual status (off, on, or jammed condition) for the various conveyor sections. The panels and electrical components are reviewed in more detail in Chap. 9.

CUSTOMER RETURNS, OUT-OF-SEASON PRODUCT, AND STORE TRANSFERS

The next activity of a carton handling warehouse and transportation department is the handling of customer returns, out-of-season product, and store transfers.

The customer return activity is a warehouse activity in all industries, but it occurs at a low volume in the carton handling industry. Out-of-season product and store transfers are another low-volume activity in the retail store distribution industry.

Customer Returns

The customer return activity ensures that your customer's store returned-order quantity was received at the warehouse, the product was entered into inventory and was physically returned to the pick position, and credit approval was issued to the customer.

Several measures can be taken to improve the return activity: keep the paper document simple; request that the SKU discreet number, pick position, store number, and carton quantity appear on the document and carton; and limit the returns to occur three or four times per year.

Returns occur for the wrong product ordered, out-of-date product, damaged product, pick errors, and incorrect quantity ordered.

Fixed SKU Pick Position Improves Employee Return Productivity.   A fixed SKU pick position theory specifies one pick position for the SKU. During the life cycle of a SKU, this pick position concept is flexible and allows the SKU to be moved from one pick location to another when the SKU becomes less popular. This location transfer keeps the SKU in a warehouse location that has SKUs with equal or similar popularity, thus optimizing warehouse employee productivity. This transfer occurs two or three times in the life cycle of a SKU and at specific times of your company's business year.

During the return process activity, the SKU is physically returned to the pick position. An employee identifies and attaches this pick position to the SKU. With the fixed pick position concept, where the SKU was picked with a label, then the pick position is more than likely current. If the return does not have a pick label, then the SKU pick position is located in the inventory listing and attached to the SKU. This quick pick position identification means improved return handling and processing productivity.

Later, when the merchandise is physically returned to the pick position, the employee assures that the return SKU matches the SKUs in the pick position.

Handling Customer Returns to the Pick Area.   If your warehouse operation is required to handle customer returns, then the activity is similar (reverse) to order-picking slow-moving SKUs.

Customer Returns Placed in the Active Pick Position.   Customer returns are placed in the active pick position for the SKU, with manual replenishment of the carton to the SKU pick position. This activity is considered a reverse order-pick activity that handles SKUs with a low hit density and hit concentration. With these SKU characteristics, the returns employee must cover a great travel distance between two SKU pick (return) positions, with a random sequence of return SKUs on the pallet load, which is located in any warehouse aisle that has a normal-width pick position.

Disadvantages of this method are that it is labor-intensive and expensive and the process is slow, and in a floating-slot inventory method it is difficult to obtain the active pick position.

Advantages of the return method are that SKU inventory is in one pick location and it is possible to have the oldest SKU inventory pick first for the next customer order.

Customer Returns Handled in a Special Warehouse Zone.   An alternative customer returns handling strategy is to lay out a special zone in the warehouse to handle your customer returns. In this special zone you establish SKU pick position sizes that are designed to accept small, medium-sized, or large cartons. The number of SKU pick positions is sufficient to accommodate a predetermined number of weeks of customer returns, which is based on your past return experience.

As your return employee travels through the special return aisle or zone, at the first vacant pick (return) position, the employee transfers the SKU from the vehicle load-carrying surface to the return pick position. After transferring the SKU to the pick position, the employee writes the SKU pick position identification number and the SKU identification number on the return transfer form. An alternative information transfer is the bar-code scanning method, in which the employee hand-scans the SKU identification bar code and the pick position bar code. With both methods, the SKU and pick position identification numbers are entered into the computer inventory files.

To maintain a good inventory turn and good warehouse space utilization, this return inventory in these temporary pick positions is held for a 4- to 8-week period and then is withdrawn en masse. This and the SKU return inventory are handled as merchandise that is packaged, sent to the discount or outlet store, donated to charity, or placed in the trash.

Various Inventory Control Methods for Returns

After the inventory update, you establish the return inventory priority. The first alternative is to have the return inventory handled as a new receipt of product on a LIFO rotation. With this method, when the existing inventory in the (normal) present pick position is depleted by your customer orders, the return pick position becomes the next active pick position for your next customer order. This is a floating-slot inventory arrangement for this special return area.

A disadvantage of the method is that the oldest SKU inventory is withdrawn last. Advantages of the method are that for your regular merchandise order-picking activity, order-picker productivity rate and high return-to-stock employee productivity rates are maintained.

An alternative method is to have the return inventory handled as the oldest inventory in the warehouse on the FILO product rotation. In this arrangement the computer treats the return SKU inventory with the highest priority. With the next customer order, the computer prints the pick instruction document for the SKU in the return pick position. When the inventory is depleted by the customer order, the computer transfers the next pick instruction to the normal SKU pick position.

Disadvantages of this method are low employee pick productivity during peak activity due to increased travel time through the return area and SKU inventory in two locations. Advantages are reduced quantity of return inventory at the end of the period, which has the potential for a write-off and removes the oldest SKU inventory from the warehouse.

Out-of-Season Product

Out-of-season product activity is a temporary warehouse holding function for retail store merchandise that did not sell at the retail outlets. With top management approval, the merchandise has been returned to the warehouse for temporary storage and awaits top management decision on how to handle the merchandise. To assure accurate and efficient out-of-season product handling, each carton's contents are identified on the exterior along with a packing slip.

Store Transfers

Store transfers consist of overstock product from one retail store that is transferred to another retail store that has high sales of the product. Store transfers with the proper paperwork and labeled cartons are handled as an across-the-dock merchandise at the warehouse.

NONCONVEYABLE CARTON PICK

It is the focus of this section to review the various nonconveyable carton pick concepts, operational characteristics, and disadvantages and advantages.

In the past years, the carton warehouse and distribution industry has implemented highly mechanized or automated order-pick systems to reduce picking expenses, increase the throughput volume, and improve order-picking accuracy. These systems are very common in the retail store distribution industry. One segment of the business that is not handled by these order-pick systems is the nonconveyable SKUs. These SKUs are characterized as odd (irregular)-shaped, small in size, high-cube, or heavy-weight. These SKUs account for 5 to 25 percent (varies per retail type of business) of the volume that is handled by the operation.

Customer Order Handling Methods

Of the various nonconveyable pick methods, two customer order handling concepts are available to your manager:

1. The SKUs (product) that are on the carts or mobile load-carrying surfaces and the customer order staging locations are on both sides of the vehicle path. This arrangement permits single customer order handling or batched pick of customer orders. Batched pick is a customer order handling concept that has more than one customer (four to eight customers) order-picked per wave (trip).

2. Customer orders are located on the carts or load-carrying surfaces, and the product (SKUs) are located on the sides of the vehicle path. This concept handles only single customer orders or a batch of two customer orders.

Various Nonconveyable Pick Methods

There are two groups of nonconveyable pick methods: manual and mechanized. The various methods are

● Manual type

   Manual pushcart, pallet truck, or platform truck

   Electric pallet truck with floor-stack or rack positions

   Order-picker or side-loading truck

Manual push overhead trolley

- Mechanized type

  Powered conveyor

  Inverted power and free conveyor

  S.I. Cartrac

Order-picker productivity in these nonconveyable SKU pick methods is generally lower than that in the conveyable pick methods because it is difficult for an employee to build a unit load or the small number of SKUs (units) per unit load. The latter increases the employee travel time and unit-load handling time.

Various Manual Nonconveyable Pick Methods

The first group of nonconveyable pick methods consists of the manual methods, which require the employee to physically move or steer a vehicle to the pick location and perform the pick transactions there.

Manual Pushcart, Pallet Truck, and Platform Truck Pick Methods. The first manual nonconveyable pick method employs the manual pushcart, pallet truck, or platform truck. An employee pushes or pulls the mobile device with a load-carrying surface through the warehouse aisles. On each side of the aisle are located the floor-stacked or rack pick positions. As required by the customer order, the employee transfers the appropriate SKU from the pick position to the vehicle load-carrying surface.

The manual pushcart, pallet truck, or platform truck method is best for a single-order handling system but provides the lowest productivity of the various methods. In a dynamic or high-volume business, these vehicles are not preferred because they travel only short distances, have low vehicle load-carrying capacity, and low employee productivity (45 to 60 units per hour). The economics of this method is low on a per vehicle basis.

Electric Pallet Truck Method. In the electric pallet truck method, the employee controls an electric powered vehicle with a load-carrying surface through the warehouse aisles. At the appropriate pick position, the employee stops the vehicle and transfers the required SKU to the pallet truck load-carrying surface. The vehicle is available in several alternative models: (1) walkie vehicle, where the employee walks with the vehicle and controls its movement by the operator's handle; (2) walkie-rider vehicle, with the vehicle controls available to the employee from the operator's platform, permitting the employee to walk or ride on the vehicle through the warehouse aisles; and (3) remote-controlled vehicle with a remote-control device attached to the employee. It allows the employee to walk in the aisle between the pick position and the vehicle load-carrying surface and control the forward movement of the vehicle. This feature permits the vehicle to be controlled from any side of the vehicle and improves the safety features of the activity. With over four vehicles per warehouse, special consideration is given to the remote-control system.

Another feature of this method is that the powered pallet jack is available as a single-pallet vehicle with a step platform with a double-pallet load-carrying capacity. The step platform permits easier access to the elevated pick levels. The double-pallet truck transports a heavier or longer load.

As the employee travels through the pick area, the pick activities are characteristically the same as those of the manual platform truck method. In a single customer order handling operation, the electric pallet truck permits the employee to handle heavier SKUs, travel greater distances, and achieve a productivity rate of 60 to 70 units per hour. The economics is medium on a per vehicle basis.

Powered Tugger with a Train of Carts Method. In the powered tugger with a train of carts method an employee travels with a tugger and a train of carts through the warehouse aisles. As required, the employee stops the vehicle and transfers the appropriate SKU quantity from the pick position to the cart.

Various tugger models are used in the nonconveyable pick operation: (1) gasoline or diesel powered tugger models that are used outdoors and (2) electric powered tuggers that are powered by a rechargeable electric battery. Within this group, there are manually and remote-controlled tuggers: a manually powered tugger which requires an employee to have physical access to all forward and reverse movement controls and a remote controlled tugger that has the same operational characteristics as the remote-controlled pallet truck.

The unique feature of the tugger models is the ability to batch-pick customer orders. With a train of carts, each cart or shelf of a cart handles one customer order. When compared to the other manual order-pick concepts, this feature improves order-picker productivity of 70 to 80 units per hour. The economics is medium on a per vehicle (tugger) basis.

High-Rise Side-Loading Truck Pick Method. In the high-rise (multilevel) order-picker truck or side-loading truck method the employee controls horizontal and vertical truck movement in a guided aisle between two rows of racks. These racks have pick positions from the finished floor level to a height of 25 ft and contain SKUs that appear on the customer order. As required, the operator's platform or load-carrying surface is elevated to the required position for the employee to perform the transaction.

With the ability to utilize the entire air space between the finished floor and the ceiling, this method improves space utilization and has the capacity to handle carts or pallets. With a pallet, the concept handles a single customer order. With a divided cart and small customer order quantities, it handles batched customer orders. The economics is high on a per vehicle basis, in addition to the high cost of the guidance system.

Pallet Flow Rack Pick Methods. The pallet flow method has two alternative designs.

1. *SKUs on the pallet flow lane.* In this design, the SKU (pick positions) are on the pallet flow lane with the customer staging locations on each side of (facing) the pallet flow lane path (Fig. 7.17a). A lift truck places the SKU (pallet board) on the flow lane, and an employee pushes the pallet board forward on the flow conveyor (rails). Per the picking instruction, the employee stops the pallet board on the flow lane at the appropriate location and transfers the required SKU quantity from the SKU pallet board to the customer staging location.

2. *Customer location on the pallet flow lane.* In this arrangement the customer location is on the flow lane with the SKU pick positions on the sides of (facing) the pallet flow lane path. In this design, the order picker places an empty pallet board on the flow lane and pushes the pallet board on the pallet flow lane. The lane is between single-high floor-stack or rack positions. Per the pick instruction, the order picker

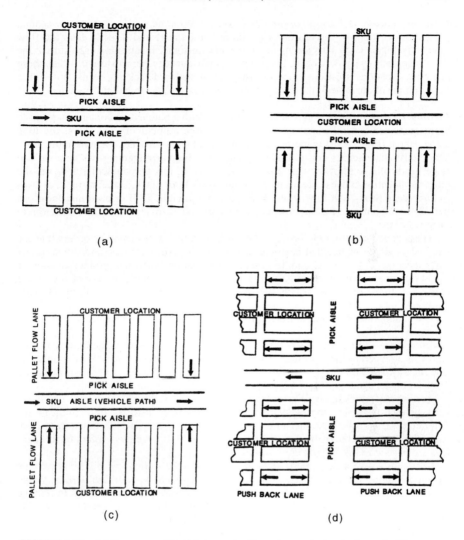

FIGURE 7.17 (*a*) Nonconveyable pick concept with customer staging in the rack; (*b*) nonconveyable pick concept with SKUs in the rack; (*c*) customer locations parallel to the pallet flow lane path; (*d*) customer locations in the loop arrangement.

stops the customer pallet board at the appropriate location and transfers the required SKU from the pick position to the pallet board.

These designs have the same components. The first component is a pallet flow lane (rails) with a slight pitch that is fixed to the floor. For best results, flanged conveyor wheels or a slave pallet board is used. The second component consists of the SKUs or customer locations that are perpendicular to the flow lane or face an aisle that is perpendicular to the lane.

Customer Locations Parallel to the Flow Lane.   With the single customer or a small number of customers, the customer locations are parallel and directly face the flow lane path (Fig. 7.17c). This arrangement has one customer location on each side of the flow lane path every 4 ft on centers. For a small quantity per customer order, these customer locations are floor-stack or single-deep standard rack positions; but with a high volume of customers, these customer staging positions are pallet flow lanes.

Customer Locations in a Loop Arrangement.   With a batched customer order (six to eight customer orders) handling method or a large number of customers, the loop arrangement (Fig. 7.17d) is preferred because this layout provides the maximum number of customer facings per middle flow lane. The loop arrangement has an aisle that is perpendicular to the flow lane path. On the sides of this perpendicular aisle are the customer locations. These locations are single-deep floor-stack, standard rack, or two- to three-deep push-back flow lanes. There are four to eight SKU or customer locations per 28-ft-wide module for the two-deep flow lane and 36-ft-wide module for the three-deep flow lane.

These concepts allow the order picker to handle SKUs or customer orders that are handled as single or batched customer orders. With permanent customer locations or SKU locations close to one another, there is a reduction in the nonproductive walk or travel time, reduction in selection errors, and handling of a wide variety of SKUs and various customer order sizes. The anticipated employee productivity is 90 to 100 units per hour.

The economics of the pallet flow lane method is low on a per foot basis.

Manual Pushed Overhead Trolley Pick Method.   In this arrangement a trolley device is attached to the overhead track. The trolley device has two sets of spools that ride on the track and has a load-carrying surface. The track path is between two rows of pick positions and is a closed-loop path through the two warehouse aisles.

As the order picker pushes the trolley through the aisle which is between floor-stack or rack positions, the pick instruction for a single customer order indicate the required SKUs on the customer order. At each location, the SKU is transferred from the pick position to the trolley load-carrying surface. All completed orders or full trolleys are pushed to the staging area for unloading.

In an alternative arrangement SKU is placed onto the trolley and distributed to the customer staging locations on the sides of the trolley path. This design permits single customer order picking from both sides of the aisle, but requires a lift truck to remove the full pallet from the trolley load-carrying surface, and the picker productivity is 75 to 80 units per hour. The economics of the trolley concept is medium on a per foot basis and medium on a per carrier basis.

Mechanical Nonconveyable Pick Methods

The next major group of nonconveyable pick methods is the mechanical group. In these methods the SKUs or the customer pallet load are mechanically moved to the pick location for an employee to perform the required pick transactions.

Powered Pallet Accumulation Conveyor Pick Methods.   The first pick method in this group is the powered pallet accumulation conveyor method, which consists of motor-driven rollers on close centers. The rollers propel the pallet forward until they stop, then the rollers reduce the forward pressure on the pallet. At the end of the conveyor is a gravity conveyor runout or an end stop. The conveyor path is directly in front of the pick positions, which are floor-stack or rack positions. The pallet board on the pallet conveyor is the SKU or customer location. With the required elevation of the conveyor, this concept has one-side order-pick positions.

If the SKU is on the pallet conveyor, then the order picker stops the pallet with a bar and in accordance with the pick instruction, transfers the required SKU quantity from the pick position to the customer location. The pick location faces the conveyor or is in a loop arrangement.

If the customer location is on the pallet conveyor, then the order picker stops the pallet with a bar and transfers the required SKU from the pick position to the customer location on the conveyor.

With the pallet on the powered conveyor, employee effort is needed to move the pallet, but with the required conveyor elevation, only one-side order-picking activity is permitted. The picker productivity is 75 to 85 units per hour. The economics of the concept is high on a per foot basis.

Inverted Power and Free Conveyor Pick Method. The inverted power and free conveyor system is a motor-driven endless chain that pulls a series of load-carrying surfaces. These surfaces have a set of wheels that ride on a rail and are programmed to momentarily stop in front of the pick position. When the inverted power and free carrier is stopped, the employee performs the order-pick transactions.

The pick position layout and arrangement in this method are the same as those for the powered pallet conveyor method. Picker productivity is 75 to 85 units per hour. The economics of the concept is high on a per foot basis.

S.I. Cartrac Pick Method. This mechanized nonconveyable order-selection method, the S.I. Cartrac pick system, is designed with a series of carts with load-carrying surfaces that travel on a fixed-path track. The fixed cart path is a rotating shaft that propels the load-carrying surface. The method is designed to make 90° or 180° turns. By pressing a foot pedal at the pick position the employee stops the cart and completes the pick transactions.

The S.I. Cartrac is designed to handle the SKU on the load-carrying surface. The S.I. Cartrac system moves the carts with a SKU from the replenishment area to the order-selection area. As the cart arrives at the pick position, the employee stops the cart that has the required SKU for the customer order. The order selector transfers the cartons from the cart to the customer staging location. On completion of the order-selection activity, the cart is released to the next pick location.

When the S.I. Cartrac load-carrying surface carries the customer order, the order picker stops the cart and transfers the cartons from the pick position to the cart load-carrying surface. The cart path is from the order-pick area to the outbound staging area. The employee productivity is 75 to 85 units per hour.

The economics of the S.I. Cartrac concept is high.

Mechanical Nonconveyable Pick Design Parameters

Prior to the implementation of one of these nonconveyable SKU pick concepts, you review these design parameters, which include customer order (information) transmission method, paper flow, pick instruction method, labor productivity and availability, SKU characteristics, SKUs per customer order (mix and hit density and concentration), throughput volume (customer demand average and peak), material handling equipment, storage and replenishment methods and procedures, facility layout, number of customer orders, customer order and delivery cycle, number of pickers per customer order, picker routing pattern, and available capital investment.

CHAPTER 8

ARRANGING KEY PALLET (UNIT LOAD) FUNCTIONS TO CONTROL OPERATING COSTS, IMPROVE INVENTORY CONTROL, AND MAXIMIZE STORAGE SPACE

The objective of this chapter is to review equipment applications, methods, and technologies that improve pallet warehouse and distribution facility employee productivity, storage capacity, and product flow. Improvements in these product storage and movement functions increase your company's profit and provide quality service to your customers. The pallet load section of a carton or single-item warehouse is the main reserve (storage) section of the warehouse. In a pallet load operation, the pallet load positions are the pick and reserve positions of the warehouse facility.

This chapter reviews the key warehouse and distribution functions of pallet load order pick and storage. This includes manual, mechanized, and automated storage and pick concepts as well as analysis of each system's operational characteristics, design parameters, and disadvantages and advantages.

PALLET LOAD WAREHOUSE OBJECTIVES

The objective of a pallet (unit load) distribution facility is to ensure that the right SKU is in the correct inventory location, is available at the right time, is in the correct condition, is in the correct quantity, is withdrawn (order-selected) on schedule, and is delivered to the required location to satisfy the customer's or the manufacturing workstation's order.

To achieve this objective, the pallet load distribution facility is designed to handle the pallet load throughput and to provide the best storage density, unit-load accessibility, product rotation, the required SKU openings per aisle, and the lowest operating costs.

Computer Simulation of Storage Area Transactions. When you use a computer simulation to design your storage area, then you provide the SKU inventory (unit of storage) quantity for average and peak daily projected inbound and outbound transactions, the employee transaction productivity rate, and the number of unit storage positions per aisle. Per your SKU allocation and inventory rotation methods, the computer assigns the SKU storage units to storage positions. The computer program projects the number of deposit and withdrawal transactions per aisle, which can be used to project the number of required storage vehicles to handle the expected business.

FACILITY DESIGN PARAMETERS

The common design parameters for any of the pallet load-handling concepts include the pallet load dimensions (length, width, height, and weight), overhang of the support device, the pallet load bottom support device dimensions (length, width, height, and weight) with complete description (sketch) of the top and bottom, shape of the product, receiving and/or shipping quantities and method of delivery, number of storage units (average and peak) that are handled by each function, the peak number of storage units in inventory, and fire protection and/or temperature requirements or unique storage requirements.

If your inventory does not have uniform design features, then you place SKUs with individual characteristics in a separate storage area. In these separate storage areas, the product is placed in the properly sized storage position and is handled by the required specialized pallet handling and storage equipment.

To achieve an efficient, accurate, and on-time pallet load distribution facility, you arrange the facility rack and aisle layout, schedule the pallet-handling equipment, establish procedures and schedule, and organize the employees and delivery vehicles to handle the projected volume. These resources optimize the following pallet load warehouse functions:

Activities preceding pick and storage

- Unloading
- Receiving and checking
- Identification and inventory control
- Internal transportation
- Deposit

Storage activity

Withdrawal (order pick) from storage

Postwithdrawal (post-pick) activities

- Internal transportation
- Staging
- Securing
- Manifesting and shipping
- Customer returns and out-of-season product

NEED FOR SUFFICIENT CLEAR FLOOR SPACE AND AISLES

In order for warehouse functions to be performed on schedule with as little product and/or equipment damage as possible, sufficient clear floor space and adequate clear-aisle dimensions between products are required. In addition, all warehouse activities should be performed by a trained employee who controls the vehicle that moves the storage unit.

In the prepick activity areas, the unloading dock area has an aisle between the dock leveler (plate) and the unit-load staging area. The unit-load staging area that is immediately behind the dock door has sufficient space to stage the entire load of a delivery vehicle. This space is designed to handle a delivery truck, container, or railroad car quantity. In this area, there is sufficient space for unit-load identification and unit-load stabilization activities.

An aisle exists between the staging area and storage area. This aisle permits an efficient internal transportation of storage units from the receiving area to the storage area and efficient, uninterrupted transfer of lift trucks between aisles.

In the storage area, the clear aisle width between the storage positions (product to product) must be sufficient for the lift truck to perform the storage transactions.

The postwithdrawal (postpick) activities of shipping require the same space as the receiving activities.

THE FOUR MOST IMPORTANT RACK AND BUILDING DIMENSIONS

When you design a pallet load storage facility, the four most important dimensions are (1) the space between building columns, (2) the space between the finished floor and the bottom of the lowest ceiling obstruction, (3) the complete dimensions of the unit load and the bottom support device, and (4) the clearances (open space) between pallets and from pallets to rack structural members. Note that in a high-rise building this includes the upright frame baseplate.

Clear Space between Building Columns

The clear space between the building columns is the open space between building columns. This dimension indicates the floor space between the four building walls that is available for your storage concept. When you design a storage concept inside of the building walls, the open (clear) space between two building columns is an important dimension which determines the number of pallets or racks that will fit between the columns.

Verify the clear dimension between all building columns because some facilities have intermediate columns for mezzanine support. When it is required to fit the building on a site, then some building architects vary the spacing between the outside columns.

Clear Space between the Finished Floor and the Lowest Ceiling Obstruction

The second dimension is the open space between the finished floor and the bottom of the lowest ceiling obstruction. This dimension determines the number of unit loads in a storage stack. This space includes the unit load dimensions, required operational (lift truck and fire protection) clearances, and storage of structural members. Also, when an exceptional situation occurs, such as a heater, the exception is calculated in the storage capacity.

Pallet Load Dimensions

The next dimension is the pallet load dimension. This dimension has two important parts: the unit load and the unit-load bottom support device. The first comprises the length, width, and height of the unit load. The second is the dimension for the bottom support device. If the product (unit load) overhangs the bottom support device, then the overhang length and width become the storage unit-load dimensions. The weight of the unit load and the number of unit loads per rack bay determine the upright frame and load beam thickness. The number of unit loads and the rack weight determine the rack baseplate. The entire unit-load weight, the baseplate size, and the lift truck wheel size with the lift truck weight determine the floor thickness. The floor construction cost is a function of floor thickness.

The height of the unit load determines the number of unit loads in a vertical stack. The width determines the number of unit loads that will fit between two building columns or between two upright rack posts. The unit-load depth determines the depth of the storage lane or the standard rack upright frame depth. Along with clearances and vehicle aisle transaction requirements, the unit-load depth determines the number of unit loads and aisles between two building columns.

Pallet Load Bottom Support Device

The length and width of the pallet board (unit-load bottom support device) are important dimensions that determine the location of the rack structural components, such as load beams, load arms and rails, and upright frame posts. In most facilities, the pallet board overhangs the load beam (extends into the aisle or flue space) by 2 to 3 in. The unit-load bottom support device determines the length and type of lift truck forks or attachment.

Other Important Clear Spaces

18 in for Sprinklers. When a storage system is designed, there is at least 18 in of clearance between the ceiling sprinklers and the top of the highest pallet load. This clearance reduces sprinkler damage and accidental discharge of water that damages the product and permits the water spray to cover the area.

3 to 6 in between Pallet Loads and Racks. Another important clearance is the horizontal open space between two unit loads and between the rack structural members. The open space between two unit loads or rack members is 3 to 6 in. This space varies per lift truck vehicle and provides the open space that permits transactions to be performed with low product and equipment damage and high employee productivity.

5 to 6 in between Pallets and Racks for a Straddle Truck. When the storage facility uses a straddle lift truck with pallet boards on the floor, then the clearance between the pallet loads or pallet load and rack structure is at least 5 in, or the width of the straddle plus 1 in. This open space permits the lift truck straddle to fit between the straddle and rack structure and pallet load.

Rack-Supported Building Means Wider Rack Baseplate.   In a rack-supported building or tall rack building, the larger upright frame baseplate spreads the load, and the baseplate is considered in the rack opening clearance, or else the bottom load level is raised.

6- to 12-in Flue Space between Back-to-Back Racks or Walls.   When racks are designed as back-to-back rack rows, then there is 6 to 12 in of open space between the unit loads. This open space is defined as the *flue space*; it permits sprinkler pipe installation between two rack rows or between the rack row and the wall.

2 to 6 in between Floor Stack Loads.   When the floor stack (block storage) of pallet loads, stacking frames, or containers is designed in a storage area, with a conventional lift truck there is 2 to 3 in of open space between adjacent storage stacks. A straddle truck requires at least 5 in between adjacent loads. This open space permits the unit-load transactions to be performed without damage to the adjacent or rear unit-load stack. If slip sheets are used in the storage operation, then the slip-sheet tab (nominal 6-in dimension) extension direction is added to the space between two adjacent unit loads.

For a Tall Building Allow for Baffle Levels and Additional Sprinklers.   With most taller warehouses, the vertical unit-load storage stacks require fire baffle(s) or additional sprinkler levels within the vertical stack. This dimension is 6 to 12 in in height at the required locations.

18-in White Space from the Wall in Food Warehouses.   When the facility is used for a food warehouse and distribution facility, then an 18-in minimum open space between the wall and storage stacks is maintained along the walls. This space is required to ensure proper sanitation and is painted white.

All Conveyors at Least 6 to 12 in from Building Obstacles.   When a facility is designed, all equipment paths must have a 6- to 12-in minimum clearance from any building column, obstacle, other fixed building, or product-handling equipment.

OTHER FACTORS THAT AFFECT EMPLOYEE PRODUCTIVITY

The pallet load-handling components that affect pallet load facility employee productivity and costs are adequate floor space for staging and storage, type of pallet load-handling vehicle, type of pallet load storage system, type of pallet load bottom support device, method used to secure the pallet load, pallet locator system and inventory control method, pallet load withdrawal method, pallet load position identification method, pallet load identification method, pallet load put-away (deposit) system, pallet load put-away and withdrawal verification system, pallet load transaction instruction method, and type and shape of product.

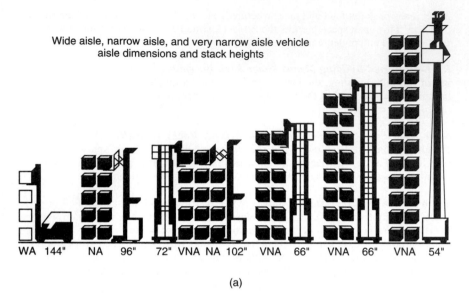

Wide aisle, narrow aisle, and very narrow aisle vehicle
aisle dimensions and stack heights

| WA 144" | NA 96" | 72" VNA NA 102" | VNA 66" | VNA 66" | VNA 54" |

(a)

FIGURE 8.1   (*a*) Various vehicle comparisons (*Courtesy of Raymond Corp.*).

Unit-Load-Handling Vehicle

The first pallet and container handling component is the type of unit-load-handling vehicle used in your operation. (See Fig. 8.1*a*.) The unit-load handling vehicle makes the unit-load deposits and withdrawals (transactions to and from the storage positions). The unit-load vehicle is the key factor that determines the warehouse aisle widths and storage heights. The aisle width and the storage height influence the warehouse cube utilization and required land.

Five Vehicle Types

The pallet load-handling vehicles are separated into five groups, with each group name determined by the basic operating characteristics of the vehicle:

- Wide-aisle (WA) group
- Very narrow-aisle (VNA) group
- Mobile-aisle (MA) or transfer car (T-car) group
- Narrow-aisle (NA) group
- Captive-aisle (CA) group

Wide-Aisle Vehicle Group.   The first unit-load-handling vehicle group is the wide-aisle group. This group gets its name from the fact that the vehicle requires a 10- to 13-ft-wide clear aisle between products. This distance allows the vehicle to make a

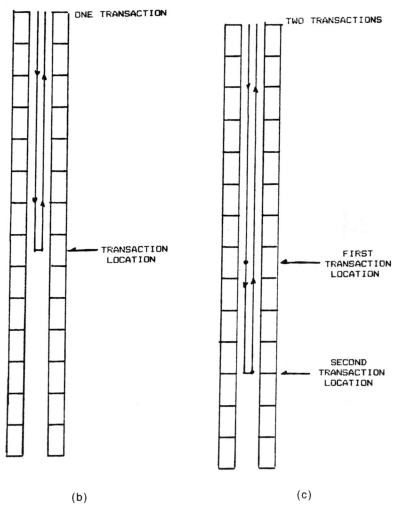

(b)   (c)

FIGURE 8.1   *(Continued)*

right-angle (stacking) turn and to perform a transaction (deposit or withdrawal from a storage position). Also it allows the operator to achieve the desired productivity with minimal product and equipment damage.

These WA vehicles are mobile-aisle equipment. Most WA vehicles are used to transport pallet loads between two warehouse activity areas. These vehicles are powered with an electric rechargeable battery, gasoline, liquid propane gas, or a diesel engine. The diesel- and gasoline-powered vehicles are preferred for outdoor activities.

The vehicle is equipped with a set-forks, slip-sheet device or other attachment to an elevating mast. The forks lift a unit load to a reserve position 20 ft high. Tilt and side-shift options improve operator productivity and reduce product and equipment damage.

The most popular wide-aisle vehicle is the three- or four-wheel counterbalanced lift truck on which the operator sits on a seat or stands on a platform in front of the vehicle control panel or steering wheel.

The disadvantages of the wide-aisle vehicle are that it requires a larger building, that it limits the lifting height, and that operators who sit tend to become lazy.

The advantages of the wide-aisle vehicle are that it is used at the dock to load and unload delivery vehicles, with free lift used under a mezzanine, it handles a slip-sheet or other attachment, it is easier to operate, it operates on a conventional warehouse floor, it requires little employee training and low capital investment, it travels up and down grades, it is used as a transport vehicle, it can reach faster travel speeds, and it interfaces with all storage concepts except two-deep and AS/RS (automatic storage and retrieval system).

Narrow-Aisle Vehicle Group.   The second unit-load-handling group is the narrow-aisle group. An NA vehicle operates in an aisle that has 7- to 10-ft clearance from product to product. The vehicle requires the clear width to make a pallet load transaction with good operator productivity and minimal product or equipment damage. Operator productivity is improved and product and equipment damage is minimized with side-shift capabilities and 5 to 6 in between two loads.

The raised load beam concept permits the vehicle to operate in an aisle that has a narrower dimension of 7 ft but requires the first storage level to be raised 12 in above the finished floor. The elevated load beam concept permits the straddle to travel under the load beam and into the rack position. This transaction is performed without the straddle striking the pallet board.

The narrow-aisle vehicle is mobile aisle equipment that travels between two warehouse aisles. The vehicle is powered by an electric rechargeable battery and has a set of forks that elevate on a mast which raises the pallet load to the storage position height. This pallet position height is 25 to 28 ft. Most models are reach types (they extend the forks outward) that permit the vehicle to handle unit loads wider than the opening between the two straddles. Some models are used in a two-deep storage system. Other models are straddle types that raise and lower the unit loads but do not reach.

All narrow-aisle vehicles have steering and fork controls that are positioned in front of the operator, who stands under an overhead guard.

The disadvantages of the narrow-aisle vehicle are slower travel speeds and increased operator fatigue from standing, and it is used in all storage systems except AS/RS.

The advantages of the narrow-aisle vehicle are that it requires a smaller building, reaches higher storage positions, travels up a slight grade, operates under most mezzanines with free lift, and travels between two aisles.

Very Narrow-Aisle Vehicles.   The third pallet load-handling vehicle is the VNA group. These vehicles operate in a 5-ft 6-in to 7-ft 6-in wide aisle that gives these vehicles their name. Operating in these aisles to make deposits or withdrawals, the vehicle places the pallet load directly into the reserve position without turning in the aisle. This feature permits the narrow aisle width.

Captive- or Mobile-Aisle Vehicle.   These vehicles are mobile- or captive-aisle equipment. Travel of a captive-aisle vehicle is restricted to one warehouse aisle. However, with the proper rack arrangement, a captive-aisle vehicle is transferred between warehouse aisles with a transfer car (T-car). Some VNA vehicles transfer aisles under their own power. A VNA vehicle is rail- or wire-guided, and its power source is either an electric direct current or electric battery or batteries. The equipment has forks or platens that lift the pallet load to a storage height ranging from 30 to above 80 ft. Some vehicles require the bottom unit-load storage level that is elevated 16 to 34 in above the finished floor.

Since the overhead guard and mast structural members require clearance between the ceiling and top unit-load position and when there are different unit-load heights in a facility, in the vertical rack design the tall unit-load storage positions are on the highest levels and the short unit-load storage positions are on the lower levels. Note that the high storage positions are the warmest, and temperature-sensitive product is stored on the lower levels.

The machine is controlled manually or by computer to perform its storage-pick transaction. In manually controlled equipment, all the controls are in front of the operator, who is located under an overhead guard. In some models the operator is inside a cab. In most models, the operator is raised with the pallet load. But in some models the operator is not raised with the load.

The disadvantages of the VNA vehicle are that it requires a large capital investment, an F75/F100 level floor/wide transfer aisle, a guidance system, and an end-of-aisle pickup/delivery station.

The advantages of the VNA vehicle are that it provides maximum storage height, requires a small building, requires the fewest employees, and ensures the least product and equipment damage.

In Chap. 10, the vehicles in the three groups are discussed in more detail.

TRANSACTION INSTRUCTION METHOD

Another component of a unit-load distribution facility is the unit-load transaction (deposit/withdrawal) instruction method. The transaction instruction method directs the lift truck operator to deposit or withdraw a unit load at a position.

The transaction instruction method must match the type of pallet load operation. The various pallet load instruction methods that are available for implementation in a new or existing warehouse operation are (1) a manually controlled vehicle, human, or machine-printed paper document or label and (2) an automated vehicle (computer network or cards).

In Chap. 5, each of these transaction instruction methods is reviewed in more detail.

Various Unit-Load Transaction Types

The next factor that affects pallet load warehouse efficiency is how the unit-load transactions are handled in the storage area. This factor determines the number of lift trucks in the storage aisles and the number of unit loads handled per hour. The two alternative transaction-handling methods are the single-command and the dual-command (cycle) methods.

Single-Command Method. With the single-command transaction-handling method, on each trip (down-aisle travel) the lift truck operator makes one storage transaction (a deposit or retrieval transaction). (See Fig. 8.1*b*.) One transaction is performed on one trip in and out of the warehouse aisle.

Dual-Command Method.   With the dual-command transaction handling method, first on each trip (down-aisle travel) the lift truck operator makes two storage transactions. On the trip into the aisle, the operator performs a deposit transaction; on the trip out of the aisle, the operator performs a withdrawal transaction. During one trip, these two independent transactions are performed by the lift truck operator. To perform dual commands, the storage area must have an end-of-aisle pickup/delivery (P/D) station (racks or conveyors) at the end of each aisle. After deposit of an outbound unit load, the P/D station permits the lift truck to pickup an inbound load. Second, to achieve dual cycles, the inventory control system has the capability to review the batched customer orders and to assign inbound unit loads to a storage position in an aisle that has a unit load required on a customer's order. A third criterion is sufficient inbound and outbound unit-load staging area in the combined receiving and shipping dock area.

In comparing the two transaction methods, the dual-cycle method performs 25 transactions per hour. This is more productive than the single-cycle method which does 15 to 20 transactions per hour. The dual-command cycle does not double the single-command cycle because additional transactions require additional slowdown and start-up speeds and fork extensions and withdrawals.

CUSTOMER ORDER HANDLING METHODS

Another customer order handling factor is the number of lift truck operators that handle the customer order. The order handling method is matched with the type of storage operation. Some customer order handling methods are preferred in a manual operation, and others are best in a computer-controlled operation. The various customer order handling methods and the type of operations are matched in Fig. 8.2.

In Chap. 5, these customer order handling methods are reviewed in more detail.

INVENTORY CONTROL SYSTEM

The next important variable in a pallet load facility is the type of inventory control system which includes the product locator system, product identification method, and storage position identification method. The preferred locator system in a pallet load warehouse is the random (floating) storage position system.

In Chap. 12, the product and storage position methods are reviewed in more detail.

Type of operation with multiple aisles	Customer order handling method
Manually controlled lift truck	Single customer order with one lift truck
	Single customer order with multiple trucks in zones (aisles)
Computer-controlled lift truck	Single customer order with multiple trucks

FIGURE 8.2   Storage transaction handling methods and operation.

Four Components of the Storage System

A unit-load storage system has four major components: facility (building), mobile- or captive-aisle vehicle, inventory control system, and storage equipment (racks).

Your facility or building ensures that the right protection is provided for company assets. The mobile- or captive-aisle vehicle ensures that the product is transferred from one location to another and that there is excellent labor productivity. The inventory control system ensures through the updating of records that the right product is assigned to the right storage position, is transferred at the right time, and is in the correct quantity. The storage equipment (racking) ensures that there is proper product accessibility and that there is maximum utilization of space.

VARIOUS UNIT-LOAD STORAGE-PICK CONCEPTS

The various unit-load storage-pick concepts include

- Floor stack and block storage (90° or 45°)
- Tier racks and stacking frames
- Double-deep and two-deep racks
- Standard pallet racks
- Bridge
- Drive-in rack
- Gravity-flow and airflow racks
- Car-in-rack
- Drive-through rack
- Push-back rack
- Cantilever rack
- High-rise rack

Storage-Pick Concept Is Key to Satisfying Objectives

The storage-pick concept is a key factor that determines the unit-load distribution facility's ability to satisfy the company's storage-pick objectives. The design parameters that influence the preferred storage-pick concepts include the type of operation, average and peak units of storage volume, unit-load dimensions and weight, type of product rotation, allowances for clearances, number of pallet loads per SKU, number of total pallet loads, type of product bottom support device, facility (building) design parameters, type of SKU and shape, type of material-handling equipment, and required storage conditions and building codes.

The objective of this section is to develop an understanding of the various unit-load storage-pick concepts with their alternative designs and operating characteristics. This chapter looks at each method's ability to satisfy the distribution facility's storage-pick objectives and to indicate the required material-handling equipment as well as the required storage concept utilization factor. The storage concept utilization factor is a percentage factor that increases the design pallet load positions to accommodate the imbalance between inbound pallet loads and withdrawals and storage position occupancy.

Floor Stack (or Block Storage)

The first pallet storage-pick method is a dense-storage concept. Floor stack or block storage has two alternative designs: the 90° stack and the 45° stack.

90° Floor Stack. The most common floor stack system is the 90° stack. This system provides the greatest number of SKU openings and storage positions per aisle.

45° Floor Stack. The 45° stack allows a minimum right-angle lift truck turning aisle, but fewer openings per aisle.

Characteristics of Floor Storage. In the floor stack storage, pallet loads or containers are placed on the floor. This arrangement provides a maximum of 6 to 10 pallet loads deep per storage lane. A storage lane is a single lane or back-to-back lanes. Because of the leaning of pallet loads and the variance of pallet load placement on the floor, longer (deeper) storage lanes reduce lift truck operator deposit and withdrawal productivity. To utilize the space, additional pallet loads are stacked onto the floor pallet loads. This requires that the SKU in the storage-lane aisle position be the SKU for the entire storage lane. In stacking unit loads, the pallet load or case of goods on the floor must be capable of supporting the stacking weight. Floor stack storage, when full, offers the highest storage density with lowest investment cost but poor pallet load accessibility.

A 60 percent utilization factor is due to honeycombing (vacant unit-load positions in the vertical stack) and vacant unit-load positions in the storage lane. See Fig. 8.3. These occur from normal lift truck pallet load transactions (withdrawals) from the lane.

In floor stack storage, to make a pallet load deposit or withdrawal transaction, from aisle A the lift truck enters the storage lane, travels to the lane pallet load position, performs the required transaction, and backs out from the storage lane to the same aisle A.

When a floor stack storage method is designed, the number of pallet loads deep per storage lane is varied to bury building columns in or between the storage lanes. The concept has a LIFO (last in, first out) product rotation, handles a high throughput volume, and interfaces with wide-aisle and narrow-aisle lift trucks.

In a floor stack warehouse, if the unit-load bottom support device is a slip sheet on a pallet board, then the storage unit load of two slip-sheet unit loads on one pallet board is possible. This practice improves employee productivity. When compared to two normal pallet load height, there is a 5-in decrease in the height of the new two-high slip-sheet unit load because there is one less pallet board. This two-high slip-sheet arrangement permits better storage utilization (addition of one unit load per stack) in your facility.

Portable Container, Tier Rack, or Stacking Frame

The second storage method involves portable containers, tier racks, or stacking frames (portable racks) which are considered a dense-storage method. When the SKU that is

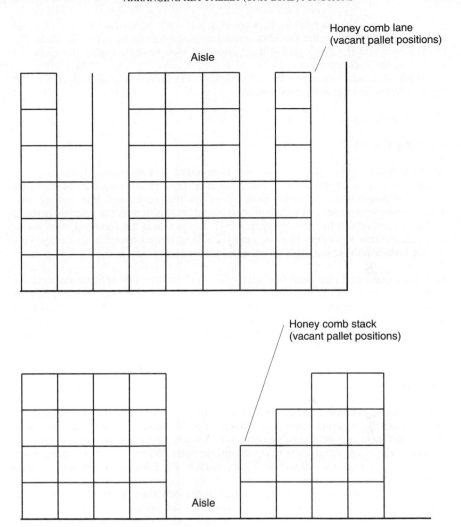

FIGURE 8.3 Floor stack honeycombing.

placed in floor or block storage is crushable (is not self-supporting or is not square), then the container, tier rack, or stacking frame makes stackable and uniform unit loads that optimize the cubic (air) storage space. When you use these storage devices, common practice has one SKU per unit-load lane and stack. Local fire protection codes are reviewed for storage height, sprinkler requirements, and depth restrictions.

Tier or Stacking Rack

The tier rack or stacking rack has a standard wood pallet board as its base. The tier rack has permanent or removable legs. After the bases of the legs are attached to the pallet board and the structural legs are connected in an X pattern at the top, the product is placed onto the pallet board to a predetermined height. A lift truck places the full tier rack into the floor stack storage position. Two or three loaded tier racks are stacked on the floor-level tier rack.

To add structural stability to the tier rack, nails are driven through each of the four leg-base holes into the wood pallet board.

Stacking Frame

Stacking frames have four legs that are permanent or removable. These legs are attached to a solid or ribbed base. There are two types of stacking frames. The first type is the nonnesting type, which means there are four permanent legs that are connected on the four corners with the top cross members that connect in a square pattern.

The second type is the nestable type, which means that it has three top cross members with the rear legs offset to allow empty stacking frames nested in its cavity. This nesting feature reduces the required storage area for empty frames.

Stacking Frame and Tier Rack Options. Optional features that enhance the stacking frame or tier rack storage capabilities include wire-mesh and solid removable base to reduce ripples in cartons, rubber straps or shrink wrap across the four sides to keep cartons from falling, and full-length fork sleeves and stir-ups to prevent the metal frame from sliding on the lift truck's metal forks.

Container

The container is a storage device made of wire mesh, wood, metal, or plastic material. The container has a permanent bottom surface or pallet board. The four sides are capable of supporting one or several containers in a stack, and the base has openings for lift truck forks. Some containers have collapsible sides. When they are not being used in the storage area, their collapsible feature reduces the floor space required to store empty containers.

The operational and design characteristics of the container, tier rack, and stacking frame storage are the same as those of the floor storage system factors with a storage lane utilization of 60 percent. The concept has a LIFO product rotation and handles a medium volume due to lower lift truck operator stacking productivity. This low productivity is due to the slow stacking transaction speeds. The storage concept provides high storage density but poor unit-load accessibility.

Standard Pallet Rack

The third unit-load storage-pick method is standard pallet racks. One, two, or three pallet loads are placed into the rack opening, which is determined by the pallet board width and load beam length. Usually pallet loads in the first vertical opening are placed on the floor, and pallet loads in the other vertical openings are placed on a pair of load beams. There is 3 to 6 in of clearance between unit loads and uprights and between unit loads and load beams. If a straddle truck is used in the pallet rack with three unit loads or if the pallet load is placed in the rack opening with its long dimension (48 in down-aisle), then the bottom rack opening is raised above the floor onto a pair of load beams. This feature provides clearance for lift truck straddles to travel under the load beams. If the straddle truck is used with the unit load on the floor in the 40-in dimension, then at least 5 in is allowed between unit loads and rack members.

A rack bay consists of two vertical upright frames and two pairs of load beams that are attached to the upright frames. At the minimum, one pair of load beams is at the first level, and a second pair of load beams is at the top of the upright frame. This feature provides good stability. Many rack installations are three or four levels high. In a tall building, they are at least four or six levels high.

The upright frames are designed by the manufacturer to hold all the rack bay's pallet load weight plus the rack weight. The pair of load beams is designed to hold the rack opening's pallet load weight. It is common for the upright frame and the load beam connection method to permit 2- to 3-in adjustability of the rack-opening vertical height.

Diagonal and Horizontal Bracing Intersection Faces the Aisle and Other Installation Options. During installation, the intersection of the upright frame's bottom diagonal and horizontal support structural members should face the aisle. If there occurs a lift truck impact against the frame, then this arrangement allows the bracing to withstand the impact. To meet stability requirements and seismic conditions, the upright frame's baseplate is anchored to the floor, back-to-back ties are attached to back-to-back upright frames, and single rows along a wall are tied to the wall. With heavy loads, the upright frame baseplate widens to disburse the weight. When a standard pallet rack is installed in your facility, the first front upright frame post is anchored at the start of each rack row, and the next anchor is on the rear of the second upright frame post. This anchor pattern ensures rack stability.

Using the Open Space between the Bottom Deck Boards to Lock the Pallet Board in the Rack Bay. In your warehouse operation, if you are concerned about the pallet load becoming dislodged from the rack bay, then one solution is to have front-to-rear members installed in each storage position (see Fig. 8.4a). These members span the open space between the two load beams and provide support for bottom exterior deck boards.

An alternative is to use the load beams and the bottom exterior deck boards to lock the pallet board in the rack bay (Fig. 8.4b).

To use the open space between the bottom exterior deck boards to lock the pallet board in the rack bay, the depth of the upright frames and load beams is equal to the pallet load depth less the depth of two exterior bottom deck boards. After the lift truck operator places the pallet load in the rack, the two bottom exterior deck boards are on the outside of the two load beams. This arrangement has two or three stringers of the pallet board rest directly on the two load beams. Since the bottom exterior deck boards are at least $\frac{1}{2}$ in high, in the rack bay this reduces the pallet board forward and reverse movement. This arrangement does not interfere with the performance of the lift truck storage transaction because the pallet board fork openings are maintained at the standard height.

STANDARD RACK OPTIONS AND ACCESSORIES

Rack options and accessories are designed to increase the rack's structural stability and the lift truck operator's ability to efficiently handle SKUs. These are the options (Fig. 8.5a):

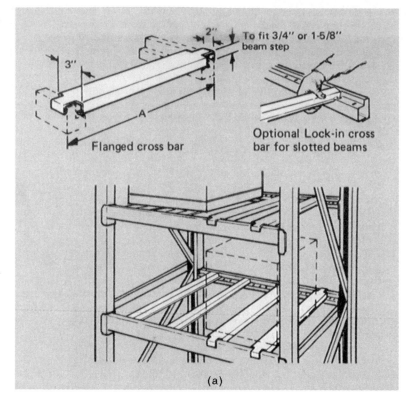

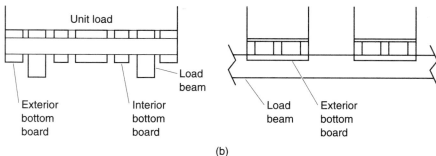

FIGURE 8.4 (*a*) Front-to-rear members (*Courtesy of Interlake Materials Handling.*); (*b*) load locked on the load beam.

Column protectors   Skid channel

Rigid row protectors   Cross bars (front-to-rear members)

Carton stops   Antidebris column

Splice plate   Wood-filled column

Reel holder   Double column

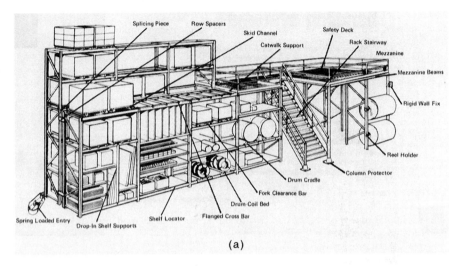

(a)

FIGURE 8.5   (*a*) Rack options. (*Courtesy of Unarco Materials Handling Corp.*)

Safety deck	Hanging-garment rail
Fork clearance bar	Lift truck stops
Catwalk support bar	Drum cradle
Drum coil bed	Stairway
Decking (solid metal, wood, wire mesh)	Entry guide and guiderail
Railcar door opener	

Other Standard Rack Characteristics

In a distribution facility, the standard pallet rack is designed with single-rack rows or back-to-back rack rows. The pallet load position utilization factor is 85 percent. With access to all pallet positions, the pallet load rotation is FIFO (first in, first out), and it handles a medium to high volume. Whenever possible, all building columns and fire sprinklers must be in the flue space (8 to 12 in of open space) between the back-to-back load beam rows or walls. If required, the building columns can be designed in a rack bay but not in the aisle. With an aisle between each rack row, most conventional standard rack storage arrangements have low storage density but excellent pallet load accessibility. The standard rack interfaces with most lift truck types.

Bridge-across-the-Aisle Method

The standard pallet rack that bridges a warehouse aisle is a storage method that takes advantage of the total available storage space in the warehouse. This method has the same design and operational characteristics as the standard pallet rack. Pairs of load beams with front-to-rear (cross) members span the aisle and are connected to the upright frames. This arrangement forms a bridge across the aisle. For each rack bay, the bridge across the aisle provides one or two additional pallet openings. A very important bridge design characteristic is to allow sufficient space for the lift truck mast (collapsed) clearance between the floor and the bottom of the bridge load beam. The storage density and pallet load accessibility are the same as for the pallet rack.

Before we look at other storage racks, we mention that the upright frame and the upright post are common structural members used in other rack storage systems.

Upright Frame

The upright frame (Fig. 8.5b) has two designs, the straight and the cant-leg, which provide the support for the load beams.

Straight Upright Post. The first upright post type is a straight type, and it is very common type in the warehouse industry. With the straight design, the front (aisle) upright post is a straight structural member from the baseplate to the frame top.

Cant-Leg Upright Post. The second upright post type is the cant-leg (set-back) upright post design. This design has the aisle post from a location interior to the rack position extend on an angle upward and outward toward the aisle and connect with the straight upright frame post. In a wide-aisle lift truck warehouse, the cant-leg design provides additional clearance which allows the lift truck operator to make right-angle turns without hitting the upright frame leg with the unit load or the counterbalanced lift truck's rear chassis.

Upright Frame Post and Load Beam Designs

A metal upright rack frame post (Fig. 8.5c) has three basic designs: the C, round, and structural. The three metal load beam designs (Fig. 8.5d) are the step, structural, and rectangular types. With decking, the step load beam is preferred for a decking level because the deck surface extension into the rack opening is minimized by the step. Each manufacturer has a series of holes that are punched into the upright post for load beam connection and possible future rack opening height adjustments. The various manufacturers' hole designs are unique and prevent you from mixing one manufacturer's upright post with another rack company's load beam.

Load Beam Connection Methods. Two load beams provide the unit-load support structure. There are three methods to connect load beams to the upright frame: the bolt, nut, and washer; the clip-on; and the safety lock.

Important Rack Dimensions

In the design of a rack storage-pick area, several important rack dimensions can ensure that your desired storage design fits your building and that your lift truck oper-

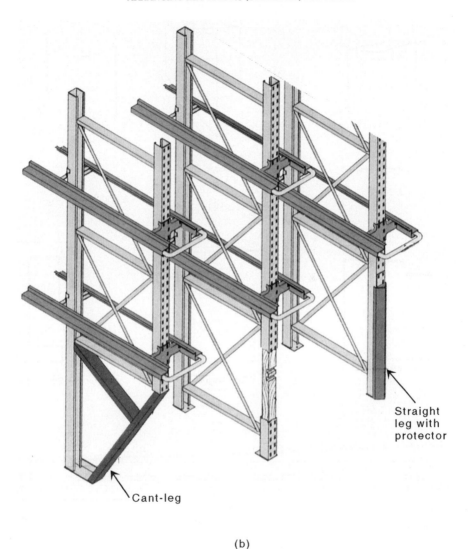

Straight
leg with
protector

Cant-leg

(b)

FIGURE 8.5   (*Continued*) (*b*) Rack leg types. (*Courtesy of Unarco Materials Handling Corp.*)

ators meet the projected transaction productivity. The pallet load dimensions are a known design parameter. These rack dimensions are the rack depth, rack height, and rack length.

Rack Depth.   The first important rack dimension is the rack depth. The rack depth is the dimension that holds the storage unit (pallet load). The rack depth is determined by the unit load and the unit-load support device depth.
    In a standard or two-deep rack, this dimension is the upright frame depth, which is 4 to 6 in less than the pallet load dimension.

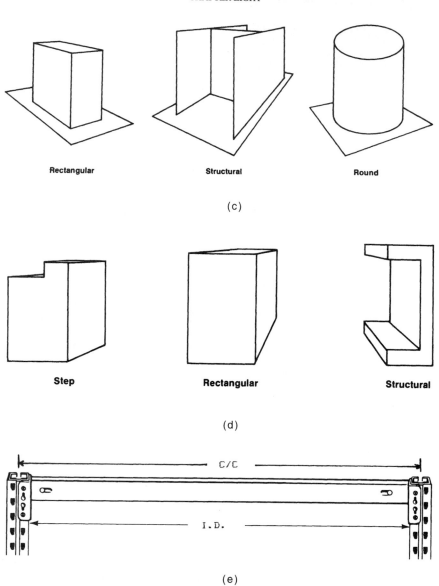

Rectangular   Structural   Round

(c)

Step   Rectangular   Structural

(d)

(e)

FIGURE 8.5 (*Continued*) (*c*) Various upright frames; (*d*) various load beams; (*e*) two load beam lengths.

In dense-storage rack concepts, this dimension is a combination of the upright frames and the upright posts that are equal to the number of pallet loads per lane.

Edge Load.   *Edge load* arises in a warehouse rack storage situation when the edge of the pallet load rests on the edge of the two load beams. This is not considered good practice because it requires the lift truck operator to exactly place (deposit) the pallet load into the storage position. This requirement decreases the lift truck operator's productivity and increases the possibility of the pallet load falling into the rack bay.

Rack Height.   The second important dimension is the rack height. It is the upright frame height or post height that is sufficient to support the top unit load and to permit sufficient horizontal and diagonal bracing members.

In a standard rack or a two-deep rack, this upright frame height extends 2 to 6 in above the top of the top load beam level. This additional upright post height serves as a guide for the lift truck operator in placing the pallet on the top load beam level.

In some hand stack racks (short-depth racks that are tall), the aisle-side upright posts extend upward to a height that permits lift truck passage with a full extended mast. At the top of these upright posts, an overhead tie (cross-aisle tie) is attached to both posts. This arrangement provides additional stability to the rack structure and permits the lift truck to pass through the aisle with the operator performing transactions at the highest load beam level.

In drive-in or drive-through rack designs, the front (aisle) upright posts are extended upward and are tied at the top. The overhead tie has sufficient height to permit a unit-load transaction at the top level. This attachment provides stability to the rack structure.

Rack Length.   The rack length has two important dimensions (Fig. 8.5*e*). These dimensions are the internal dimension (ID), or load beam length, and the centerline-to-centerline (C/C) rack opening length.

The load beam length (ID) dimension is the rack length that handles the combination for the width (fork opening) of the storage pallets (include overhang) and the required clearance. Another definition of the internal dimension is the open (clear) space between two upright posts or frames.

The centerline-to-centerline dimension is the rack length that is the combination of the load beam length and the upright post and frame length. When you calculate the warehouse aisle length, you use the centerline-to-centerline dimensions plus the length of one upright post. To determine the warehouse aisle, you count the number of rack bays (openings) and multiply the number of bays by the centerline-to-centerline dimension plus a nominal 3- to 4-in post width.

If you calculate the warehouse aisle with the load beam length (internal dimension or ID), then the actual warehouse aisle is short by the sum of the number of upright post widths in the aisle length.

Two-High Unit Loads on the Floor

When the standard, drive-in or drive-through rack design has two-high unit loads, the two-high unit load is set on the floor and the rack levels are one unit load high. This rack feature requires your rack manufacturer to specify the racks. With this arrangement, the upright frames and upright posts are manufactured with additional strength for the increased bottom-level vertical height. Also this installation design is approved by the rack manufacturer. Rack post protectors are placed on the aisle side of the upright frame post to reduce lift truck damage.

Two-Deep (Double-Deep) Rack

The fourth unit-load storage system is the two-deep or double-deep pallet rack, which is considered a dense-storage method. The two-deep pallet rack's components and design characteristics are similar to those for the standard pallet rack, with a few exceptions. The first exception is that the rack upright frame has two alternative designs. The selection of the upright frame design depends on the weight of the pallet load and the desired flexibility to reuse the upright frames at another facility that has single deep-lift trucks.

Long Frame. The long- or one-frame system has a long upright frame with four load bars and two load beams. See Fig. 8.6a.

Two Standard Upright Frames. The second frame configuration consists of two standard upright frames with four standard load beams. See Fig. 8.6b.

Two-Deep Rack Characteristics

The two-deep rack has a utilization factor of 85 percent for the interior pallet positions and 70 percent for the exterior pallet positions. The double-deep rack row design requires that one rack bay be placed directly behind the other rack bay. The space between the exterior and interior pallet positions varies between 1 and 4 in. This space is a function of the lift truck stroke or fork extension. The double-deep rack design allows the rack opening to hold four conventional-sized pallet loads. The second unique feature is that the two-deep rack concept requires a double-deep reach lift truck that has a stroke (forward extension of the forks) of 43 to 51 in.

In the two-deep concept, the SKU first unit-load deposit is made to the interior rack position, and the SKU second unit-load deposit is made to the aisle (exterior) rack position. The SKU first unit-load withdrawal is made from the aisle (exterior) unit-load position, and the SKU second unit-load withdrawal is made from the interior rack position. This feature and the time required to extend the forks provide a LIFO product rotation and the ability to handle a medium volume. With four unit loads per pallet position (two deep), double-deep rack storage provides medium storage density and fair unit-load accessibility.

The two-deep or double-deep frame design has two arrangements.

Up-and-over or the Raised-above-the-Floor Arrangement. In the first arrangement, the bottom rack opening is raised above the floor (see Fig. 8.7a). The open space permits the lift truck straddles to pass under the bottom load beam and to turn in a narrower aisle. Also this feature allows easy pallet load deposit and withdrawal at any rack level because the lift truck straddles are not required to straddle the floor-level pallet board. This raised bottom rack level (up and over) concept is required for smaller pallet loads or for pallet loads with the 40-in dimension (depth) into the rack. Also this concept requires a narrow rack opening, which means a greater number of positions per aisle but increases the rack investment and rack and building height.

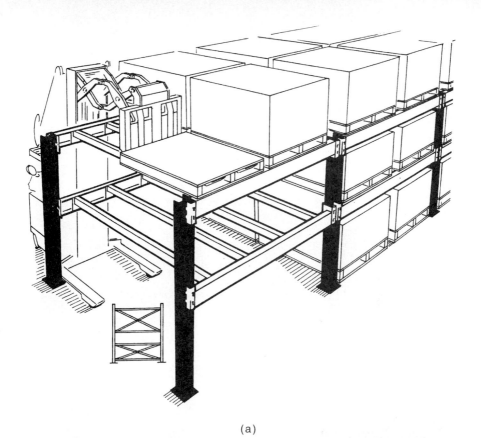

(a)

(b)

FIGURE 8.6 (*a*) Long-frame two-deep racks (*Courtesy of Frazier Indl. Co.*); (*b*) two-deep standard racks (*Courtesy of Unarco Materials Handling Corp.*).

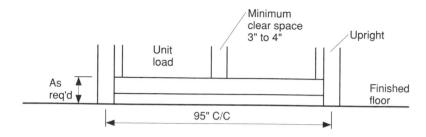

(a)

FIGURE 8.7 (a) Up-and-over system (*Courtesy of Raymond Corp.*); (b) on the floor (*Courtesy of Unarco Materials Handling Corp.*).

Bottom-Load-on-the-Floor Arrangement. In the second arrangement, the bottom pallet load is on the floor (Fig. 8.7b). This concept requires that all unit loads in the elevated storage positions be in direct alignment with the bottom pallet load. This feature reduces the lift truck operator's transaction productivity due to the additional

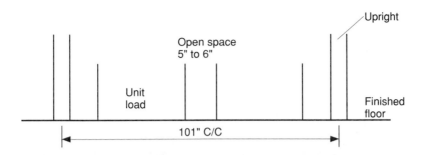

(b)

FIGURE 8.7   (*Continued*)

time required to line up the straddles between the pallet boards. Also this concept reduces the rack investment and height of the building, but it provides a wider rack bay (load beam length) and fewer loads per aisle.

Long Load Beam

If your warehouse facility is a tall warehouse (above 30 ft) and you expect not to change the vertical height of the unit load, then in your rack specification state that a load beam can span two and three rack bays (openings). This unique load beam design has the load beam attached to the face of the upright post. For stability, the load beams are staggered (the front load beam spans two bays, and the rear load beam spans three bays, next the front load beam spans three bays, and the rear load beam spans two bays). Compared to standard upright frame depths, the upright frames are narrower. In comparison to standard rack installation costs, this arrangement has the potential to reduce rack material and installation costs. With the staggered load beam feature, if the rack opening vertical height is adjusted, then the entire row requires adjustment.

Tips and Insights on Installing Used Rack

When you are considering installing used storage rack in your facility, these guidelines can reduce potential installation problems.

Determine What Is Required in Your Facility

After you have completed the storage rack area layout design and drawings, from the rack layout determine the various quantities of upright frames and load beams that are required for the storage area. If you have a mix of new and used racks, then on the layout drawing the appropriate rack bays (frames and load beams) should be classified as new or used equipment.

Develop Written Specifications for the Rack

In addition to the above storage area information and drawings, you must develop a complete storage rack written specifications package which is sent to rack vendors. In addition to the regular storage rack specification information, you identify who is responsible for taking down the existing rack equipment and how the anchor bolts are to be removed and the holes filled in level to the finished floor. This includes the material to fill in the holes. Then who is responsible to bundle the upright frames and load beams and other rack component equipment into bundles which are handled by a conventional lift truck? And who is responsible to load and/or unload the equipment? Who is responsible for transporting the equipment?

Communicate with Rack Vendors

You must develop a rack vendor list that contains both used-equipment and new-equipment dealers. Prior to sending a bid package to these vendors, phone them to assess their willingness to participate in your used-storage-rack bid. To the vendors who are willing to participate, send the used storage rack layout and drawings and written specifications.

Obtain the storage rack vendor list from your company's past vendor list, industry directories, or industry newspapers.

Review the Equipment

After you receive the completed used-storage-rack bids, determine which vendor provides the rack that satisfies your specifications and is most economical. Prior to equipment purchase, obtain (if available) a complete set of drawings of the existing rack system, upright frame and load listing, description, pictures, and written history of the equipment. This includes the stated capacity for the upright frames and load beams. If the stated capacity is not available and you know the original rack owner and/or manufacturer, then contact the rack manufacturer because most manufacturers have records of this information. If you visit the site or location of the existing rack, review the upright frames and load beams for damage, accumulated dirt, discoloring, and rust. It is good practice to take pictures of the racks.

Installation and Start-up Tips

To complete your rack layout, it is possible that you could mix various manufacturers' racks. This situation means that your future storage rack layout will have mixed colors of racks. If top management accepts this color situation, that reduces the installation project problems. If top management requires one color for the upright frames and load beams, have the upright and load beam components painted prior to installation in your facility.

Also based on centerline-to-centerline dimensions and if you are required to mix different racks within one row, then the row length increases by the dimension of each new manufacturer's upright post.

If the existing rack has accumulated dirt or rust, then the rack components should be cleaned by air pressure or water and, as required, sanded and/or coated prior to installation in your facility.

Drive-in Rack

The fifth storage concept is drive-in rack that handles non-self-supporting unit loads (see Fig. 8.8a). Drive-in rack consists of rack storage lanes and is a dense-storage method. The rack components are required upright frames, upright posts, support arms, guiderails, support rails, and required side, top, and back bracing. The drive-in rack bottom storage lane has the bottom unit loads set on the floor, and the elevated storage lane unit loads set on the rack structural members. A drive-in rack lane is

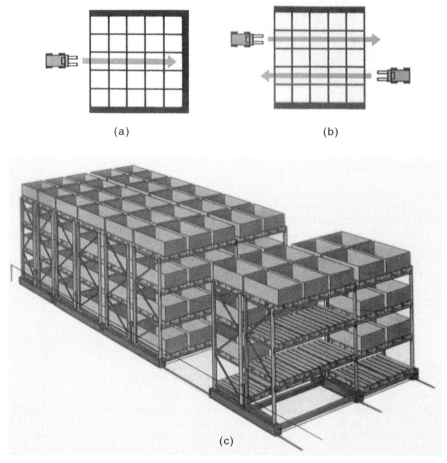

(a)   (b)

(c)

FIGURE 8.8 (*a*) Drive-in rack (*Courtesy of Frazier Indl. Co.*); (*b*) drive-through rack (*Courtesy of Frazier Indl. Co.*); (*c*) mobile (sliding) rack (*Courtesy of Spacesaver.*).

designed two to ten unit loads (positions) deep and three to four levels high per storage lane. This rack layout provides medium to good storage density and poor unit-load accessibility; the product rotation is LIFO.

Drive-in rack is designed with a single row or back-to-back rows. The SKU that is in the storage lane floor-level (bottom) aisle position is the SKU that is in all unit-load positions of the rack storage lane. This feature provides a unit-load utilization factor of 66 percent.

In designing drive-in racks, the distance between the rack second-level storage lane and the structural members must be sufficient to permit a lift truck overhead guard and, as required, straddles to enter and exit the floor-level storage lane.

For the drive-in-rack, the number of unit loads deep per storage lane is varied to place building columns within the flue space of the back-to-back drive-in rack rows. The drive-in rack storage lanes are best designed between building columns. This feature means that the building columns are not in the drive-in rack storage positions or lanes. In all drive-in rack down-aisle designs, structural members require fire sprinklers and lift truck mast and overhead guard clearances.

Most drive-in racks are designed with the fork opening side of the pallet board facing the aisle. This arrangement has the maximum number of faces per aisle and provides excellent unit-load stability in the rack position. If possible, in a carton-handling (pick) operation, the rack has the 48-in (stringer) dimension or side of the pallet board facing the aisle.

To make deposits and withdrawals in a drive-in rack storage concept, a lift truck enters the floor-level storage lane from aisle A, deposits or withdraws a unit load, and backs out of the storage lane into the same aisle A. Because of this operational characteristic, a drive-in rack can handle a medium volume and has a LIFO product rotation.

Drive-through Rack

The next storage method is the drive-through rack that handles non-self-supporting unit loads (Fig. 8.8*b*). Drive-through racks have the same rack components, design characteristics, and utilization factor (66 percent) as drive-in racks except that there is no back bracing. But there is a requirement for top bracing on the rack. With no back bracing, drive-through rack is designed as a stand-alone rack row with a lift truck aisle on both sides. This means that it is not designed with back-to-back rows. This arrangement handles a medium volume, and the product rotation is LIFO or FIFO.

With LIFO product rotation, the lift truck enters the storage lane from aisle A, deposits or retrieves the unit load, and backs out of the storage lane into aisle A.

With FIFO product rotation, the lift truck enters the storage lane from aisle A, deposits the unit load in the unit-load position, and backs out (from the storage lane) into aisle A without a unit load from the storage lane. An alternative procedure in FIFO product rotation is to exit from the storage lane by driving without a unit load through the storage lane and exit into aisle B. To retrieve a unit load, the lift truck enters the storage lane from aisle B, retrieves the unit load, and backs out of the storage lane into aisle B.

Drive-through rack has medium storage density and poor unit-load accessibility. The dimension of the pallet load position has the same characteristics as that of drive-in racks.

Mobile (Sliding) Rack

The next storage system involves mobile (sliding) rack, which is similar to the standard rack (Fig. 8.8*c*). The exceptions are that it is considered a dense-storage method. It requires fewer aisles, and the rack rows move to create a lift truck aisle. Mobile racks are standard single-deep pallet rack rows or back-to-back rack rows that are placed onto movable bases. All rack rows are placed in a 90° angle or are perpendicular to the main traffic aisle.

Mobile rack is designed with nominal 6-in sections of back-to-back movable rack rows, one lift truck aisle, and at each end of these movable sections a single-deep standard rack row. The mobile rack has a ratio of 4 to 5 unit loads high to 1 load deep. If sprinklers are required in the racks, then the mobile rack manufacturer is notified of the design requirement, and a specially designed sprinkler system is used in the rack.

For access to the required unit-load position, a mobile rack row moves to the side and creates a lift truck transaction aisle between the required rack rows. The lift truck enters this aisle, performs the required storage transaction, and exits into the main traffic aisle. After the transaction, as required, the mobile rack sections are moved to create a new aisle between different rack rows. Sensing devices on the movable base bottom sense the existence of an object or employee in the path and stop the mobile rack movement. This feature prevents the movable rack and equipment from being damaged and prevents injury to personnel.

With one access aisle, mobile rack provides high storage density and good unit-load accessibility. Mobile racks have a utilization factor of 85 percent and handle a low volume due to the slow rack movement for lift truck access. With a good inventory control program and batched storage transactions per row, a medium to high volume can be handled.

Gravity Flow Rack

The next storage rack is gravity flow or flow-through rack (see Fig. 8.9a). This storage method is designed as a single or stand-alone rack that has one aisle for unit-load deposit and a second aisle for unit-load withdrawal.

Unit-load gravity flow racks consist of upright frames, upright posts, bracing, brakes, end stops, and skatewheels or roller conveyors that make individual flow (storage) lanes. In a conventional warehouse, unit-load gravity flow storage lanes are 3 or 4 high. In a hybrid or high-ride facility, the storage lanes are designed with 7 to 8 storage levels high. The weight and height of the unit load determine the slope and pitch of the flow rack system. The unit-load height/length ratio is 3 to 1. If unit loads exceed this ratio, there is a potential for uneven flow-through or hang-ups in the storage lane. The gravity flow rack system is designed with 3 to 20 unit loads per storage flow lane. In most unit-load flow rack systems, to ensure a smooth flow through the storage lane, the unit load is placed onto a captive or slave pallet board. In some flow rack systems, the conveyor rollers have flanged wheels that act as guides for the unit load as it flows through the rack. To prevent rack damage, lane entry guides, upright post protectors, lift truck wheel stops, and sufficient lift truck turning-aisle widths are recommended at the entry and exit positions.

The pallet gravity flow rack functions with a lift truck in aisle A that places a unit load onto the skatewheel or conveyor at the deposit (entry) end of the storage flow lane. Gravity and the unit-load weight on the rollers allow the unit load to flow through the storage flow lane to the withdrawal (exit) end of the storage flow lane. The unit load is removed by a lift truck in aisle B from the exit end of the storage flow lane. This activity permits the next unit load in the flow lane to move (index) forward to the withdrawal position. In long systems to reduce line pressure and product damage, unit-load brakes and at the exit position a load separator are installed in the lane.

The unit-load gravity flow system indexing movement of the unit loads from the deposit (entry) position to the withdrawal position allows each storage flow lane to accommodate one SKU per storage lane. This feature permits unit flow racks to have a utilization factor of 85 percent and to handle a high volume. Gravity flow storage with two aisles has high storage density and fair unit-load accessibility.

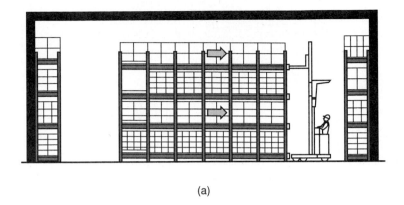

(a)

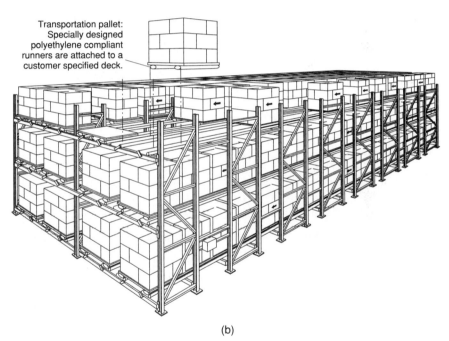

Transportation pallet:
Specially designed
polyethylene compliant
runners are attached to a
customer specified deck.

(b)

FIGURE 8.9   (*a*) Gravity flow rack (*Courtesy of Webb-Triax.*); (*b*) airflow rack (*Courtesy of Sail Rail Enterprises.*).

Airflow Rack

Another type of unit flow-through storage concept is airflow racks (Fig. 8.9*b*). Airflow racks are very similar to unit-load gravity flow racks except that the unit-load racks are indexed through the storage flow lanes by air which is forced through tiny holes in the pallet rails. This feature requires a higher pitch or slope to the storage lane and an air compressor with its associated piping network.

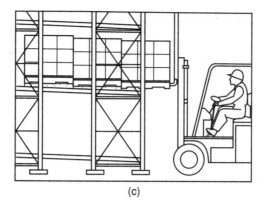

(c)

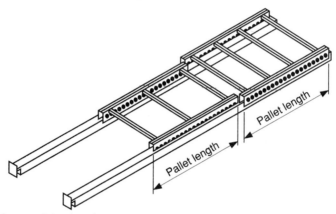

Three-deep push back rack

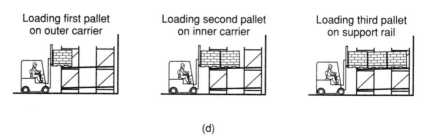

| Loading first pallet on outer carrier | Loading second pallet on inner carrier | Loading third pallet on support rail |

(d)

FIGURE 8.9 (*Continued*) (*c*) Standard push-back rack (*Courtesy of Frazier Indl. Co.*); (*d*) telescoping push-back rack (*Courtesy of Interlake Materials Handling.*).

Push-Back Rack

The push-back rack is a stand-alone concept with one aisle. This unit-load storage method consists of the same rack structural components and design characteristics as for the pallet gravity flow rack with few exceptions.

A push-back rack is designed as a single-rack row that is installed along a building wall, in a building location that permits one aisle for a lift truck to perform all storage transactions, or as back-to-back rack rows. The concept is designed to hold 3 to 4 loads deep and 3 to 4 loads high.

To deposit unit loads, a lift truck places the inbound unit load against the existing load that is in the storage lane aisle position. When the existing unit load is sufficiently pushed back into the storage lane, it creates the required new unit-load storage space. To withdraw a unit load, the lift truck slowly raises the unit load 2 to 3 in high from the storage lane position and backs out into the aisle. As the unit load is removed from the storage lane, gravity moves the next unit load into the storage lane exit (aisle) position. This feature permits a 66 percent storage lane utilization factor and a LIFO product rotation. The push-back rack provides good storage density and fair unit-load accessibility and handles a low to medium volume, and each lane handles one SKU.

Standard Conveyor Push-Back Rack

In a standard pallet flow rack, the push-back flow rack does not require brakes but requires end stops on both ends of the flow lanes.

Telescoping Push-Back Rack

In a telescoping carriage push-back rack, the push-back concept consists of unit-load carriages that ride on a set of tracks (Fig. 8.9d). The carriage design permits the carriage that travels into the interior unit-load position when not in use (without a unit load) to nest over the empty carriage that is adjacent to the aisle.

Car-in-Rack

The next storage method uses the car-in-rack which is designed with one or two aisles (see Fig. 8.10a). It consists of upright frames, posts, bracing, support arms, car rails, unit-load support rails, a car-in-rack track vehicle, slave pallets, and a deposit/retrieval vehicle. Sufficient space is designed between the unit load and rack members (storage lane) for fire sprinklers and a car with an elevated unit load to travel in the storage lane. The car-in-rack concept is designed with 3 to 8 storage levels high and with 10 to 20 unit-load positions per storage lane. Each storage lane has one SKU. A very narrow-aisle vehicle carries the car-in-rack vehicle between the rack rows to the required storage lane and performs the storage transaction.

One-Aisle Car-in-Rack

In a one-aisle arrangement, the car-in-rack vehicle leaves the host vehicle from aisle A, enters and travels into the storage lane, arrives at the required unit-load storage lane position, performs the unit-load transaction (deposit or retrieval), and exits the storage lane to the same aisle A onto the host storage vehicle. This feature provides a LIFO product rotation and unit-load utilization factor of 66 percent and handles a low

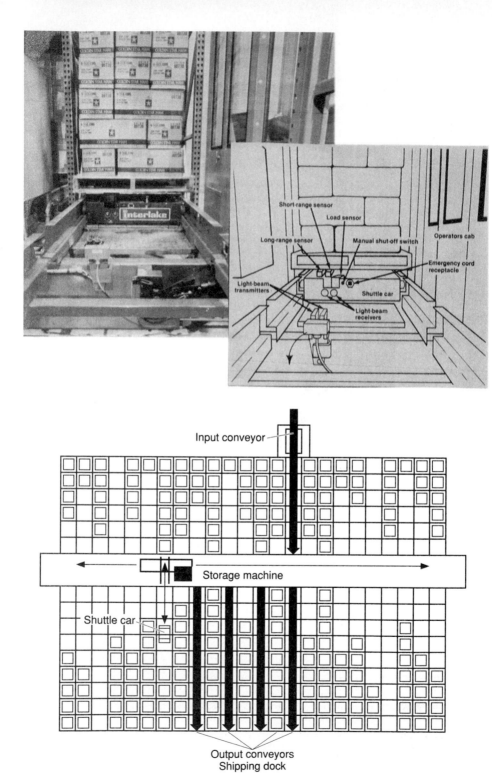

Short-range sensor
Load sensor
Long-range sensor
Manual shut-off switch
Operators cab
Emergency cord receptacle
Light-beam transmitters
Shuttle car
Light-beam receivers

Input conveyor

Storage machine

Shuttle car

Output conveyors
Shipping dock

FIGURE 8.10 (*a*) Car-in-rack. (*Courtesy Modern Materials Handling Magazine.*)

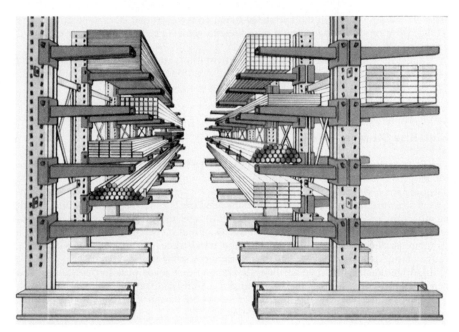

FIGURE 8.10 *(Continued)* (*b*) Cantilever rack. (*Courtesy of Unarco Materials Handling Corp.*)

volume. One-aisle car-in-rack storage provides high unit-load storage density, poor unit-load accessibility, and excellent security.

Two-Aisle Car-in-Rack

The two-aisle car-in-rack requires two host storage vehicles and two car-in-rack vehicles. In a two-aisle concept, from the inbound aisle A all deposits are made by an aisle A host storage vehicle with its car-in-rack vehicle. From an outbound aisle B, all withdrawals are performed by the aisle B storage vehicle with its car-in-rack vehicle.

The two-aisle car-in-rack concept provides a FIFO product rotation, excellent security, high unit-load storage density, and poor unit-load accessibility.

If unit loads in the storage lane are indexed forward by a car-in-rack vehicle toward the withdrawal aisle, then the two-aisle car-in-rack concept has an 85 to 90 percent utilization factor and handles a high volume.

Cantilever Rack

The next storage concept is a cantilever rack which is designed to handle long unit loads such as piping or lumber (see Fig. 8.10*b*). Cantilever rack consists of upright posts, support arms, legs, and bracing. At several levels sufficient space is designed between the unit loads and metal members for fire sprinklers. With solid or wire-mesh decking on the support arms, cantilever rack is designed for the storage of cartons and furniture.

The cantilever rack is designed as a single-arm row or as double-arm rows. The arms extend outward from the upright post and create unit-load storage positions. The storage positions are serviced by a narrow-aisle vehicle, high-rise order-picker truck, or wide-aisle vehicle.

Cantilever rack permits a FIFO product rotation, excellent storage density, and excellent product accessibility. The unit-load position has a utilization factor of 85 percent and handles a high volume.

High-Rise Storage Rack

The final storage system is the high-rise storage facility that uses the vertical airspace for unit-load storage rather than the horizontal footprint. These storage methods are designed for conventional buildings or in a rack-supported structure. The rack-supported structure has the racks' upright frames and posts as the support and attachment members for in-rack sprinklers, the walls (skin), and the roof. All these storage systems require a dead-level or F75/F100 level floor and a vehicle guidance system. The structure is designed for snow, seismic and wind forces, fire protection (including walls, barriers, and sprinklers), and mast clearances at the top of the rack.

In an automated high-rise facility, all inbound unit loads pass through a size and weight station to verify that the unit-load dimensions and weight conform to the system design standards. The high-rise storage racks are designed with single-deep racks, two-deep racks, and flow racks in the structure.

High Rise Pallet Rack Openings

The single-deep rack openings have two alternative designs. The rack opening matches the storage and retrieval vehicle load-handling device. There are two types of rack openings.

Standard Pallet Rack Openings.   The first type of rack opening is the standard rack bay that has two unit loads set on a pair of load beams. This rack opening requires that the storage/retrieval vehicle have a set of forks. To complete storage transactions, the forks enter the pallet board fork openings as a normal lift truck. In the typical high-rise facility, the first unit load is set 30 to 36 in above the finished floor. Compared to the alternative single-deep concept, the standard rack concept has these advantages: fewer upright posts per aisle, greater number of facings per aisle, and shorter building.

Four-Arm Pallet Rack Opening.   The second rack opening has four load arms and two unit-load rails that support one unit load per opening. This rack opening requires a captive (slave) pallet board and the storage/retrieval vehicle to have a set of platens that go under the unit load to perform the transaction. The first rack level is approximately 17 to 24 in above the finished floor, but at each rack level an additional 3 to 6 in of clearance is designed for the platens between the top of the unit load and the support arms of the next storage level.

Various Unit-Load Bottom Support Devices

An extremely important factor in a pallet load distribution facility is the method used to support the unit load. The unit-load support method determines the storage concept and the unit-load-handling equipment.

The various methods to support the bottom of a unit load are reviewed in greater detail in Chap. 11, but the methods are as follows:

- Slip sheet
- Pallet board
- Skid
- Slip sheet on a pallet board
- Two slip sheets on one pallet board
- Container

Various Unit-Load Securing Concepts

An important feature of unit-load storage is the stability of the unit load on its support device. The method used to secure the cartons or product on the support device prevents product damage and ensures an efficient operation. Methods include

- No method
- Tape
- Shrink-wrap
- Netting
- ti × hi (pattern of cartons)
- Stretch-wrap
- String
- Glue
- Band (metal/plastic straps)

In Chap. 11, these methods are reviewed in detail.

When to Consider a Rack-Supported Facility

When your company is confronted with a business situation to increase pallet (unit-load) storage positions for its warehouse operation, there are two basic storage facility design options as solutions: (1) new construction of a conventional facility with free-standing (pallet) storage racks while continuing with the present operation and lift trucks and (2) new construction of a rack-supported structure with aisle-guided storage vehicles.

Conventional Warehouse Characteristics

The conventional storage warehouse alternative consists of a building with the following characteristics: (1) a 20- to 30-ft clear ceiling height from the floor, (2) square or rectangular building, (3) a normal (level) warehouse floor for a counterbalanced or straddle reach lift truck operation, (4) no heavy (high) point loads (therefore, it does not require a thick floor), (5) freestanding pallet racks and floor storage method, (6) building with a bay size that has a range of 38 to 42 ft on center, (7) building columns that support the walls and roof, and (8) building depreciated (noncash business expense that reduces gross income and the gross asset book value) as a building which occurs over a longer accounting period.

Alternative Conventional Building Storage

The alternative floor stack or freestanding pallet rack storage methods for a conventional building are many. When you consider the alternative freestanding pallet rack storage concepts, the alternative unit-load storage concept and storage vehicle estimated investment is shown in Fig. 8.11.

Disadvantages of Conventional Building

The disadvantages of the conventional building with floor stack or freestanding storage racks are that it requires a larger floor building area and limited stacking height; increases the travel distance between two warehouse locations; creates lower lift truck operator productivity, longer construction period, and higher total facility (land, building, and equipment) investment; and increases building, equipment, and product damage.

Characteristics of Rack-Supported Facility

A rack-supported facility consists of a structure that has a 40-ft minimum clear ceiling height from the finished floor. The facility shape is a rectangle or oversized rectangle

Storage method	Investment*			
	Storage equipment[†]		Storage vehicle	
	Cost, $		Type	Cost, $[‡]
Floor stack	-0-		C/S	20–30
Tier rack and stacking frames	150–200		C/S	20–23
Standard pallet rack	50–75		C/S	20–30
Double-deep rack	50–75		DD[1]	30–35
Drive-in, drive-through	150–200		C/S[2]	20–30
Gravity flow and push-back rack	200–250		C/S	20–23
Car-in-rack	250–300		C[3]	50–60
Cantilever rack	75–125		C/S	20–30

*Does not include building costs.
[†]$100 per unit load position.
[‡]$1000 per vehicle.
[1]Conventional wide aisle.
[2]Straddle reach.
[3]Deep reach.

FIGURE 8.11

Storage method	Storage rack,* $	Investment[†] storage vehicle	
		Type	Cost,[‡] $
Pallet rack	50–75	VNA	80–110
		H	175–200
		AS/RS	250–300
Gravity flow rack	200–250	Same as above	
Car-in-rack	250–300	Same as above	

*$100 per unit-load position.
†Does not include building costs.
‡$1000 per vehicle.

FIGURE 8.12

that has F-75 to F-100 level warehouse floor in the storage rack area as defined by the Face Company. The facility requires a VNA guided storage vehicle. The rack structure provides support for pallet loads, roof, walls, and facility utilities (lights, sprinklers, HVAC, and other systems). During a unit load transaction heavier rack loads and increased weight on one lift truck wheel means a thicker floor and specific depth for rack anchor bolts. With wire aisle vehicle guidance system it means that all metal objects (re-bar and drain covers) are required to be a specific distance from the wire path. With this rack supported structure no columns exist in the building except structural-steel roof support members along the walls and in the runout areas (vehicle turning aisles). This means a greater number of unit loads per square foot which is depreciated (noncash business expense that reduces gross income and gross asset book value) as a piece of equipment which occurs over a shorter accounting period.

When we consider a rack support facility, the pallet rack storage concept and storage vehicle investment requirement are as shown in Fig. 8.12.

Characteristics of Rack-Supported Facility

The remainder of the section looks at the design features and operational characteristics of the rack-supported facility (Fig. 8.13).

The rack-supported facility (structure) has many similar design components to the conventional building with freestanding rack. Yet when you look in detail at each component, there are differences between the conventional building and the rack-supported facility in rack members. These differences are the ground preparation and floor design; roof and wall supports and runout areas; rack structure and baseplates; storage vehicle and overhead clearances; increased seismic, snow, and wind forces due to facility height and implementation schedule and enclosing the rack support frame skeleton.

Floor Facility Design Factors. The rack-supported facility floor design has a heavier psi (pounds per square inch) value due to the following: (1) The storage rack point loads are heavier because there is an increase in the number of unit loads and rack members per bay, roof, wall, and other supported building related material and equipment. (2) When compared to the conventional lift truck weight, the additional VNA

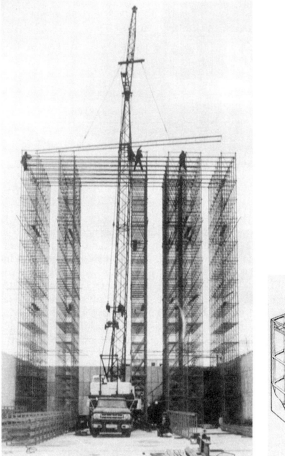

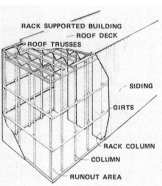

FIGURE 8.13   Rack-supported facility. (*Courtesy of Materials Handling Institute and Frazier Indl. Co.*)

lift truck weight is a nominal 12,000 lb. In completing a pallet load storage transaction with some lift trucks, approximately 56 percent of the lift truck and pallet load weight is assumed on one lift truck wheel. (3) The VNA lift truck aisles are a nominal 5 ft 4 in to 6 ft 4 in between rack posts, and the wide-aisle vehicle lift truck has a 12- to 13-ft-wide aisle between rack posts. (4) The snow, ice, wind, and seismic loads are transferred to the floor through the rack baseplate. After soil borings and testing and if the soil is not compact or does not satisfy specific standards, then pilings are required in the ground. (5) In the storage area, the floor has minimal pour joints. The concrete floor is poured in the aisle direction for the entire length of the storage area. This practice reduces the number joints and lift truck wheel damage and creates a greater possibility of a level floor. (6) To minimize rack bay and lift truck fork misalignment at the higher rack levels, the concrete floor preparation and pour must meet the Face Company specifications and standards.

Compared to the conventional building square-footage cost, the above floor design characteristics increase the floor cost of the rack-supported facility. But they ensure an excellent and level base for a rack-supported facility with a lower shim baseplate requirement. This means a faster rack installation which reduces the potential wind damage to the exposed rack. A level floor ensures proper upright post (plumbness) of the elevated rack positions to the finished floor. This situation means excellent alignment of the load beam and lift truck fork to complete a storage transaction at the elevated storage levels. An uneven floor can cause a lift truck fork to deflect at elevated levels, which means that the lift truck cannot perform the required storage transaction.

To minimize rack-supported facility costs, your architect requires accurate facility and vehicle design data. These data include the maximum unit-load storage and rack member weight per rack bay, the location of each rack post and each rack post baseplate size, load weight per storage unit, lift truck weight (with maximum unit-load weight per wheel), all building items (lights, sprinklers, mechanical units, HVAC, and other building items attached to the roof and walls), and location of the wire guidance and anchor bolts and their depth.

Roof and Wall Characteristics.   The second design difference lies in how the roof and wall (skin) components are attached to the structural support members (skeleton). The majority of the upright posts of the rack-supported structure are the support members for the roof (purlins) and wall (girts) components. In the rack-supported facility, runout areas (end of rack vehicle turning aisles) and some walls have structural-steel members (columns). Structural-steel members are used in these areas due to the long, clear spans (vehicles' turning aisles) that are required by most VNA vehicles. These vehicle turning-aisle widths range from a minimum of 18 to 30 ft.

Design Factors of Uprights.   The third difference lies in the rack structure (skeleton), upright post, and baseplate. The rack upright post and bracing components assume the entire unit-load weight, facility support items (sprinklers, mechanical units, lights, etc.), and the external forces. Some of these forces are seismic, snow loads, and wind loads. When stating the external forces to the rack manufacturer, you specify the geographical location of the facility. This provides the rack manufacturer with sufficient information to research these engineering facts and to design the rack upright posts and baseplates for your proposed facility.

Compared to conventional (standard) pallet rack, to assume these loads, the rack upright facility rack frame posts are larger and have a heavier gauge. Also, to spread the load, your rack post baseplate is large and requires a greater number of anchors per baseplate. Since the baseplate and upright post connection support a heavy load, proper attachment of the post to the baseplate is required. The proper upright post and baseplate attachment is performed per the manufacturer's engineering specification standards and by the rack manufacturer, at the manufacturing facility. It is not wise to permit baseplate attachment to the upright post in the field (at the installation site).

With some VNA vehicles, the rack baseplate cannot protrude into the aisle and interfere with the storage vehicle travel in the aisle. If the rack baseplate and anchor bolts extend into the rack bay, then your alternatives are to raise the bottom storage level or have the baseplate extend into a wider-than-normal rack bay. To raise the bottom load beam eliminates the problem but increases the load beam (quantity) investment, upright frame height, and building height. The wider-than-normal load beam eliminates the problem, but increases the aisle length, which increases the facility cost and operator travel distances.

During the rack-supported facility implementation phase, after the pouring of the concrete floor, the most critical installation is the rack (skeleton) installation and wall (skin) attachment to the rack. This phase of the installation project is most critical to success. If the rack is standing with no walls and a strong wind arises in the area, there is a good chance of the wind collapsing the rack structure (skeleton) due to the sail effect.

Rack Storage Concepts.   The various types of storage rack concepts that are considered for a rack-supported facility include (1) a standard pallet rack with load beams that provides two standard pallet positions for VNA vehicles with forks and (2) a combination of three-bay-long load beam and two-bay-long load beam design which lowers the rack material and installation labor cost. In this design the load beams are attached to the aisle side of the upright frame. Compared to the standard upright frame depth, this frame depth is narrower and reduces the ability to adjust unit-load storage opening height. They also include (3) a rail pallet rack on cantilever support arms that provides one pallet position per bay for VNA vehicles with a platen load handling, (4) a pallet flow rack, and (5) a car-in-rack.

In most of these rack-supported structures, the insurance underwriter and local building and fire authorities require additional in-rack or ESFR (Early Suppression Fast Response) sprinklers, barriers, and walls.

Storage Position Design Factors.   Design your rack position to improve operation product handling and to reduce product and equipment damage. With these design features, all short unit loads are allocated to the lower storage positions of the rack structure. This feature increases the structure stability, and all tall unit loads are allocated to the top storage positions of the rack structure. This arrangement reduces additional clearances for the lift truck overhead guards and masts. With the warmest storage positions at the highest rack positions, all heat-sensitive product is allocated to the lower rack storage positions, and this allows a minimum of 3 to 6 in of open (clear) space between unit loads and rack structures.

Vehicle and Aisle Design Characteristics.   The next important aspect of a rack-supported facility is the unit-load handling vehicle that performs storage transactions in the rack positions. In a rack-supported facility, these vehicles are required to operate in very narrow aisles. These aisles are between two rows of racks with a rail or wire guidance system to ensure controlled vehicle travel in the aisle. The controlled vehicle travel minimizes rack and equipment damage.

Other Options.   Other features that ensure a highly productive operation with minimal product or equipment damage include (1) aisle entry guides with a rail guidance system; (2) with wire guidance systems an *uninterruptible power source* (UPS) which during a power failure has sufficient time to permit all vehicles to exit the storage aisles; (3) two front-to-rear (cross) members per pallet position or these rack components at least above the sprinkler locations; (4) on operator-down vehicles, a shelf locator system to improve storage transaction efficiency and reduce product and equipment damage; (5) end-of-aisle P/D (pickup and delivery) stations that permit the lift truck operator to perform dual commands.

To reduce rack damage and improve operator efficiency, the P/D stations have the following characteristics: (1) sufficient width between two rows' upright frames for the in-house transport vehicle to perform a transaction, (2) if possible, freestanding upright posts, (3) rub bars full length between the upright frame posts to protect the anchor bolts and baseplate, (4) backstops throughout the full length of the rack bay opening, and (5) with a pallet truck in-house transportation system, highway guarding at the main-aisle side of the P/D station.

Vehicle type	Type of storage
Order-picker truck	Hand stack
Swing-reach or turret truck	
Operator-up	Hand stack or pallet load
Operator-down	Pallet load
Free-path, ranger or hybrid truck	Hand stack or pallet load
Storage/retrieval vehicle	
AS (Automated Storage)	Pallet load
MS (Manual Storage)	Hand stack or pallet load
Swing-mast truck	Pallet load

FIGURE 8.14

The various narrow-aisle vehicles that are used in a rack-supported facility are shown in Fig. 8.14.

When your rack-supported facility has a side-loading storage vehicle that turns the unit load to perform a storage transaction, then you must decide to turn or not turn the pallet load in the aisle. The decision to turn the pallet load in the aisle requires an aisle that is a nominal 6 in wider than an aisle that does not permit turning of the pallet load. A narrower aisle means a smaller floor area and lower floor cost, but with a slight decrease in operator productivity because at the P/D station one-half of the unit loads are turned (rotated) in the proper orientation for the storage transaction.

In a rack-supported facility with a guided aisle vehicle, consider extra cab lighting on the vehicle to reduce the permanent aisle lighting to a minimal lighting level.

In a multiple-vehicle operation, consider the end-of-aisle slowdown concept. According to your vehicle type, the slowdown concept involves different-color rack members or electric or magnetic devices embedded in the floor and sensing devices on the undercarriage of the vehicles.

The final item to ensure your rack-supported facility is on schedule is the project implementation schedule. The preferred project schedule is a bar chart that has each major activity of the project identified with an amount of construction time, start and completion dates, and who is responsible for the completion of the activity.

CHAPTER 9

WHICH HORIZONTAL OR VERTICAL TRANSPORTATION METHOD CONTROLS INTERNAL TRANSPORTATION COSTS AND HANDLES THE VOLUME AND PRODUCT*

The objective of this chapter is to review the various methods to perform your horizontal and vertical transportation warehouse function. The chapter identifies and evaluates the various transportation systems, equipment applications, and technologies that are used to move product between two horizontal or vertical warehouse function areas or workstations (levels). These in-house transportation methods are designed to save handling time and effort as product is moved between two locations. In most operations, these locations are between inbound dock areas and storage ares, through the warehouse to various locations or levels, or from the picking area through the packing area to the outbound dock areas.

TRANSPORTATION LINKS TWO WAREHOUSE FUNCTIONS

When you consider transportation equipment or systems, you are considering the link between warehouse function areas or stations, facility floors or levels, and warehouse and manufacturing workstations.

The employee- or computer-controlled transportation equipment and concepts ensure that the right product is moved on schedule from one location to the next correct (assigned) location.

The various transportation equipment and concepts that move small items, cartons, or pallet loads are similar, but have differences. The concepts are similar because they move product, and the differences between them involve

---

*This chapter is based on an unpublished article that is accepted by *Plant Engineering Magazine*.

- Power source
- Weight capacity
- Required space and path
- Volume handled
- Unload and load ability

VARIOUS TRANSPORTATION CONCEPTS

Various transportation methods have two major classifications:

- Horizontal transportation

  Above-floor (nonpowered or powered)

  Overhead (nonpowered or powered)

  In-floor (powered)

- Vertical transportation (nonpowered or powered)

Transportation Design Parameters

Your transportation design parameters are clearly defined prior to the purchase and implementation of a horizontal or vertical transportation concept. These include the product dimensions (length, width, height) plus weight, shape, and gap between SKUs; type of product bottom support device; throughput volume (surges or continuous flow; average or peak); individual customer delivery amount (space); crushability and fragility of the product; transportation distance; travel path (fixed or variable, horizontal or vertical); clearance; building obstacles; other product-handling equipment; delivery on schedule or on demand; loading and unloading method; and number of pickup and delivery points.

Computer Simulation of the Transportation Function

When you use a computer simulation to design your transportation concept, you must provide the following information: in a unit-load warehouse, the number of pickups and deliveries for SKUs that are assigned to each warehouse location; and in a single-item or single-carton warehouse, the SKUs that have been assigned in the pick aisle to determine the number of SKUs transported from each aisle.

With this SKU activity information, the computer program determines the number and frequency of pickups and deliveries per station, a projection of the potential queues and the single-item or single-carton pallet withdrawals per aisle, the potential accumulation locations, and activity at merge and transfer locations.

SMALL-ITEM HORIZONTAL TRANSPORTATION

The first section of this chapter reviews the small-item and hanging-garment transport methods. Horizontal transportation is separated into the following groups:

1. Above-floor, nonpowered
   Human carry, pushcart, skatewheel, and roller conveyor with containers
2. Above-floor, powered
   Human-controlled burden carrier; automated AGV; powered skatewheel, roller, and strand conveyor with containers and belt conveyor
3. Overhead, nonpowered
   Gravity skatewheel and roller conveyor with container
   Trolley (basket and hanging garment)
4. Overhead, powered
   Powered conveyors, powered trolley, and pneumatic tube

Above-the-Floor Nonpowered Horizontal Transportation

The horizontal transportation methods move merchandise on one floor (level) between two locations. These transportation concepts are above- or in-floor and are powered or nonpowered.

Human-Carry Method.   The first above-floor nonpowered horizontal transportation system is the human-carry concept. With this variable-path concept, an employee picks up an armful, apron, sack, or tote of small items and physically carries the merchandise to the required location. The method handles a low volume and is restricted to the weight and capacity that can be handled by an employee. This method is considered the basic concept that handles one SKU or multiple SKUs; it requires a low investment.

Manual Pushcart and Pallet Truck with Shelves.   The second horizontal method is the employee pushcart, a variable-path concept (Fig. 9.1a). The manual pallet truck with containers that are stacked on a pallet board has operational characteristics and features similar to this concept. The pushcart is a four-wheel manually pushed vehicle with one or more load-carrying surfaces and push handles or bars. Swivel casters or wheels on the front and rigid casters or wheels on the rear provide excellent steering and ease of turning at the end of the aisle.

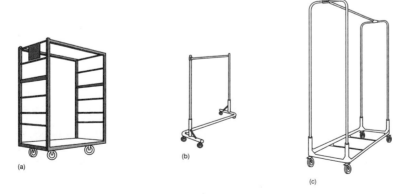

FIGURE 9.1   (*a*) Manual pushcart; (*b*) Z frame cart; (*c*) regular frame cart (*Courtesy of Railex Corp.*).

To operate, an employee places merchandise on the load-carrying surface, pushes the cart onto the next location, and unloads the merchandise. The method can handle a low volume of loose, individual packaged or containerized merchandise, and the system requires a low capital investment. Containers, shelves, and dividers permit separation of merchandise or customer orders on the load-carrying surface.

Z-Shaped Bottom Frame Cart Permits One Employee to Haul Two Carts. If your product is hanging garments, then a rolling rack is used for transportation. This vehicle has four swivel wheels, a hang bar, and support members that are connected at the wheels (Fig. 9.1*b*). Some rolling racks have a Z-shaped bottom support members that enables two rolling racks to be hooked together, so that one employee can pull two rolling racks (Fig. 9.1*c*).

Skatewheel and Roller Conveyor. The next above-floor nonpowered horizontal transport method involves the skatewheel (Fig. 9.2*a*) or roller conveyor (Fig. 9.2*b*). The merchandise must be in a container that has a conveyable bottom surface. This fixed-path concept requires an employee to place the container onto the conveyor. The employee pushes or moves the container via gravity to the next workstation. At the discharge location there is an end stop that stops the container flow. The method can handle a medium volume, and separators inside the container allow multiple SKUs or customer orders in one container.

The conveyor concept consists of a series of ball-bearing tubular rollers or skatewheels that are carried and turned on a shaft which is between two side rails or beds. The bed is 10 to 12 ft long and from 12 to 48 in wide, and it is supported by stands. To direct the container flow and prevent damage, either side guards are attached to the bed or the rollers and skatewheels are set low (below the bed) (Fig. 9.2*c*).

Above-Floor, Powered Horizontal Transportation

The above-floor powered horizontal transport methods use an electric or power motor that propels the small parts carrying surface from one location to another.

Burden (Personnel) Carrier. The next concept in this group is the burden (personnel) carrier or powered cart, a variable (flexible) path concept. The powered pallet truck with containers stacked on a pallet board and the tugger with a cart that has separated shelves are similar. These vehicles have three or four wheels. This sit-down or stand-up rider vehicle is powered by LP gas, gasoline, diesel, or electricity. The most common indoor vehicle is the electric powered type. Other vehicle features are an operator's platform with all controls and a steering device and a load-carrying surface with containers or shelves to separate the merchandise and customer orders.

This operation has the same characteristics as the pushcart; however, there is less employee fatigue and improved productivity due to the fact that a power source moves the vehicle. With containerized merchandise, individual packaged or loose merchandise, the method handles a medium volume.

Miniload AGV. The next transportation method is a fixed-path, closed-loop concept: the miniload *automated guided vehicle* (AGV), which is a self-controlled or host-computer-controlled, programmable, bidirectional vehicle. The mini-AGV has a static (one-level or shelved) or container elevator load-carrying surface. For safety, the vehicle has a protective bumper and an underside guide-sensing device. The power source is several rechargeable batteries, and the vehicle rides on three to four wheels with at least one as a steering or drive wheel.

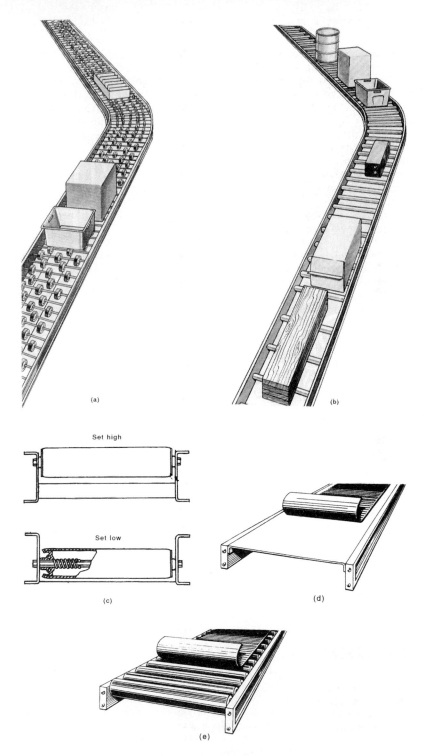

FIGURE 9.2 (*a*) Skatewheel conveyor; (*b*) roller conveyor; (*c*) roller set high and low; (*d*) slider flatbed conveyor; (*e*) slider flatbed conveyor (*Courtesy of Shuttleworth Corp.*).

After being programmed and placed onto the wire guide path, the AGV transports the merchandise from the pickup (dispatch point) station to a drop-off (delivery point) station. At any activity station, the solid deck AGV requires an employee to load and unload the merchandise. Additional stations or feet of an inductive (floor-embedded wire) or optical (chemically treated paint or tape) guide path are easily added to the system up to a nominal 5000 ft. The guide path transmits an electric signal to the AGV sensing device which directs its travel. The most common guide path is the inductive method with a UPS (backup electric power system).

When the travel distance is great and over a fixed path, the AGV is considered an economical and operationally feasible transport method. The AGV's main travel path consists of straight runs, 90° to 180° turns, slight grades, and at the workstation spur lanes (sidings) which are parallel to the main travel path. When an AGV is picking up or delivering merchandise at a workstation in a multivehicle method, the sidings permit an uninterrupted flow of vehicles on the main line.

Powered Belt, Roller, Skatewheel, or Strand Conveyor.   The powered belt, roller, skatewheel, and strand conveyor are very commonly used to transport merchandise. The economics of a powered conveyor is a function of its length. When a tremendous volume of merchandise is transported between two locations, the powered conveyor is the preferred transport method. If the merchandise is bulk-picked, individually packaged, not fragile or crushable, and destined for a sortation induction station, then the merchandise is transported en masse and loose on a belt conveyor. If the merchandise is single customer-picked, loose, fragile or crushable, and destined for several workstations (packing), then the merchandise is transported in discretely identified containers (totes) that ride on a powered roller, skatewheel, strand, or smooth-top belt conveyor. For best results, the container travels with its length in the direction of travel.

A conveyor consists of a conveyor bed, motor or drive, pulley, conveying surface, guarding, operational and emergency controls and devices, and power and control panels. The conveying surface ranges from 12 to 48 in wide.

Belt Conveyor

The belt conveying surface is the simplest form of powered conveying surface. This conveying surface is turned by a center or end motor or drive and turns on a head-tail pulley. If your operation handles loose merchandise, a belt conveyor is a very versatile and economical transportation method. By changing the sprocket or motor size, the conveyor speed can be changed to increase the speed with a small sprocket and decrease the speed with a large sprocket. As required by the order-pick concept, packing (workstation) concept, or building, a conveyor is designed with multiple loading and unloading points, fixed or movable sections, and automatically or manually controlled sections.

With a motorized conveyor, the conveyor is separated into 100- to 200-ft sections per drive motor. For merchandise transfer between conveyor sections, there are two systems.

In the first method, there is no elevation change between conveyors. The powered belt turn does not redirect the merchandise flow and creates minimal merchandise damage. The belt product transfer method is preferred for cartons or containers. Both conveyor sections are at the same elevation. The carton or container is transferred from the first conveyor to the second conveyor section. The carton or container length does not fall into the small gap (nominal 1 to 2 in wide) that exists between the two conveyor sections.

When cartons or containers make right-angle (90°) turns or merge onto another conveyor, each location requires flow and jam controls. These controls ensure a smooth flow without damage to the carton.

In the second method, the first conveyor section overlaps (waterfall) the second conveyor. At a waterfall right-angle turn or additional load point, the first discharge conveyor is elevated and overlaps the second (take-away) conveyor. The conveyor's flow and gravity force transfers the merchandise from the intersecting conveyor onto the 90° take-away conveyor. In most waterfall applications, at the end of the discharge conveyor, a sheet-metal slide is preferred to reduce merchandise fall (impact) on the second conveyor. The second belt propels the merchandise away from the discharge conveyor's belt return. This feature reduces product damage from merchandise clinging to the return belt as the belt turns on its pulley. At each waterfall, special attention is given to side guard height and jam controls. To improve employee safety at these locations, many companies have installed personnel guarding and E (emergency) stop pull cords. When belt conveyor is used, the E stop pull cord quickly stops the conveyor and all conveyors in the line of sight.

Slider Flatbed Belt Conveyor.   A slider flatbed conveyor moves over a solid, flatbed of wood, metal, or plastic. This bed provides a solid surface under the belt. It is used for short runs that handle light loads and merchandise that has irregular shapes.

Slider Roller-Bed Conveyor.   The slider roller-bed belt conveyor is designed with the conveying surface moving over a series of rollers. These rollers are similar to gravity rollers. The roller spacing is on centers to have at least three rollers to support the shortest individual item with no undue belt deflection between rollers. With less drag resistance, the slider roller-bed conveyor transports larger volumes of merchandise and heavy items for greater distances.

Other Belt Conveyor Features.   Other features of belt conveying systems are as follows: If two-way traffic is required, then the conveyors are separate units that are stacked (see Fig. 9.3*a*). If two-way traffic is desired, then a transporter conveyor is used which has one belt conveyor with two large pulleys or four pulleys. Divert devices prior to each pulley transfer product from the top level onto workstations and on the return (bottom level) to the other workstation. For best performance and minimal belt wear, the belt should be nominal 1 to 2 in from the side guard. For short runs, manual takeups are used to tighten and loosen the conveyor belt. On long runs, automatic belt takeups are the preferred option. Belt conveyors transport loose merchandise or containers and totes. The motors are sized to pull a stopped conveyor surface with the surface 100 percent loaded with product. Crown and laggered pulley helps track the belt conveyor.

Powered Roller, Skatewheel, and Strand Conveyor.   The powered roller, skatewheel, and strand conveyor requires the merchandise to have a conveyable bottom surface or to be in a conveyable container or tote. For good traction, there are three rollers under an individual container or tote. Rollers on close centers are ideal for short containers, and rollers on longer centers are ideal for long containers. The container has sufficient weight to permit the rollers to move the container. Also the height of the container activates the sensing devices and permits side guards to guide the container travel. In transporting containers that have slanted lead or tail ends, uncontrolled line pressure can cause the containers to buckle, jam, and jump from the conveyor surface.

Powered Skatewheel Conveyor.   The powered skatewheel conveyor transports uniform and sturdy containers of merchandise between two locations (Fig. 9.3*b*). The powered skatewheel conveyor is similar to the gravity skatewheel conveyor except that the container rides on the skatewheels and is moved by a powered belt which is in

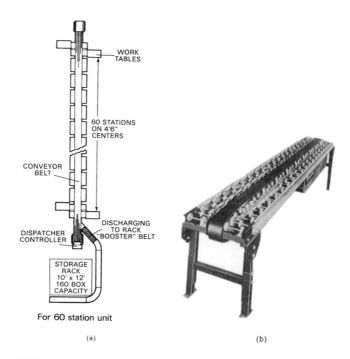

For 60 station unit

(a)   (b)

FIGURE 9.3   (*a*) Transporter (*Courtesy of Diamond Transportation Co.*);
(*b*) powered skatewheel (*Courtesy of Rapistan Demag Corp.*).

contact with the container bottom. The belt rides on a series of specially designed wheel cam devices that activate and deactivate the load-carrying powered belt. The 6-in-wide powered belt is in the middle of the bed and rides on these cam devices. As the lead container travel is stopped, the cam device lowers the cam wheels. This action eliminates the belt force on the container bottom, and the other containers behind the lead container accumulate on the conveyor.

Powered Roller Conveyor.   The powered roller conveyor is very common. The container is propelled forward by rollers which are turned by one of several motor-drive-pulley arrangements. There are four arrangements. (1) In this continuous-chain concept (Fig. 9.4*a*), a full-length chain travels over each roller sprocket. The chain path is on one side of the conveyor and is driven by the pulley, and the sprocket turns the rollers. (2) In the roller-to-roller arrangement (Fig. 9.4*b*) adjacent sets of rollers are driven by a complete chain drive. (3) In the line shaft method (Fig. 9.4*c*), each roller is driven by an elastomeric belt that is connected with bushings to a drive shaft which runs the full length of the conveyor. The shaft is turned by a motor. (4) In the roller on belt method (Fig. 9.4*d*), the rollers are turned by a motor-driven belt that runs underneath the roller for the full length of the conveyor. To turn the sprocket drive for long runs at high speeds, a flat belt is preferred; in short runs at low speeds, a V belt is preferred. When the roller conveyor makes a 90° or 180° turn, the tapered roller is preferred, to maintain container orientation.

Strand (Chain-in-Channel) Conveyor.   In the strand (chain-in-channel) conveyor (Fig. 9.5*a*), an endless looped chain transports uniform cartons or containers on two skatewheels across a fixed path. The fixed path goes from the charge end to the dis-

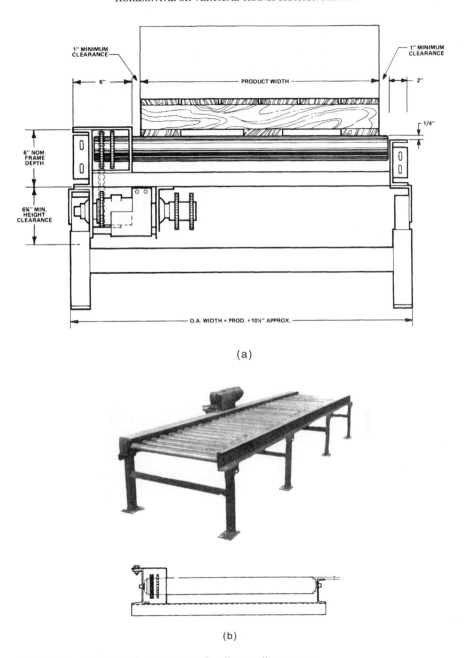

(a)

(b)

FIGURE 9.4 (*a*) Chain-driven conveyor; (*b*) roller-to-roller conveyor.

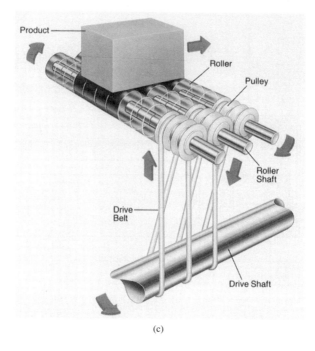

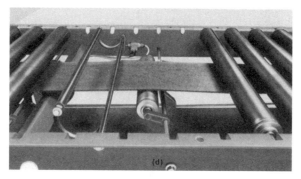

FIGURE 9.4 (*Continued*) (*c*) Line shaft conveyor; (*d*) belt-driven conveyor.

charge end. As required ,the path inclines and declines at a maximum of 30° and moves a tote at a speed of 70 to 120 ft/min.

The components of the strand conveyor are the C channel for the chain with a length up to 1200 ft, drive motor and sprocket, idle end assembly, straight or curved sections, conveying surface of two skatewheels of nylon-edged aluminum with stabilizer runners, steel-hardened chain, and side guards on each side.

After the carton, container, or tote is placed onto the conveyor skatewheels, it rests on the chain and two skatewheels. As the bottom of the tote comes in contact with the chain, the forward movement pulls the tote across the skatewheels. At the discharge station, a pusher-diverter or an employee removes the tote from the conveyor surface. As required, the strand conveyor accumulates totes prior to a workstation.

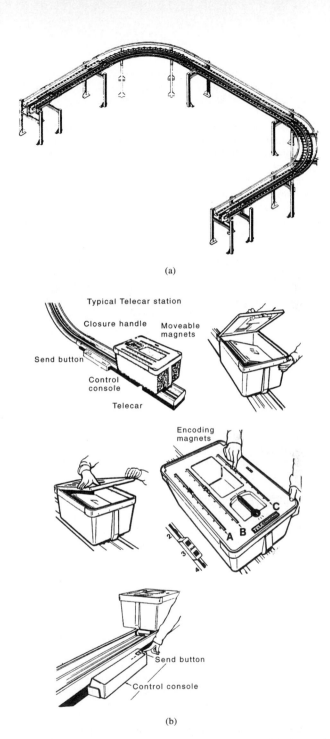

(a)

Typical Telecar station

Closure handle   Moveable magnets

Send button

Control console

Telecar

Encoding magnets

A B   C

Send button

Control console

(b)

FIGURE 9.5 (*a*) Strand (chain-in-channel) conveyor (*Courtesy of John Burton Machine Corp.*); (*b*) TELECAR, small load (*Courtesy of TeleEngineering, Inc.*).

9.11

The disadvantages of the strand (chain-in-channel) conveyor are that it requires a uniform product and difficult-to-control merge flows. The advantages are the low capital investment, ease of relocation, fewer moving parts, and low maintenance.

Conveyor Options.   In roller conveyor systems, pop-up rollers or stops serve as accumulators (temporary stops). When activated by a blocked photographic eye or wire prong, the conveyor roller accumulator extends upward above the conveying surface and stops the container flow on the system. During accumulation, line (built-up) pressure between containers is created by the pressure of the roller's forward movement on the bottom of each container. To reduce the line pressure, at these locations the roller conveyor system is designed with low- or zero-pressure conveyor sections.

To ensure smooth container flow and few jams, prior to the workstation or conveyor merge location, brake belts or brake modules are used to control container flow.

To access workstation, gravity conveyor lift gates are placed at appropriate locations. Prior to the lift gate, powered belt conveyors ensure controlled flow across the gate. Also handles on both sides permit easy lifting of the gate, and an elevation change between the charge and discharge ends ensures container flow across the gate. Divert devices at workstations transfer the container from the main travel line onto a spur (branch line). The conveyor surface type, merchandise volume and type, packing station, building features, and container orientation at the workstation determine the divert device.

How to Determine the Conveyor Path.   When you use powered conveyor to transport product between two locations, you should follow certain guidelines to ensure on-time, continuous product movement with minimal product damage.

There are seven important conveyor design parameters: (1) incline/decline slope (angle) with a power tail and nose, (2) over-conveyor support surface (ceiling or floor), (3) radius for turns, (4) conveyor speed, (5) conveyor frame and surface width, (6) drive and sprocket locations (end, middle, side, or top), and (7) emergency stops and start-stops.

Magnetic Conveyor.   The next powered small-item horizontal transportation method uses the magnetic conveyor, a fixed-path conveyor. The magnetic conveyor transports metal parts and consists of a polyester web that is laminated with a tough urethane surface which slides over a magnetized heavy-duty steel slider bed conveyor surface. The belt is driven by an electric motor, and the belt take-up is pneumatic to ensure proper belt tension. The drives are located at the center or end pull the 12-in-wide belt in any desired length up to 100 ft per module. The speed has the same range as a standard belt conveyor.

After the metal product is placed onto the belt, the metal part is stationary on the belt surface which slides across the magnetized steel bed. When the product arrives at the end of the first conveyor section, the product is transferred to a second conveyor module.

The disadvantages of the method are that it is limited to transporting metal products and that the transport distance for one module is 100 ft. The advantages of the magnetized conveyor are that it transports metal parts and with additional conveyor sections can transport more than 100 ft.

TELECAR Small-Load Electric Track Conveyor.   The TELECAR electric track transports small loads in a sealed container between two warehouse locations (Fig. 9.5*b*). The conveyor consists of (1) the four-wheel and dc powered cart with an attached lockable cover (the cover address panel consists of three sets of magnets, each with zero to nine possible locations), (2) electric powered track that secures the cart's four wheels to the track and, as required, switch-track devices to divert or merge carts between the main travel path and loading or unloading spurs, and (3) a send/receive console station.

The TELECAR electric track has the capability of transporting a car (container) with a 20-lb load (2600-in^3 volume) over a horizontal or vertical path at 120 ft/min. As required, a load-carrying self-leveling device is used to maintain a level load on the incline-decline transport path. After the car arrives at a send-to station, if it is not handled in a specific time, then an alarm sounds to have an employee unload and send the car to the next station.

There are three send/receive station designs. (1) In a reentry station, the carrier enters and departs over the same single track with a 3- to 5-carrier staging capacity. At the station, the cars are handled on a first-in first-out (FIFO) basis. (2) In a magazine station, a switch device has a separate departure track (spur) that holds 8 to 10 carriers, which are handled on a FIFO basis. (3) In a through station that is a high-capacity continuous station, the carrier is handled on a FIFO basis.

The horizontal transport path is a straight line or curve, and the vertical path has a bend to make a transition from the horizontal to the incline or decline. The travel path requires minimal clearances (1 to 6 in) above the carrier top. This feature allows the transport path to be installed on the ceiling. If the travel path passes through a fire barrier, then the manufacturer designs a fire deluge or a doghouse to protect the fire wall–floor penetration. Another travel path option is to use a switch that transfers a car from one lane to another.

To give management additional control, the mimic display panel shows the status of the cart's travel path.

After the operator loads the merchandise onto the carrier, the lid is closed and locked onto the container sides. With the lid secured on the container, the dispatch (send-to) address is entered on one of or all three address magnets. When the address magnets are set, on the control console the operator presses the send button. This dispatches (sends) the carrier on the travel path (track) to the receiving station. Arriving at the receiving station, the carrier activates a receiving station full alarm. At the receiving station, the operator unloads the merchandise and follows the dispatch instruction to send the container to the next station.

The disadvantages of the TELECAR are the fixed travel path, possible queues created by single track path, and high capital investment. The advantages are that the travel path is above the floor level, it is expandable, the carrier is secured with the option of having a leveling device, there is a reusable container-address code, there are many address locations, each location can have a bypass spur, and the travel speed is fast.

Novasort Method.   The Novasort method is a modular tilt-tray sortation method that has a closed-loop fixed travel path. It consists of a train of three to four tilt trays which have a 500-lb load-carrying capacity each, tilts to the left or right, a programmable controller for each train of tilt trays, a fixed travel path, and an electric bus bar and on-board sensing devices to detect objects or an employee in the travel path.

The Novasort vehicle travel path can be on a ceiling or floor and travels up slight grades and over 90° to 180° curves. The path directs the tilt tray train past an induction station and a delivery station. At the induction station, the induction employee places the product onto the tray and assigns the tray's tilt location in the programmable controller. At the delivery station, the programmable controller has the appropriate tray to tilt and discharge the product onto the assigned station.

With the on-board programmable controller, if required at the workstation, a train of trays is unloaded or loaded by the workstation employee. After all trays are tilted at the required stations, the train of tilt trays travels on the path and accumulates prior to the induction station or continues to the maintenance spur. On this maintenance spur, the tilt tray carriers wait for the next assignment.

As an option, the tilt tray has a fixed load-carrying surface and has the capacity to handle single-tray items, small items, and dual-tray items. As required to satisfy expansion of the work area, additional track is added to the existing travel path.

The design parameters and operational characteristics of the train of tilt trays are similar to those for the tilt tray sortation. Induction is manual or automatic with a tray that travels on a fixed closed-loop rail path between two locations.

The disadvantages of the method are that it handles a medium volume and has a fixed closed-loop path, the travel path requires an employee or equipment path to access or egress the workstation, single travel path could create queues, and there is a high capital investment.

The advantages are that the product is transported as required, it is a ceiling- or floor-supported concept, it is unloaded or loaded at the workstation, and expansion of the travel path is easy.

Overhead Nonpowered Horizontal Transportation

The unique characteristics of these transportation concepts are that the transport path is above the floor and requires an employee or gravity to move the merchandise between two locations.

Gravity Roller or Skatewheel Conveyor.   The gravity roller or skatewheel conveyor requires merchandise in containers or totes, an end stop, an elevation change (maximum of 5°) between the infeed (charge) end and the outfeed (discharge) end and side guarding. With gravity, the container flow is uncontrolled and has uncontrolled line pressure on the accumulation containers. To control product flow, the bed is tilted to the guide rail, or a roller turning is retarded, which slows the container speed.

Tubular Rail with Manual Trolley or Carrier.   The second fixed-path transportation method involves the tubular rail and manual trolley or carrier. A trolley or carrier consists of 6-ft 4-in high tubular (traffic) rail network, switches and end stops, and a trolley with a carrier.

The rail is nominal $1\frac{5}{16}$-in-diameter galvanized tubular steel sections that are connected with inserts or a V-shaped track that is welded together. The rail is supported on 6-ft centers from the ceiling or floor. It is designed with main traffic rails that have switches to transfer a trolley onto a branch (spur) line or from the branch spur onto the main line. This rail network allows a carrier with merchandise to move between two locations. At the discharge location, the rail is slightly sloped down for gravity to move the trolley and has an end stop that prevents trolley movement. As required, trolley-retaining side bars have the rail network make 45°, 90°, or 180° turns.

The 45- or 48-in-long trolley consists of two sets of metal or plastic wheels or heads with spools that are attached at the lead end and the tail end of the trolley. A load bar is supported between the base of the heads. The wheels or spools ride on the tubular rail or V track network, and the load bar holds the garments or flatwear carrier.

After merchandise is placed on the carrier, the trolley carriers are accumulated and an employee pushes or pulls the trolley train to the required workstation. To assist with this transport effort, a trolley pull bar is used to apply force on the rear trolley. As required, the trolley is manually diverted to the appropriate workstation.

There are two types of trolley rail, tubular and flat stock rail.

Tubular J Hook Rail.　　The tubular gravity trolley traffic rail network is supported by a J hook every 6 ft on centers (Fig. 9.6*a*). The entire trolley rail concept is a floor-supported or ceiling-hung structure.

The J hook is clamped and then screwed to the support structure with the open side of the J hook facing one side of the rail. Since most employees are right-handed, the J hook offset side of the direction of travel is on the left side, which allows the employee to easily pull trolleys on the rail.

FIGURE 9.6　　(*a*) Tubular rail (*Courtesy of Omni Flow, Inc.*); (*b*) V stock rail (*Courtesy of Omni Flow, Inc.*); (*c*) standard conveyor (*Courtesy of SDI.*); (*d*) programmable conveyor (*Courtesy of SDI.*); (*e*) trolley identifiers (*Courtesy of W & H Systems.*); (*f*) empty-trolley transporter (*Courtesy of Railex Corp.*); (*g*) carrier extension (*Courtesy of SDI.*).

V Bar Stock Rail.   A second slick rail type is the bar stock system with a V travel lane (Fig. 9.6b). The second trolley concept has a four-wheel trolley that rides on bar stock which has the bottom portion (track) in the shape of an inverted "V". The structure is floor-supported or ceiling-hung.

Rail Components.   Both rail systems have the *top of rail* (TOR) 6 ft 4 in above the finished floor or, as required, a slight slope that permits the trolley to travel on the gravity (slick) rail by gravity or manual power.

The other components of a gravity manually powered rail consist of switches and stops that direct or control the trolley travel on the gravity rail system. These components are (1) fixed end stop, (2) spring switch, (3) master switch, (4) open or closed stop, (5) level or drop switch, and (6) special switches.

Special-Purpose Switches.   The next group of trolley switches is the special-purpose switches. These switches include the crossover switch, cross-through switch, hinged knife switch, removable bar, overhead door switch, parallel stacking switch (trolley trap), telescoping swivel switch, telescoping removable docking switch, sectional loading boom, and extendible boom.

Other Trolley Design Parameters.   When two-way hanging-garment rails are parallel, the centerline-to-centerline dimension (distance) between the two rails is 24 to 27 in. This dimension is determined by the type of garment being transported on the rail system. Another design feature with a standard trolley is that the main and branch lines are 45°, 90°, or 180° turns. A standard trolley with two traffic rails that have trolley flow in different directions or make a 180° turn has the two rails on 6-ft centers. When long trolleys (48 in) are used on the system, then the 180° turn is 9 ft on centers. When baskets are used on the system, to prevent the basket from jumping the rail on the curve, trolley antijumping bars are installed on the curves.

Trolley Components.   The trolley is the basic travel mechanism for hanging garment and basket transportation, and there are three components (Fig. 9.6c). (1) Two heads with two sets of spools (the spools are metal or plastic wheels) have a space between the wheels equal to the rail diameter. (2) The load bar is attached to the two heads that have standard lengths of 24, 30, 36, or 42 in. The load bar length is equal to the end of the spools, but does not extend beyond the spools. The load bar is the load-carrying surface for hanging garments or basket. Most trolleys have three to five pegs that stabilize the garments. When the trolley travels on inclines or declines, to keep garments from falling off the bar, pegs are recommended on the load bar. Also the pegs are used to separate SKUs or customer orders. (3) A second trolley type is used on the V track system that has four wheels. These wheels ride on the V lane. This type of trolley does not require master switches.

For trolleys on an automatic trolley transportation system that has automatic diverts the lead neck to the trolley head must be long enough (neck length) to accommodate code identifiers (Fig. 9.6e).

In a hanging-garment transportation system, there are two options that increase employee productivity: the empty-trolley transporter and a carrier extension for short garments.

Empty-Trolley Transporter.   The empty-trolley transporter is a four-wheel structurally supported vehicle that has several hang bars for empty trolleys (Fig. 9.6f). At the packing station, an empty-trolley transporter provides a device that accumulates empty trolleys and is manually pushed with empty trolleys to another location.

Carrier Extension. The second device is a carrier extension which is used in a hanging-garment system for short garments (Fig. 9.6g). The carrier extension is a rigid or folding load bar with two hooks that are placed on the trolley's load bar. With short garments less than 28 in long, the device is attached to the load bar and approximately doubles the load bar's carrying capacity.

Various Trolley Basket Carriers. In trolley small-parts flatwear SKU transportation, there are three basic carriers: the metal folding basket, circular basket, and canvas bag.

Folding Metal Basket. The folding metal basket is 36 in high by 36 in long by 18 or 24 in deep with two hooks that permit the basket to be attached to the trolley's load bar. The basket load-carrying surface has a slight slope to the rear which helps prevent the product from falling to the floor. The basket interior is separated into small compartments for carrying multiple SKUs or customer orders. When not in service, the basket is folded for storage or remains on the trolley, which is on a branch spur. Mesh netting with hooks or velcro reduces incidences of the product falling to the floor.

   The metal basket has several disadvantages: the weight of the basket, its carrying capacity is approximately 75 percent of the space, and its one-side-fill capability.

Metal Circular Basket. The metal circular basket is hooked onto the trolley and is supported in the center. There are several levels to the basket. Each basket level is divided and handles multiple SKUs or customer orders. Compared to the other baskets, the circular basket is a two-side fill type and carries a smaller capacity.

Canvas Bag. The folding canvas bag is a carrier that provides a load-carrying surface with a solid bottom and sides. The canvas bag has steel structural members and two hooks. The two hooks are used for attachment to the trolley. There are two types of canvas carriers, the top-loading type and the front-loading type. Both devices are lightweight and are attached to a trolley or rolling rack. The front-loading type has a slight slope that prevents product from falling to the floor.

Rope and Clips. The next overhead, nonpowered small-item transportation method is a fixed-path concept. Double clips are attached to a rope. The rope is an endless loop around two pulleys. The product is attached to the other end of the clip.

   To operate at one workstation, the clip is attached to the rope, and the product is attached to the clip. The operator pulls the rope that propels the product to the next workstation. At the new station, the employee removes the product and clips from the rope.

   The disadvantages of this method are that it handles a low volume and lightweight product. The advantages are that it is easy to operate and requires minimal investment.

Overhead Powered Horizontal Transportation

These transportation methods have similar characteristics to the overhead nonpowered transportation methods. But these methods move product by a load-carrying surface that is propelled by an electric motor. These concepts are supported from the ceiling steel with structural members (racks or stands) from the floor.

Some Powered Conveyor Overhead Protection Requirements. The powered belt, roller, skatewheel, or strand conveyor is most common. In a facility, for code and safety reasons, all overhead transportation methods require the following: (1) An elevation above 7 ft 6 in, side guards, and underside protective netting or solid structure are needed. (2) All belt-driven conveyor belt returns below 7 ft 6 in must have protective underside guarding. (3) All fire wall penetrations must be protected. (4) All conveyor rollers above 7 ft 6 in must be self-locking.

Powered Chain and Trolley or Carrier.   The powered chain and trolley or carrier has an endless chain of links contained in a C-type channel. A trolley pusher is attached every 10 ft on centers of the chain and extends downward toward the rail. The first pusher type is a hard-pusher pendant (dog), which is a metal device. The pusher engages the trolley lead head, which pulls the trolley along the rail system. The second pusher is a soft-pusher (flexible) pendant, which is a rubber device. On a horizontal run, the rubber device has sufficient rigidity to engage the trolley lead head and to move the trolley along the rail. When the trolley is halted on the rail, then the rubber device has the flexibility to slide over the trolley head and not move the trolley. This action provides trolley or carrier accumulation without line pressure.

A powered trolley chain has the motor or drive located at the highest elevation and the manually operated chain take-up at the 180° turn at the lowest elevation that is immediately past the highest elevation. When the chain has completed the trolley transportation path, then the last divert location is a castoff. A castoff has the powered chain rise above the trolley heads, which eliminates the forward movement on the trolley. The path for the chain with the pendants is in a different direction. If the chain does not push trolleys, then the required chain travel path (vertical space) is minimal because the clearance is for the chain and pendants.

In a powered trolley, the trolley is automatically diverted from the main traffic line at a 30° to 45° angle onto a branch spur (line). When the spur has accumulated trolleys, the branch line has full line controls, which does not permit the divert device to operate. In this automatic method, the trolley lead head requires a divert location identifier that is a reflective tab, sliding tab, bar-code label, or wire prong. When the identifier is read or detected by a code reader, this activates the divert device to transfer the trolley onto the branch line.

Pneumatic Tube.   The next transportation method involves the pneumatic tube, a fixed-path method. This method consists of a container that travels through an air-pressurized or vacuum metal tube network. There can be a single (simple) receiving location or multiple-send (complex) multiple-receive locations. In a two-way system, only one container travels at a time. The container traffic flow through the tube system is two-way or one-way. If the system is one-way, then there are separate sending and receiving tubes or an alternative transportation method for empty-container return to the sending station. The container's diameter ranges from 3 to 6 in, the container is $12\frac{5}{8}$ in long, and it has the capability to carry 3 to 5 lb at speeds from 15 to 20 ft/s.

To operate the pneumatic tube, the loose or individually packaged merchandise is placed into the container. The container is sealed at the sending station. The desired receiving station is programmed (set) electronically on the sending station box, and the container is placed inside the tube. Air pressure or vacuum is automatically turned on and pushes the container through the metal tube for a maximum distance of 2000 ft. When the container arrives at the desired (set) receiving location, the air pressure is automatically shut off. This decrease in air pressure discharges the container at this receiving station box, which was previously set at the sending station.

CARTON OR PALLET LOAD HORIZONTAL TRANSPORTATION

The second major horizontal transportation method involves carton, pallet load, container, or skid transportation.

The various horizontal transportation methods include the following:

- Above-floor nonpowered

   Human carry

   Manual push or pull

   Cart

   Dolly

   Roller, skatewheel, or strand conveyor

   Platform truck

   Hand truck

   Roller pallet

   Semilive skid

- Above-floor powered

   Burden carrier

   Conveyor

   Pallet truck (single or double pallet)

   Tractor (tugger) train of carts

   In-floor towline

   Inverted power and free

   S.I. Cartrac

   Automatic guided vehicle (AGV)

   Air cushion pallets

- Overhead nonpowered trolley basket
- Overhead powered trolley basket, conveyors, monorail

Above-Floor Nonpowered Horizontal Transportation

The first horizontal transportation group is the above-floor nonpowered group. The equipment in this group moves product via manual power between two locations on one floor.

Human-Carry Method.   In this variable-path method, an employee lifts, carries, and deposits the carton between two locations. It is the basic concept in a carton-handling warehouse and is used to transport cartons a short distance.

The disadvantages include low employee productivity, the great number of employees required, the potential for injuries, and low volume (carton) handled. The advantages of the human-carry method are that no capital investment, a narrow path, and a variable-path system are required.

Two-Wheel Hand Truck.   The two-wheel hand truck transports one carton or several cartons. It is designed with a metal or wood H frame that has (1) two handles on the upper end of the H frame, (2) horizontal bracing between the frame legs, (3) two wheels on an axle at the H frame's lower end, and (4) a load-carrying surface that has a lip or nose ahead of the wheels.

To operate this hand truck, an employee picks up or places the product on the load-carrying surface, to balance the truck and to push the truck between the two locations.

Compared to the human-carry method, the disadvantages are the short distance traveled, the small load-carrying surface, and the low volume handled.

The advantages are low capital investment, little employee training needed, a heavier load carried over a greater distance, and less employee fatigue.

Dolly and Four-Wheel Platform Truck.   The third and fourth manual transportation vehicles are considered as one group. This group consists of a dolly and a four-wheel platform truck that transports cartons, pallet loads, or skids. Both are variable-path vehicles. The platform trucks come in nontilt or tilt types.

Dolly and Platform Truck, Nontilt Type.   The nontilt type has all four wheels resting on the floor under the four corners of the platform. The dolly has three wheels that are equally spaced in a triangular pattern under the load-carrying platform. The dolly with four wheels has the wheels under the four corners of the load-carrying surface.

Platform Truck, Tilt Type.   The tilt type is a specially designed four-wheel platform vehicle. It has a wheel in the center of each side under the rectangular load-carrying platform. The two wheels under the rectangular long sides of the platform are higher than the two wheels under the short sides of the platform. This feature provides the tilt characteristic to the truck.

Dolly and Platform Truck Characteristics.   The dolly and platform vehicles have at least two steering wheels or casters. The load-carrying surface is a solid or slat load-carrying surface. The dolly does not have a push handle, and the platform truck has a push handle or bar.

To move the dolly between the two locations, an employee pushes against the exterior surface of the product. This feature with the small vehicle wheels that have swivel casters enables the dolly to handle lightweight and rigid product. This wheel-caster feature makes the dolly difficult to steer.

The standard platform truck has large wheels located in a four-corner pattern under the rectangular platform. The front two wheels have rigid casters, and the rear wheels have swivel casters. The platform truck has a push bar and is of more rugged construction, which enables it to handle a wide variety of loads up to 4000 lb. The truck's carrying surface is designed with various structures to handle specific loads.

To operate these vehicle, an employee places the product onto the vehicle's load-carrying surface. To move the truck, the employee pushes against the product on the dolly or against the handle on the platform truck. To assist in the pulling and steering effort of the dolly, a pull handle is used to pull the dolly.

The disadvantages of the dollies and platform trucks are the same as those of the two-wheel hand trucks. In addition, it is difficult to travel up grades, and to load or unload the vehicle requires another labor activity, which lowers the total facility productivity. An additional disadvantages of the dolly is that the lack of a push bar or handle limits the type of product that can be handled on the load-carrying surface.

Compared to the two-wheel hand truck, these vehicles enjoy additional advantages. First, there is less employee fatigue, and the employee travels longer distances. Second, the platform truck handles larger and heavier loads.

Human Pushcart.   The next horizontal transport vehicle is the employee pushcart, which is considered a variable-path vehicle. The pushcart is a four-wheel manually powered vehicle with a load-carrying surface or several surfaces with side guards and a push bar or push handles. The pushcart handles one carton or several cartons.

Carts that carry loads of 750 lb and have a short load-carrying surface length of 4 ft or less have the swivel wheels or casters on the front and the rigid wheels or casters on the rear. This caster-wheel arrangement provides excellent steering and easy turning at the end of the warehouse aisles. Carts with a longer load-carrying surface that can carry a heavy load (750 lb or more) have the swivel wheels or casters in the rear and the rigid wheels or casters in the front. This wheel-caster arrangement is designed for easy steering of the cart.

Unlike the dolly and platform trucks, the pushcart is designed with shelves or dividers. This permits the cart to transport different SKUs or customer orders.

Hand-Operated Walkie Pallet Truck.   The next manual transport vehicle is the hand-operated nonpowered walkie pallet truck (jack). The hand-operated pallet truck is designed with an operator handle that activates the hydraulic fork-elevating mechanism, a hand- or foot-operated hydraulic pressure release level, a wheel brake, one or two steering wheels, and two load-carrying wheels under a set of forks. To reduce cartons falling from the pallet board, a backrest is attached to the vehicle. The hand-opeated pallet truck handles loads up to 3000 lb. For best results, the pallet board or container has fork entry openings with chamfered bottom boards, which reduce wheel hang-up.

When a pallet board is used on a manual pallet truck, for easy pallet board pickup and reduced product damage, the pallet truck fork ends have pallet entry wheels and the pallet board outer bottom deck boards are chamfered on both edges. The chamfered deck boards are shaved at the top for the wheels to travel over the deck boards. When skids that have a higher fork opening are used in the warehouse, the manual pallet truck has a skid adaptor which increases the height of the pallet truck forks.

The vehicle picks up an empty pallet board, sets the hydraulic level in the elevate position, and elevates the pallet board. As the employee pulls or pushes the vehicle through the aisles, he or she arrives at the assigned location and deposits the pallet board in the area.

The disadvantages of the hand-operated pallet truck are that it is a slow transport method, it requires an empty pallet board, tape, string, or other material is used to secure the load, and there is increased operator fatigue from pushing and/or pulling the vehicle.

The advantages of the pallet truck are that it is the best economically per case of the nonpowered transport methods, it requires little employee training, with a pallet board it handles a wide variety of SKUs, it requires a low capital investment, and facility productivity is increased because the product is on pallet boards.

Roller or Skatewheel Conveyor.   The next above-floor nonpowered carton, pallet load, skid horizontal transport method involves the roller or skatewheel conveyor, which is considered a fixed-path system.

A conveyor consists of conveyor sections that are secured together. The conveyor section is a series of tubular rollers or skatewheels that revolve on a shaft connected to the bed. The conveyor bed is supported by adjustable metal stands. To guide cartons or containers across the conveyor bed, side guarding is installed above the conveying surface, the wheels or rollers are flanged, or the conveyor surface is set below the bed section.

Gravity or an employee provides the force that moves the product between two locations.

When short cartons are transported, the roller or skatewheels are located on close centers. The roller size varies according to the carton's characteristics. When pallets or skids are transported, the conveyor components are heavy-duty rollers. These rollers span the full width between the side beds, or narrow rollers (wheels) are located under the two exterior pallet board stringers (pallet or skid outer edges). If the product is transported more than 25 to 30 ft, a slave pallet board and side guides or flanged wheels ensure positive flow between the two locations.

The disadvantages of this system are the fixed path, a conveyable SKU, and the human or machine load and unload. The advantages of the concept are the low capital investment, little employee training needed, easy operation to manage, and medium volume handled.

Semilive Skid. The next product transport method employs the manually powered semilive skid, which is considered a variable-path system. Product is transported between two locations over a short distance. The semilive skid consists of a two-wheel jack with a stud and a skid platform. The platform has two rear rigid casters or wheels, two legs in the front, and an eyehole coupler.

The skid platform has a load-carrying surface made of wood slats. These wood slats are modified with steel posts on the sides, removable sidings, and posts. The carrying surface has a deck as small as 24 in wide by 36 in long up to 36 in wide by 72 in long, and the two rear wheels are 2 in wide and 6 to 7 in high. The two support legs are in the front corners of the deck underside and are 5 to 6 in high. In the middle of the deck at the front is attached a coupling plate with an eyehole.

The two-wheel hand jack has a hand bar, two fixed casters or wheels that are 6 to 7 in high (higher than the legs), and a stud that fits into the coupler eyehole.

To operate the semilive skid, the hand jack stud is inserted into the coupler. As the hand jack is declined to the pull position, the hand jack supports the platform by raising the front of the skid and legs above the floor. In this position, the employee pulls and steers the semilive skid between locations.

The disadvantages of the semilive skid are that it is not flexible (used in other warehouse functions), travel up grades is difficult, it is more expensive than a pallet board, and it is difficult to steer.

The advantages of the semilive skid are that it handles a heavier load, travels over a greater distance, and requires less employee effort.

Above-Floor Powered Horizontal Transport Methods

The above-floor powered horizontal transport method is next. To move product between two locations, these transport methods rely on electric power that is supplied to a conveyor chain or that is an engine-driven vehicle.

Above-Floor Powered Conveyor

The powered conveyor is one of the most common transport methods in the warehouse industry. The factors that make conveyor transport popular are the multiple load and unload locations, it handles a large volume, few employees are required, and transport is over a long distance.

The cost of a conveyor is a function of the length and type of conveyor. These are some conveyor types:

Powered roller	Powered skatewheel
Strand conveyor	Air conveyor
Magnetic	Apron
Drag chain	Flight
Sliding chain	Slat
Vacuum	Belt

CONVEYOR OPTIONS AND ACCESSORIES

Conveyor options and accessories are designed and installed on a conveyor to increase employee productivity, improve operator safety, and increase the flexibility of the workstation. These options and accessories include (1) fixed end stop; (2) hand case stop; (3) case counter; (4) gate section; (5) guardrails that are channel (single or double), solid, skatewheel, or continuous (adjustable or flared); (6) safety signs; (7) hold-down strips; (8) traffic cop; (9) 45° sweep rail and sweep plate; (10) guide drums and guide wheels; (11) roller-activated belt (roller slave sleeve); (12) merge belt and table; (13) gap plate or conveyor set at different elevations with a flat gap or ribbed gap plate; (14) adjustable stop; (15) slave-driven turn; (16) elevation change on the turn; (17) turntable; (18) brake belt or brake module; and (19) ball top transfer.

How Employees Cross the Conveyor Path

When you design a conveyor system to transport product between workstations or as a pick-to-belt system, then it is likely that your employees will work within a particular area. If this situation exists in your work area, then you must provide a means for employees to cross the conveyor path. You must provide an access or egress path for employees at the workstation, and this will ensure an emergency exit from the workstation for employees.

Painted lines on the warehouse floor show employees where to cross a conveyor path. There are three ways to cross the conveyor path: (1) The employee walks up and over the conveyor via stairs and a platform bridge or a stile with a bridge or ship's ladder with slats between the rollers. (2) The employee walks through a liftgate. (3) The employee walks on incline and decline belts to form a bridge.

The method that you choose is determined by these factors: (1) What is the purpose of the bridge and crossover? (2) What is the conveying surface? (3) What type of product is on the conveying surface?

Incline and Decline.   The conveyor path incline and decline create a bridge between
the finished floor and the conveying surface (Fig. 9.7*a*). This arrangement requires an
incline conveyor belt from the floor, an extended nose-over straight conveyor section
that creates a bridge, and a decline belt conveyor to the floor with a powered tail.

(a)

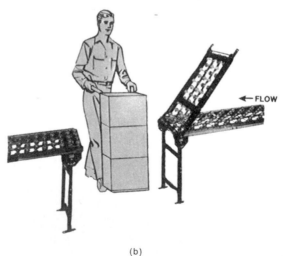

◄—FLOW

(b)

FIGURE 9.7   (*a*) Incline and decline over the aisle (*Courtesy of Modern
Materials Handling Magazine.*); (*b*) liftgate (*Courtesy of Rapistan Demag
Corp.*).

The incline and decline belt handles loose merchandise or merchandise in cartons, totes, and containers. It provides a path for a large number of employees or for mobile warehouse pallet handling equipment. The concept requires a runout for the incline and decline belts.

Liftgate. In the second system, the employee passes through the conveyor via a lift-gate in the conveyor path (Fig. 9.7*b*). A liftgate section is a 3-ft-long gravity skate-wheel conveyor section. In the open position, it gives an employee access to a work-station; in the closed position, it permits carton flow across the conveyor section. For best operation, the liftgate section is spring-counterbalanced with handles on both sides. To ensure continued product flow across the liftgate, conveyor design should follow these guidelines: (1) When the gate is used on a gravity conveyor system, the liftgate lead end is 1 in higher than the discharge end. (2) When the gate is used on a powered conveyor system, prior to the liftgate is a 4-ft-long powered belt section, and the liftgate has a switch that controls the electric power to the belt or roller section. The powered belt or roller section feeds cartons onto the liftgate. If the liftgate is in the open position, the switch turns off the powered conveyor. Without power there is no product travel onto the liftgate. In the closed position, the lift gate permits power to the conveyor which allows product to flow across the liftgate. (3) When the gate is in the up position, a minimum of 36 in of clear space is needed for the gate.

The liftgate is low-cost and provides a path for a few employees to cross the conveyor path.

Stairs and Platform. In this system, the employee walks up and over the conveyor path (Fig. 9.7*c*). There are three variations.

First is the stairs and bridge platform. A series of stairs and handrails lead to or decline from a platform with handrails and kickplates. The stairs and platform consist of a welded structure whose exterior metal is coated and that meets local codes. These components provide a bridge for employees to walk over the conveyor path.

The stairs and bridge platform permit the conveyor to handle loose merchandise or product in cartons, totes, containers, pallets, and carts. Compared to the other systems, the stairs and platform have a higher cost but handle a large number of employees.

Stile. The stile is similar to the stair and platform except that the stile is a pre-designed coated-metal structure (Fig. 9.7*d*). Also the stairs to the platform are at a steeper angle.

The stile provides a means for employees to cross a conveyor that handles loose merchandise or merchandise in cartons, totes, containers, pallets, or carts. The stile costs little and handles a small to medium number of people.

Ship's Ladder. The ship's ladder with slats between the rollers on the conveyor consists of metal-coated handrails (Fig. 9.7*e*). Between the two handrails, as required, are slats between two rollers. As an employee walks across the conveyor, these slats are where employees put their feet. Before crossing the conveyor path, the employee makes sure that there is no product in or approaching the crossover path. If there is no product on the conveyor, then the employee crosses the path. If there is product on the conveyor, then the employee waits for the product to pass or stops the product and crosses the conveyor. For best results the conveyor should be a low- or zero-pressure conveyor.

The ship's ladder costs little and is used on a roller conveyor that transports product in containers and handles a small number of people.

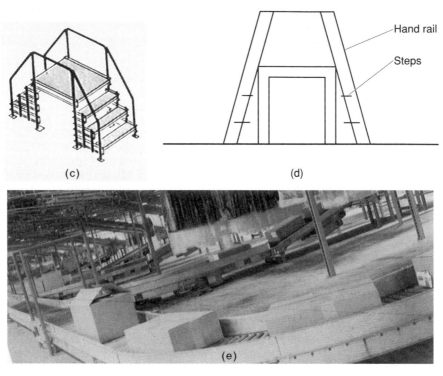

(c)   (d)

(e)

FIGURE 9.7 *(Continued)* *(c)* Stairs and platform; *(d)* stile; *(e)* ship's ladder *(Courtesy of Modern Materials Handling Magazine.)*.

DETERMINING THE CONVEYOR PATH

When you use a conveyor to transport product between two locations, following certain guidelines ensures on-time and continuous product movement with minimal product damage.

These guidelines are similar but vary for each conveyor type: nonpowered roller or skatewheel; powered roller, skatewheel, belt, or strand; and overhead chain.

For powered and nonpowered roller and skatewheel conveyors, follow these guidelines: (1) SKU characteristics are the length, width, and height plus weight and conveying surface. (2) Determine the operating environment. What is the temperature (freezer or normal), and how dusty is the operating area? (3) When the frame or roller and wheels could be touched by product or an employee, a galvanized coating reduces the transfer of dirt and reduces rust. In most large systems, coating the frame with paint is standard practice. (4) To ensure smooth, continuous product flow, there should be at least three rollers or axles for the skatewheel conveyor under the shortest product. To determine the proper roller-axle spacing, use this formula:

$$\frac{\text{Load length}}{3} = \text{maximum roller-axle spacing}$$

(5) The roller or wheel capacity is the maximum weight that a roller or wheel (axle) supports as product is moved across the rollers or wheels (axles). The formula is

$$\frac{\text{Unit-load weight}}{\text{No. rollers or wheels (axles)}} = \text{weight per roller}$$

If you exceed the roller or wheel (axle) capacity, then increase the number of rollers or wheels under the load.

(6) The conveyor frame capacity is the maximum weight that a roller or wheel conveyor frame section can support between the two stands or hangers. (7) Curve section is the method used to convey product around curves. There are three types of roller curves plus the belt curve. These roller curves are single straight roller, double-roller differential, and tapered roller. With a single roller, a square or rectangular product does not track on single straight roller curves, but tends to slow at the exit of the curve. With a double-roller differential curve, the rollers reduce the twisting of the carton but skew at the exit of the curve. With tapered rollers on the curve, the roller at the interior of curve is smaller or narrower than the wider or larger part of the roller at the outside of the curve. This roller design keeps the loads in the same position during travels around the curve. With a belt curve, there is a complete belt on the curve, and it is used to control or track the product on the conveyor system. (8) Since most conveyor manufacturers have precalculated curve charts, with the basic product information (length, width, and desired clearances) you determine the curve width. (9) The conveyor straight path is determined by the product width, the clearance between product and guide rails, and the decision to use guardrails. (10) If the product is conveyed without guardrails, then design the frame width for the widest product plus a reasonable overhang ($\frac{1}{2}$ in) from the frame's side. If product is conveyed with guardrails, then design for the product width plus the desired clearance from the guardrails. Add 1 to 2 in to the frame width. (11) The incline or decline of the conveyor path determines the pitch or slope of the conveyor for the elevation change between the elevated end and the lower end. These slopes range from $\frac{5}{8}$ to $\frac{1}{4}$ in/ft and depend upon the product and the conveying surface. Most manufacturers have predetermined charts that show the plan view length for the conveyor run including a powered tail and nose-over. (12) Conveyor supports are used to support the conveyor frame. There are three kinds: portable floor supports, permanent floor supports, and ceiling-hung supports. The portable floor supports are tripods or casters or wheels under each end of the conveyor frame. Typically these tripods or casters are used to unload and load delivery vehicles. Each conveyor frame has a hook at one end and a stud at the other end for attaching one frame to the other frame. The permanent floor supports are anchored to the floor and rack and are secured to each frame section with bolts and nuts. The permanent floor support method is used for conveyor sections that move product between two warehouse workstations and require employees to manually transfer product between the conveyor and the workstation. The ceiling-hung supports are attached to the joists and have threaded rods that permit connection of each conveyor section to another conveyor section. There is approximately 12 in of thread rod on both sides of the conveyor-connected locator to permit future adjustment. In all ceiling-hung installations, the architect is provided with the dead and live loads to ascertain if the ceiling structure can handle the weight. This conveyor hanging method is very common in dock areas or over processing areas of the warehouse. When the conveyor ceiling is hung from the building joist, then the conveyor support members are attached at the panel points. The panel point is at the intersection of the diagonal members and the bottom straight joist member. (13) With a belt conveyor, use crown and laggered pulleys to help ensure belt tracking. In crown pulleys, the center of the

pulley is higher than the ends. *Laggered* means that the pulley has rough grooves in the exterior. When the conveyor is above 7 ft 6 in from the finished floor, then it has underside guarding. When the conveyor path travels past an employee workstation, it should have an E stop pull cord, at potential jam locations. The E pushbuttons are located for easy employee access. When the conveyor path penetrates a floor or fire barrier, the opening must be protected. Allow a minimum of 6 to 12 in of clearance between the top of the product and other items.

Design Guidelines for Overhead Chain Conveyor

When the product movement system uses an overhead chain conveyor, following these guidelines can reduce operational problems. (1) Determine the SKU characteristics (length, width, height, and weight) and the number of SKUs. (2) Determine the number of carriers required to handle the product volume, and calculate the travel speed of the conveyor by these formulas:

$$\frac{\text{No. parts} \times \text{parts per carrier}}{60 \text{ min}} = \text{carriers per minute}$$

$$\frac{\text{No. carriers} \times \text{6- or 5-ft spacing}}{60 \text{ min}} = \text{ft/min}$$

(3) Determine the carrier design and clearances. The carrier attaches the load to the chain conveyor and ensures that the load arrives on the carrier at the assigned location. During the travel, the carrier balances the load on the conveyor, provides stability, and ensures that the lead end of the carrier (trolley) arrives at the workstation. It is also important to have sufficient clearance from all building obstacles and other carriers. The neck of the carrier has space for the placement of the destination code device. (4) Locate the carrier horizontal path, decline and incline paths, and divert or loading spurs.

(5) Two other important factors are the carrier weight method used to attach the carrier to the conveyor and how the product is loaded and unloaded from the carrier. The two types of carrier attachment to the chain conveyor are the one-point and two-point suspension methods. The one-point suspension method is in the center of the carrier and is used for lightweight SKUs or carriers. The two-point suspension method is attached at both ends of the carrier and is used to handle heavy SKUs or carriers. A general rule of thumb, as the carrier travels on a curve, is to allow 6 in of clearance between two carriers. This clearance prevents two carriers from hitting (jamming). To determine the space between two carriers on a 180° curve, sketch to scale two carriers on a curve, and scale the open space (curve clearance between the two carriers)—or use the manufacturer's standard distance between conveyor paths. When 6 in of clearance is allowed on a straight conveyor section, then it is defined as the straight clearance or open space between two carriers.

(6) When the conveyor path makes an incline or decline between two elevations, the carriers require clearance. When the overhead chain or carrier makes a decline or an incline, special attention must be paid to incline-decline and head clearances. The chain incline or decline is the open path that the chain requires to change elevation, the carrier (load) clearance or open space between two carriers. This clearance is affected by the carrier suspension method. When a single-point suspension method is used, as the carrier declines or inclines, the clearance between the two carriers shrinks. With a two-point suspension method, the clearance between two carriers shrinks, but

as the carrier travels across the top and bottom of the incline or decline transmission point, it becomes a critical location which requires a curved transmission area to prevent hang-ups.

(7) Other important chain conveyor features include the following: Take-ups are located on a 180° curve at the low point of the conveyor path. The drive is located at the highest point of the conveyor path. At potential jam locations, E stop buttons are located on the floor level for easy employee access. When the chain passes a workstation, the E stop pull cords should be located for easy employee access. When the chain conveyor is more than 7 ft 6 in from the finished floor, underside guarding is needed. When the chain penetrates a floor or fire barrier, the opening requires protection.

Directing the Flow of Carton Travel on a Roller or Skatewheel Conveyor.　When you design a roller or skatewheel conveyor transport system, the conveyor system requires that your cartons, totes, and containers be directed to one side of the conveyor path. This requirement is determined by the divert, scanner, label, or other device prior to a workstation.

Carton Travel on a Conveyor System

Various concepts used to direct carton travel on a conveyor system include (1) skewed rollers, (2) sleeve-wrapped (taped) rollers, (3) an angled deflector, and (4) a tilted conveyor section.

Skewed Rollers.　The first method of controlling carton travel on a belt-driven roller or conveyor system is to skew (angle) two ore three rollers on the conveyor section. The skewed rollers create an angled path on the conveyor surface. As the carton, tote, or container travels across the skewed rollers, which revolve forward and are directed toward one side of the conveyor, the skewed rollers direct the carton travel to the required side of the conveyor.

The skewed roller system can affect the tracking of the conveyor belt, but maintenance employees can install the system on the conveyor and reduce tracking problems.

Sleeve-Wrapped (Taped) Roller.　The sleeve-wrapped (taped) roller is used on any type of roller conveyor system. Prior to the location where it is desired to have the cartons on the side of the conveyor, a specific number of rollers are wrapped in a spiral manner with girt-covered tape. The tape on the roller pattern is from the far side of the conveyor to the near side of the conveyor. With each progressive roller, the spiral tape location moves to the near side of the conveyor. With the girt tape surface, as the carton travels across the tape, the bottom comes in contact with girt tape and directs its travel from the far side to the near side of the conveyor surface.

Tape is a low-cost application that is used on any type of roller conveyor system.

Angled Deflector.　The angled deflector consists of a solid metal plate, skatewheel guard conveyor section, or channel guardrail. The deflector is located on the far side of the conveyor and projects outward toward the near side of the conveyor.

As the product travels on the conveyor surface and comes in contact with the deflector, the travel speed, movement, and deflector angle of the carton move it from the far side to the near side of the conveyor.

The angled deflector is a low-cost method used on any type of roller or skatewheel conveyor surface.

Tilted Conveyor.   The fourth way to control the carton travel path on a conveyor system is to tilt the conveyor section. This approach is used on a gravity roller or skatewheel conveyor section. The far side of the conveyor section is slightly higher than the near side. When the carton travels across this pitched conveyor surface, the pitch causes the carton to move from the far side to the near side of the conveyor surface.

CHANGING THE DIRECTION OF TRAVEL

When the carton label does not face the workstation employee, then you must change the carton's direction of travel. This change of travel direction enables the carton label to face the employee at the workstation. If the sortation conveyor cannot handle the divert on the return loop section, there are four alternatives: (1) Manually change the travel direction; (2) mechanically turn the table; (3) divert onto a curved belt conveyor; and (4) use a curved roller conveyor with inverted tapered rollers.

Manual Change of Travel Direction

By the manual method, at the end or prior to the workstation the employee physically lifts and turns the carton as it travels on the conveyor. This action rotates the carton so that the label will face the employee at the workstation.

The disadvantages are that it handles a low volume, it requires an employee, it is a slow activity, and it increases the possibility of injuries. The advantages are that it requires no capital investment and does not require additional floor space.

Mechanical Turning of Conveyor Section

The mechanical turntable conveyor section is an automatic powered conveyor section (table) with sensing devices that have the ability to accept a carton, turn the carton 180°, and power the carton off the conveyor section. As the carton travels on the powered conveyor table, the label does not face the workstation. When the carton leaves the powered conveyor table, the change in the travel direction of the carton causes the label to face the workstation.

The disadvantages of the method are that it increases capital investment, requires electric power, and increases maintenance. The advantages are that it handles a high volume, requires no employees, and reduces employee lifting injuries.

Divert onto a Curved Belt Conveyor

The next way to change a carton's direction of travel is to divert the carton from one conveyor onto a 90° or 180° curved belt conveyor. The divert device must be a pusher (straight-line) type.

In this method, the carton travels directly in a straight line from the sortation conveyor onto a 90° or 180° curved belt conveyor. As the carton enters the charge end of the belt conveyor, the label does not face the workstation. As the carton arrives at the discharge end of the curved belt conveyor, the label faces the workstation. When the carton arrives at the discharge end of the belt conveyor, it is transferred onto another conveyor.

The disadvantages of this method are that it requires additional capital (conveyor) investment, a sortation system, additional floor space, and electric power. The advantages of the method are that it handles a medium volume of 25 to 30 cartons per minute, does not require an employee, and reduces employee lifting injuries.

Inverted Tapered Roller.   Inverted tapered rollers on a curve have a specially designed 90° (wide) curve between the infeed (divert) location and the outfeed (takeaway) conveyor. To design this turn method, the design parameters are the typical carton design parameters for a conveyor system.

On this wide 90° curve are a series of inverted and standard tapered rollers. The width of the curve must be able to handle the diagonal length of the carton since it turns on the curve.

After the carton is diverted from the sortation conveyor, it travels across a 45° curve onto a straight conveyor section. From the straight conveyor, the carton travels in the long direction with the label facing one side onto the curve. As the carton travels across the curve, the inverted tapered rollers start to turn the lead end of the carton. When the carton completes the curve, it is in the long direction of travel but faces the other side. After the carton completes the curve, the original lead end of the carton becomes the tail end of the carton. When the carton leaves the 90° curve to assure singulation, angled skatewheel guide rails direct the carton onto the standard straight conveyor.

The disadvantages of this method are that it requires approximately 20 ft between the sortation conveyor and the take-away conveyor, requires a higher conveyor investment, and handles a standard size and shape of carton or tote. The advantages are that it handles a high volume, requires the fewest employees, and reduces lifting injuries.

TIPS AND INSIGHTS ON INSTALLING USED CONVEYOR

If you are considering installing used conveyor equipment in a facility, then following these guidelines can reduce potential problems: (1) Determine what is required in your facility. (2) Develop conveyor specifications and drawings. (3) Communicate with the conveyor vendors. (4) Review the equipment. (5) Follow installation tips and start-up tips.

Determine What Is Required in the Facility

After you have completed the conveyor layout design and drawings, determine the various conveyor types (powered or gravity; transportation or accumulation; incline or decline) and the required quantity for the system. These conveyor components and associated quantities are classified as new or used equipment and identified as to type, width, side guard height, floor-support or ceiling-hung, and other information.

Develop the Conveyor Specification

In addition to the above information, you must develop a complete conveyor specification package which is sent to selected conveyor vendors. In addition to the regular specification, you must identify the following: who is responsible for taking down the equipment, cutting or removing the anchor bolts, and filling in the boltholes? Who will bundle the equipment into bundles that are handled by a conventional lift truck? Who is responsible for loading and unloading the equipment? Who is responsible for transporting the equipment? Who will supply the necessary safety signs and provide the necessary material to bring the system to acceptable safety standards?

Communicate with Conveyor Vendors

Develop a conveyor vendor list that contains both used- and new-equipment dealers. Before you send a bid package to these vendors, telephone to assess their interest and on-hand inventory. Send conveyor drawings and written specifications to the interested conveyor dealers.

Create the conveyor vendor list from your past vendor lists, manufacturers' directory, and used-equipment directory or industry newspaper.

Review the Equipment

After you receive the completed used-conveyor bids, you must determine the best vendor to provide the equipment that satisfies your specifications and has the best economics. Prior to equipment purchase, obtain (if available) a complete set of drawings of the conveyor system, conveyor component listing, description, pictures, and written history of the equipment.

With this information, visit the location of the equipment. During the visit review the equipment for damage to the various components. If the equipment is standing in the facility and it is possible to have the conveyor system turned on, watch the conveyor system move product. Then review the maintenance records. Another good practice is to take pictures of the system.

Installation and Start-up Tips

Color of the Equipment. To complete your conveyor system, you could mix conveyor equipment from several conveyor manufacturers. Then the conveyor system in your facility will have a mixture of coatings (colors). If top management accepts this mixed-color situation, then this reduces the installation project. If top management requires one color for the conveyor beds and stands, then the conveyor components must be painted prior to installation in your facility.

Change Belts. If the used-conveyor system has not operated for a long time, to reduce start-up and operational problems, replace the old belts with new ones.

Motor Replacement. If the conveyor system has not operated for a long time, keep spare motors on site to replace bad motors. The motors could be bad because the motors have been idle and some components have not been oiled.

Equipment Testing.   To reduce operational problems, let the equipment operate without product for several days or a week. This practice has the best potential to identify conveyor problems. Generally, these problems occur within the first several days of operation.

Electric Motors and Electric Draw.   When you relocate a conveyor system from one geographic location to another, make sure that the electric power in the proposed facility matches the existing motor's electrical requirement. If these do not match, then there may be motor problems or slow belt speeds.

Identify Existing Equipment on Proposed Conveyor Layout.   After you receive the existing equipment layout drawings, identify each piece of equipment with discrete alphanumeric characters on these drawings. From your new layout drawings, this same identification is placed on each piece of equipment.

Prior to Installation, Clean Equipment.   Prior to the takedown or transfer of the used equipment from the existing facility, to ensure clean equipment and fewer maintenance problems in the future, have the equipment cleaned with air pressure. This practice keeps the facility clean.

Burden (Personnel) Carrier.   The next above-floor powered carton transport vehicle is the burden (personnel) carrier which is powered by LP gas, gasoline, or an electric battery and carries a rider. This vehicle has a load-carrying platform and an operator control area. Compared to other transport vehicles, the burden carrier takes a small load and is a variable-path vehicle. For these two reasons, the burden carrier is not considered a prime vehicle for the transportation function.

Lift Truck and Order-Picker Truck.   The next horizontal transportation concept is the powered lift truck or order-picker truck. These are manually controlled vehicles that travel over a variable path and handle unitized or carton loads. With the ability to make lift truck transactions in the staging area, these vehicles do not require another lift truck to perform an unload or load or set-down activity.

When the product has a short travel distance, is large, is bulky, or requires a special handling device, then the sit-down four-wheel counterbalanced lift truck is the preferred transport vehicle.

Since the maximum safety load capacity is one unit load, the lift truck transport productivity is increased by towing a train of carts. To pull a train of carts, the lift truck's rear chassis requires a hitch. A hitch on the lift truck increases the aisle width for a right-angle (stacking) turn.

The order-picker truck is similar to the lift truck. But it has a long right-turning-aisle requirement which makes the order-picker not the preferred vehicle for the transportation function.

Electric Single-Pallet Truck

The next transportation vehicle is the electric single-pallet truck (jack) that has one drive (steering) wheel and two load-carrying wheels. These load-carrying wheels are under the set of forks.

VARIOUS TYPES OF TRUCKS

The various types of pallet jacks are the end rider, midcontrol, and walkie.

The rider trucks have an electric battery and an operator's platform located in the middle or at the front end of the vehicle. In the operator's area are complete controls for the vehicle, and in the rear there is a set of forks with a backrest. The forks elevate and carry the load. The truck requires a pallet board, container, or skid and has the capacity to carry a 2000- to 4000-lb load. Typically, these vehicles have a 12- or 24-V battery power source. Some trucks do not have an operator's platform; therefore, the operator walks and controls the vehicle with the vehicle's hand controls. This vehicle has slow travel speeds; therefore, it is not preferred as a transportation vehicle. Backrests, fork entry wheels, and chamfered pallet boards reduce product falling from the pallet board and improve employee productivity.

The advantages and disadvantages of the electric truck make it a good vehicle to move product short distances in the dock area. In the dock area, the vehicle is used to load or unload delivery vehicles.

Manually Operated Rider Electric Double-Pallet Jack

The next transportation vehicle is the manually operated electric rider double-pallet truck (jack). This truck has two designs, the end rider type and the midcontrol type. The design and operational characteristics are similar to those for the manually operated electric single-pallet truck. The difference between the two vehicles is that the double-pallet truck can carry two unit loads, or 6000 lb. These unit loads are located behind the operator's platform. The pallet truck power source is a 12- or 24-V electric battery.

The disadvantages and advantages of the double-pallet truck are the same as those of the powered single-pallet truck. Additional disadvantages are that it requires a wider intersection (aisle), it is difficult to travel up grades, and it requires more employee training.

An additional advantage is that the vehicle is more productive because it transports two unit loads, which makes it one of the most productive and low-cost manually controlled transport vehicles.

Manually Operated Powered Rider Tugger

The next vehicle is a manually operated powered rider tugger, a variable-path vehicle. The vehicle's power source is LP gas, gasoline, or an electric battery. The electrically powered tugger is best used indoors. The tugger capacity is a 10,000-lb rolling load or 750-lb normal drawbar pull for a train of carts. Typically, a train has four to five carts each of 2000 lb.

When the travel distance is over 300 ft and the volume is low to medium with a varied travel distance and several load and/or unload points, then the tugger or train of carts is a cost-effective system.

The most commonly used cart in a trailer train is the nontilt type that has all four wheels on the floor.

CART TRAIN

When you are considering a tractor (tugger), an AGV (automatic guided vehicle), and lift truck cart trains, cart (trailer) steering and the hitch or coupler are two important factors. These factors determine the number of carts which determines the volume of product transported between two locations, employee productivity, and turning-aisle requirements.

The two possible concepts for a cart as a transport vehicle in a warehouse and distribution facility are as a manual push/pull cart or as a train of towed carts. Prior to cart purchase, the specific material-handling role of the carts does have a direct bearing on the type of steering for the cart. When a cart is manually pushed or pulled in a facility, the cart is easily moved between two locations. When a cart is used in a train that is towed by a powered vehicle, it must have excellent trailing characteristics.

In a warehouse and distribution facility, the cart steering consists of the wheel or caster arrangement and location of wheels or casters on the cart's underside. These two factors control the cart's direction of travel and its ability to make 90° and 180° turns.

CART STEERING

A cart is designed with one of these steering concepts:

- Caster steering
- Two-wheel steering
- Knuckle steering
- Four-wheel steering
- Fifth-wheel steering (single or double)

Caster or Wheel Steering

In caster steering, there are four wheels under the cart platform which has a fixed drawbar. The centers of these wheels are arranged in a rectangular pattern on the four corners of the underside of the platform. The front two casters or wheels are the swivel type and are the casters or wheels that turn the cart. The rear two casters or wheels are the rigid type. This arrangement of swivel and rigid casters provides the proper lead and avoids any whipping action of the cart train.

Caster steering is the most common steering in the warehouse industry. It provides easy manual movement and excellent trailing characteristics for a 5000-lb load.

Knuckle Steering

In knuckle steering, four wheels are arranged in the four corners of the rectangle on the underside of the cart platform. These wheels are spaced on equal centers. This steering concept is sometimes referred to as the *Ackerman steering concept.*

Knuckle steering has two designs: two-wheel and four-wheel.

Two-Wheel Knuckle Steering.   In two-wheel knuckle steering, there are four wheels under the four corners of the rectangular cart platform. The two rear wheels are on rigid casters. The two front wheels are connected by a bar. The wheel-connecting bar is connected to the drawbar. The drawbar is turned by the towed vehicle, and this simultaneously turns the two front-wheel knuckle mechanism. With the two wheels turned, the cart turns in the direction of the tow vehicle.

Four-Wheel Knuckle Steering.   Four wheels are spaced on equal centers on the four corners of the rectangular pattern under the platform. The front two wheels have a knuckle mechanism as the two-wheel knuckle steering type. The rear two wheels have the knuckle steering assembly that is connected with a bar to the front knuckle assembly. The connections between the two wheel assemblies are on opposite sides of the tow bar center.

In this arrangement, the front two wheels turn in the same direction as the tow vehicle, and the rear wheels turn in the opposite direction to the tow vehicle. This wheel combination provides four knuckle steering with accurate trailing characteristics and load stability. But it is difficult to manually maneuver the carts between locations.

Fifth-Wheel Steering

Single Fifth-Wheel Steering.   Four wheels are equally spaced on centers under the four corners of the rectangular platform. The rear two wheels on centers are the rigid type and are on a fixed axle. The front two casters or wheels are rigid and are connected to a front axle that turns. The axle is attached to and is supported by a fifth-wheel (turntable) assembly. The tow vehicle drawbar is connected on both sides of the fifth wheel.

As the tow vehicle turns, it exerts a force on the front fifth-wheel axle, which turns the cart in the same direction as the tow vehicle.

Double Fifth-Wheel Steering.   The front wheel arrangement is the same as that for the single fifth-wheel design. The rear wheels are supported on an axle that is attached to a second fifth-wheel turntable. The front and rear axles are connected via a bar. This connecting bar is on opposite sides of the drawbar. In this arrangement, the front wheels turn to follow the tow vehicle, and the rear wheels turn in the opposite direction.

Double fifth-wheel steering does not provide accurate trailing characteristics, requires a relatively large turning radius, and is preferred for heavy loads of 8000 lb plus.

CART COUPLER (HOOK)

Two devices are used to couple (hook) the carts to the tow vehicle or to another cart: a hitch and a coupler.

The coupler is attached to the tail end of the tow vehicle or cart. The hitch is attached to the lead end of the cart and is secured to the coupler.

Manual Hookup

Manual hookup requires an employee to physically move the lead (tow) vehicle in front of the trailing vehicle. This requires the employee to insert the tow vehicle's coupler hook, pin, and top lever into the eye or hole of the hitch. There are several means of manual hookup: hook and eye, spring-loaded pin or clevis, pin and clevis, and pintle hitch.

Manual Hook and Eye.   In the hook-and-eye manual hookup, a drawbar is secured to the lead end of the cart. The drawbar is 18 to 19 in long, is raised and lowered, and has a hook that is inserted into the coupler's eye. The coupler is secured to the middle rear end of the tow vehicle or cart. The coupler has an eye of sufficient size to accept the hook. One option is a safety latch on the hook.

This combination has the ability to handle a weight range of 2500 to 5000 lb.

To operate, an employee moves the tow vehicle's (cart) rear chassis (coupler) close to the lead end of the cart hook. The employee raises the drawbar, moves the cart, and inserts the hook into the eye. Then the employee moves the cart train with the powered vehicle.

Manual Pin and Clevis.   The drawbar has a closed-loop hitch on the lead end of the cart. The rear (chassis) end of the tow vehicle or cart has the pin and clevis. The clearance between the pin and the vehicle (cart) and the space between both clevis members accept the drawbar. To ensure that the pin is available, the pin is secured by a chain or string to the tow vehicle.

The clevis is in the form of C metal component which is round and is secured to the tow vehicle. Each round member of the clevis has a diameter that will accept the pin.

To operate, the vehicle should be close to the cart's drawbar. The operator removes the pin from the clevis and inserts the drawbar between the two members of the clevis. After the drawbar loop is inside the clevis, the pin is inserted into the clevis hole, and a vehicle moves the cart.

Spring-Loaded Pin and Clevis.   The spring-loaded pin and clevis have design features and operational procedures similar to those of the pin and clevis except that a second clevis holds a spring-loaded pin.

Manual Pintle.   The pintle hitch consists of a lower and upper member that are secured to the rear of the vehicle (cart). There is a gap between the two members. The lower member is fixed. The upper member is attached to the pintle base which is raised and lowered onto the lower member. The gap between the two members has the clearance to accept the drawbar loop. The upper member has a safety latch.

To operate, the employee raises the upper member and places the tow bar into the lower member's cavity. After the loop is in place, the upper member is lowered and locked onto the lower member. In this position, the cart is moved to a new location.

Manual Safety Latch Cap and Ball.   The ball is secured to the lead vehicle, and the safety latch cap (drawbar) is secured to the trailing vehicle. The safety latch cap fits over the ball. With the safety latch in the raised position, the cart is released.

The lead vehicle is moved close to the cart. With the safety latch cap in the unlocked (raised) position and a raised drawbar, the cap is lowered onto the ball. When the cap is on the ball, the safety latch is secured and the cart is moved to a new location. The load capacity is 2500 lb.

AUTOMATIC COUPLER (HOOKUP)

The automatic coupler (hookup) requires an employee to move the tow vehicles or cart bail onto the cart jaw. When the cart is to be released, this arrangement does not require the employee to leave the tow vehicles.

There are two types of automatic coupler: the spring-loaded jaw and spring-loaded bail and the spring-loaded jaw and counterweighted bail. Each has a foot pedal release mechanism that permits the trailing cart to become detached from the lead vehicle.

Spring-Loaded Jaw and Spring-Loaded Bail

The spring-loaded jaw and spring-loaded bail are two components. The jaw is secured to the trail vehicle's drawbar. The distance to the jaw's interior is 5 to 7 in. The jaw has fixed upper and lower members. The lower member has a spring-loaded lever that locks the bail to the interior. By applying pressure, the spring-loaded bail is secured to the lead vehicle's rear (chassis) end, and the lever is pressed down. The length of the bail to the interior ranges from 8 to 10 in, and the bail has a fixed bracket. With the bracket, there is a spring-loaded loop that moves forward and backward.

To operate, the employee backs the tow vehicle up to the cart's spring-loaded jaw and continues its reverse movement until the bail has passed the spring-loaded jaw lever. In this position, the cart is towed to a new location. To disengage the cart, the jaw's pedal lever is depressed and the cart or vehicle can be moved.

Spring-Loaded Jaw and Counterweighted Bail

The spring-loaded jaw and counterweighted bail comprise the second automatic coupler (hookup). The spring-loaded jaw is the same as the jaw, but its length is approximately 7 in. The counterweighted bail is secured to the tow vehicle or lead cart. It consists of a fixed bracket with a bail that is raised and lowered onto the fixed bracket. The distance to the interior side of the bail is 10 in.

To operate, the operator makes sure that the bail is in the lowered position and backs the tow vehicle toward the cart. This direction is continued until the bail passes the spring-loaded lever. In this position, the lead vehicle pulls the trailing cart to a new location. At this location, the operator depresses the foot pedal that releases the spring-loaded jaw. This action releases the trailing cart from the lead cart or towed vehicle.

AUTOMATIC GUIDED VEHICLE

The next above-floor powered horizontal transport method involves the automatic guided vehicle (AGV). The AGV is an electric, battery-powered, driverless vehicle that has a load-carrying surface or towing ability. The AGV follows a fixed closed-loop path between two locations. Stop (address) locations and spur (branch) lines are segments from the main traffic line and run parallel to the main traffic path. The return line has a separate path or runs parallel to the main traffic path. The travel path spur and return line design permit a vehicle to unload or load at a stop while other vehicles travel in the opposite direction. All AGV types are equipped with safety bumpers, and multiple vehicles travel on one guide path. You can extend or change the guide path, as required, up to a nominal 5000 ft. The unique features of the AGV are as follows:

- Weight and how AGV moves: light transporter (mini-AGV), towing, pallet truck, unit load
- Guidance of AGV: inductive (wire-guided), optical (paint or tape), laser beam
- How the AGV performs tasks: one-directional, two-directional, four-directional

The AGV is the preferred method where and when the following criteria are met: (1) A transport activity is performed by a lift truck or pallet truck or tugger with a train of carts. (2) Delivery is between two locations. (3) The delivery path is fixed. (4) Deliveries are frequent and on schedule. (5) The load is loaded or unloaded automatically and is lift-truck-assisted. (6) The load is within the design parameters (cartons or pallet load from 2000 to 12,000 lb). (7) There is a smooth or level floor (grades are gradual), (8) The delivery route is a long distance or is a closed loop.

Various AGV Control Methods

The AGV is driverless. Control of AGV movement for travel is programmed by three different methods: basic controls, advanced controls with microprocessor, or microprocessor controls to interface with another system.

The basic control method is the simplest control method and is implemented in a system that has one or two vehicles, has less than four address locations, and has a total guidance path of less than 5000 ft.

The advanced control method with microprocessor controls is used for more complex systems that have a guidance path of more than 5000 ft, a vehicle travel path that is more complex or requires blocking, if the activity requires more than four vehicles and more than four automatic unload and load stations.

The microprocessor control method is the most sophisticated control method, and it interfaces with another automatic system. The transportation vehicle performs pickups and deliveries of a load to and from another automated system. These pickups and deliveries require on-time performance and have at least the same operational design parameters as the advanced control method.

Various AGV Dispatch Methods

There are two ways to dispatch AGVs: the fundamental human dispatch method and via remote communications to a host computer (automated system).

To manually dispatch an AGV, there are three different ways to program the AGV controls: through the toggle switch, thumbwheel switch, or pushbutton numeric pad.

With AGVs there are several basic mechanical and electrical features that improve vehicle efficiency and safety. The first is a safety bumper on the vehicle's lead end. When the bumper comes in contact with an object or employee, this stops the vehicle.

Running lights are illuminated when the vehicle is traveling on the guidance path. This is a colored light that is readily noticed by employees near the travel path.

An AGV system has a control panel and a mimic display panel that shows the operational status of each vehicle on the guide path and permits you to start or stop the system. These panels indicate where each AGV is located on the guide path and the safety stops, and they contain other devices and controls of the AGV system.

An off-line stop device is an emergency device that reduces damage to product or material handling equipment and the facility. When an AGV deviates (strays) from the guide wire by more than 2 in, the device is activated and the vehicle is immediately stopped.

An UPS (uninterruptible power source) provides temporary electric power to the AGV wire guidance system during a brownout or electric power company failure. It provides power long enough to ensure that all AGVs on the travel path are returned to the main dispatch station. This reduces damage to the product, material handling equipment, and building.

AGV Anticollision Methods

The most sophisticated AGV systems have vehicle anticollision controls which are considered a blocking system. The blocking system permits several vehicles to travel on one guide path. The anticollision system is on board the vehicle or built along the guide path. The two methods are the optical anticollision method and the zone blocking methods (point-to-point blocking, continuous blocking, and computer zone blocking).

Optical Anticollision Methods.   The optical controls on each vehicle consist of a light beam (source), receiver, and reflective target. On the front end of each vehicle, a light device produces a light beam at a fixed level and directed to the front of the AGV. A receiver on the front of each vehicle is at a fixed level to receive the reflected light. A reflective target at the same elevation as the light beam is located on each vehicle's rear end. On a multivehicle transport system, when the second (trailing) vehicle light beam is returned (reflected) from the first (lead) vehicle's target, then the on-board controls of the second vehicle stop the second vehicle's travel. This anticollision system is used with unit-load, miniload, and towing AGV vehicles that have a structural member on the rear or a cart for the attachment of the light source, light receiver, and reflective target. The system is not preferred on a driverless pallet truck because there is not always a reflective target on the rear of the vehicle. If a reflective target is temporarily attached to the rear of the unit load, then the anticollision system functions as a normal system.

Zone Blocking Method.   In the zone blocking method, the guide path is divided into various zones (segments) of sufficient length to contain a vehicle (train of carts) plus a margin of safety. The anticollision control program permits only one vehicle per zone at a time; therefore, if a vehicle occupies a zone and another vehicle enters that zone, the entering vehicle must wait until the first vehicle exits the zone.

The three zone blocking methods are point-to-point, continuous, and computer.

Point-to-Point Blocking.   The point-to-point zone blocking method consists of zone-sensing devices that are mounted on physical facility objects (columns) adjacent to the guide path. As a vehicle passes from zone A into zone B, the vehicle actuates the sensing devices in the zone B control package to prevent another vehicle from entering zone B and permits another vehicle to zone A

Continuous Blocking.   The continuous blocking method consists of an on-board blocking signal transmitter, on-board receiver, and auxiliary multi-loop wire that is buried in the floor under the travel path. Each looped wire section becomes a zone. As each vehicle travels on the guide path, it sends a signal into this looped wire that is detected by the second vehicle's blocking devices. When a vehicle is in one zone wire loop and a second vehicle sends a signal on the loop wire, then the second vehicle's sensing device picks up the signal and stops (queues) until the first vehicle leaves (clears) the zone or travels from the loop wire.

Computer Zone Blocking.   In the computer zone blocking system, each vehicle is equipped with an on-board microprocessor and on-board transmitting and receiving antenna and has the ability to communicate with a host computer or another smart vehicle. The floor area under the travel wire guide path contains magnets or plates that identify each zone. As the vehicle passes over the magnet or plate, it sends a signal through the guide wire or by FM radio link. This method links the vehicle zone status to a second vehicle and to the host computer. When one vehicle is in a zone, a second vehicle that wants to enter the occupied zone is restricted by its on-board microprocessor or host computer.

AGV Design Parameters

When you are considering an AGV transportation system in your warehouse facility, you are required to define these fundamental design parameters. (1) There must be good traction for the wheels. (2) The floor is hard. (3) Metal is not within 2 in of the wire. (4) The guide wire path avoids expansion joints. (5) If expansion joints are passed, the wire is looped. (6) All grades are 10 percent or less in slope. (7) There must be a UPS for the system. (8) Determine and identify the number of turns in the system that are uncompensated (tangential) turns which have a shorter width to turn, that are compensated turns which have a wider width to turn, and that are mitered or 90° (right-angle) turns.

(9) Calculate the vehicle travel requirements. Basic is used for a one- or two-directional vehicle system with a simple manual dispatch method. Advanced is used for two-directional smart vehicles with a host computer or automated dispatch system. Simulation is used for the sophisticated system with a host computer or automated dispatch method that interfaces with another automated system. (10) Calculate the number of required vehicles on the vehicle loop travel time (travel from a start location and return to the start location), an 80 percent allowance for equipment utilization, and dispatch and on-line placement times (include load and unload time). (11) Determine the guidance system which is inductive. It has a floor-embedded wire that carries a low (40-V) electric current and sensors on the vehicle's undercarriage. This is the most common guidance method: Chemically treated tape or paint and an ultraviolet light beam from the vehicle underside stimulates fluorescent particles in the path and transfer the light beam back to the vehicle's underside sensor. Or the laser beam that has red light sent from a light on the mast of the vehicle is reflected back to the vehicle from strategically placed reflective targets.

Various AGV Types

The various AGV types are miniload, towing, driverless pallet truck, and unit load.

Light-Load (Miniload) Transporter.   The light-load (miniload) transporter is more commonly known as the miniload AGV. Compared to the other AGVs, the miniload AGV has a travel speed of 100 ft/min, a narrow travel path, and a narrow turning radius. The load-carrying surface is designed for small loads (cartons or containers), and the majority of the guidance systems are wire guidance systems.

AGV Towing Vehicle.   The second AGV transportation vehicle is the towing vehicle (Fig. 9.8a). When you have frequent deliveries (high volume) over a long travel distance (more than 300 ft) and several carts (loads) are assigned to one warehouse location, then a towing AGV should be considered. Also, another important feature is that one AGV with a train of carts can drop or pick up loads at multiple locations. This feature requires an employee to uncouple or couple the carts and to program the AGV towing vehicle for travel to the next location. If a towing AGV with a train of carts is to make a turn, consideration must be given to the train-of-carts turning requirement. A second consideration is that the loading or unloading spur (siding) must be long enough to accommodate the vehicle and train of carts. This design feature permits other towing AGVs on the main travel path to pass the AGV that is on the siding. A third consideration is that the vehicle is manually controlled for cart hookup and unhooking.

Driverless Pallet "Stop-and-Drop" Truck.   The driverless pallet truck carries one or two pallet loads to one or two drop (warehouse) locations (Fig. 9.8b). This vehicle has the name *stop and drop* and is a driverless vehicle that travels in one direction. It can be manually controlled. Manual control is required for the pallet truck to travel in the reverse direction for pickup of two pallet loads. After the loads are on the pallet truck forks, the employee enters the dispatch code or codes in the vehicle pushbutton control panel and dispatches the vehicle. Arriving at the assigned warehouse location, the driverless pallet truck travels onto the siding, stops, and drops the pallet loads at the location. The driverless pallet truck travels on the main travel path to perform, if one load was dropped, another drop function or continues to the dispatch location.

The vehicle is battery-powered and has an operator control area, safety light, and safety bumper.

Unit-Load AGV.   In the automated warehouse, the unit-load AGV is the most common mobile vehicle used to transport unit loads between an AR/RS pickup/delivery station and other warehouse locations (Fig. 9.8c). The unit-load AGV has a battery compartment, three to four wheels (one at least for steering), safety bumper, control device, path sensor device, and load-carrying surface. The three important nonvehicle components of an AGV system are the vehicle guidance system, warehouse location of the pickup/delivery stations, and control or mimic panels.

On-board rechargeable batteries power the AGV on a guide path. The most common guide path uses wire embedded in the floor. This is the inductive guidance method in which an electric charge travels through the wire. The sensor devices under the AGV maintain travel control via electric impulses that direct the AGV travel on the guide path. The other guidance systems are the optical method, which uses chemically treated tape or paint that is on the floor, or the laser beam method, which uses a light beam sent from the vehicle to reflect from reflective targets.

1 The Stop & Drop vehicle is manually loaded by the operator who backs into two pallets and lifts them off the floor.

2 The operator programs the vehicle for two destinations, one for the Stop & Drop spur and the second for its return to a loading area spur. The vehicle is driven to the guide path and dispatched. This completes the manual portion of the pallet delivery cycle... the rest is all automatic and requires no driver.

3 The vehicle arrives at its destination spur, stops, lowers its forks and deposits the pallets on the floor, pulls out of the pallets and returns to the main guide path—all automatically.

4 The pallet delivery cycle is completed when the empty Stop & Drop vehicle automatically returns to the programmed loading area spur. It is now ready to be manually loaded again.

(a)   (b)

FIGURE 9.8   (a) Towing AGV (*Courtesy of Barrett Indl. Trucks, Inc.*); (b) stop and drive (*Courtesy of Barrett Indl. Trucks, Inc.*).

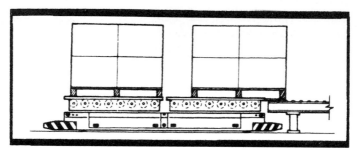

(c)

FIGURE 9.8   (*Continued*) (*c*) Unit-load AGV. (*Courtesy of Litton Indl. Automation.*)

With three or four wheels, the AGV travels in the forward or reverse direction. In some newer AGV models, the load-carrying surface is designed to handle a wide variety of unit loads.

The next important unit-load AGV feature is the load-carrying surface, or how the AGV handles the unit load. The first method is to tow the unit load on a cart, and it was reviewed earlier in this section. The second method is to carry the unit load.

Various AGV Load-Carrying Surfaces

The AGV unit load carrying surface is equipped with any type of device that is placed onto a manual pallet or lift truck. Some of the most common unit load carrying surfaces are (1) fixed-position single-unit-load carrier that requires a lift truck (device) to deposit or lift the load from the vehicle's carrying surface, (2) lift and lower the single-unit-load carrying surface for unassisted loading or unloading of unit loads, (3) gravity roller surface for side-assisted loading and unloading unit loads, (4) powered roller surface for automatic loading and unloading of unit loads, and (5) a set of forks to elevate unit loads from the floor to a raised position on a platform.

The AGV pickup/delivery (P/D) station is an important design criterion and is matched to the AGV unit load carrying surface and travel direction capability. The majority of the distribution facilities have elevated P/D stations and are designed to permit the AGV to pick up and deliver a unit load.

Various AGV P/D Stations

There are several types of P/D stations. (1) In the manual type, a lift truck lifts or deposits the unit load. (2) The AGV automatically loads or unloads onto a structural stand, which requires the AGV to turn and back up to the stand. (3) The conveyor stand has a gravity or powered roller conveyor interface with the AGV. (4) In the slave-driven stand, the AGV provides the power to transfer the unit load. (5) In the floor or rack station, the unit load with fork openings is placed within a specific marked floor location or in a rack position.

Air Cushion Pallets

The air cushion pallet is a specially designed solid pallet board that forces air between the bottom of the carrying surface and the finished floor surface. As the air is forced from the bottom carrying surface through a plenum and through holes onto the finished floor, an air bubble forms. The bubble raises the load-carrying surface and the pallet board is moved between two warehouse locations.

For best results, the load weight should be in the center of the pallet board's load-carrying surface. The air cushion pallet board handles heavy and difficult-to-handle loads over a flat, smooth floor. With no load-carrying wheels (such as lift truck wheels), the load capacity of the floor is lower than that of the normal lift truck floor. But the capital investment per air cushion unit is higher than that for the standard wood pallet board and lift.

ABOVE-THE-FLOOR POWERED CARTON OR PALLET HORIZONTAL TRANSPORT

The next above-floor powered carton or pallet horizontal transport methods are as partial or in-floor systems. These are fixed-path systems that have control panels to start and stop the system, E stops (emergency stops to prevent employee injury or equipment or building damage), and a mimic display panel to show the status of the system.

These transport methods include in-floor towline, inverted power and free conveyor, and S.I. Cartrac.

In-Floor Towline

The in-floor towline is very common and travels at 60 to 90 ft/min (Fig. 9.9a). It is used to transport a high volume of goods (hand-stacked or unitized) between two locations.

The in-floor towline is considered a fixed-path system that is designed with multiple automatic divert and infeed locations (spurs and sidings) and that is installed in a facility with inclines or declines having a 10 percent grade or less slope. Also, towline carts are transferred from one chain to another. Throughout the towline system, a tow cart is manually removed or placed onto the chain.

The in-floor towline has two major components: a towline chain and towline cart.

There are two types of towline chain: the conventional heavy-duty towline and the low-profile towline.

The conventional towline chain is in a pit 7 in deep, 6 in wide at the top, and 3 in wide at the bottom with cleanout pits that are 4-in-deep holes.

These cleanout pits are periodically checked and cleaned out. Also, along the chain path there are automatic oilers at various locations to ensure proper lubrication of the chain.

(a)

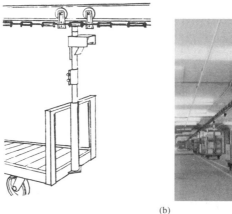

(b)

FIGURE 9.9 (*a*) In-floor towline (*Courtesy of Rapistan Demag Corp.*); (*b*) overhead tow conveyor (*Courtesy of Plant Engineering Magazine.*).

The low-profile towline chain is in a pit 3 in deep and $2\frac{1}{2}$ in wide. The low-profile chain has a lower installation cost and is installed in a thin or average-width floor. If required, this towline can be designed for installation of a mezzanine. If the mezzanine has a metal graded floor, then the cart travel path has a solid metal surface.

With a towline, there should be a good preventive-maintenance program. And employees should be prevented from sweeping trash into the slot trench for the chain.

Inside of these chain paths are bottom and top wear bars which are strips of hardened metal. These wear bars are the tracks for the towline chain. As the chain travels across the wear bar, it prevents excessive chain wear.

The towline chain is pulled by a motor-driven sprocket that is located on a straight run and is in a pit. In this pit is a takeup device that allows the towline slack chain to be taken up. Slack chain results from wear and weather changes. A tight chain ensures good performance from the towline.

Chain Links and Links with a Cavity (Dog).   The towline consists of a series of motor-driven chain links that travel in a track, and it is designed to travel over 90° or 180° turn or to incline or decline over grades of approximately 10° to 15°. It is composed of numerous chain links, and every 20 ft there is a link with a cavity (dog). For long-lasting wear, these chain components are hardened steel. At every 20 ft or as determined by the cart length, a dog is located; this is the towline chain component that pulls the cart. The dog is a specially designed link with a cavity to hold the cart's tow pin. The motor, sprocket, and take-up device are located in a pit. The motor and sprocket pull the chain through the system, and the towline is designed to pull the maximum number of carts on a chain. The take-up device provides the means to increase or decrease the chain length, which adjusts the chain tension.

The spur is a left or right nonpowered track that intersects the chain at 45° angles. It is considered a branch line that permits nonpowered cart travel to a warehouse location. The chain sensor device (pad) is in the floor prior to the spur. After the spur is activated by the cart's coded sensor, the spur diverter (floor-mounted lever) slides across the chain path. The lever being in this position and the chain pulling the cart forward together force the cart pin to become disengaged from the dog and to flow into the spur's open path. The spur allows the diverted cart to clear the main travel path, which permits continuation of other cart traffic on the main line. All spurs or branch lines are designed for a specific length and have full line sensors. When activated by a cart, these sensors do not allow the diverter mechanism to divert additional carts onto the spur. Therefore, when you purchase used carts, be sure that the carts match the towline system specifications.

Towline Cart

The towline cart is a four-wheel vehicle that has a load-carrying surface and a code bar on its front end. The two standard carts are 6 and 10 ft long and 3 or 4 ft wide. The most common load-carrying surface is a flat wood surface with a metal border; however, the load-carrying surface is designed to accept all loads handled by a manual or powered lift truck or manual pushcart.

The three unique features of the towline cart are the tow pin; selector pin rack, selector pin, or bar-code label or scanner; and location and types of wheels under the cart. These features make the towline cart different from the tow tractor, lift truck, manual, and AGV carts.

Tow Pin.   To prevent excessive wear, the tow pin is a hardened steel rod that is set inside a sleeve. The pin sleeve is attached to the lead end and the center of the cart. As required, it allows the tow pin to be removed from a tow cart pin sleeve. The tow pin has two positions, nonengaged and engaged.

In the nonengaged position, the tow pin is in the secured position that keeps the pin in a raised position above the floor. This nonengaged position is maintained by a cap (retainer) on the end of the tow pin. This prevents the pin from mistakenly being placed in the tow chain. When the cart is not being used in the system, this is the preferred position to prevent pin damage.

The engaged position is in the lowered predetermined position. The exact pin length is a specification based on the track and dog depth. The rod retainer or cap permits the tow pin to reach this predetermined depth. When the tow pin is in the engaged position and in the track, as the empty dog arrives at the cart, the pin automatically slides into the dog (open link) and the tow chain pulls the cart to the assigned location.

Selector Rack, Pins, or Bar-Code Labels or Scanners.   The selector rack and selector pins are at the lead end of the cart, and in combination these two devices allow an employee to assign a divert address to the cart. The selector rack(s) is (are) on each side of the tow pin (center of the cart). A series of holes (12) permit the selector pin to hang in a vertical position which is approximately floor level. Each position has an alphanumeric character in the front. The pin's hanging arrangement from the specific selector rack hole corresponds to a discrete series of sensors in the spur (floor-mounted) sensor device.

One selector pin rack handles up to 12 address locations, two selector pin racks can handle up to 150 locations, and three pin racks can handle several hundred locations.

The selector pin (probe) is a metal rod with a flexible or spring-loaded magnetized tip and metal cap (retainer). To prevent pin damage, the pins are attached to the selector pin rack with a string or chain. When not in use, the pins are placed in a tray that is behind the selector pin rack.

The metal cap and selector pin rack permit the pin to hang vertically at specific heights to have the magnetized tip at approximately floor level. This is the proper height to activate the floor spur sensing device which activates the diverter to transfer the cart from the towline.

An alternative to the selector pin is the bar-code label on the lead (front) side of the cart. The bar code is read by a scanning device that communicates with a controller. The controller triggers the divert device to direct the cart from the main line path to the spur to another line.

Wheels of Towline Cart.   The towline cart has four wheels. The two rear wheels are rigid casters or wheels and are located on the two corners under the rectangular load-carrying surface. The two front casters or wheels are the swivel type and are located under the two front corners of the rectangular load-carrying surface. However, compared to the rear wheels, with an underside plan view of the cart, the front-wheel locations are interior. This feature reduces drag as the cart turns the corners of the towline path.

During towline operation, to prevent injuries to personnel and damage to the material handling equipment or the building, a cart is not pushed or towed by another cart that is pulled by a dog.

To operate a towline, the employee places the cartons or pallet load onto the cart's load-carrying surface and manually pushes the cart on the towline track. At the towline track, the selector pin is inserted into the assigned selector rack hole. The employee waits for an open link (dog) on the towline (every 20 ft) and pushes the cart over the track. In this position, the tow pin is released and rides on the top of the track, and the dog automatically engages the tow pin. When this occurs, the cart is moved by the tow chain to the new warehouse location that corresponds to the selector pin arrangement.

As the cart approaches the assigned spur, the magnetized selector pin passes over the floor-mounted sensing device (the bar code passes a reader) and activates (trips) a floor-mounted divert device (lever) which extends across the cart path. When the tow pin of the cart strikes the lever, the tow pin slides from the dog's mouth (cavity) and the cart travels onto the spur away from the path of the next cart. At this location, the employee removes the cart from the system, removes the pins from the engaged rack position, and transfers the product.

Inverted Power and Free System

In this fixed-path system, there are multiple load and unload locations. The inverted power and free method consists of an in-floor electric chain that is in a track and has a series of load-carrying surfaces. These surfaces travel past SKU transfer locations.

Most material handling professionals recognize that the power and free method is a conventional overhead power and free conveyor which is on the floor.

S.I. Cartrac

The next in-floor powered horizontal transportation is the S.I. Cartrac method. It is more commonly considered an order-pick system, but in many distribution facility layouts, it provides the transportation link between two locations. S.I. Cartrac is a fixed-path method that has a closed-loop track and a series of pallet load-carrying surfaces. The load-carrying surface is stopped and started by the employee at any warehouse location. S.I. Cartrac links the receiving area to the storage area and the storage area to the pick area.

OVERHEAD NONPOWERED HORIZONTAL TRANSPORTATION

The next carton or pallet horizontal transport method is overhead and nonpowered. This consists of the manually pushed trolley and hanging basket, which handles cartons. The design and operational characteristics are the same as those for the single-item manually pushed trolley.

OVERHEAD POWERED HORIZONTAL TRANSPORTATION

The unique feature of this group is that the product is conveyed above the floor between two locations. This permits the floor below to be used for other warehouse activities. When these transport methods are used in the warehouse, the elevated transport vehicle must have side guards, underside guards, E stop buttons, and rollers or skatewheels secured to the bed section. The various means of transportation are (1) the trolley and basket, (2) conveyors (belt, pusher bar, roller, skatewheel, apron, slat, bucket, and strand), (3) monorail, (4) powered and free, and (5) overhead flow conveyor.

Powered Trolley and Hanging Basket

When the trolley and basket are used in this situation, the overhead trolley basket requires protective underside netting or wire mesh.

Powered Conveyors

The overhead carton or pallet load conveyor has the same design parameters and operational procedures as the floor-level conveyor. When these conveyors are used overhead, they require underside guarding, side guards, and rollers that are secured to the channels (conveyor beds).

Overhead Free Conveyor

In this fixed-path system, a series of individual trolleys ride on a track and are propelled by a motor-driven chain. The load carriers are suspended from a second trolley track that runs on a free (second or independent) track. This second track is parallel to and below or beside the chain-carrying power track. The power chain dog mates with a similar extension on the load-carrying trolley. The mating of these dogs and extensions permit the carrier to be pushed forward on the track. As required, the carriers are accumulated and disengaged from the chain onto the spur.

In the most common application, the carrier is suspended from two trolleys that are coupled by a towing bar. A retractable dog on the front trolley is engaged by a pusher on the powered chain, which moves the carrier. When the carrier dog is retracted, the pusher and dog clear each other and the carrier slides to a slow stop. Speeds of 40 to 60 ft/min are achieved.

Self-Powered Monorail

The self-powered monorail is a fixed path system. The monorail consists of an electric or air self-powered carrier that rides on an overhead track. The monorail carrier can travel at speeds of 220 ft/min on an electric track that is a beam.

These self-powered carriers are programmable and are diverted onto sidings (spurs); they have anticollision devices; and when empty, they can incline or decline between elevated levels on a spiral track.

Overhead Tow Conveyor

The overhead tow conveyor is a fixed-path system. The overhead tow conveyor consists of an overhead motor-driven chain and trolley. The trolley can pull a pallet truck, dolly, or cart. The trolley is equipped with a pusher dog that has a sling or strap which extends downward and engages a four-wheeled vehicle's handle or rigid mast. The overhead trolley is equipped with a code device, and the track is equipped with a sensing device and a divert mechanism that switches the trolley from the main line to the spur.

The overhead powered conveyor has the same design parameters and operational characteristics as the light-duty in-floor towline.

VERTICAL TRANSPORTATION SYSTEMS

Many existing and future facilities have multiple levels (floors). The use of multiple levels takes advantage of the airspace above or below ground level. In these facilities the product is transported between the floors. As with all product transportation systems, the material-handling equipment used in the facility is designed for the product handled, available space, and operational factors. Compared to overhead powered systems, the additional safety features are the elevated floor penetration that has handrails, kickplates, and fire protection.

The objective of this chapter section is to identify the material-handling equipment applications and technologies that are available to move product between warehouse levels.

The remainder of this chapter separates the vertical transport methods into three groups: single-item product, carton product, and pallet load product.

Single-Item Vertical Transportation

Single-item vertical transportation is separated into nonpowered and powered. The vertical single-item transport methods include human carry or slide and chute, gravity skatewheel or roller conveyor, belt conveyor, miniload powered chain with trolley or carrier, chain system with fixed hooks, powered squeeze conveyor, air conveyor, magnetic conveyor, continuous vertical lift, AS/RS, vertical reciprocating conveyor, vertical carousel, pneumatic tube, screw conveyor, vertical trolley lift, sandwiched conveyor, lapped stream conveyor, gripper conveyor, whiz lift, freight elevator, power and free conveyor, and ramp with mobile powered equipment.

Nonpowered Vertical Transportation

Nonpowered product transport methods require human power or gravity to move product between two levels. When gravity transports product, it is an uncontrolled product flow. To reduce product damage, on the floor it requires a longer runout distance and jam controls. Whenever possible, gravity flow is the preferred method because of the low cost.

Human-Carry. An employee carries product and walks on a ramp or stairs between floors. This method is basic but is not preferred because of the physical effort required to move product between locations.

Slid or Chute. The slide (chute) is a material handling device that transports loose, individually packaged, or containerized product by gravity from a higher to a lower level. Because of line pressure buildup from jams or product accumulation, light, fragile, crushable, and heavy products are not mixed in the same chute. A slide consists of sheet-metal or coated-metal, plastic, or wood material that has a charge and discharge location. The typical slope for a slide is 10° to 30°. A greater slope is required in humid or outdoor environments.

Since product has uncontrolled flow in the slide, jam control devices are installed through the underside or sideguards of the slide. When the jam control device is blocked, the jam control system shuts down the infeed or product flow at the charge end of the chute. When a slide transfers product to a workstation (packing station), the discharge end requires an end stop (door). To reduce line pressure and provide additional product accumulation in the slide, a step is designed in the bottom of the slide. The length of the step is nominal one-half of the slide length and 6 to 8 in deep or the depth for combination of two of the tallest packages. If the slide transfers product to another conveyor, then at the discharge end there is a transmission slide to a powered take-away system.

Flat-Bottom Slide. A flat-bottom slide is preferred for short travel distances and containerized product. When irregularly shaped, loose merchandise travels on the slide and accumulates, there is a high probability of jams.

Concave-Bottom Slide. The concave-bottom slide is preferred for transporting irregularly shaped (loose) or individually packaged merchandise. The concave bottom reduces jams.

Spiral Chute. The spiral chute is sometimes referred to as a *helical chute,* and it is designed around a center support. This design requires a small footprint. The spiral chute consists of overlapping coated-metal panels or a continuous concave plastic member.

Gravity Conveyor. The gravity conveyor transfers containerized product from an elevated level to a lower level. When a gravity roller or skatewheel conveyor is used, the lower level (discharge end) requires an end stop, transmission area to a powered conveyor or gravity conveyor runout of sufficient length for accumulation. The pitch or slope of the conveyor is at least 5°, and the conveyor requires sideguards.

Hanging-Garment Slide Rail. The slide rail uses gravity to transport hanging garments from a higher to a lower level. The slide rail consists of a tubular rail between two levels at a 20° to 30° slope. The exact slope is determined by the garment weight and type of hanger. The charge end is on the higher level, and the discharge end is on the lower level. An end stop at the horizontal runout end of the slide rail stops the hanger flow and permits hanger accumulation.

To operate, on the mezzanine level an employee places a garment on a hanger onto the elevated (charge) end of the slide rail. Gravity provides the power to move the hanging garment from the higher level to the lower level. At the lower level, hanging-garment flow is stopped by the end stop, and garments accumulate on the slide rail. As required, an employee removes the garments from the slide rail. To keep the hangers from falling to the floor, a metal cap runs the entire length of the slide rail.

The disadvantages of the system are that it handles one garment per trip and handles low weight. Its advantages are the low capital investment and ease of operation.

Powered Vertical Transportation

In most warehouse facilities that transport product to an elevated level, powered transport is preferred because it ensures continuous, controlled product flow between levels. All powered vertical transport methods are considered fixed-path systems. The various systems employ a lift truck, vertical carousel, pneumatic tube, belt conveyor, miniload, vertical trolley lift, powered chain or trolley carrier, chain system or fixed hooks, sandwiched conveyor, lapped stream conveyor, gripper conveyor, whiz lift, freight elevator, power and free conveyor, screw conveyor, powered squeeze conveyor, air conveyor, magnetic conveyor, continuous vertical lift, AS/RS, vertical reciprocating conveyor, and ramp with mobile powered equipment.

Powered Chain with Trolley.   When the trolley is used to transport hanging garments, the trolley load bar has at least two pegs. There are as many as seven pegs on the load bar to keep the hangers from sliding on the load bar. During the incline and decline, garments in the space between two pegs are stopped (by the pegs) from nesting against the trolley neck and falling to the floor.

If the trolley basket travels an incline or decline of 18° or more with a maximum of 30°, then a basket leveler is preferred. The leveler is connected to the trolley load bar by a set of basket hooks. As the trolley inclines or declines, the leveler keeps the basket's load-carrying surface level to the floor. In designing the rail system on the horizontal with a basket leveler, the top of the rail (TOR) must be at least 8 in higher than the normal height of 6 ft 4 in; therefore, the TOR is 7 ft to compensate for the leveler.

When a trolley leaves a horizontal travel area and is elevated or lowered, then the trolley is indexed and singulated forward onto the incline or decline conveyor section. This conveyor section is the hard push section of the chain that has a hard-push dog every 5 to 6 ft on centers. There are three methods to ensure that trolley lead head is properly engaged by the hard-push dog: manual, chopper, and hold and release.

Manual Trolley Infeed Method.   The manual infeed method consists of a manual push trolley. An employee pushes the trolley on the rail until the spool head merges with the incline or decline powered chain. At this location, the manual push rail is slightly inclined to ensure that the index of one trolley onto the powered chain pusher which engages the trolley head.

To operate, an employee observes an empty powered chain pusher dog approaching the infeed location and pushes the trolley until it is engaged by the hard-push dog.

This method handles a low volume and requires an employee to operate, but it requires no capital investment and has a low probability of jams.

Chopper Infeed Mechanized Method.   The chopper method consists of a chopper device and a gravity rail section that declines to merge with a powered chain hard-push section. Prior to the chopper, the manual (gravity) rail section inclines and declines to accumulate trolleys and is long enough to hold two or three trolleys up to a maximum of six.

The chopper device is an air-operated bar with trolley stops on both ends that operates in a seesaw manner. As the powered chain pusher dog passes a series of prong sensing devices, it activates the chopper device. This action permits the trolleys on the gravity accumulation section to index forward. The trolley travels on the gravity rail, and the trolley's lead spool is captured by the pusher dog. After the pusher dog has engaged the trolley, the trolley is moved forward by the chain and trolley head which activates another sensing device. This permits the next trolley on the line to index forward.

Holdback-and-Release (Automatic) Method.   The holdback-and-release method consists of a rail section that is a double-action device. This device is located prior to the incline or decline conveyor section and at the end of the soft-push powered chain section. The holdback-and-release device in the up position ensures that the trolley's lead spool is engaged by the push dog. This is an air-operated device that moves in an up-and-down path. In the down position, it temporarily holds one trolley. This up-down movement is activated by a pusher dog and trolley activating a sensing device.

In this method, as the trolley travels on the soft-push (trolley accumulation) rail section, the trolleys are accumulated by the holdback-and-release device which functions as a temporary trolley stop. When the powered chain hard push (dog) passes a sensing device, it activates the holdback-and-release device to raise up and permit the trolley's lead spool to become engaged in the pusher dog. The pusher dog pulls the trolley on the rail to the incline or decline section. As one section of the holdback-and-release is raised, the second section is lowered, which permits a trolley placed in the lower position to wait for the next hard-push dog to activate the holdback-and-release device that is raised up.

The holdback-and-release device requires an investment, but it handles a high volume and does not require an employee.

Screw Conveyor.   The screw conveyor is a fixed-path system that transports single hanging garments between levels. The screw conveyor is a motor-driven helix (thread) that turns around a tubular rail. Some models have a U cap over the helix to keep a hanger from falling off the rail. As the motor turns the helix, the shaft revolves and the hanger between the revolving helix couplings is propelled forward. Some screw conveyors are designed with the helix component in a U trough. These conveyors transport loose granular product between warehouse locations.

To operate a screw conveyor, an employee places an individual hanging garment on the infeed slide rail, which is on the left side of the screw conveyor. The revolving screw conveyor picks up the hanger. After the hanger is between the helix revolving coupling, it is moved forward. At the discharge end, the hanger is automatically discharged onto a slide rail, on the right side of the screw conveyor.

The disadvantages are that it handles a low volume, one-way transport, and a limited weight capacity. The advantage is that it is a simple operation.

Chain with Fixed Hooks.   The next vertical transport method is designed to convey hanging garments or medium-sized baskets between two levels. In this fixed-path electric motor powered chain system, a series of fixed carriers are attached to the chain links. The chain is an endless loop that travels at 45 ft/min and can carry 10 lb per carrier. There are two models of the chain system with fixed hooks.

The first model is a single-hanging-garment or basket system that is side-loaded and -unloaded at the two warehouse locations. In this system, the carrier is side-mounted to the chain link. In the second model, a carrier hangs below the chain and that has multiple hooks which increase the carrier's capacity. The carriers transports multiple hanging garments or a single basket to the warehouse location.

To operate, an employee places the hanging garment or basket onto the hook. The chain transports the product to the second level of the warehouse. On that level, another employee removes the product from the carrier.

The disadvantages are that it handles a low volume, handles a limited weight capacity, and requires an employee at both levels. The advantage is that it requires a low investment.

Vertical Trolley Lift.   The vertical trolley lift is a fixed-path concept. It is designed to lift a fully loaded trolley of hanging garments between two levels. There are a motor-driven chain, rail bar that moves between the two warehouse levels, automatic trolley entry and exit stops to secure the trolley on the rail, and at both levels automatic adjustable rail bar lift stops. It is designed to elevate a trolley between 10-, 15-, or 20-ft-high levels.

To operate, an employee pushes a trolley onto the vertical lift load bar and ensures that the trolley is in the correct position. The employee places the controls in the elevate or decline position, and the load travels to the transfer station at the next warehouse level. At this new warehouse location, the chain automatically stops the rail bar at the proper level. This allows the trolley transfer to travel on the rail to the new location.

The disadvantages are that it handles one trolley per trip, handles a low volume, and requires employees at both levels, and the maximum lift is 20 ft. The advantages are the quick installation, ease of operation, and automatic location of the load bar on the rail.

Powered Squeeze Conveyor.   In this single-item vertical transport fixed-path system, two (dual) belt conveyors are designed to have the low-profile product squeezed between two separate belt conveyors. These belt conveyors vertically transport the product between two levels.

The bottom conveyor is the carry conveyor, and the top (cover) belt conveyor secures the product for the incline or decline transportation. To ensure that the product is secured between the two belts, offset full-width rollers are located at specific intervals. This arrangement acts as a ladder for the product to climb between the two levels.

To operate, the product is placed onto the infeed station of the bottom conveyor. As the product moves forward, it becomes squeezed between the two belt conveyors. During the vertical climb or decline, the full-width rollers, spaced at specific intervals and offset, secure the product as it travels between the two levels.

Powered Sandwich Conveyor.   The next small-item vertical transport method involves the powered sandwich conveyor, a fixed-path system. There are two (dual) belt conveyors (FIg. 9.10a). The product is sandwiched between the two separate belt conveyors which vertically transport product between two levels. The bottom conveyor is the carry belt, and the top (cover) belt conveyor secures the product for the incline and decline transportation. During vertical transportation, to ensure that the product is secured between the conveyors, the two belt conveyors pass through several side rollers that form the contour of the belt. The belt sandwich conveyor handles many materials such as granulates and low-profile product. The belt width is 24 to 72 in.

To operate this conveyor, the product is dumped onto the horizontal runout of the bottom belt conveyor which travels over a series of trough idlers that force the product to the middle of the belt. The product moves forward and the top (cover) belt conveyor comes over the sandwiched product which forms a sandwich. The conveyor with the product starts to incline or decline between the two levels. During its travel, it passes specially designed rollers that form the belt contour. These rollers are placed along the conveyor sides at specific intervals until the product arrives at the next level. On this new warehouse level, the belt has a horizontal runout. At the desired location, the top belt stops, and the product is moved forward by the bottom belt conveyor.

The sandwich belt conveyor path follows a straight-line, L, Z, or C configuration. The disadvantages are that it handles low-profile product and that product is singulated. The advantages are that it handles a high volume, it inclines and declines in a short space, it handles a wide variety of low-profile products, and operation is easy.

Air Conveyor.   The fixed-path air conveyor transports lightweight flat-surface containers (empty plastic or metal cans or bottles and chipboard or corrugated cartons) between two levels. It consists of a perforated conveyor surface plate, plenum (passageway beneath the conveyor), and blower. The airflow generated by the air blower flows through the plenum beneath the conveying surface of the plate. As the air escapes through vertical jets, it provides the force that lifts the product upward. The

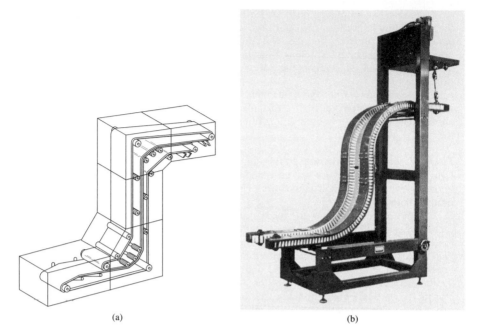

(a)

(b)

FIGURE 9.10 (*a*) Sandwich conveyor (*Courtesy of C. Tech, Inc.*); (*b*) gripper (*Courtesy of Arrowhead Conveyor.*).

air emerging from the lateral jets provides the force to move (thrust) forward on the conveying surface. This combination of air circulation provides rapid and gentle movement of small items over the conveyor.

The disadvantages are that the product must have a flat bottom (conveyor surface) and that the product must be lightweight. The advantages are that it transports in a small area, it transports several item types, and it has few moving parts.

Lapped-Stream Conveyor. The lapped-stream conveyor, a fixed-path conveyor, transports low-profile items (newspapers) between two levels. The lapped-stream conveyor consists of two conveyor surfaces (series of coated cords or narrow belts) that squeeze the product between them. These conveying surfaces face each other. As the electric motor-driven belts incline or decline to the next level, the pressure is maintained on the product by the two conveying surfaces passing over a series of rollers.

The lapped-stream conveyor handles low-profile product (3 in high and 17 in wide) at a travel speed of 315 ft/min.

Magnetic Conveyor. The magnetic conveyor, a fixed-path conveyor, is used to transport metal parts. It consists of a polyester web that is laminated with a tough urethane surface which slides over a magnetized heavy-duty steel slider bed conveyor. The belt is motor-driven, and the belt take-up is pneumatic to ensure proper belt tension. The drives are center- or end-located and pull the 12-in-wide belt at any desired length up to 100 ft per module at any comparable speed to the standard belt conveyor.

After the metal product is placed onto the belt, the metal part is stationary on the belt surface which slides across the magnetized steel bed. When the product arrives at the end of the first conveyor (110-ft) section, which is the lower conveyor, the product is transferred to a second conveyor. This second conveyor is above the first conveyor. This transfer of product is achieved because the two magnetized conveying surfaces face and overlap each conveyor section.

Gripper Conveyor.   The fixed-path gripper conveyor (Fig. 9.10*b*) transports a large volume of lightweight (approximately 20 lb/ft) items with the same width between two levels. It consists of two plastic strands of conveyors designed so that the product is gripped between the two conveyors and vertically transported to the next level. These two plastic conveyors are a series of continuous endless chains of side grippers that face each other and form a narrow product path between them. The product path width between the conveyors is changed to accommodate different products. At both the infeed and outfeed sections of the conveyor are horizontal runouts that have additional bottom conveyors 21 in above the finished floor. This elevation ensures that the product is singulated and indexed forward to avoid jams.

The conveyor speed is 250 ft/min, and it handles a product with a maximum width of 8 in; the elevation change is 88 in between levels.

To operate, as the individual product is conveyed to the horizontal infeed runout, two strands of side gripper conveyor grasp the product. On this horizontal section, these grippers secure the product and move the product up or down the incline or decline conveyor section onto the discharge horizontal runout section. As the product is moved forward on the outfeed horizontal conveyor runout, the side grippers release the product. At this point, the product movement remains singulated and is under the flow control of the bottom conveyor.

The disadvantages of the side gripper are that a higher capital investment is required, product must be singulated, the product's maximum width is 8 in, flexibility to change product width is limited, and it handles various-sized product on a continuous flow.

The advantages are that it handles a high volume, singulates the product, and inclines or declines in a short space.

Continuous Vertical Lift Conveyor.   The continuous vertical lift conveyor handles individual containers or large individually packaged items of merchandise (Fig. 9.11*a*). The continuous vertical lift consists of two motor-driven sprockets that propel the two endless chain loops. Attached to the chains are several load-carrying platforms (flights or groups of narrow slats). The chains follow a path through a guide channel, and the entire mechanical system is contained in a protective enclosure.

As the sprockets turn the chains, the load-carrying platforms travel past the powered conveyor infeed (charge) or outfeed (discharge) stations. To feed in product, the floor-level horizontal powered conveyor moves the container forward onto the load-carrying platform. The load-carrying platform movement (rotation) is in the same direction of travel as the container is pulled on the load-carrying platform. To discharge the container, the load-carrying platform moves the container onto the floor-level horizontal powered conveyor.

The first continuous lift design has a Z configuration similar to the letter Z. This continuous lift conveyor receives the container on one side and discharges the container on the opposite side. The continuous lift path does not change the container's direction of travel. The second continuous lift design has a path similar to the letter C. The container path enters from one side and exits on the same side. This continuous lift path does change the direction of container travel.

Whiz Lift.   The whiz lift handles a wide variety of individually packaged merchandise or containers of product (Fig. 9.11*b*). During the vertical travel up steep inclines, to ensure the merchandise is retained in the container, for best results, the container

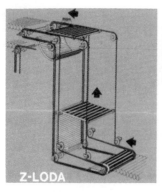

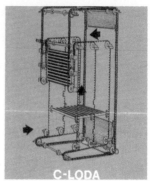

Z-LODA UNITS...
accept material in a
given direction, lift or
lower and continue flow
in same direction with
discharge on the
opposite side.

C-LODA UNITS...
receive material at one
level, lift or lower, to
discharge on the same
side at a different level.

(a)

FIGURE 9.11 (*a*) Continuous vertical lift (*Courtesy of Z-Loda Systems Engineering.*); (*b*) whiz lift (*Courtesy of W & H Systems, Inc.*); (*c*) vertical reciprocating conveyor (*Courtesy of Piflow Industries, Inc.*).

should have a cover. The whiz lift consists of a steep inclined powered conveyor (belt type), powered infeed and outfeed stations (conveyors), and a closed-loop motor-driven snug elastic metal-ribbed shroud envelope. The snug elastic ribbed shroud is as wide as the inclined belt conveyor, and its path is directly above the belt conveyor. The ribbed shroud channel almost touches the top belt conveyor.

As the merchandise travels on the infeed conveyor, the top of the merchandise comes in contact with the elastic shroud and is pulled forward onto the inclined powered belt conveyor. Under the power of the incline powered belt conveyor and the shroud, the metal ribs engulf the container. Then the metal ribs and shroud hold the container, and the merchandise travels between floors to the elevated outfeed station. At the station, the container leaves the engulfed (restricted) position between the top shroud and the belt conveyor, which permits travel onto the powered take-away conveyor.

Automated Storage/Retrieval or Lift Truck.   The automatic (AS/RS) or manual (S/RS) storage/retrieval system or a lift truck is considered a fixed-path system that handles pallet loads from the ground level to above floor level to 80 ft high and performs a deposit/withdrawal transaction at any unit-load position level. Therefore, the deposit position is an infeed station or a warehouse level. As required, on the elevated-floor transaction there is an opening in the wall or handrail. This opening is protected by a door, double-acting safety gate, or corral.

Freight Elevator.   The freight elevator is a pallet load fixed-path transport system. The elevator is a manually operated vehicle that travels within a shaft which runs the entire height of the facility. Additional runouts are below the ground level and above the roof. The freight elevator's load-carrying surface can take 4 to 6 pallets; it stops at each floor and is an electric motor-driven chain or hydraulic (oil) system.

Vertical Reciprocating Conveyor.   The vertical reciprocating conveyor is a fixed-path system (Fig. 9.11c). It consists of an electric motor-driven load-carrying surface, guiderails, protective cage, and controls.

To operate, a mobile pallet load handling vehicle or pallet conveyor transfers the pallet load onto the load-carrying surface. After the safety and operating controls are set, the unit-load-carrying surface carries the load up to the appropriate level. At this level, the control devices stop the upward travel of the platform. An employee opens the safety gate, and the pallet load is removed from the carrying surface. The employee sends the reciprocating conveyor to another level.

The reciprocating conveyor makes multiple stops at several floor levels, carries a load up to 10,000 lb, and travels at a speed of 25 to 30 ft/min. The load-carrying platform is 8 ft by 10 ft. The vertical reciprocating conveyor is a stand-alone system. It requires a 6-in-deep pit and roof runout that is determined by the pallet load height. If an existing facility has an elevator shaft, then the vertical reciprocating conveyor can be installed in the shaft.

Power and Free Conveyor.   The vertical power and free conveyor has the same design parameters and operational characteristics as the horizontal power and free conveyor. The incline and decline slope of the system does not exceed 15°.

Ramp with Powered Vehicles.   The last pallet load vertical transport concept is the ramp designed for self-powered vehicles or carts which are towed by a chain. This fixed-path system requires a solid surface that has an incline or decline not exceeding 10° to 15°. The various vehicles are in-floor tow chain, overhead tow chain conveyor, AGV, pallet truck, and lift truck.

CHAPTER 10

THE BEST LIFT TRUCK AND MOBILE WAREHOUSE EQUIPMENT FOR AN OPERATION*

The objective of this chapter is to identify and evaluate lift trucks and other warehouse equipment applications, methods, practices, and technologies used to move product between the storage and pick areas of a facility. These factors make the key storage function a more efficient and effective operation.

The storage function has four components: storage positions, inventory control, and building and product handling equipment. Storage positions, inventory control, and the building area were reviewed in earlier chapters.

The product (unit-load or pallet load) handling equipment that is operator- or computer-controlled ensures that the right product is transferred on schedule from the right storage place (position) and to the right place (workstation).

The various product handling equipment types for use inside or outside a distribution facility are similar but not identical. They all move product; the differences lie in their power source, aisle width, capacity, storage height, and controls. The various product-handling equipment groups are (1) lift trucks and storage/retrieval vehicles, (2) HROS (High Rise Order Selection) trucks, (3) rough-terrain (outside or yard) trucks, (4) bridge cranes, (5) pallet trucks, (6) tractors (tuggers), and (7) AGVs.

DESIGN PARAMETERS

Prior to the purchase and implementation of product-handling equipment, you must clearly define the operational parameters: (1) battery charging, dc electric or fuel storage requirements; (2) unit-load dimensions with overhang (length, width, height, plus weight); (3) average and peak inbound and outbound transactions per hour; (4) unit-load storage position type and clearances; (5) unit-load bottom support method (pallet board type, skid, slip sheet, or container) and attachments on the lift truck; (6) stack-

---

*This chapter is based on "Comparison of Unit Load Handling Vehicles," published by *Plant Engineering Magazine*, May 1991.

ing height and storage area's environmental conditions; (7) condition of storage area floor (indoor or outdoor) and surface (including depth and metal locations); (8) aisle width (product to product) and length; and (9) equipment complies with the latest ANSI Standard B56.6.

The first type of truck examined is the lift truck. To achieve maximum lift truck operator productivity in a high-volume operation, at P/D stations the lift truck operator stays in the storage aisle and limits the truck travel outside of the aisle as a transport vehicle. However, the exception arises in a low-volume operation that uses a counterbalanced lift truck in other key warehouse functions. This equipment flexibility improves the total facility productivity and return on investment.

To achieve maximum productivity from the lift truck operator, the operation or design manager should ensure that there is sufficient clearance between building obstacles, rack components, and unit loads; understand the volume and characteristics of the product; reduce lift truck travel distances; and implement in the warehouse an excellent SKU inventory control program.

Introducing a New Lift Truck to the Existing Warehouse Fleet.   When you introduce a new manufacturer's lift truck to the warehouse operation, to improve the lift truck driver's acceptance, you must "sell" these new trucks to your employees. Require the lift truck manufacturer and your managers to set up training sessions. This reduces the employee's resistance to the new lift truck.

Keep Uniform Loads in One Area.   If your distribution operation has all pallet loads with identical dimensions, weight characteristics, and bottom support devices (wood pallets), this reduces operational problems. If your operation has different support devices, then to reduce operational problems, each pallet should have its own storage area or aisle.

Lift trucks are used in a warehouse operation to handle a unit load that includes product. The product sets on a pallet board, on a skid, on a slip sheet, or inside a container. The forks on lift trucks are the chisel type and can adjust to different pallet board, skid, or container fork openings. A slip sheet requires a special attachment on a counterbalanced truck or low-lift pallet truck. These attachments are reviewed in greater detail later in this chapter.

VARIOUS LIFT TRUCKS

Different unit-load storage vehicles that are used in a warehouse operation have a wide range of capabilities. In this section, we consider the unit-load storage/retrieval vehicle and lift truck as a vehicle capable of picking up a unit load, with the unit load traveling through a warehouse aisle (down-aisle) and performing the unit-load deposit or withdrawal transaction. There are three basic warehouse storage vehicle types, defined by the stacking (right-angle turn) requirement. Within each major group subgroups are identified by the vehicle's power source or ability to travel outside the storage aisle. The three basic lift truck classifications are *wide-aisle* (WA), *narrow-aisle* (NA), and *very narrow-aisle* (VNA).

In this section, the objective is to develop an understanding of the various product storage vehicle types. We identify each vehicle's power source, type of controls, right-angle stacking width, lift height, throughput capability, and captive- versus mobile-aisle ability.

Wide-Aisle Lift Trucks

The first group is the WA lift trucks (vehicles) that operate in 10- to 13-ft-wide aisles.

Walkie Stacker Truck. The first vehicle (Fig. 10.1*a*) is the walkie stacker truck. It is a manual push or electric powered drive wheel vehicle that handles a maximum load of 1500 lb up to a stacking height of 10 to 12 ft. The second vehicle (Fig. 10.1*b*) is the electric powered straddle walkie stacker which has two stabilizing straddles that extend forward from the vehicle's base and a space between the two straddles for a pallet load. Some models are reach types. The third vehicle (Fig. 10.1*c*) is the counterbalanced walkie stacker. It has the unique characteristics of a long battery compartment and a long counterbalance weight area.

A walkie stacker has a set of forks on a telescopic mast that elevate and lower pallet loads at 10- to 12-ft-high storage positions, two load-carrying wheels, one drive or steering wheel, and an operator's handle. The operator's handle contains all vehicle movement controls. The operator who walks behind the vehicle has easy access to the fork movement controls. It receives its power from a rechargeable electric battery.

Straddle or Counterbalanced Straddle Truck. The electric powered straddle or counterbalanced walkie stacker requires a minimum 6- to 12-ft-wide aisle for a right-angle stack or transfer aisle. It has slow travel speeds and has a low throughput capability. The stacking height of a 1500- to 3000-lb load is 10 to 12 ft. Free lift allows the vehicle to operate in a low ceiling area. In free lift, the forks elevate before the mast elevates. The vehicle requires a normal warehouse floor surface and receives its power from a rechargeable electric battery.

Sit-Down Counterbalanced Truck. The fourth vehicle is the sit-down counterbalanced lift truck (Fig. 10.1*d*). With a low mast and overhead guard, it is a very maneuverable and versatile pallet-handling vehicle. The truck is used as a dock, transport, and storage area vehicle, and it requires a normal warehouse floor surface. Also it can travel up a 15° grade or ramp. The vehicle is designed with a set of load-carrying forks that elevate and lower on a telescopic mast and an operator's area with a seat. From the seat, the operator has access to all horizontal vehicle and vertical fork movement controls. Some models have see-through masts that provide the operator with a complete view of the vehicle's path. The vehicle counterweight is located in the rear chassis, and the vehicle has a long wheel base. These features provide the vehicle with the ability to offset the pallet load weight.

The most common indoor truck is powered by a rechargeable electric battery. When the lift truck is used as an outdoor vehicle, the power source is diesel, LP gas, or gasoline. These alternative power sources are common on the counterbalanced vehicle. The vehicle has three or four wheels, with one wheel as a drive and steering wheel. The wheels are fitted with pneumatic tires for outdoor use and cushion, solid polyurethane, or rubber tires for indoor use.

These counterbalanced vehicles are available with one-, two-, three-, or four-stage masts that provide the vehicle with a stacking height of 16 to 18 ft. Counterbalanced lift trucks are used to deposit or withdraw a 2000- to 4000-lb pallet load from a floor stack or any rack position except a double-deep rack position. Some trucks have the counterweight and wheelbase to handle heavier loads at lower storage positions. Fork side shift and mast tilt devices are options that increase operator productivity and reduce product and equipment damage. An experienced operator makes 20 to 25 transactions per hour. All pallet load deposits and withdrawals are controlled or eyeballed by the operator.

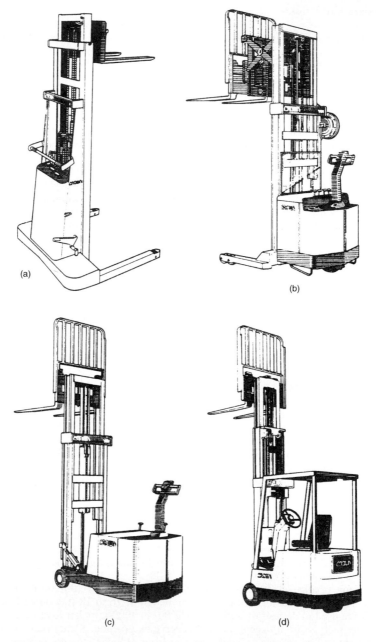

(a)

(b)

(c)

(d)

FIGURE 10.1 (*a*) Manual walkie stacker; (*b*) electric powered straddle stacker; (*c*) electric powered counterbalanced stacker; (*d*) sit-down counterbalanced lift truck (*Courtesy of Crown Equipment Corp.*).

The sit-down-rider counterbalanced lift truck with a low mast and low overhead guard is the most maneuverable and versatile pallet-handling vehicle. The vehicle with its wheelbase has good grade clearance that allows it to easily load or unload delivery vehicles, to transport pallet loads through the facility (including towing a train of carts), and to perform the standard floor and rack storage activities. With limitations, a standard counterbalanced lift truck handles a wide variety of unit loads over a normal warehouse floor surface or operates on a flat, hard outdoor surface.

The four-wheel counterbalanced lift truck with its long wheelbase and chassis weight is fitted with various attachments to perform other warehouse tasks. However, for a given lift truck, verify with the manufacturer that the lift truck will accept the attachment weight and new center of load and will reach the elevation.

Lift Truck Stability

When we use the term *lift truck load capacity with stability,* it means that when a counterbalanced lift truck picks up a unit load, then the fulcrum (balance point) for the counterbalance action is the centerline for the front wheels. The unit-load weight plus any attachment is counterbalanced by the truck's counterbalance weight and wheelbase. The truck's weight includes the battery weight and the counterbalance weight.

Lift Truck Capacity

The lift truck *capacity* is the amount of weight that it safely lifts. Each lift truck has a capacity which is stated by the manufacturer. This weight capacity decreases as longer loads are handled. The lift truck handles shorter loads up to the limit specified by the manufacturer. Therefore, as the length of the unit load increases from the standard 24-in center, the lift truck can handle less weight. If the unit load is heavier than or over the standard dimensions, then a bigger, heavier, and more expensive lift truck is needed.

When a load is heavier or longer than the lift truck's capacity and the operator performs a storage transaction at a high storage level, the lift truck has a tendency to tilt forward on its front wheels.

Lift Truck Mast

The second most important lift truck component is the type of mast. There are two mast specification parameters. The first is the overall extended (lift) height that allows the vehicle to deposit or withdraw a unit load from the highest floor or rack storage position. The second is the lowered (down) or collapsed height of the mast. This height is the clearance for the lift truck to pass through the facility doorways, enter delivery trucks, and enter dense storage rack lanes.

In considering these specification parameters for maximum lift height, the vehicle's lift height is influenced by the backrest height or unit-load height to the lowest ceiling (building) obstruction. The clearance height for the lift truck is affected by the overhead guard height which may be higher than the mast. In this situation the truck's ability to pass through a low-ceiling area, doorway, or into a rack lane is restricted.

Seven types of masts are available for a lift truck: (1) single mast, (2) two-stage nontelescopic mast, (3) standard two-stage mast, (4) two-staged mast with high free lift, (5) three-stage mast, (6) four-stage mast, and (7) rigid mast.

Single-Stage Nontelescopic Mast.   The single-stage nontelescopic mast lifts a unit load to a 7- to 8-ft-high floor or rack storage position. In today's warehouse operation, the single-stage mast has very limited application and is uncommon in a company's lift truck fleet.

Two-Stage Nontelescopic Mast.   The two-stage nontelescopic mast has little or no free lift. No free lift means that as the forks start to rise, the mast starts to rise. With this mast, the lift truck deposits or retrieves pallet loads at storage positions that are 12 to 13 ft high.

Two-Stage Standard Mast with Low Free Lift.   With this mast, the lift truck deposits and withdraws a unit load from a storage position that is 15 to 20 ft high. This mast feature of low free lift conserves energy and allows the lift truck to be used as a multipurpose vehicle in a warehouse with a low or high ceiling.

Two-Stage Mast with High Free Lift.   This free-lift mast permits the pallet load to be raised to almost half the mast height without elevating the mast. This feature permits the lift truck to operate as a dock vehicle that loads or unloads delivery vehicles. Also the lift truck deposits and retrieves unit loads from storage positions 20 ft high, and it can operate in an area that has a low ceiling.

Three-Stage Mast with Full Free Lift.   The mast feature allows the lift truck to deposit or withdraw unit loads from 20- to 25-ft-high storage positions. Also the full free lift permits the lift truck to operate in facility areas with low ceiling clearance.

Four-Stage Mast with Full Free Lift.   This mast feature permits the lift truck to deposit or withdraw unit loads from the 30- to 40-ft-high positions.

Rigid Mast with or without Free Lift.   The rigid mast permits the unit-load deposits and withdrawals from an elevated rack position 40 to 60 ft high. The rigid mast is associated with the hybrid or storage/retrieval system vehicles.

Lift Truck Maneuverability

The next most important factor is the lift truck's maneuverability, or ability to make a right-angle stacking turn or a turn in an intersecting aisle.

Right-Angle Stacking Turn.   A right-angle stacking turn is the distance required for the lift truck to turn and make a deposit or withdrawal from a storage position (Fig. 10.2*a*). This distance is the aisle width. An easy way to determine the aisle width is to take a conventional pallet load length (stringer) plus the distance from the face of the forks to the centerline of the truck drive wheel plus the outside turning radius. For excellent lift truck operator productivity and reduced product and equipment damage, 6 to 12 in is added to the above calculation. If there is unit-load overhang, the overhang dimension is added to the stringer length.

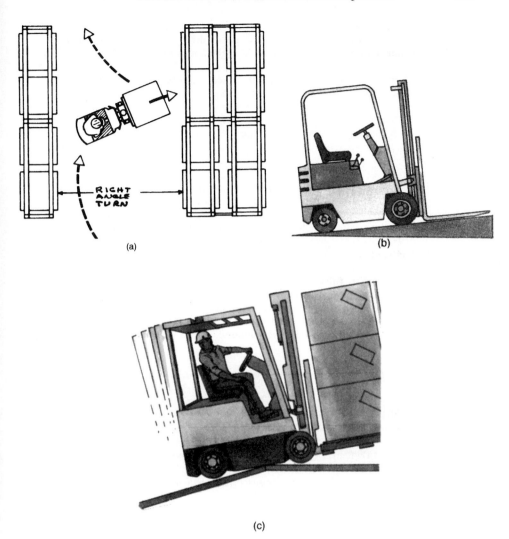

(a)

(b)

(c)

FIGURE 10.2 (*a*) Right-angle stack (*Courtesy of Plant Engineering Magazine.*); (*b*) lift truck gradeability (*Courtesy of Kelley Co., Inc.*); (*c*) lift truck clearance (*Courtesy of Kelley Co., Inc.*).

In planning a warehouse aisle with the lift truck manufacturer's aisle width recommendation, for good lift truck operator productivity and low product and equipment damage, the storage aisle width should be that from product to product plus 6 to 12 in.

Lift Truck Wheelbase. In theory, the counterbalanced lift truck wheelbase determines the aisle width. The wheelbase is the distance between the front and rear wheels. A lift truck with a short wheelbase requires a narrow right-angle turning aisle. Therefore, a lift truck with a short wheelbase makes the truck very maneuverable. A truck with a long wheelbase requires a wide right-angle turning aisle, but has an improved ride, steering traction, and stability.

The second maneuverability factor is the lift truck's ability to climb a ramp (gradeability) and its underclearance.

Lift Truck Gradeability. The gradeability of a lift truck is the steepest percent grade (ramp) that a lift truck climbs with a unit load (Fig. 10.2*b*). An electric lift truck's steepest grade is 15 percent, and the grade for a gasoline, LP gas, or diesel truck is 20 to 25 percent.

Lift Truck Clearance. The lift truck's grade clearance is the steepest grade percent that the lift truck climbs without underclearance problems (Fig. 10.2*c*). Underclearance problems occur at the top of the ramp (dock plate) as a hang-up of the lift truck bottom or at the bottom of the ramp with the load (forks) hitting the floor.

The lift truck underclearance is the distance between the floor and the lowest part of the lift truck undercarriage. The important underclearance locations are midway between the wheelbase and the bottom of the mast because of the mechanical operating parts in these areas. In considering a vehicle that will climb ramps, enter and exit delivery vehicles, and vehicles for outdoor operations, gradeability and underclearance are important, to avoid future repair expenses and lift truck downtime. In theory, a lift truck with a short wheelbase and high underclearance is the best for ramp, dock, and outdoor work.

Stand-up-Rider Counterbalanced Lift Truck. The stand-up-rider counterbalanced lift truck (Fig. 10.3*a*) is considered part of the wide-aisle (WA) group. The vehicle has similar design features and operational capabilities to the sit-down-rider counterbalanced lift truck. The major difference is that the operator stands up on the operator's platform. This feature allows a shorter wheelbase. With this shorter wheelbase, the vehicle makes right-angle turns in a 9- to 10-ft-wide aisle and places a 2000- to 3000-lb pallet load into an 18-ft-high floor or rack storage position. The 24- or 36-V electric rechargeable battery powered mobile-aisle vehicle is very maneuverable and versatile, performs all activities on a normal warehouse floor, and unloads and loads delivery vehicles. Some models, in addition to the tilt and side shift option, have an option for a retrackable overhead guard which permits easy delivery truck entry.

Narrow-Aisle Lift Trucks

Narrow-aisle vehicles operate in a 7- to 9-ft-wide aisle.

Stand-up-Rider Straddle Lift Truck. The stand-up-rider straddle truck is powered by a rechargeable electric battery (Fig. 10.3*b*). The truck does not enter delivery trucks or travel up grades. The vehicle is designed with two load-carrying wheels and one drive or steering wheel, a set of forks that elevate and lower on a telescopic mast, an overhead guard, stabilizing straddles (outriggers), and an operator's platform. The operator's platform has both horizontal truck and vertical fork movement controls. To elevate, lower, and transport a pallet load, the outriggers assume the majority of the pallet load weight. Since the lift truck straddles (outriggers) engulf the majority of the

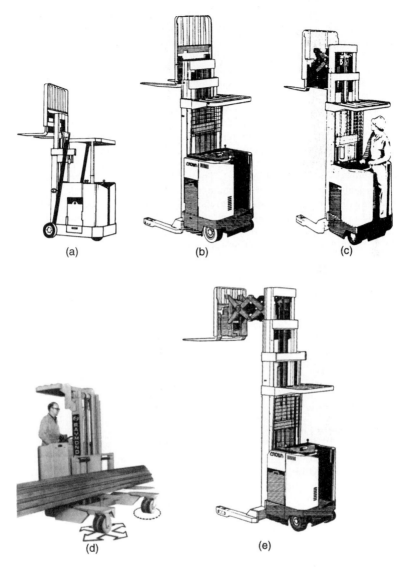

(a) (b) (c)

(d) (e)

FIGURE 10.3 Stand-up-rider (*a*) counterbalanced lift truck (*Courtesy of Crown Equipment Corp.*); (*b*) straddle lift truck (*Courtesy of Crown Equipment, Inc.*); (*c*) straddle reach lift truck (*Courtesy of Crown Equipment, Inc.*); (*d*) four-directional lift truck (*Courtesy of Raymond Corp.*); (*e*) double-deep lift truck (Courtesy of Crown Equipment Corp.).

pallet load between the two stringers (outriggers), the straddle lift truck operates in a 7- to 8-ft-wide aisle. On a normal level warehouse concrete floor, the straddle lift truck handles a 2000-lb load up to a 20-ft-high floor or rack storage position, and the operator makes 18 to 20 transactions per hour.

Straddle Reach Lift Truck.   The stand-up-rider straddle reach lift truck (Fig. 10.3c) has similar design features and operational characteristics to the straddle truck, except that its forks are attached to a pantographic reach device. The pantographic reach device allows the forks to extend out beyond (forward of) the outriggers to pick up and retract a unit load. This pantographic device permits the reach truck to handle a pallet load wider than the interior space between the outriggers. This reach feature increases the right-angle stack aisle requirement of 8 to 10 ft and allows an operator in a normal aisle length to make 15 to 18 transactions per hour.

Four-Directional Lift Truck.   This unique stand-up-rider straddle reach truck (Fig. 10.3d) has similar operational features to the regular trucks, but all four truck wheels turn to the direction of travel. One load wheel hydraulically shifts to forward or reverse or lateral for travel (sideways). The other load wheel is the free-swiveling wheel. This feature gives the truck its name, the four-directional vehicle.

While traveling in the main aisle with the long load in front and for entry into the narrow aisle to perform the storage transaction, the operator stops and shifts the load wheel direction. This permits the truck to travel sideways into the narrow storage aisle to perform a transaction. The narrow storage aisle is a normal 8-ft 6-in wide aisle.

Stand-Up Two-Deep Lift Truck.   The stand-up-rider double-deep (two-deep or deep-reach) lift truck is the next vehicle (Fig. 10.3e). Its basic design features and operational characteristics are the same as those of the straddle reach lift truck except that the pantographic reach device has a greater extension. It extends forward out into the second deep interior pallet rack position of the two-deep rack system. For best results, this extension (stroke) is a fixed number of inches (43 or 51 in). This means that all pallet boards must have the same dimension. Operators can make 13 to 16 transactions per hour. With a 9- to 10-ft-wide aisle, the two-deep system does increase the storage density per aisle.

Very Narrow-Aisle Lift Truck

The very narrow-aisle (VNA) lift truck operates in a 5-ft 6-in to 6-ft wide aisle. When VNA vehicles are used in an operation, for maximum employee productivity and low product and equipment damage, they require an aisle guidance system and are kept in the warehouse aisle the maximum amount of time. The most common aisle guidance concepts are the inductive (wire is embedded into the floor) or rail concept. Other aisle guidance methods are the optical (chemically treated tape or paint) and laser beam.

Level-Floor Requirement and Other Options.   Several options are recommended: a shelf locator control system to achieve the correct fork elevation for a transaction, a dead-level or F-75 to F-100 floor that is permitted to vary $\pm \frac{1}{8}$ in 10 ft, and a pallet load pickup and delivery station at the end of each aisle.

Lift Truck Load Handlers.   There are three load handlers for VNA trucks. The first is the *L fork,* which is similar in design to conventional lift truck forks. The forks are side-attached to a device that elevates or declines and traverses left or right on the mast to perform a transaction. In the second design, the forks are top-attached and cantilevered from the mast to the elevating or declining device that slides left or right to make a transaction. With the top-attached and cantilevered device that has a set of forks extending down and outward, this fork configuration resembles a C or J shape. The third load handler design is referred to as the *track-and-scissor forks.* These forks are bottom-attached to outriggers that rise and lower on a mast to the unit-load position. The forks are rotated on a turntable or flipped in the desired direction to perform a transaction.

Aisle Widths.   With side-loading vehicles there are two aisle widths in which the vehicle can perform storage transactions: an aisle width to turn the unit load and an aisle width not to turn the unit load.

Aisle Width to Turn the Unit Load.   When you design a side-loading vehicle to turn the unit load in the aisle and perform a storage transaction, the aisle width must be a nominal 6 in wider than the diagonal length of the unit load. This dimension permits the side-loading vehicle to rotate (turn) the unit load as the vehicle travels in the storage aisle. This feature does improve operator productivity because the unit load has the correct orientation for a lift truck storage transaction at the P/D station or storage position. But the method requires a wider warehouse aisle and facility.

Aisle Width Not to Turn the Unit Load.   When you design the sideloading truck not to turn the unit load in the aisle, then the aisle width must be a nominal 6 in to 8 in wider than the length of the unit load. This dimension does not permit the lift truck operator to rotate (turn) the unit load as the truck travels in the aisle. To turn the unit load, the lift truck operator does so at a vacant storage position or at the end of the aisle. This feature requires disciplined lift truck operators and a narrower building but results in slightly lower productivity.

Mast Types.   The two basic masts on these trucks are the two-stage masts for elevated heights at 25 ft and three-stage masts for elevated heights from 22 to 40 ft high.

Other NA Vehicle Characteristics.   In considering the implementation of a side-loading vehicle in a new or existing plant, note that the floor is designed to accept the vehicle wheel loads of approximately 56 percent of the vehicle's weight as the lift truck performs a transaction.

The term *side-loading lift truck* describes the normal direction that the lift truck faces to perform a transaction and how the pallet load is taken on board at the P/D station or at the storage position in the aisle. The majority of these vehicles can perform front transactions in a very wide aisle.

Rider Side-Loading Lift Truck.   The rider side-loading lift truck (Fig. 10.4*a*) has a set of load-carrying forks that are attached to a pantographic device which elevates and lowers on a telescopic mast. It has an overhead guard and an operator's area. The operator's area is at one end of the vehicle with full machine controls and a steering device. The vehicle has two steering wheels and two load-carrying wheels that are under the two decks (outriggers or straddles). There is sufficient width between the decks for the pallet load which is taken onto the forks into the well. When long loads (lumber or pipe) are handled, the forks extend forward and pick up and raise the load to clear the decks. Prior to travel the unit load is lowered and rests on the front and rear decks. The two decks become the stabilizing platform for the unit load.

The side-loading vehicle takes the pallet board onto the forks with the pallet opening facing the rack position. Since the forks face one side of the aisle, all transactions are performed from one side of the aisle; therefore, without a computerized inventory letdown program, the number of transactions per hour is 17 to 19 in a normal aisle

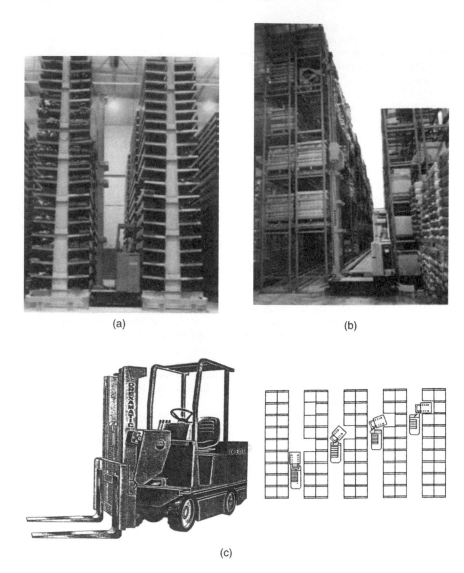

(a)

(b)

(c)

FIGURE 10.4 (*a*) Rider side-loading lift truck (*Courtesy of Raymond Corp.*); (*b*) rider side-loading double-deep lift truck (*Courtesy of Raymond Corp.*); (*c*) rider swing-mast lift truck (*Courtesy of Drexal Industries, Inc.*).

length. This feature allows the lift truck to operate in a 5- to 6-ft-wide aisle. The side-loading vehicle handles a 2000 to 4000-lb pallet load at a 30-ft-high rack position.

Stand-up-Rider Double-Deep Lift Truck. The stand-up-rider double-deep side-loading vehicle (Fig. 10.4*b*) has the same design features and operational characteristics as the side-loading lift truck except that the pantographic device can perform transactions from a two-deep rack storage position. This vehicle requires a 6- to 7-ft-wide aisle. Due to the double-deep transaction at high rack positions the operator performs 15 to 17 transactions per hour.

Operator-Up Side-Loading Lift Truck.   The operator-up side-loading vehicle has the same design and operational characteristics as the standard side loader except that the operator elevates and lowers with the unit load to the rack storage position. The vehicle performs 1 to 2 additional transactions per hour more than the standard side loader.

Sit-Down-Rider Counterbalanced Side Loader with Swing Mast.   The next vehicle is the sit-down-rider counterbalanced side-loading and front-loading lift truck (Fig. 10.4c). The design features and operational characteristics of this lift truck are the same as for the regular counterbalanced lift truck. However, to make transactions into a rack or floor storage position, the telescopic mast swings 90° to the right side of the aisle. The mast movement gives the lift truck its nickname, which is the *swing-mast lift truck.* An electric rechargeable battery is the power source, and the truck operates in a 6- to 8-ft-wide wire-guided aisle with a minimum aisle width which is a nominal 1 ft wider than the pallet load. The swing-mast truck makes 3000-lb pallet load transactions from floor or rack positions 25 to 30 ft high. In a warehouse aisle of normal length, the operator's throughput capability is 17 to 20 pallet loads per hour for a dual cycle. But when the transactions are not on the same side of the aisle (single-cycle), then productivity is 15 to 17 transactions per hour because of the need for two down-aisle trips. The truck requires a dead-level floor and a 10- to 12-ft intersecting aisle.

Compared to other VNA lift trucks, the swing-mast truck can function as a transport vehicle and as a vehicle to service standard floor and rack positions from a wide aisle. Able to travel with its forks facing the side of the aisle, the swing-mast truck transports long unit loads.

Stand-Up Outrigger Side-Loading Rising Cab Lift Truck (Turn-a-Load or Rack Loader).   The unique feature of this truck is that the load-carrying forks are placed on a set of outriggers. This feature permits the vehicle to pick up and deliver the pallet load from the truck's side. As the forks rise, the operator's cab rises with the pallet load. In one model, at the rack position, the forks rotate on a turntable to perform a transaction from either side of the aisle. In a second model, at the pickup station the forks are flipped to the proper side to make the transaction at the rack position. This feature gives the trucks their nickname of *turn-a-load* or *rack loader.*

These vehicles are powered with an electric rechargeable battery. They operate in an aisle that is 10 to 20 in wider than the unit load. The bottom 2500-lb unit-load rack position is required at 14" to 16" above the floor and the top rack position is 20 to 21 ft high. An operator performs 17 to 18 activities per hour.

Operator-Down Counterbalanced Side-Loading Lift Truck.   The operator-down sit-down rider counterbalanced side-loading lift truck is the next vehicle. On this electric rechargeable battery powered lift truck, the forks are attached to a telescopic mast. The forks can move from the right side to the left side of the aisle. This permits the lift truck to handle a 2000- to 4000-lb pallet load at 30- to 40-ft-high rack positions. These lift trucks have side shifts to assist the pallet load deposit or withdrawal activity. The lift truck's nickname is the *turret* or *swing-reach truck.* Aisles are a nominal 20 in wider than the pallet load or 6 in wider than the diagonal dimension of the pallet to turn the pallet in the aisle as the vehicle travels down-aisle.

The operator load throughput capacity is 20 to 23 transactions per hour in a guided aisle of normal length. With the 40-ft-high stacking capabilities, the mobile-aisle lift truck requires a dead-level warehouse floor. The intersecting aisle must be a minimum of 17 ft wide. For best productivity, these vehicles require an end-of-aisle pickup and delivery station.

Side-Loading Turret Lift Truck with Operator Up or Down. On this vehicle the forks are attached to the top of the carriage. The carriage is attached to a telescopic mast that permits the forks to handle pallet loads from rack positions on either side of the aisle. For best operational results, the lift truck requires a guidance system, end-of-aisle pickup and delivery stations, and a dead-level floor. The truck travels in a 5-ft 6-in to 6-ft 6-in wide aisle that is a nominal 1 ft wider than the pallet load. To change aisles, it requires a 17- to 18-ft-wide intersecting aisle. The lift truck receives its power from an electric rechargeable battery. The vehicle has the operator-up feature.

Counterbalanced Side-Loading Truck with Rising Cab and Auxiliary Mast Lift. This multimast vehicle (Fig. 10.5a) carries the pallet load in front of the operator's cab. The operator's cab has all controls for vehicle movement and fork travel. As the cab moves up or down, the pallet load on an auxiliary mast moves with the cab. The unique feature of this truck is the auxiliary mast attached to the front of the cab. This feature provides the truck with 60 in of free lift that allows a 2000- to 3000-lb load to be deposited or withdrawn from a 40-ft-high rack position with minimal damage to the product or equipment because the operator has a complete view of the transaction. Also, in a carton storage or order-pick operation, the operator can hand-stack cartons. The truck requires a 16- to 20-in clearance from the bottom of the ceiling. Compared to other lift trucks, operator productivity is increased by 1 to 2 unit loads per hour.

When to Consider Operator-Down or Operator-Up Lift Trucks. If the warehouse stacking height is three to four unit loads, consider an operator-down vehicle. For a stacking height above three to four loads, the operator-up vehicle has become more popular in the warehouse industry. With the operator-up vehicle with its operator platform and environmental control of the cab, warehouse utilities are kept to a minimum.

Outrigger Stand-Up (Sit-Down) Truck with Rising Cab and Fixed Mast. This vehicle (Fig. 10.5b) is considered a hybrid, something between the AS/RS machine and a side-loading lift truck. These vehicles handle a 2000- to 2500-lb load at a 60-ft-high rack. With some vehicles, the forks are attached to the front of the cab, and the cab elevates and declines with the pallet load on one rigid mast to the pallet position. Other trucks have dual masts with shuttles or forks between them. The first storage position is a nominal 16 in above the finished floor. Some vehicles carry the pallet load between the outriggers and are guided by a rail in the middle of the warehouse aisle. Still other vehicles have guide wheels that are mounted on both sides of the truck. In the aisle the truck obtains its power from the ceiling dc system and recharges the battery. When it is slowly transferring aisles, the power source is the electric rechargeable battery. Normal aisle transfer time ranges from 1 to 2 min, and the transfer aisle width ranges from 24 to 27 ft. Due to the electric hookup, top rail guidance system, and wide transfer aisle, in most designs the lift truck enters from one end of an aisle. With proper electric devices at the top of the dual-mast vehicle, lift truck entry is designed at both ends of an aisle. The storage aisle is 4 to 6 in wider than the pallet load. To perform a transaction at the end of the aisle, an 8- to 10-ft runout is required at the other end of the aisle. Operators perform 25 to 30 transactions per hour in the normal warehouse aisle.

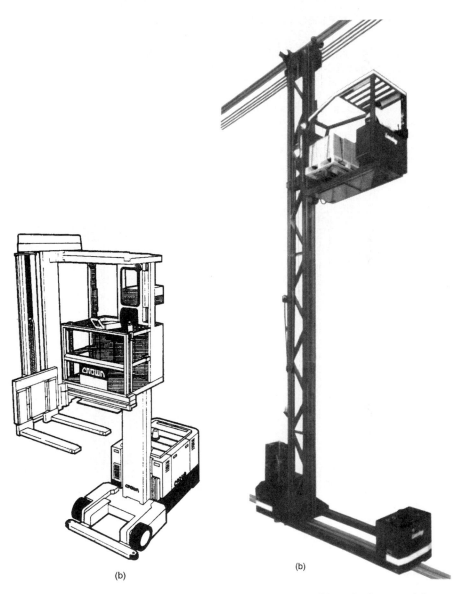

(b)

(b)

FIGURE 10.5   (*a*) Counterbalanced rising cab with auxiliary mast lift truck (*Courtesy of Crown Equipment Corp.*); (*b*) outrigger rising-cab fixed-mast lift truck (*Courtesy of Lansing Bagnel Ltd.*).

STORAGE/RETRIEVAL VEHICLES

Storage/retrieval vehicles (Fig. 10.6*a*) require an aisle that is a nominal 3 to 6 in wider than the pallet load. The vehicle is designed with one or two masts (Fig. 10.6*b*) and as a manual or completely automated vehicle that deposits or withdraws 2000- to 4000-lb pallet loads from 40- to 80-ft-high rack storage positions. These vehicles require a dead-level floor, middle rail guidance system, a first storage level 34 in above the floor, and clearance at the ceiling for the mast and ceiling-attached dc system. All storage and retrieval vehicles require a runout of 15 to 20 ft to perform transactions at the end rack storage positions. The runout is determined by the vehicle's length.

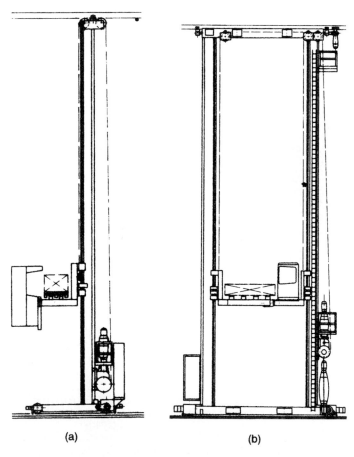

(a)   (b)

FIGURE 10.6   (*a*) Single-mast AS/RS (*Courtesy of Rapistan Demag Corp.*); (*b*) dual-mast AS/RS (*Courtesy of Rapistan Demag Corp.*).

Manually Controlled Vehicles

In the manually controlled storage and retrieval vehicle, the operator's cab rises with the pallet load. For best results, the vehicle requires an end-of-aisle pickup and delivery station.

Load-Handling Devices (Forks or Platens)

The pallet load is carried on a set of platens or forks that lift the pallet load to complete a transaction on either side of the aisle. When platens are used, the bottom of the pallet load requires a slave pallet board, and the rack opening consists of two rails with four support arms. As an alternative, the vehicle can be equipped with a set of forks that allow it to handle a pallet board in load beam rack positions.

Single-Command Mode

The typical storage vehicle performs 18 to 24 single commands per hour. In a single-command mode, the vehicle makes one down-aisle trip to deposit or withdraw one pallet load. This activity represents one unit-load transaction.

Dual-Command Mode

In the dual-command mode, the vehicle makes one down-aisle trip to deposit one pallet load and to withdraw one unit load during the return trip. This represents two unit-load transactions per trip, and operator productivity is 22 to 26 dual commands per hour. The dual-command mode does not double the single-command productivity rate because of the additional travel time and fork activity time.

Dual Shuttles

Some vehicles are designed with two sets of shuttles (platens or forks), which increases the throughput by 20 to 25 percent.

Other Operational Parameters

The automated storage and retrieval vehicle operates in a controlled environment with minimal utilities. In this environment, the manual vehicle requires an operator's cab. In an automated system, all unit loads must pass a size and weight station to verify that the unit-load dimensions and weight meet design parameters.

Captive-Aisle Vehicle

In a captive-aisle vehicle, there are multiple vehicles in the warehouse operation. In the captive-aisle method, one vehicle is allocated to one aisle. In this aisle, the vehicle performs all storage transactions. The projected pallet transaction activity has each storage vehicle performing at its maximum activity rate, and the storage transactions are evenly distributed to each aisle.

Mobile-Aisle Vehicle

A mobile-aisle vehicle is used when there are multiple vehicles in the warehouse operation. Vehicles can perform storage transactions in any aisle. With this vehicle allocation concept, the projected pallet transaction activity has each storage vehicle performing at its maximum storage transaction rate, but the storage transactions are randomly distributed to the various warehouse aisles. The mobile-aisle vehicle requires at least one vehicle transfer aisle (some have two vehicle transfer aisles) which increases the building costs but reduces the warehouse equipment costs.

T-Car

The storage and retrieval vehicle is a captive-aisle vehicle. But with a T-car (transfer car), the vehicle is transferred between aisles (Fig. 10.6c). The transfer car takes the AS/RS vehicle on board and transfers the machine between aisles. The transfer aisle is 25 to 40 ft wide according to the storage/retrieval machine length.

Automatic Wire-Guided Stacking Vehicle

The next vehicle is the automated wire-guided stacking vehicle, a programmable vehicle (Fig. 10.6d). The vehicle is either a counterbalanced or straddle type. The power source is several electric rechargeable batteries. The vehicle has three to four wheels (at least one is steering), a fixed mast, a set of load-carrying forks, a manual or remote-controlled on-board control device, and a wire guidance system. The vehicle is guided on a wire, tape, or painted path that requires a level warehouse floor.

The vehicle picks up a pallet load, travels through the warehouse, and deposits the pallet load on the floor or into an elevated rack position. The 2000- to 3000-lb load is placed into the pallet rack storage position which is 180 in high. Since the vehicle is used primarily as a transport vehicle, the pallet load storage throughput is 10 to 12 pallets per hour. The right-angle stacking aisle requirement for the counterbalanced type is 10 to 12 ft, and for the straddle type it is less.

Due to the instability of the floor-stacked pallet loads, the majority of VNA vehicles stack pallets one high on the floor. It is very rare to use these vehicles to stack two pallets high on the floor.

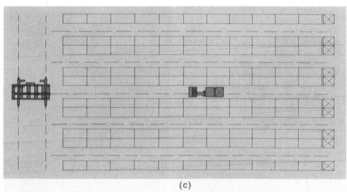

(c)

FIGURE 10.6   *(Continued)* (*c*) Transfer car (T-car). (*Courtesy of Rapistan Demag Corp.*)

HIGH-RISE ORDER-PICKER TRUCK

The high-rise multilevel order-picker vehicle is the next vehicle group. The power source is a rechargeable electric battery. There are two types of vehicles: a counterbalanced truck and an outrigger (straddle) truck. The counterbalanced truck handles pallet boards, and the straddle truck handles pallets or long carts. For best results, these vehicles travel in a guided aisle. These trucks have an operator's platform with com-

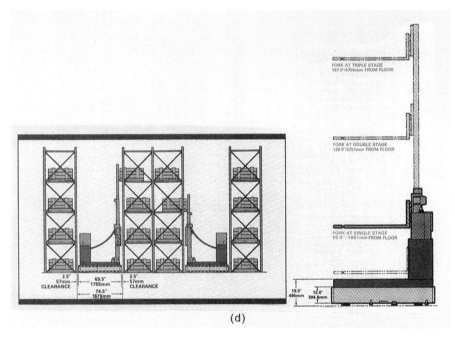

(d)

FIGURE 10.6   (*Continued*) (*d*) AGV stacking vehicle. (*Courtesy of BI Systems.*)

plete controls and a set of load-carrying forks that elevate and decline with the platform. These vehicles have four to five wheels with one wheel as a drive or steering wheel. With a pallet board securing device, the operator can hand-stack cartons between the rack positions and the vehicle's carrying platform. If these vehicles are used to make pallet load transactions in a floor or rack storage area or as transport vehicles, then employee productivity will be low because it is a poor use of the vehicle.

The counterbalanced narrow-aisle rider platform truck has the same design characteristics and operational features as the fork truck models. The exception is that the load platform is a solid deck instead of a set of forks.

ROUGH-TERRAIN (YARD OR OUTDOOR) LIFT TRUCK

The next major vehicle group is the rough terrain (yard or outdoor) lift truck group. These trucks are manually controlled from an operator's platform area which has a seat. From this seat the operator has access to the steering device and the fork controls for elevating, declining, and forward or reverse.

The rough-terrain vehicle comes in six major types: sit-down counterbalanced truck, variable-reach truck, side-loading truck, straddle carrier, van carrier, and straddle hoist carrier.

These trucks are human-controlled, are designed to operate outdoors, and can handle a wide variety of products. The key functions performed by these trucks are to unload and load flat-bed trailers, to stack and unstack palletized loads, to deposit and withdraw product from storage racks (standard or cantilever type), and to transport product between workstations. The truck's wheelbase, underclearance, wheel types, tires (pneumatic type), counterbalance weight, and vehicle weight permit it to operate effectively and safely over rough, soft, and slippery surfaces. The majority of the vehicles have diesel engines.

Due to the wind forces, the storage of unit loads outdoors is limited to an approximately 20-ft-high stack. This stacking height improves storage density, inventory control, and security. The warehouse operation stacks a load on another load, using stacking frames, pallet racks, or cantilever racks. The selected storage method is determined by the SKU stackability.

Sit-Down Counterbalanced Vertical Mast Lift Truck

The first outdoor vehicle group is the sit-down counterbalanced vertical mast truck (Fig. 10.7a). This truck elevates and lowers loads similar to the conventional indoor counterbalanced lift truck. It has a high degree of flexibility, handles a wide variety of loads, and performs other key warehouse functions. These trucks deposit and withdraw pallet loads from 20-ft-high rack positions. The truck operates in a 12- to 30-ft-wide aisle. The aisle width varies according to the type of product and weight handled by the truck. Acute-angle (45°) floor stacking reduces the aisle width to make a right-angle stack. Crab-steering can improve the truck's maneuverability. Crab steering is multidirectional steering, either turning both wheels at one end or the other of the truck or turning all four wheels of the truck.

These trucks are fitted with attachments to handle special tasks. Some attachments are special carriages, a set of forks, fork extensions, clamps, rams, booms, winches, and load rotation devices.

VARIOUS VERTICAL LIFT VEHICLES

These vehicles are powered by gasoline, LP gas, or diesel engines. Within this group, there are four types of trucks: vertical mast two-wheel steering two-wheel-drive, vertical mast skid steering four-wheel drive, vertical mast articulated-frame steering four-wheel drive, and vertical mast two- or four-wheel steering four-wheel drive.

The vertical two-wheel steering two-wheel-drive type of outdoor truck is considered a general utility truck. It handles pallet loads from 4000 to 20,000 lb.

The vertical mast skid steering four-wheel-drive vehicle is a highly maneuverable outdoor vehicle. It handles pallet loads from 1500 to 4000 lb and is used indoors as required.

The vertical mast articulated-frame (connected by joints) four-wheel drive truck is a highly maneuverable outdoor vehicle. It handles unit loads from 3000 to 15,000 lb. The unique feature is that the truck has excellent flotation and traction.

The vertical mast two- or four-wheel steering four-wheel-drive truck gives the best performance on rough terrain. The truck handles pallet loads from 4000 to 20,000 lb and has superior stability, flotation, and traction.

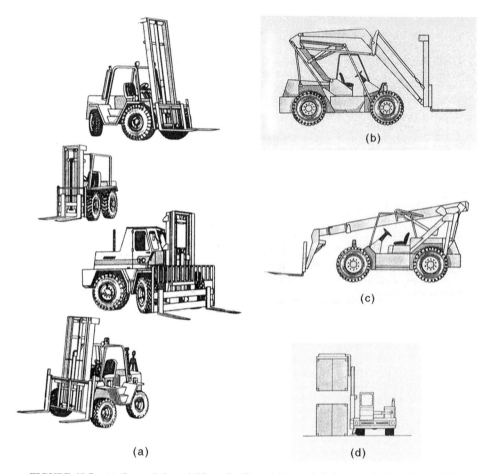

(b)

(c)

(a)

(d)

FIGURE 10.7 (*a*) Counterbalanced lift truck; (*b*) variable-reach linkage mechanism; (*c*) variable-reach telescopic boom; (*d*) side loader (*Courtesy of Modern Materials Handling Magazine.*).

Various Front-Load Variable-Reach Lift Trucks

The second outdoor truck group is the front-load variable-reach group. The variable-reach truck handles pallet loads from 5000 to 10,000 lb up to a height of 35 ft. These trucks require a wide aisle because the mast configuration extends from the rear of the vehicle to the front. The truck is flexible and is used for two-deep storage or loading/unloading of pallet loads from the side of a flatbed highway truck or railcar. The forward extension of the forks is 10 to 26 ft. Within the variable-reach lift truck group, there are two types, the linkage mechanism truck and the telescopic boom truck.

Linkage Mechanism Lift Truck. The variable-reach truck with the linkage mechanism (Fig. 10.7*b*) lifts a 5000- to 8000-lb pallet load straight in front of the truck. This feature gives the operator a clear view of the unit load. The truck has two- or four-wheel crab steering and two- or four-wheel drive.

Telescopic Boom Lift Truck.   With the telescopic boom, the vehicle is a front-load-handling type and allows the fork extension forward farther than the linkage type of truck. (See Fig. 10.7c.) This ability to extend forward allows the truck to pick up loads that are below ground level. The truck lifts pallet loads that weigh from 5000 to 10,000 lb and has two- or four-wheel or crab steering and two- or four-wheel drive.

Side-Loading Lift Truck.   The side-loading lift truck (Fig. 10.7d) is considered the VNA vehicle of the outdoor group. It has a pantographic device attached to a tele-scopic mast. The mast is located in the middle of the truck between the two decks. These features allow the lift truck to handle unit loads from 2500 up to 10,000 lb at rack heights of 10 to 15 ft. The required aisle width is a few inches wider than the truck and load.

As the vehicle travels down-aisle with a unit load, to stabilize the unit load, the operator allows it to rest on the two decks (outriggers).

Straddle Carrier Lift Truck.   The straddle carrier lift truck (Fig. 10.8a) has a series of shoes (short forks) that are suspended from the arch or top support member. When the load is deposited in the storage position, a bottom support device is needed to allow the shoes to be removed or inserted under the load. The vehicle handles loads from 72 in wide that weigh 12,000 to 60,000 lb and to a height of 12 ft. The loads are stored in a 90° position to the storage aisle. The row layout requires an additional 36 to 50 in between the storage lanes (objects) to allow the carrier wheels to travel between the loads.

Van Carrier Lift Truck.   The van carrier (Fig. 10.8b) has a lifting frame that attaches to the container under the arch and stacks the container at a maximum of 26 ft or three containers high. Each storage container row has a 36- to 50-in clearance (open space) between containers for van carrier travel.

Straddle Hoist.   The straddle hoist (Fig. 10.8c) has a cross beam on each arm sup-port. On the beam is attached an electric trolley and hoist which has a set of arms, sling, or magnet that becomes the support member of the unit load. The vehicle stacks loads up to 45 ft high. Storage loads weigh between 30,000 and 150,000 lb. The stor-age row layout allows for 44 to 50 in of wheel clearance between lanes.

Design Parameters for Outdoor Vehicles.   When you are considering an outdoor truck, you need exact specifications. In addition to the normal lift truck specifications, the outdoor vehicle specifications include (1) definition of the equipment application (floor surface, weather conditions, load weights and dimensions, stacking and for-ward-reach requirements and heights, clearances, and aisle widths), a statement that the lift truck complies with ANSI Standard B56.6, and fuel storage requirements.

BRIDGE CRANE

A bridge crane is a fixed-path unit load handling piece of equipment with an electric hoist. The bridge crane is designed for indoor or outdoor applications. The hoist is

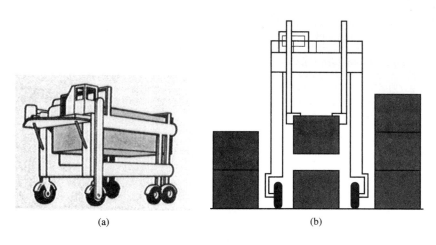

(a)　　　　　　　　　　　　　　　　(b)

(c)

FIGURE 10.8 (a) Straddle carrier (*Courtesy of Modern Materials Handling Magazine.*); (b) van carrier; (c) straddle hoist (*Courtesy of Shuttlelift, Inc.*).

suspended from a single or double girder with electric end trolleys that travel on a path supported by roof trusses and columns.

Various Types

For heavy loads, the two types of bridge cranes are the top-running and the bottom-running types. In designing a bridge crane concept, the main features are the span, headroom, end space, and hoist lift. A bridge crane is floor-operated with an electric hoist or has an operator's cab that is suspended between the strands of the girder.

The top-running bridge crane has cross girders on the top end of the trolleys. The bottom-running crane has the cross girders suspended from the bottom end of the trolley.

The variety of lifting devices for a bridge crane include grabs and hooks, magnets, vacuum lifting pads, counterbalanced hooks, and a set of forks.

A recent model of a bridge crane has a set of forks suspended from cross girders that are above standard storage racks. With a floor-operated system, the forks elevate, decline, or extend to deposit or withdraw a unit load from one of the rack storage positions.

IMPROVING TRUCK OPERATOR PRODUCTIVITY AND REDUCING PRODUCT AND EQUIPMENT DAMAGE

This section of the chapter examines procedures, equipment applications, and ideas to improve the lift truck operator's productivity, reduce product and equipment damage, and reduce the number of accidents. These factors are equipment options, enhancements, or modifications made to an existing lift truck, to the facility, or to a new lift truck.

In other chapters we indicated good warehouse practices to make an operation more productive and efficient. The factors reviewed in other chapters include (1) unit-load identification program, (2) clear and understandable reserve-pick position markings, (3) a stock locator system, (4) good housekeeping (keep the aisles clear), (5) adequate lighting and good maintenance, and (6) adequate aisle width clearances.

Shelf Selector

The first item is a lift truck option. This option is an electric or mechanical shelf selector (level finder) system that aligns and cycles the travel of a lift truck forks to the required elevation for a unit-load transaction from a rack storage position. The shelf locator system at higher rack positions improves employee productivity and reduces damage to the product, rack, and lift truck.

Notched Mast

The second factor is modification of existing equipment to reduce the need for lift operator to eyeball the fork alignment to make a unit-load transaction. Notch or mark (paint) the mast. Each notch or mark represents a forks-elevated position that corresponds with a storage rack level. The benefits are similar to those of the shelf selector with no investment.

Painted Lines on the Floor

The third idea is to paint lines to the width of the storage unit plus clearance on the facility floor. Lines are painted on the receiving/shipping dock floor area, warehouse aisle border areas, and the floor of the dense static storage areas.

In the receiving/shipping dock staging area, parallel lines are painted on the floor to create staging lanes. The distance between the staging lanes (lines) is the dimension of a pallet board plus clearance, and the length of the lines is equal to the predetermined number of pallets. This practice improves operator productivity because the pallet boards are oriented in the proper direction for easy and quick pickup. And because there is space between the parallel staged pallets, the pallets do not extend (project) beyond the staging lane into the adjacent lane or into the travel aisle, which reduces damage.

Lines painted on the warehouse floor aisle borders indicate to all employees the travel aisle and pedestrian workstation or rack. Also diagonal lines between the aisle border lines indicate a pedestrian crosswalk. The border lines ensure that the required product-to-product aisle width is maintained in the aisle. This permits unit-load deposit and withdrawal transactions to be made without damage to product or equipment and ensures high lift truck driver productivity.

Lines painted on the floor within a static dense-storage lane are a guide for the lift truck driver entering or exiting the dense-storage lane. This practice improves driver productivity and reduces damage to the product or rack post.

Strip (Paint) Load Beams and Floor

Another practice is to paint a strip on each load beam and floor pallet position of a rack bay. This indicates to the lift truck operator the desired pallet board location on the load beam and floor. In an operator-down operation, these paint marks on the load beam and floor help the driver locate the proper aisle to make a transaction, which improves driver productivity and reduces damage to product or equipment.

Sprinkler Protection

Most warehouse facilities have in-rack sprinklers, which are required by the fire insurance underwriter and local authorities.

To prevent product damage caused by the sprinkler's being accidentally hit and to ensure good lift truck driver productivity, tuck the sprinkler in the flue space between the back-to-back rack structures or near the upper portion of the cross beam. When you design the rack opening, allow sufficient clearance between pallets and rack members to permit the lift truck driver to make transactions.

See-through Masts

See-through masts give the lift truck operator an unobstructed view. The operator has a clear view over a one-high unit load as the truck travels in the aisle. Clear view of the travel path reduces damage to product or equipment and injuries to personnel.

Electric Heater for Lift Truck Cab and Mast

An option for lift trucks that operate in a freezer or cold storage environment is a heated cab to keep the operator warm and heaters to warm the hydraulic oil. These features allow the lift truck operator to perform work in the freezer for longer periods.

Fans and Lights in Operator's Cab or Platform Area

Fans and lights in the operator's area improve driver productivity.

Highway Guarding

Highway guarding is a warehouse construction item that provides a barrier between the fork lift truck's travel aisle and workstations, stationary equipment, and door frames. These barriers keep the lift truck from entering a restricted area or hitting a building structure, which reduces equipment and building damage and improves safety.

Sufficient Aisle Widths

When you design a warehouse aisle between the unit-load storage position, the dimension is product to product plus 6 to 12 in. This aisle width provides sufficient clearance for the lift truck operator to make a unit-load deposit or withdrawal transaction without damage to the product or equipment. When the operator is convinced that she or he will not hit the rack structure, that operator is more productive.

Fuel or Energy Indicator

Another important equipment option is the lift truck fuel or energy indicator which shows the energy remaining in the power source. An accurate indicator avoids the nonproductive time for fuel replacement in the middle of the warehouse or operating with a low battery.

Standardize the Fleet

The next important material handling equipment feature is the standardization of the lift truck fleet in your distribution facility. Equipment that has common operating controls, operator platform areas, and warehouse aisles and clearances designed for a particular truck allows an employee to operate a replacement lift truck with no extensive training. Also the maintenance training and spare-parts inventory are reduced because all the equipment is the same. All these facts make your employees more productive and reduce operating expenses.

Lift Truck Rodeo or Rally

The next factor is an employee activity that improves employee productivity and morale, reduces damage to product or equipment, and reduces injuries. A fork truck rodeo or rally is a two-part program that achieves the above-mentioned objectives. A prerequisite for lift truck participation in the lift truck rodeo is an accident-free year.

First the lift truck operator takes a written examination on lift truck operations and safety. Then the lift truck driver performs a vehicle safety check and drives the truck through a course that tests the driver's skill and safe-driving ability.

LIFT TRUCK ATTACHMENTS

A lift truck attachment permits a lift truck (usually a counterbalanced type) to perform additional functions or to perform the original function more efficiently and effectively.

Most lift truck attachments are load-handling devices other than the conventional set of forks, and they are attached to the front of the mast. The attachments are permanently or temporarily attached to the elevating mechanism of the counterbalanced lift truck.

The benefits of lift truck attachments are that they improve employee productivity, improve space utilization, enhance use of the lift truck, and reduce damage to product.

Economic Justifications

If there is a potential application for a lift truck attachment, consider these factors: (1) Is the unit-load weight uniformly distributed? (2) Determine the number of times that the attachment is used in the operation. (3) Estimate the purchase and installation cost of the attachment; if required, include lift truck modification. (4) Estimate the economic savings (labor and product damage).

Items to Review

Prior to the purchase of the lift truck attachment, you must figure out whether the existing lift truck can handle the attachment and the new unit-load weight. Most attachments are located in the front of the forks and plate and extend the load more forward—to the front of the lift truck fulcrum—than the normal load. The new weight position and the attachment weight move the center of gravity forward of the fulcrum, which decreases the lift truck manufacturer's stated weight capacity. To use the attachment in your operation, decide with the lift truck manufacturer whether to increase the existing lift truck's counterbalanced weight or to purchase a new lift truck with a higher rate capacity that will handle the new load weight and center. When the counterbalanced weight is changed on your lift truck, have the lift truck manufacturer do it.

Various Attachments

There are numerous types of lift truck attachments. These attachments hold or manipulate a unit load. This makes the lift truck driver more productive, increases control of the unit load, and reduces product damage.

Lift truck attachments are classified into five major groups: attachments for (1) better pallet load handling, (2) slip-sheet handling, (3) a unit load with no base, (4) better handling of a specific item, and (5) special odd-job service or specially designed attachment.

Pallet Load Handling Attachments.   The pallet load-handling attachments are side shifters, fork positioners, and load stabilizers.

Side Shifter.   The side shifter quickly and precisely moves the pallet load or forks to the left or right. This movement allows the operator to make a pallet load deposit or withdrawal with little equipment or product damage and increases operator productivity. In today's new distribution facilities, the side shifter is considered a basic device on most lift trucks that perform the key warehouse functions of unloading and loading delivery vehicles and storage transactions.

Fork Positioner.   The fork positioner is a hydraulically controlled device that resets the lift truck's fork spacing. The fork spacing is the interior distance between the two forks. The ability to quickly and easily reset the open space between the forks allows the lift truck operator to handle notched pallet boards, skids, or containers with various fork-opening widths. Suppose a distribution facility handles a 48- by 40-in four-way pallet board and in the rack pick position requires the 40-in dimension into the rack, and in the storage area the pallet board has the 48-in stringer into the storage lane. Then during the pick position replenishment transaction, the fork positioner increases operator productivity.

Load Stabilizer.   The load stabilizer is a device that holds the top and bottom of a pallet load together. The attachment applies hydraulic pressure to the top stabilizer, which has a rubber face pad. The downward pressure of the pad holds the cases of a pallet load together. When the pallet load is being transported and cartons might slip from or bounce off the unit load, the load stabilizer reduces product damage, reduces cleanup time and expenses, and improves the driver's productivity.

Slip-Sheet Attachments.   These devices handle a unitized load that is on a corrugated board, solid fiber board, or plastic slip sheet. There are three basic types of slip-sheet handlers.

Platen Push-Pull Device.   The first is the push-pull device (Fig. 10.9a), and it is the most common type. This device consists of two wide and thin metal platens that are connected together at the carriage end and a pantographic device that has a gripper and a unit-load backrest. The slip-sheet lip extends outward from under the load by 6 in and is secured by the gripper. The lift truck operator raises the unit load above the floor. During this action, the lift truck operator moves the platens forward between the raised unit load and slip sheet and the floor. This combination of pushing the platens and pulling the slip sheet brings the entire unit load and slip sheet to rest on the platens. At this time the lift truck driver places (pushes) the unit load onto a pallet board or in a floor position.

Tine Push-Pull Device.   The second slip-sheet handler (Fig. 10.9b) is very similar to the push-pull device, but it has a series of narrow chisel tines instead of two platens.

Wide-Platen Push-Pull Device.   The third slip-sheet handler is very similar to the push-pull device except that there is one wide platen instead of two platens.

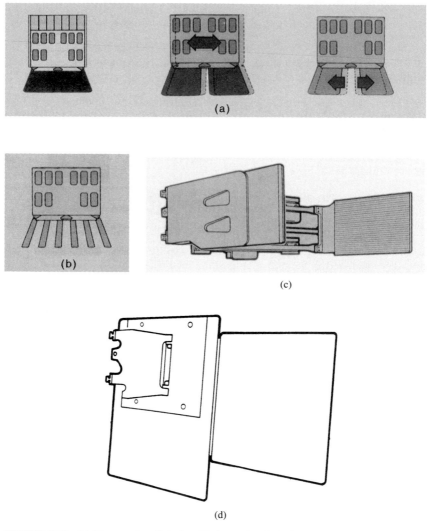

FIGURE 10.9   (*a*) Platen push-pull device; (*b*) tine push-pull device; (*c*) carton clamp; (*d*) whole unit-load clamp (*Courtesy of Cascade Corp.*).

Carton Clamp Attachments

The third major lift truck attachment group is the carton clamp group (Fig. 10.9*c*). The two basic types are the whole unit-load clamp and the one- or two-layer clamp.

Whole Unit-Load Clamp.   The whole unit-load carton clamp (Fig. 10.9*d*) consists of two large, wide, and flat padded clamp arms that are adapted to pick up a unit load of corrugated cartons with no bottom support base. These arms are hydraulically pressurized to move together. This movement exerts pressure on the two sides that hold the unit load together, and this allows the lift truck to raise and lower the unit load.

One- or Two-Layer Clamp. This clamp has similar characteristics and operational features to the large padded carton clamp, except that the clamp arms have smaller height and length. This permits it to handle one layer of cartons from a unit load. The unit load ti* × hi† carton stack pattern is interlocked with no airspace. When a layer is clamped, these features reduce product damage and improve lift truck productivity.

Specific Item Handler Attachments

The fourth attachment group is the specific item handler group. These devices are designed with a special size, weight, shape, and ability to hold, manipulate, or move items that do not have a container, pallet, skid, or slip sheet.

Double-wide forks	Tilt
Fork extenders	Vacuum lifter
Concrete handler	Long-load stabilizer
Cotton bale	Die handler
Paper roll clamp	Bin load dumper
Cargo container handler	Carpet handler
Multipurpose handler	Tire handler
Single- or double-drum handler	Appliance handler
Crane boom	Rotator
Drum arms	Tow coupler or hitch
Fork rotator	Pipe clamp
Fork hooks	Railcar door opener
Remote lift/lower	Dumping hopper
Single or double basiload	

Special Odd-Job Attachments

The fifth lift truck attachment group is the special odd-job service or specially designed attachment. These devices make the lift truck a piece of equipment that performs jobs other than to manipulate, move, and hold cartons or single items. Odd-job or special attachments are the sweeper, plow, and bulk scoop.

Note that the counterbalanced lift truck is a very versatile and flexible piece of equipment, and the lift truck manufacturer can design an attachment to handle almost any warehouse task that does not exceed the lift truck's design characteristics.

---

*The number of cartons per layer on a unit load.
†The number of layers per unit load.

Lift Truck Tires

The correct wheel on the lift truck minimizes truck downtime, lowers truck (wheel) and floor maintenance expenses, and reduces the truck's power usage. The two basic types of tires are the pneumatic (air pressure) tire and the cushion tire.

Pneumatic Tire. If a truck is used outdoors or in a yard operation, then pneumatic tires are the best. The pneumatic tire provides excellent traction on wet (slippery) surfaces, ramps, and unpaved or soft surfaces. Another feature is that the tires provide some degree of load stability and cushioning as the truck travels over an uneven surface.

Compared to the other tire group, the pneumatic tire is physically larger and is less maneuverable, but it allows the lift truck to attain greater travel speeds. Not all pneumatic tires are inflatable because there is a semipneumatic or zero-pressure tire.

Cushion and Solid Tires. Cushion and solid tires are smaller and are used on most indoor lift trucks. The two types are rubber tires and polyurethane tires.

Polyurethane tires provide less rolling resistance, longer wear, fewer chucks in the tire, and greater load-carrying capacity than pneumatic tires. The polyurethane tire is a nonmarking tire; but the polyurethane tire does have a tendency to skid or slide on smooth surfaces, inclines, and wet surfaces.

Rubber tires provide better traction on ramps and wet surfaces and a more cushioned ride, and they are less expensive than the polyurethane tire. These tires are popular on the LP gas, gasoline, and diesel powered vehicles. If these tires have treads, they leave marks on the floor. These are considered nonskidding tires.

Tire Options. In choosing a tire for an indoor lift truck wheel, consider these facts. (1) For the drive tire, the best tire is a smooth rubber tire that provides maximum floor contact and stability for the high lifts. (2) Grooved tires are best for long travel distances due to minimal heat buildup. (3) Tires with biased grooves provide the best traction and braking action. (4) Tires with lugs provide the best traction and are most common on outdoor vehicles. (5) A flat tire that covers a wide surface provides maximum traction. (6) A semi- or full-crown (high-profile or narrow) tire that provides minimal floor contact is the best tire for steering the lift truck.

Lift Truck Accessories

Lift truck accessories improve employee safety and productivity, enhance employee morale, and reduce product and equipment damage. Another important benefit of accessories is the reduction in lift truck accidents.

Lift truck accessories that improve operator safety include (1) cabs for outdoor vehicles, (2) tall-load backrests, (3) backup lights and alarms, (4) dead-operator brake controls, (5) seat belts, and (6) see-through masts.

Accessories that improve operator productivity and reduce damage and maintenance expenses are (1) power source indicator or fuel gauge, (2) hour meter that shows accumulated hours of operation, (3) travel directional arrow, (4) spotlight or cab with platform lights and fans, (5) lift intercept that restricts lift with a low battery, and (6) battery compartment rollers.

Lift Truck Power Source

There are two basic power sources: electric powered and internal-combustion engine.

Electric Powered. Electric power is power source for most vehicles that are used indoors due to the low noise and emission of fumes. Within this group, power is supplied by an electric rechargeable battery of 12, 24, 36, 48, or 72 V or by a dc bus bar. The majority of the WA, NA, and VNA mobile vehicles are powered with electric powered engines. With the rechargeable-battery vehicle types, a discharged battery is recharged on the equipment or exchanged for a second fully charged battery. The discharged battery is recharged in the battery charging area. Most of the captive-aisle or T-car transferred VNA hybrid or AS/RS vehicles have a bus bar as the power source.

Internal-Combustion Engine. If the facility has good ventilation and the facility has the following operational characteristics, then an internal-combustion engine is considered best for the job. The operational characteristics are long periods of operation, long travel distance, and high travel speeds.

With the high lift truck clearance, long wheelbase, and pneumatic tires, the internal-combustion engine is best for outdoor use and to climb grades of 10 percent or greater. Compared to the electric powered group, the internal-combustion trucks supply the highest torque or amount of force (energy) available to drive a vehicle.

The two major disadvantages of the internal-combustion engine truck are the need for an approved location for fuel storage and good facility ventilation. Most local codes require storage of the extra fuel supply to be in a secured area outdoors or in a location that has sufficient air rotation to reduce the fumes to an acceptable level.

Lift Truck Batteries. The rechargeable battery contains lead cells that are engulfed in a sulfuric acid–water mixture. These cells are connected via wire cables to the battery. The cables connect the battery to the lift truck or battery charger.

The battery's *capacity* is determined by the number of hours that it is expected to power the lift truck. The battery capacity is expressed in ampere hours.

Different Battery and Charger Connections Can Save Mischarge. When your electric battery-powered lift truck fleet is comprised of equipment with different voltage requirements, then each group of trucks with a specific voltage has its own type and color for the battery-charger connector. This feature reduces the mischarging of the batteries that destroys the life of the battery.

VNA Storage or Order-Picker Truck Guidance. To maximize the existing or new warehouse facility product storage-pick positions, consider facilities that have tall rack structures. Tall rack structures require a lift truck to perform SKU storage-pick transactions at elevated levels and operate within a very narrow aisle. These vehicle aisles have minimum clearances (nominal 6 ft 4 in) between two rows of racks. To obtain the anticipated savings and return on investment, these vehicles require an aisle guidance system.

The focus of this section is to describe the alternative very narrow-aisle (VNA) guidance concepts and their associated disadvantages and advantages.

The purpose of aisle guidance is to reduce the operator's effort of steering the vehicle down-aisle (traveling) in a very narrow aisle between two rows of rack. In most systems the clearance between the vehicle and product is 4 in on each side; but with some vehicles that handle long loads at high levels, the clearance is 8 in on each side.

VARIOUS AISLE GUIDANCE METHODS

The two alternative vehicle aisle guidance systems are rail (mechanical) and wire, tape, paint, and laser beam (electric). The mechanical guidance systems are used with some modifications on existing vehicles or in an existing or new facility. If a guidance system is considered for an existing or new facility, careful attention must be given to the floor (surface, loadings, location of metal objects) prior to purchase and implementation of the system. This is especially true for a wire guidance system.

Rail Guidance

The basic rail vehicle aisle guidance system has one or two angle-iron rails; some rack manufacturers have rails. The angle iron is 4 in by 3 in or 4 in by 4 in; manufacturer rail has a 3-in base extension which is 1 or 2 in less. These bases are anchored to the floor on both sides for the full length of each warehouse aisle. In one rail guidance system the guiderail is on right side of the aisle as the vehicle travels through the aisle. The rail is matched to the vehicle guide rollers, which are attached to the side of the vehicle, or a roller device is attached to the right side of the vehicle. The vehicle rollers ride against the rail or rails, which provides down-aisle vehicle guidance. To assist the vehicle operator with aisle entry and to reduce rack and vehicle damage, end-of-aisle angled-entry guides are installed on both openings to the aisle entrance. These occupy a portion of the main traffic aisle.

Aisle Entry Guides.   At the main traffic aisle the angled-entry guides have a wider opening (nominal at 7° to 10° from the aisle) and become narrower as each entry guide meets the aisle guiderail. The various aisle entry guides (Fig. 10.10*a*) are rounded, spring-loaded, curved, and straight.

With an entry the extension outward from the rack (Fig. 10.10*b*) creates a 27-in open space. When this area is covered with a deck, it provides a good location for trash containers or a document control table. This can improve facility appearance, safety, and housekeeping.

Rail Installation Parameters.   In most rail applications, the 4-in side of the angle iron extending upward (leg-up) comes in contact with the vehicle's guidance rollers. To secure the rail onto the floor, in the typical installation the rail is anchored on 8- to 9-in centers for the entry guides and the first 12 ft into the aisle. After the first 12 ft of the aisle, the anchors do not receive as much pressure due to the alignment of the lift truck. This permits the anchors on nominal 24-in centers. To provide smooth travel in the aisle and to reduce guide roller wear, at the locations where two pieces of angle iron are joined, the angle iron is ground smooth. Then the angle iron is painted with an OSHA-approved safety-colored coating.

The various rail guidance systems are

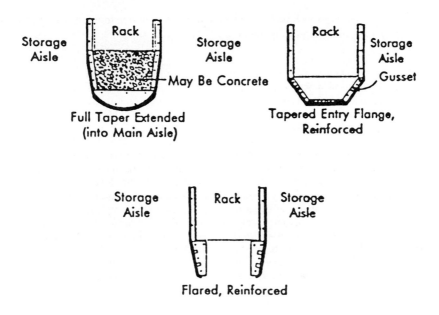

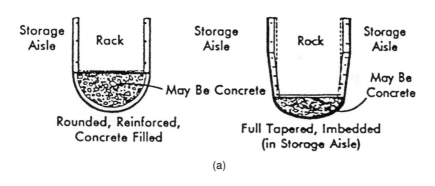

(a)

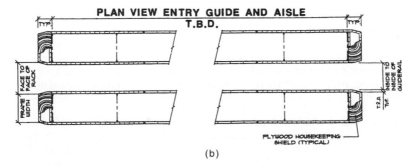

(b)

FIGURE 10.10   (*a*) Entry guides (*Courtesy of Raymond Corp.*); (*b*) entry guide cover.

• Double-rail systems—elevated, floor-mounted, and rack-mounted rails
• Single-rail systems—floor-mounted or rack-mounted rail

Double-Rail Systems.   When you consider a double-rail guidance system, it has a nominal 4 ft 6 in to 4 ft 8 in of clear space between the two rails. But the required clear space between the rails is a function of the vehicle's roller width, unit-load dimensions, and transaction height. Specific dimensions are calculated by the lift truck manufacturer for each application.

Elevated Double-Rail System.   The elevated-rail aisle guidance system consists of two rack rows on an elevated floor section (Fig. 10.11a). The aisle side of each elevated floor section has a full-length angle-iron face. The vehicle's two side-mounted guide rollers ride against the two metal sides. This method is used for storage or pick systems that handle pallet loads, cartons, and single items. Typically, it requires the replenishment vehicle to operate within the same aisle. To perform the required transaction at this elevated position above the finished floor, the vehicle turns and extends the pallet load, or the operator hand-stacks the product.

The disadvantages of the elevated double-rail system are that it is difficult to relocate, it is difficult to reuse the warehouse area, it increases capital investment, and there is the entry guide tripping hazard. The advantages are that the first position is above the finished floor, which reduces the load beam cost and reduces housekeeping problems.

Double-Rail Floor-Anchored System.   In the floor-anchored double-rail system (Fig. 10.11b), two rails are anchored to the full-length floor on both sides of the aisle. In this concept the first storage level is on a pair of load beams that are elevated above the guiderails. Replenishments are made from the same aisle or from a separate aisle, which permits the system to handle pallet loads that are set on the floor. The latter is an order-pick system for cartons and single items. It has the same operational requirements as the elevated rack system. The vehicle's guidance rollers are side-mounted, or there is a guide roller device on the sides of the vehicle.

The disadvantages are that additional investments are required for a separate replenishment aisle, load beams (with one concept), and entry guides and that a housekeeping or tripping problem is created. The advantages are that it is easy to relocate and the warehouse space can be reused.

Double Rail Attached to the Rack.   Guiderails are attached to the rack structure for the full length on both sides of the aisle (Fig. 10.11c). This system provides a 6-in open space between the finished floor and bottom of the guiderail. In a carton or single-item pick system, the vehicle has a guide roller device on both sides of the truck, or the two side guide rollers are set at a height to come in contact with the guiderails. Unit-load or hand-stack replenishment is made from the order-pick aisle with an NA vehicle or to the rear of the pick position from a separate replenishment aisle.

The disadvantages are that it requires additional load beam and rail cost and that pallet load replenishment requires a separate aisle. The advantages are that it reduces housekeeping problems, is easy to relocate, and permits good air circulation on the bottom level.

Single Rail Attached to the Floor.   A single rail that is attached to the floor for the full length of the aisle is used to guide an order-picker truck (Fig. 10.11d). The rail is on the right side of the floor and requires the vehicle to have a guide roller device on the right side. With this rail and guide roller arrangement, replenishment is a hand-

(a)

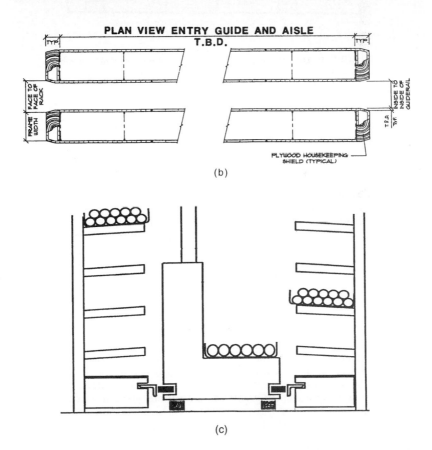

(b)

(c)

FIGURE 10.11 (*a*) Elevated floor rail (*Courtesy of Crown Equipment, Inc.*); (*b*) double-rail floor-mounted system (*Courtesy of Barrett Indl. Trucks, Inc.*); (*c*) double-rail rack-mounted system.

(d)

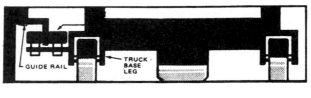

(e)

FIGURE 10.11 (*Continued*) (*d*) Single-rail floor-mounted system (*Courtesy of Barrett Indl. Trucks, Inc.*); (*e*) single-rail rack-mounted system (*Courtesy of Barrett Indl. Trucks, Inc.*).

stack activity, is performed by a VNA vehicle that extends the pallet load into the position, or is from a separate replenishment aisle.

Disadvantages of the single floor-mounted rail are that it creates a housekeeping problem, requires a vehicle with a guide roller device, and with most applications requires a separate replenishment aisle for the activity. The advantages are lower rail and load beam costs.

Single Rail Attached to the Rack. The single rail attached to the rack (Fig. 10.11*e*) is used specifically to guide an order-picker truck. The guiderail is attached for the full aisle length on the right side of the rack at the height of the first load beam level. All unit loads are replenished with the up-and-over method, performed with a stand-up-rider straddle lift truck. There is a 6-in clear open space between the finished floor and the bottom of the guiderail, and the aisle is wide enough to complete a right-angle stack transaction.

This method allows carton or single-item order-pick activities from SKUs that are hand-stacked into position or from SKUs on pallet loads. This system conserves warehouse space because there is one aisle for pallet load and hand stack replenishment and carton or single-order pick transactions.

The disadvantages are that it entails an additional load beam cost, the vehicle has a guide roller device, rail is attached to the rack, and it requires a straddle lift truck which has a low lift of 14 to 18 ft. The advantages are that it requires less warehouse aisle space, reduces housekeeping problems, and permits good air circulation on the first pallet load level.

Electronic Guidance Group

The second major type of VNA vehicle guidance is electronic. Within this group, there are four systems: wire, magnetic tape, magnetic paint, and laser beam.

These trucks require the aisle guidance wire, tape, or paint to extend into the intersecting aisle, which provides the necessary distance for the vehicle to pick up the guidance signal. Entry guides are optional for trucks that are guided by one of these systems.

Wire Guidance.   Wire guidance (Fig. 10.12*a*) is the most popular system of the electronic group. A wire is buried (nominal $\frac{3}{8}$ to $\frac{5}{8}$ in) in a saw-cut path in the center of the aisle. After the wire is embedded in the floor, the saw-cut path is filled in with an approved substance. The wire runs the full length of the aisles and is a closed loop for an approximate length of 4000 ft to 5000 ft. The loop starts and ends at a line driver (electric impulse creater). At floor joints and especially expansion joints, the wire is looped to compensate for movement. These electric impulses are picked up by the vehicle sensing device. This sensing device is the second component of the system and is attached to the vehicle undercarriage. On the truck, to prevent equipment damage and for safety, most systems have a short-duration UPS capability and an off-wire stop feature.

After the vehicle is on the wire and the system is activated, if the vehicle leaves the wire by a nominal 1 to 2 in, its travel is stopped. Note that certain floor levelness, metal object location, content of metal hardener, and guide wire to re-bar (wire mesh) depth must meet specific standards to ensure good operation.

When you design the vehicle path, sufficient aisle runout (wire extending beyond the P/D station) is in the layout for easy truck pickup of the wire.

Disadvantages of wire guidance are that (1) it creates slurry when the floor is cut, (2) specific floor tolerances are satisfied, (3) it increases vehicle investment, (4) it is difficult to add length, and (5) the capacity length is 4000 to 5000 ft per line driver. The advantages are that (1) it reduces housekeeping problems and tripping hazards, (2) floor positions are used, (3) it is used by both storage and pick vehicles, and (4) it is preferred for a multivehicle multiaisle system.

Tape Guidance.   A magnetic tape strip is applied to the surface of the floor for the entire length of the vehicle path.

A sensing device is located on the vehicle undercarriage and directs a light beam onto the tape. The light is reflected back from the reflective tape to the sensing device, which ensures that the vehicle is on the guide tape.

The disadvantages and floor-level requirements of tape are similar to those of wire guidance. Two additional concerns with tape are (1) that it has low durability because tape tears easily and wears with cross vehicle traffic, which becomes a problem in the main aisle for vehicle aisle entrance, and (2) that in a carton or single-item pick system, to replenish pick positions, it requires a separate replenishment aisle or a vehicle that extends the pallet load into the pick position, or else the product must be handstacked in the pick position.

The advantages are that it is easy to add length and is less expensive.

Magnetic-Paint Guidance.   This system has similar design and operational features to tape.

Laser Beam.   This new technology consists of a light beam source from the vehicle and reflective targets (Fig. 10.12*b*). These targets are located strategically in the aisle to reflect the light back to a receiver on the vehicle. If the vehicle does not receive the light beam, then aisle travel is stopped.

The disadvantages are that it increases capital investment, the vehicle requires a line of sight, and it requires the in-aisle replenishment vehicle to have the same guidance system or to use an alternative guidance system or separate replenishment aisle.

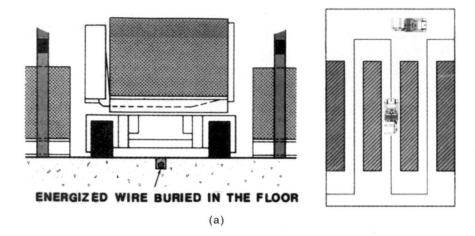

ENERGIZED WIRE BURIED IN THE FLOOR

(a)

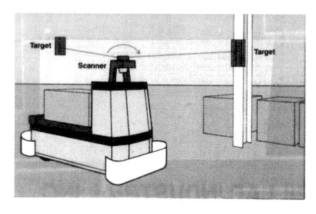

(b)

FIGURE 10.12   (*a*) Wire guidance system (*Courtesy of Hyster Co.*); (*b*) laser beam guidance system (*Courtesy of Caterpillar, Inc.*).

When to Use Rail or Wire

If building floor conditions can accept either system, which system do you choose for implementation in your warehouse? For the wire guidance system, the downtime and investment are factors that influence the decision. If the operation is in a remote area and the factory-qualified wire guidance and vehicle service center is not close to the operation, which creates an unacceptable downtime, then choose rail guidance. If the operation is close to a factory-qualified wire guidance and vehicle service center, then choose wire guidance.

With the mobile vehicle system (one truck per two or more aisles), wire guidance represents the lowest investment. If the vehicle is a captive-aisle vehicle or is a mobile

| Type of | Number of trucks | | | | | | | | | | |
guidance concept	1	2	3	4	5	6	7	8	9 +	Total	%
Rail	66	26	15	3	1	1	-	1	2	109	33%
Wire	127	55	12	7	7	6	1	4	2	221	67%
Total	187	81	27	10	8	7	1	5	4	330	100%

(a)

Type of guidance concept	1984	1985	1986	1987	1988	1989	1990	Total	%
Rail	5	9	8	11	20	48	10	111	35%
Wire	6	31	25	24	45	51	30	212	65%
Total	11	40	33	35	65	99	40	323	100%

(b)

FIGURE 10.13 (a) Number of VNA trucks per facility; (b) number of facilities with VNA vehicles per year.

vehicle between two aisles, then rail guidance offers the best return on investment (Fig. 10.13a and b).

END-OF-AISLE VEHICLE SLOWDOWN DEVICES

The focus of this section is to describe the alternative concepts to slow a VNA vehicle's travel speed or to stop the vehicle at the end of the aisle. Also we look at each system's disadvantages and advantages.

The majority of the newly constructed or remodeled warehouse facilities have tall racks with very narrow storage aisles. These tall rack structures require lift trucks that perform SKU storage-pick transactions at elevated positions and operate within very narrow aisles. These vehicles have minimum clearances between the two rows of storage racks. Some employees refer to the aisle travel as traveling in a tunnel. To achieve the anticipated savings, return on investment, and required employee productivity, these vehicles travel at maximum travel speeds and give the operator access to all positions. In many facilities, to meet these objectives, these vehicles transfer between storage aisles and while in the storage aisle they require a guidance system.

Per the storage area design, the vehicle must perform the following activity: In a storage (pick) area which has two turning aisles, prior to an aisle transfer to obtain a very slow travel speed. This feature reduces potential accidents. Also in this situation the vehicle can complete an end-of-aisle storage transaction at an end-of-aisle storage position. In a storage-pick area with dead-end aisles (one turning aisle) prior to aisle transfer or to perform an end-of-aisle storage transaction, the vehicle stops or slows down. This feature reduces potential accidents and permits completion of a storage transaction at an end-of-aisle storage position or in a dead-end aisle stop to avoid hitting the building wall.

The end-of-aisle vehicle travel stop or slowdown provides the operator with a signal or stops the vehicle to prevent uncontrolled travel into a wall at the end of a dead-end aisle or exits the aisle in a high travel speed with the potential to hit another vehicle or employee.

The various VNA vehicles that require a travel stop or slowdown system are:

- Order-picker trucks (counterbalanced, straddle, platform, and decombe)
- Storage vehicles (turret operator-down, turret operator-up, side loader, four-directional, swing mast, hybrid fixed mast, and AS/RS)

Two Alternative Concepts

The two means of vehicle slowdown or stop are the rail and rack (mechanical) and the electromagnetic (automatic). These systems are used with some modifications on an existing vehicle, for an order storage-pick system, or in a new facility. If the electromagnetic system is considered for an existing facility, careful attention must be given to the floor (surface, loadings, depth, metal object location) prior to purchase and implementation of the system.

Various Slowdown Devices.   The various means of vehicle stop or slowdown are (1) operator-controlled, (2) operator-controlled with end-of-aisle rack bays painted a different color, (3) rail and bumper stop, and (4) electromagnetic stop.

It is the premise of this section that these VNA storage-pick vehicles have a vehicle aisle guidance system. This system reduces the operator's effort of steering the vehicle down-aisle (traveling) at maximum speeds in a very narrow aisle between two rack rows. Also after the load-carrying forks or operator's platform is elevated above a specific point, the vehicle's travel speed is not maximum.

Manually Controlled Slowdown

In this system, the operator has complete control of the vehicle's travel speed and determines vehicle arrival at the end of the aisle. At the end of the aisle, the operator proceeds very slowly from the very narrow storage aisle to the main traffic aisle or stops at the end of a dead-end aisle.

The disadvantages of the method are additional employee training, low employee productivity, potential equipment damage, potential vehicle accidents, potential employee injuries, and lack of a system to control the vehicle.

The advantages of the concept are that there is no additional equipment expense or investment and it is implemented in an existing building.

Rack Bays Painted Different Colors.   In this operator-controlled method, end-of-aisle rack bays and load beams are painted a different color from the other rack bays (Fig. 10.14*a*). The operator has complete control of the vehicle's travel speed and with the rack color determines the vehicle's arrival at the end of the aisle.

Prior to the end of the aisle, the upright frame posts and load beams of the last two to four rack bays are painted a different color from the other rack bays of the aisle. These different colors are a signal to the operator that the vehicle has reached the end of the aisle. With these signals at these aisle locations, the operator is trained to stop travel at the dead-end aisle or to slow the vehicle for entry into the main aisle.

The three standard manufacturer's colors are green for high travel speed in the

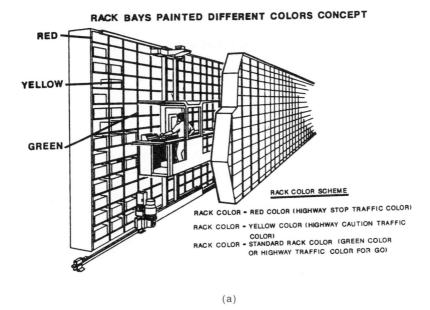

(a)

FIGURE 10.14 (*a*) Rack bays painted different colors. (*Courtesy of Sims Consulting Corp.*).

middle of the aisle, yellow for slow to medium travel speed prior to the end of aisle, and red for very slow travel or stop at the end of the aisle

The disadvantages of the colored-rack system are additional operator training, lack of a system to control the vehicle, and lower employee productivity. The advantages are that it is low-cost, no additional expense or investment is required, it is implemented in an existing facility or on existing equipment, and it is used in combination with other systems.

Rail Bumper. The third VNA vehicle stop or slowdown system is a mechanical rail and bumper (Fig. 10.14*b*). The rail consists of a 4-in by 4-in or 4-in by 5-in angle iron that is secured with the leg up on 8- to 9-in centers to the warehouse floor and is across the storage aisle between the storage racks. With some AS/RS vehicles, the stop device is a bumper which is secured to the warehouse floor at the end of the travel rail. When the vehicle strikes this angle iron or bumper at slow travel speeds, vehicle travel is stopped. The rail or bumper is painted with an OSHA safety color. This method is used in the storage aisles that have dead-end aisles.

The disadvantages of the method are that it creates equipment and floor maintenance problems, is an employee tripping hazard, and is used on dead-end aisles. The advantages are that it is low-cost, it ensures the vehicle will stop, and it is installed in existing building or system.

The rail bumper is used in conjunction with the color rack. This combination provides the advantages of both concepts and reduces equipment and floor maintenance problems.

RAIL ACROSS THE FLOOR CONCEPT

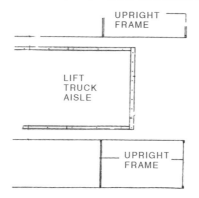

BUMPER AT THE END OF THE RAIL CONCEPT

(b)

FIGURE 10.14 (*Continued*) (*b*) Rail and bumper across the aisle. (*Courtesy of Interlake Materials Handling.*)

Electromagnetic Guidance

In the electromagnetic system, a sensing device is attached to the vehicle undercarriage, and magnets are set in the floor on both sides of the aisle at a specific distance to the end of the aisle. As the storage vehicle travels over the magnets, the sensing device network automatically slows the vehicle and warns the operator of approaching the end of the aisle. With this slow travel speed, the vehicle leaves the storage aisle and enters the main travel aisle. This system is best used in a storage-pick area that has two turning aisles and a wire guidance system for the vehicle.

ELECTRO-MAGNETIC CONCEPT

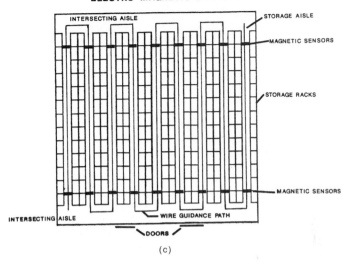

(c)

FIGURE 10.14   (*Continued*) (*c*) Electromagnetic system.

The disadvantages of the electromagnetic system are the additional equipment investment and operator training. The advantages are that it is not operator-controlled, it reduces the potential for equipment damage or accidents, and it ensures vehicle slowdown.

When the electromagnetic system is installed with colored racks, then the advantages of both systems are enjoyed.

Before deciding which end-of-aisle slowdown or stop system to install in your facility, review the facility layout, floor quality, and required operational procedures. These facts should be reviewed with the VNA vehicle supplier and management staff and vehicle operators.

Based upon our economic estimates for the various systems, we have drawn the conclusions shown in Fig. 10.15.

System	Investment cost, $	Results
Operator-controlled end-of-rack upright frame post and load beam painted	0	Requires training
Operator control	0	Requires training
Rail or bumper stop	100–1000	Requires training
Electromagnetic	2000–3000	Best with a wire guidance vehicle

FIGURE 10.15

PICKUP AND DELIVERY STATIONS

In this section we describe the various P/D (unit-load pickup and delivery) station concepts, review the operational characteristics, and identify the disadvantages and advantages of each system.

In the pallet (unit load) storage warehouse industry, recently many companies have increased their warehouse space utilization (storage space) by implementation of turret trucks, fixed-mast (hybrid) trucks, or fixed-mast (AS/RS) storage vehicles. In these storage systems, lift trucks perform storage transactions at elevated levels, operate within a very narrow aisle, and perform the maximum number of transactions per hour. Therefore, these vehicles remain in a particular warehouse aisle and perform the maximum number of dual or single storage transaction cycles per aisle. Since the typical warehouse inbound activity is skewed to the morning hours and outbound activity is skewed to the afternoon, an imbalance in the projected lift truck storage transactions is created. To obtain the desired transactions and return on investment, these vehicles require unit-load P/D stations at the end of each storage aisle.

The P/D stations are unit-load positions located at the end of the storage aisle. They provide temporary accumulation for inbound and outbound unit loads that are assigned for deposit in a storage position or transported to another warehouse location. In addition, the P/D station location gives the lift truck direct, unobstructed access from the storage aisle to the pallet load for the start or completion of a storage transaction. Also the P/D station ensures that the pallet board openings are in the correct orientation for pickup and delivery by the storage lift truck or transport vehicle.

The P/D station improves lift truck driver productivity, decreases product damage, decreases rack and lift truck damage, and reduces transaction errors.

P/D Station Design Parameters

In considering a P/D station for the storage system, these design parameters determine the best P/D station system:

Type of VNA lift truck

- Turret truck (floor-level or elevated pickup)
- Fixed mast
- AS/RS vehicle
- Captive-aisle
- Mobile-aisle

Type of truck guidance system

- Rail (middle or side)
- Wire, paint, tape, or laser beam

Type of pallet board (slave, two-way, or four-way)

Volume (low, medium, or high)

Type of storage area design (one or two transfer aisles)

Type of in-house transport vehicle

- Pallet truck
- Lift truck
- AGV (side and rear loading)

- Conveyor
- Monorail
- Cart

Type of storage vehicle handling device (forks/platens)

Various P/D Station Setups

The end-of-aisle P/D station is designed as a static or dynamic system. Static P/D stations are staggered or flush and have the floor, structural stand, or standard pallet rack to hold the unit load. Dynamic P/D stations have four-wheel carts, on-board transfer car, gravity flow conveyor, powered roller conveyor, or powered shuttle car that hold the unit load.

The P/D station is one component of the total storage system which includes the rack, storage vehicle, in-house transport vehicle, unit load (including bottom support device), inventory control system, and computer inbound or outbound balance system.

Staggered P/D Station.   In the most common staggered static P/D station, the pallet board fork openings face the storage aisle (Fig. 10.16*a*). This staggered (sawtooth) rack layout consists of four rack rows and two storage aisles. In a plan view of the storage area, the rack arrangement has the interior two rack rows shorter by the number of required pallet positions at the P/D station, and the exterior rack rows extend outward toward the main aisle by an equal number of P/D positions. With these extended rack bays, this rack design provides sufficient aisle width (10 to 15 ft) between the two exterior rack rows. This open space permits most in-house transport vehicles the necessary maneuvering space to perform a pallet load pickup or delivery transaction. The staggered rack system enjoys good lift truck driver productivity and is used with a rail, tape, paint, laser, or wire guided turret truck.

Flush P/D Station.   In the flush P/D station (Fig. 10.16*b*), the pallet board openings face the main or transfer aisle of the storage area. This system decreases the lift truck driver's productivity due to increased travel time, provides no accumulation (one pallet), handles a low volume, and interfaces with turret trucks. It is not considered for a dynamic warehouse operation.

Static P/D Station.   The static P/D station requires the in-house transport or storage vehicle to place the pallet board onto the floor or elevated stand or rack position. The pallet board is stationary until it is removed by the storage or in-house transport vehicle.

Floor P/D Station.   The P/D warehouse floor station is the basic P/D station (Fig. 10.17*a*). The floor area extends from the end of the rack into the main aisle. In this area, the in-house transport or storage vehicle sets on the floor the inbound or outbound pallet board. The pallet openings face the aisle.

This static system provides one-level-high P/D station which is poor space utilization above the P/D station. The system requires no investment except lines painted on the floor and a backstop to ensure proper pallet alignment at the P/D station. The floor P/D station interfaces with all guidance systems for a turret truck storage vehicle and a pallet or lift truck transport vehicle. Since the fixed-mast and AS/RS storage vehicle load-handling devices are raised above the floor, the floor P/D station is not used with

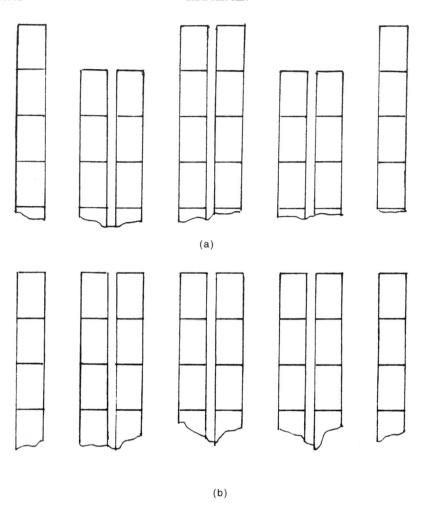

(a)

(b)

FIGURE 10.16 (*a*) Staggered P/D stations; (*b*) flush P/D stations.

these vehicles. The static floor P/D station handles two-way or four-way pallet boards, and pallet board flow through the P/D station is controlled. The floor P/D station is designed for any storage area layout and has no pallet load accumulation.

Structural Stand P/D Station. The structural stand is an extension from the rack into the main aisle (Fig. 10.17*b*). This structural stand has the capacity to hold one pallet board which is elevated above the finished floor. The pallet load has four structural members that extend to each corner of the stand or are four sets of arms with two rails.

The structural stand P/D station interfaces with all mobile transport vehicles except the pallet truck and interfaces with all storage vehicle types. The P/D station with two sets of arms with rails handles storage vehicles with platens. The four-top-member

(a)

(b)

FIGURE 10.17 (*a*) Floor P/D stations (*Courtesy of Lansing Bagnel Ltd.*); (*b*) structural stand P/D stations (*Courtesy of Interlake Materials Handling.*).

system handles two-way and four-way pallet boards, and the system with two sets of rails handles all pallet board types. In the structural stand there is a controlled flow through the P/D station. With this system there is no accumulation, and it is designed for any type of storage area.

Staggered Standard Pallet Rack P/D Station. In the staggered standard pallet rack, specific rows of rack extend outward from the storage rack into the main aisle. The standard rack consists of upright frames and load beams that are connected to create rack openings. These openings are multilevel from the floor to the maximum transaction elevation of the transport vehicle. Therefore, this system is designed with the first P/D station level on the floor or elevated to the required storage vehicle transaction height. With additional rack levels above the P/D station, additional P/D storage positions are provided. To reduce rack post damage and good lift truck transaction activity, the baseplate of the upright post is calculated in the side dimension of the rack opening.

Standard Pallet Rack P/D Station Options. Options with the standard rack include rub bars (metal strips between the upright posts), front-to-rear members, and backstops which improve pallet orientation in the P/D station and reduce damage to the rack upright posts (Fig. 10.18*a*). In a rack-supported facility, an additional upright frame between the storage racks and the P/D stations creates a freestanding P/D station. With the freestanding P/D station, if a transport vehicle damages a rack post, damage to the rack-supported facility is minimal.

The standard pallet rack P/D station interfaces with all turret truck storage vehicles and most in-house transport vehicles (pallet and lift trucks, carts, and some AGVs). The standard rack P/D station handles both two-way and four-way pallet boards. The pallet load transaction sequence through the P/D station is random. The standard rack P/D station is medium-cost.

Of all the static P/D station systems, standard rack can support the roof and walls of a rack-supported facility.

The staggered standard rack P/D station is very common with all types of turret lift truck operations with any of the vehicle guidance methods. It handles a medium volume with a random flow through the P/D station.

Various Dynamic P/D Station Systems

The dynamic P/D station requires the transport and storage vehicles to place the pallet load onto a gravity or powered conveyor section. The conveyor moves the pallet board

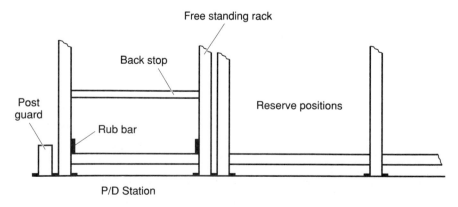

FIGURE 10.18   Standard pallet rack options.

into position for the completion of a P/D station transaction. When the pallet board is on a cart, the cart is manually moved into the P/D station lane to complete a transaction.

The various dynamic P/D station systems include the cart, on-board T-car, gravity conveyor, powered conveyor, and powered shuttle car.

Four-Wheel Cart.   In this P/D station, the pallet board is on a four-wheel cart (Fig. 10.19*a*). A cart with a pallet board on its load-carrying surface is transported by a towline, AGV tugger, or manually controlled tugger to the P/D station. At the P/D station, an employee aligns the cart underside guide pin between the cart guiderails and pushes the cart into the P/D station lane. The P/D lane is between the vehicle guiderail (path) and storage rack center post. This lane is the unit-load accumulation area; empty carts are removed from the lane or remain in the lane to receive outbound unit loads.

The pallet board is placed onto the cart with the correct orientation of the pallet fork openings. The cart's load-carrying surface is designed to handle lift truck vehicles with forks or platens. With the cart P/D station and structural rack support design, the positions above the lane are used for storage.

The cart P/D station requires additional employees to move carts between the transport vehicle's drop-off or pickup location and the P/D lane. The system requires an AGV, towline, or tugger transport system and handles all pallet load types. The pallet load transport sequence is random. With the structural rack design and cart guiderail, the system is medium-cost.

On-Board Storage/Retrieval Transfer-Car.   The on-board storage/retrieval transfer car (T-car) is used in a storage system that has one aisle for both inbound and outbound transfer of pallets (Fig. 10.19*b*). This one aisle is used for storage vehicle transfers and main aisle traffic.

A T-car transfers the storage/retrieval vehicle between two storage aisles. On each side of the storage/retrieval transfer car cavity are conveyor sections which ensure that the pallet loads flow to the proper P/D transaction location. One side (conveyor section) is used for inbound loads, and the other side (conveyor section) is for outbound loads. As required, the storage vehicle performs the P/D transactions between the storage aisle and T-car, and a manually operated lift truck handles the pallet loads from the T-car.

The on-board T-car ensures a FIFO unit-load flow through the P/D station. The concept handles a four-way or slave pallet board and is designed for a storage area with one main aisle. The additional investment beyond that for the T-car is minimal.

Gravity Flow Conveyor.   The gravity flow conveyor consists of a series of rollers that turn on a shaft (Fig. 10.20*a*). The shafts are attached to channels or are between the side guards, or the pallet has guides on the bottom that direct it through the system. This P/D station is used in a storage area with two aisles. One aisle is used for lift truck transfer activity, and the other is used for the P/D transaction activity. The gravity flow conveyor holds three to four pallet loads and is pitched from the inbound end to the outbound end. In the storage area layout, the gravity flow conveyor directs the pallet flow as an extension of the single-deep storage rack or parallel to the aisle for pallet flow rack storage system.

The gravity conveyor P/D station is designed to handle all storage vehicle types with forks or platens. It handles four-way or slave pallets. The conveyor section has sufficient accumulation, and at the P/D station the conveyor has a slight slope or is

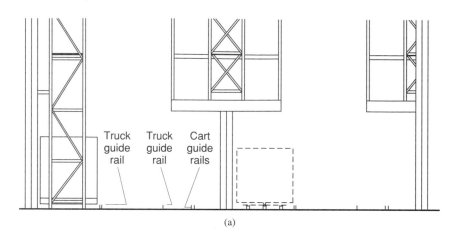

Truck
guide
rail

Truck
guide
rail

Cart
guide
rails

(a)

(b)

FIGURE 10.19   (*a*) Four-wheel cart P/D station; (*b*) on-board T-car P/D stations (*Courtesy of Interlake Materials Handling.*).

(a)

FIGURE 10.20 (a) Gravity conveyor P/D station. (*Courtesy of Lansing Bagnel Ltd.*)

flat. With long flow lanes, brakes and load separators are used to control the pallet flow and line pressure. The gravity flow P/D station is medium-cost.

Powered Conveyor P/D Station. The powered conveyor P/D station is used in a storage area that has one aisle for P/D transactions and one aisle for vehicle aisle transfer (Fig. 10.20b). In this system a series of rollers are moved by an electric motor-driven chain. The turning of the rollers on the shafts moves the pallet load between the guiderails through the P/D station area.

The powered conveyor path is an extension of the storage rack. Compared to the gravity conveyor, the conveyor length is shorter and there is a pallet load recirculation loop in front of the storage area. If there is no recirculation loop, the lift truck recirculates the pallet loads.

The powered conveyor P/D station interfaces with all storage vehicles that have forks or platens. It handles two-way, four-way, and slave pallets and provides a FIFO pallet load flow through the P/D station. The system is used for a storage area with a P/D aisle and a vehicle transfer aisle. The powered conveyor is very attractive economically.

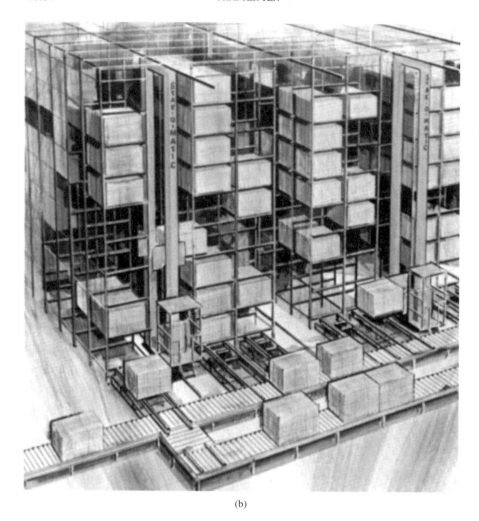

(b)

FIGURE 10.20 *(Continued)* (*b*) Powered conveyor P/D stations. (*Courtesy of Automation.*)

Powered Shuttle Car P/D Station.   The powered shuttle car is a wheeled load-carry-ing platform that rides on two rails (Fig. 10.21*a*). Two rails direct car travel between the pallet in-house transport method (conveyor, AGV, towline, or powered monorail) transfer location and the storage vehicle P/D transaction station. The load-carrying surface is designed to interface by elevating and declining for unit-load pickup or dis-charge at the transport system transfer station and the P/D station.

To use the shuttle car P/D transfer station requires a unit-load transport system that can provide empty transport devices or open space for outbound unit loads. With inbound loads, the transport system can queue or recirculate the unit load when a group of loads are assigned to one aisle.

The shuttle car is designed with two aisles in the storage area or has captive-aisle vehicles, interfaces with all storage vehicle types with forks or platens, and handles four-way or slave pallets. The economics of shuttle car are very attractive.

FIGURE 10.21   Powered shuttle P/D stations. (*Courtesy of Rapistan Demag Corp.*)

CHAPTER 11

STACKING STACKABLE AND NONSTACKABLE PRODUCTS

In this chapter we review the various pallet types, methods to protect or secure unit loads, slip-sheet types, stacking frames, and containers. For each type we review the storage space requirements, lift truck requirements, disadvantage and advantages of each method, and operational characteristics.

Your *unit of storage* is the common design factor in your warehouse. When you design a warehouse facility that has a transport vehicle, lift truck, and storage system, the common design factor is the unit (handling) load size, shape, weight, and protection requirements. The handling unit can be a single item, carton (case), container, unitized load on a cart, or palletized load on a pallet board. When a pallet board or slip sheet is used in the warehouse operation, the carton overhang of the pallet board or slip-sheet lip extension is considered part of the unit load.

UNITIZE OR PALLETIZE LOADS AS SOON AS POSSIBLE

For your warehouse facility to reduce operating costs, move product on schedule, and maximize the storage (cubic) space, the operation must unitize or palletize the product. When the warehouse facility receives unitized loads, the employees or material handling system moves product with the fewest possible number of movements, in the shortest time, and with the greatest amount of product. These material handling benefits are realized by your manufacturer, warehouse operation, and customer.

Your best opportunity to unitize product is to become involved in a unit-load unitization program with your supplier. If your product is not unitized by your supplier, then you operation unitizes the product at the receiving dock or as soon as possible before product handling increases and reduces the benefits.

To seize the opportunity to unitize loads, your operation can become involved in a pallet exchange program. In a pallet exchange program, your warehouse receives from your supplier a palletized delivery and then returns an equal number of pallets to your supplier. The pallet exchange arrangement does not create a pallet cost to your operation because of the exchange of pallet boards.

The two types of product handled at your warehouse facility are *stackable* and *nonstackable*. Stackable product is self-supporting or is enclosed in a carton that has the strength to support additional cartons. Stackable product falls into three classes: (1) individual carton and SKU, (2) palletized load on a pallet board and unitized load on a slip sheet or cart, and (3) unitized loads in containers.

Nonstackable product consists of product or cartons that are not self-supporting or cannot support an additional load. Nonstackable product characteristics include product shape, height, size, fragility, and strength of the corrugated material or product.

Humidity (moisture) in the air of the storage area reduces the strength of normal warehouse cartons (corrugated material).

STACKING DESIGN PARAMETERS

In considering the stocking of unit loads in a new or an existing storage facility, the storage (stacking) method and unit-load support device are two important design factors. Other important design factors include the following: (1) unit-load bottom support device (include weight), (2) design of underside of unit-load bottom support device and how the material handling and storage equipment handles the load, (3) warehouse floor condition (can it handle the stacked weight and imposed load of the material handling and storage equipment?), and (4) whether the material handling equipment has the lift capacity and capability to handle the new lift heights, new load centers, and weight. The weight of the special attachment is included in the lift truck specifications.

Floor Stack

The floor stack is the basic and simplest method whereby an employee or lift truck with an attachment (forks) stacks one unit of product on top of another unit of product. Typically, these products are self-supporting or in cardboard or fiberboard cartons. Some of these items are paper rolls or appliances.

The disadvantages of the method are as follows: With hand stacking there are increased employee injuries and fatigue, and employee productivity is low. In a low-ceiling warehouse, space utilization is low, the greatest number of employees and equipment is required, use of space is not maximized due to honeycombing, and there is no access to all SKUs.

The advantage of floor stack are its low cost (no rack investment), good vertical space utilization in a very low-ceiling warehouse, and the narrow aisle required with hand stacking.

Other Stocking Methods

The other stocking methods for stackable products were reviewed in Chap. 7. These methods include racks of all types, stacking frames, and portable racks.

The objective of this section is to define the pallet board terms, to develop an understanding of the various pallet board designs, and to identify pallet board materials.

PALLET BOARDS

The pallet board used in your distribution is a key factor that allows product to flow smoothly through the warehouse facility. In the warehouse and distribution industry, the pallet board is the common unit-load bottom support device for corrugated product or stackable product. Within the pallet board group, the wood pallet board is the most common device. It is handled by all lift trucks, fits into virtually every type of storage structure, and is handled in the operations of your vendors and customers.

The pallet board's popularity is due to three reasons: (1) there are common operational characteristics in your warehouse, manufacturing operation, vendor's facility, and customer's locations; (2) pallet board influences the layout, design, and specifications for storage and handling equipment; and (3) pallet board purchases and repairs are a business expense that affects your income statement.

Pallet boards are available in many sizes and shapes and are manufactured from a variety of materials. The pallet board is a rigid-base structure with stringers and blocks and top and bottom deck boards. It provides the support for the unit load and allows material handling equipment to handle the unit load. A unit load consists of cartons or cases that are stacked onto the pallet board.

The majority of warehouse facilities use a standard or a specially designed (engineered) pallet board. Both these pallet boards support a wide variety of products and are easily handled in all warehouse functions.

Three Basic Types

Four basic types of pallet boards are used in the warehouse and distribution industry (Fig. 11.1): the standard pallet board, the throwaway or exchange pallet board, take-it-or-leave-it pallet board, and specially engineered pallet board.

A *standard pallet board* is used in a warehouse operation that has manual or powered fork lift pallet trucks. Standard pallet board comes in two types: throwaway pallet board and exchange pallet board.

A *throwaway pallet board* (Fig 11.1b) is purchased from the supplier and sent to the warehouse facility. After the warehouse facility has made effective use of the pallet board, it is thrown in the trash. Given these operational characteristics, throwaway pallet board is less expensive to manufacture than exchange pallet board. Usually, these pallet boards have thin top deck boards, narrow stringers, and few thin bottom deck boards.

The *exchange pallet board* is used by your vendors, in your facility, and at your customer's locations. All these operations have the same specifications and standards for pallet boards. Upon receipt of a unit load from your vendor on an exchange pallet board program, your facility trades one of your empty pallet boards for the pallet board under the product (unit load).

The *take-it-or-leave-it pallet board* (Fig. 11.1c) is basically an exchange pallet board that is used by some vendors and warehouse facilities. It consists of a combination of a slip-sheet unit load on a pallet board that has a specially designed top deck. This unit-load combination allows your warehouse operation or your customer to handle the unit load as a slip-sheet unit load or as a unit load with a combination of pallet board and slip sheet. The customer is given the opportunity to purchase the combination of the slip sheet and pallet board or to handle the slip-sheet unit load. If the cus-

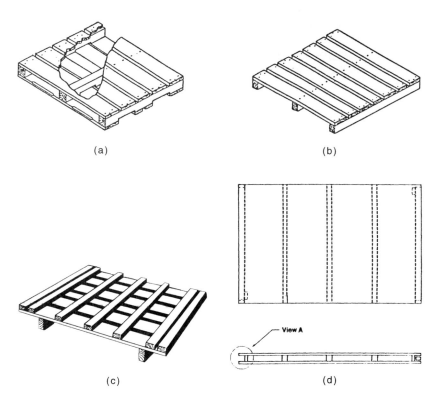

(a)   (b)

(c)   (d)

FIGURE 11.1   (*a*) Standard pallet board (*Courtesy of Litco Int'l, Inc.*); (*b*) throwaway pallet board (*Courtesy of Litco Int'l, Inc.*); (*c*) take-it-or-leave-it pallet board (*Courtesy of Modern Materials Handling Magazine.*); (*d*) specially designed pallet board (*Courtesy of Plant Engineering Magazine.*).

tomer takes the slip sheet, then with a push-pull tine device the slip-sheet load can be pushed and/or pulled from the pallet board.

The specially designed (engineered) or *slave* (captive) *pallet board* (Fig. 11.1*d*) is manufactured to your distribution facility's specifications and standards. Because of your distribution policy or the material handling equipment and system requirements for a high-quality pallet board, these pallet boards do not leave the warehouse. A high-quality pallet board meets the following criteria: (1) it does not have broken or loose top or bottom deck boards or stringers; (2) nails do not protrude above or below the deck boards; (3) it does not have crooked or cupped (bowed) members; (4) the nails are the flathead threaded (screw) type; and (5) all wood members are knot-free.

Pallet Board Materials

Pallet boards are manufactured from a wide variety of materials. The factors that determine the pallet board material for a distribution facility are (1) SKU characteristics, (2) amount of allowable board deflection (bowing), (3) weight that is acceptable to the material handling system, (4) storage and safety considerations (such as spark-free, noncorrosive freezer or high-humidity environment), (5) pallet board investment plan, (6) material permitted by local fire codes, and (7) other distribution facility characteristics. The most common pallet board material is wood with threaded screw-type flathead nails for attachment of the pallet board members.

The most widely used material to construct pallet board is wood (Fig. 11.2a): hardwood, softwood, and plywood.

Hardwood Pallet Board.   The hardwood pallet board is a class C type wood as defined by the National Wooden Pallet and Container Association (NWPCA). These pallet board components are nailed together; usually the nail holes are predrilled for easy entry. By nature hardwood pallet board ensures that the nail is held, provides the greatest support strength, resists shock or damage (splinters), and is the heaviest. Since hardwood pallet board is the heaviest of the wood pallet boards, it is preferred in a pallet load warehouse operation.

Hardwoods include white ash, white beech, red oak, birch, rock elm, hickory, hard maple, hackberry, and oak.

Because of the high cost associated with hardwood, the weight, and difficulty of drying the hardwood pallet boards, they are used less frequently in the single-item and carton handling warehouse and distribution industry.

Medium-Hardwood Pallet Board.   Medium-density hardwood is a class B wood pallet board. To prevent the wood from splitting, nail holes are predrilled. It has medium strength, dries more easily, and has medium weight. These characteristics make medium-density wood pallet boards preferred for use in a carton or single-item handling warehouse facility. Some of the medium-density woods include ash (but not white ash), soft elm, tupelo, butternut, soft maple, yellow poplar, chestnut, sweet gum, sycamore, walnut, and magnolia.

Soft-Hardwood Pallet Board.   The soft-hardwood pallet board is class A wood. Of the hardwoods, the soft-hardwood pallet board has medium support strength, a possibility of splitting, and is more lightweight. Soft hardwood pallet board is preferred for use in a carton or single-item warehouse operation because employees handle empty pallet boards. Some of the soft hardwoods are aspen, basswood, buckeye, cottonwood, and willow.

Softwood Pallet Board.   Softwood pallet boards have components made of wood that tends to split easily and are very lightweight. The wood has low support strength and dries easily. The pallet board components are nailed, stapled, or glued together.

These pallet boards are lightweight, are low-cost, and are easily handled by employees, so they are used in a carton or single-item warehouse operation or as an expendable pallet board in a pallet load warehouse. Some of the softwoods are Douglas fir (coast and mountain types), western hemlock, southern pine, and western larch.

The disadvantages of softwood pallet board are that nails and splinters damage product; it becomes dirty and infested, is difficult to use in a spark-free environment, and adds weight and height to the unit load; and empty pallet boards take up space.

The advantages are that it is repairable and/or reusable, is widely used in the warehouse and distribution industry, and interfaces with all lift trucks.

Plastic Pallet Board.   The plastic pallet board is one piece of preformed molded plastic (Fig. 11.2b). Its operational characteristics are similar to those of wood pallet

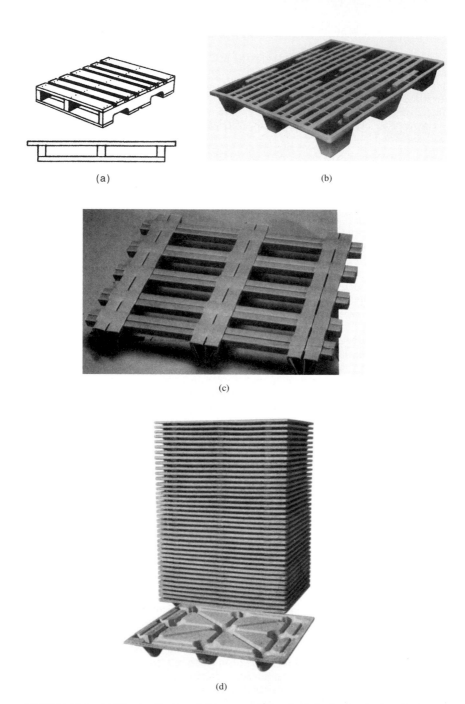

(a)

(b)

(c)

(d)

FIGURE 11.2 (*a*) Wood pallet board (*Courtesy of Litco Int'l, Inc.*); (*b*) plastic pallet board (*Courtesy of Armin Thermodynamics.*); (*c*) corrugated pallet board (*Courtesy of Modern Materials Handling Engineering Magazine.*); (*d*) pressboard and fiberboard pallet board (*Courtesy of Litco Int'l, Inc.*).

11.6

board. An option during the molding process that will increase the plastic pallet board's strength is to have metal rods placed in the plastic.

The disadvantages are that it is difficult to repair and has a tendency to bow and there may be restrictions by local fire codes. The advantages are that it is washable, has a long life, and can be used in a spark-free environment.

Corrugated Pallet Board.   Corrugated pallet board is treated heavy-duty corrugated board (Fig. 11.2c). The corrugated pallet board is preformed as one piece with its components glued or nailed (stapled) together. Some pallet boards have plastic cups for support legs.

The disadvantages are that is is not as durable as other material, it is limited to low-humidity environments, and there may be restrictions by local fire codes. The advantages are that it is less costly, is used in a spark-free area, is lightweight, and is available in a wide variety of sizes and shapes.

Pressboard and Fiberboard Pallet Board.   This pallet board (Fig. 11.2d) is manufactured in one piece of preformed molded mixture of wood fiber and synthetic resins. Some have plastic cups for legs.

The pressboard and fiberboard pallet boards have similar disadvantages and advantages to corrugated pallet board.

Rubber Pallet Board.   Rubber pallet board is made from one piece of preformed molded polyethylene rubber (Fig. 11.3a). Rubber pallet board has the same disadvantages and advantages as corrugated pallet board. These pallet boards are used in a spark-free environment. An option during the molding process that will increase the strength of the pallet board is to have metal rods placed in the rubber.

Metal or Metal-Clad Pallet Board.   The metal and metal-clad pallet board (Fig. 11.3b and c) is the last material type. The full metal pallet board has aluminum or steel members. A metal-clad pallet board has aluminum or steel deck boards with wood blocks or stringers. These pallet boards are used in a pallet load warehouse operation.

Disadvantages of the metal pallet board are that it is difficult for employees to handle, it adds weight to the unit load, and it is more costly. The advantages are that it is fire-resistant and has a longer life.

Important Pallet Board Dimensions

Pallet boards are available in a wide variety of sizes. These factors determine the pallet board size: (1) the material handling equipment, (2) SKU dimensions and weight, (3) SKU ti × hi (palletizing pattern), (4) order-picking requirements, and (5) required SKU storage-pick openings.

The pallet board's length (into rack, depth, bearers, or stringers) is stated first. The width or down-aisle dimension (lift truck opening) is stated second. The third dimension is the height.

A general rule of thumb is that the stringer length determines the depth for the lift truck or pallet jack fork length. In the U.S. warehouse industry, the most popular pallet board size is the 48-in by 40-in by $5\frac{1}{2}$-in flush, nonreversible, partial four-way entry, open-deck pallet board. The height of the pallet board lift truck opening is determined by the unit-load handling equipment and storage system requirements.

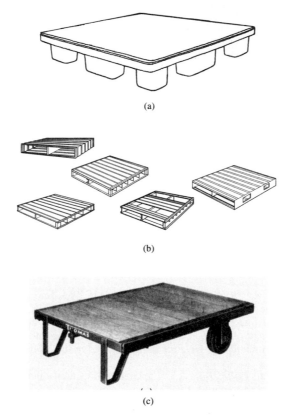

(a)

(b)

(c)

FIGURE 11.3 (*a*) Rubber pallet board (*Courtesy of Armin Thermodynamics.*); (*b*) metal pallet board (*Courtesy of Armin Thermodynamics.*); (*c*) metal-clad pallet board or skid (*Courtesy of Thomas of Rockford, Inc.*).

Generally, a counterbalanced chisel lift truck or pallet truck forks require a 3- to 5-in-high pallet board opening. Most gravity pallet flow storage systems require a solid slave pallet board that is approximately 1 to 2 in high. The pallet board lengths and widths most commonly used in the warehouse and distribution industry are listed in Fig. 11.4.

Five Basic Pallet Board Designs

The five basic pallet board designs are block, leg, solid (slave), stringer, and flue.

Block Pallet Board. The block pallet board (FIg. 11.5*a*) has equally spaced blocks along its length and width. Deck boards are attached to the blocks.

Leg (Honeycomb) Pallet Board. The leg (honeycomb) pallet board (Fig. 11.5*b*) has a solid deck with equally spaced legs along the length and width of the pallet board.

Fork opening	Stringer	Fork opening	Stringer
40 in	48 in	48 in	40 in
40 in	40 in	36 in	36 in
48 in	48 in	42 in	42 in
42 in	52 in	44 in	56 in
30 in	24 in	32 in	40 in
40 in	36 in	44 in	54 in
1165 mm	1165 mm	1100 mm	1100 mm
1000 mm	1200 mm	1200 mm	1000 mm
54 in	54 in	78 in	65 in
120 in	54 in		

FIGURE 11.4   Pallet board sizes.

These legs are attached to the solid deck. The base of the legs sits directly on the floor, and the open space between the top surface and the base is the fork opening.

Solid (Slave) Pallet Board.   The solid (slave) pallet board (Fig. 11.5c) has one solid piece of plywood or other material for the top and bottom surface. Because it does not have legs (blocks or stringers), the slave pallet board sits flat (directly) on the floor. In many storage and pallet conveyor systems, solid pallet board is used as a support device for another pallet board. Some solid (slave) pallet boards are designed with four holes in the corners of the deck for easier employee handling of the pallet board.

When a lift truck handles a slave pallet board, the slave pallet board is set on a support device that permits the lift truck's forks to extend under the slave pallet board. This feature allows the lift truck to complete the transaction.

Stringer Pallet Board.   The stringer pallet board (Fig. 11.5d) has two exterior stringers and one interior stringer for top and bottom deck board attachment. The exterior stringers either are solid or have two notches (openings) in the stringer side. The notches are additional fork entry openings that permit a lift truck to handle the pallet board from all four sides.

In a distribution facility that uses a rack storage system, the stringer pallet board is the most common pallet board.

Flue (Rippled) Solid-Piece Pallet Board.   The flue pallet board (Fig. 11.5e) is a rippled solid piece or perforated piece of material that has every other ripple sitting directly on the floor. The depth and width of the flue space between the ripples are the space or openings for the lift truck fork entry.

The two most popular pallet types are the wood stringer pallet board (Fig. 11.6a) and wood block pallet board (Fig. 11.6b). There are several pallet board components common to both.

Pallet Board Components

Top Deck Boards.   Top deck boards are the first component. These flat wood members are attached to the stringer pallet board's stringers or to the block pallet board's

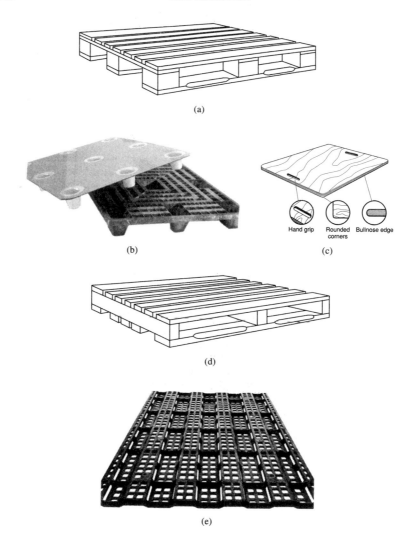

FIGURE 11.5   (*a*) Block pallet board (*Courtesy of National Wooden Pallet and Container Assoc.*); (*b*) honeycomb leg pallet board (*Courtesy of Modern Materials Handling Magazine.*); (*c*) solid (slave) pallet board (*Courtesy of Modern Materials Handling Magazine.*); (*d*) stringer pallet board (*Courtesy of National Wooden Pallet and Container Assoc.*); (*e*) flue (rippled) pallet board (*Courtesy of Armin Thermodynamics.*).

stringer board. The top deck boards provide the support for cartons or cases. The top deck boards are slats or a solid piece of material. When top deck board slats are used on a pallet board, the width of the slats (deck boards) varies from 4 to 8 in and the open space between slats varies from 1 to 3 in. The open space is called the *deck*

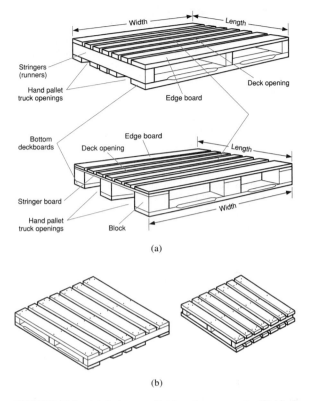

FIGURE 11.6 (*a*) Stringer pallet board components; (*b*) block pallet board components. (*Courtesy of National Wooden Pallet and Container Assoc.*)

opening. The deck opening allows air circulation and is very common on pallet boards used in a refrigerated or freezer storage facility.

Bottom Deck Boards. The bottom deck boards are attached to the bottom of the stringers or blocks. These deck boards provide a rigid and flat surface for the board to sit on the floor or rack load beams. When a pallet jack with front-load wheels is used in the operation, then the space between the bottom edge board and the next bottom deck board permits the pallet jack wheel to raise and lower the pallet board.

Edge Boards and Other Deck Boards. The top and bottom deck boards at the two normal fork-entry ends of the pallet board are the edge boards. These top and bottom boards are a nominal 6 in wide and $\frac{1}{2}$ in high. The width of the remaining top and bottom boards varies from 4 to 8 in with a nominal $\frac{1}{2}$-in height, and the space between the deck openings varies between 1 and 8 in.

Chamfered-Edge Boards. In some pallet boards, the two bottom edge boards and the bottom boards adjacent to the pallet jack wheel space are chamfered for easy pallet truck fork entry. In chamfered-edge boards the top edge of each bottom edge board is cut on a 35° angle. The angle cut is 12 in wide.

Fork-Entry Opening. With the top and bottom deck boards attached to the stringers or blocks, the overall pallet board height is 5 to 6 in with a nominal $\frac{1}{2}$-in-high deck board. This arrangement creates a 4- to 5-in-high open space that is the fork entry opening for pallet jack and lift truck forks.

Notches. When a solid exterior stringer has two additional openings for fork entry, the stringer openings are called *notches.* The notch is a nominal $1\frac{1}{2}$ in high and 9 in long. The notches permit a fork lift truck with chisel forks to handle the pallet board from the stringer side.

Stringer Board. Stringer board is 1 in thick and is used on block pallet board for attachment of the top deck boards to the blocks.

Stringer (Bearer). The stringer (bearer) is the stringer pallet board component that runs the entire depth of the pallet board. Small pallet boards have two stringers, and large pallet boards have three. The stringers are 2 to 3 in wide and serve several purposes. First, they attach at the top and bottom deck boards. Second, they serve as the spacer that helps create the fork entry opening. Third, they provide support for the load.

Blocks. The blocks are components of the block pallet board. The blocks are square or rectangular parts that are placed under the four corners and in the middle of the pallet board. The blocks serve the same purpose as the stringers. Small pallet boards have six blocks, and large pallet boards have nine. These blocks vary from 3 to 4 in in length and width.

Pallet Board Connecting Devices. The stringer or block pallet board components are attached to each other by helically threaded (spiral) nails, staples, nuts and bolts, or glue. The fastener method selected for the pallet board is determined by the cost, pallet board material, and warehouse operational requirements.

The two basic pallet board types are the flush and wing pallet board types.

Flush Pallet Board. In the flush-type pallet board, the width of the top or bottom deck boards overhangs the exterior stringers or stringer boards. The flush pallet board is the most common pallet board in the warehouse and distribution industry.

Wing Pallet Board. In the wing-type pallet board, the width of the top deck boards or sometimes of both the top and bottom deck boards overhangs the exterior stringers or stringer boards. When a warehouse rack design calls for a wing pallet board with drive-in, drive-through, or flow (without a slave pallet board) rack, then the pallet support rails are on closer centers.

Pallet Board Configurations

Both the standard flush and wing-type pallet boards are designed with one of the following design modifications.

FIGURE 11.7 (*a*) Open-deck pallet board; (*b*) closed-deck (solid) pallet board; (*c*) two-way entry pallet board; (*d*) four-way entry pallet board; (*e*) partial (four-way) or notched pallet board; (*f*) single-wing pallet board. (*Courtesy of National Wooden Pallet and Container Assoc.*)

Open-Deck Pallet Board. The open-deck pallet board (Fig. 11.7*a*) is designed with 1- to 3-in-wide openings (spaces) between the top deck board slats. The slats are 4 to 8 in wide. This open-deck design is very common in the warehouse and distribution industry.

Closed-Deck (Solid) Pallet Board. The closed-deck or solid pallet board (Fig. 11.7*b*) is designed with a solid one-piece surface as the top deck.

Two-Way Entry Pallet Board. Two-way entry pallet board (Fig. 11.7*c*) is a stringer pallet board that has two solid external side stringers.

Four-Way Entry Pallet Board. The four-way pallet board (Fig. 11.7*d*) is a block pallet board designed with the normal high fork-entry openings on all four sides.

Partial (Four-Way) or Notched Pallet Board. The fifth design is a partial (modified or notched) four-way entry pallet board (Fig. 11.7*e*). This stringer pallet board has the two normal fork-entry openings at both ends and an additional fork-entry opening on both stringers. These two additional openings are in the exterior two stringers; these notches or holes are 9 in wide by $1\frac{1}{2}$ in high which permits a lift truck with chisel fork entry to handle the pallet board.

Single-Wing Pallet Board.   In the single-wing pallet board (Fig. 11.7*f*) the width of the top deck board extends beyond the two exterior stringers or blocks. In a straddle lift truck operation, there is sufficient clearance between the floor and wing for the straddles to engulf the pallet.

Double-Wing Pallet Board.   In double-wing pallet board (Fig. 11.8*a*), the width of the top and bottom deck board extends beyond the top exterior stringers or blocks. In a straddle lift truck operation, the lift truck straddles to engulf the pallet.

Reversible Pallet Board.   This pallet board (Fig. 11.8*b*) has two solid deck boards with no pallet truck wheel space on both the two and bottom; therefore, the side that faces up provides the support surface for the product. This pallet board is not handled by a pallet jack. It is handled by a lift truck with a set of forks.

Nonreversible Pallet Board.   On nonreversible pallet board (Fig. 11.8*c*), the top deck boards provide the support surface for the unit load. In the warehouse industry, the bottom deck board has the spacing for pallet jack wheels. Nonreversible pallet board is the most common warehouse pallet because it is handled by a pallet jack.

Take-It-or-Leave-It Pallet Board.   This two-way pallet board (Fig. 11.8*d*) has the two normal pallet board fork-entry openings and has a top deck surface with ribs that run the entire depth (stringer dimension) of the pallet board. These ribs elevate a slip-sheet unit load (slip sheet and product) above the top deck and provide a series of clear spaces or openings. These openings permit a lift truck with a tine push-pull device to enter under the unit load between the rib spaces. This feature allows the unit load (slip sheet and product) to be removed from the take-it-or-leave-it pallet board.

Nestable Block Pallet Board.   This block pallet board (Fig. 11.8*e*) has a bottom deck board that spans the width of the pallet board. The open space along the pallet board's bottom length allows the unused and empty pallet boards to be nested for storage.

Specially Engineered Pallet Boards.   The specially designed pallet boards include the round pallet board, the beer keg pallet board, and the barrel pallet board.
     The disadvantages of these pallet boards are that product fits on the pallet board, they are difficult to dispose of, and costs are average. The advantages are that they are used in many industries, most are repairable and durable, they raise the product a nominal 6 in off the floor and rack, and they can be used in all types of environments.

SLIP SHEET

The slip sheet is a unit-load support device that is used primarily in the transportation activity between two facilities or warehouse functions. With a pallet board or lift truck with a push-pull device under the slip sheet, the structural or beam strength of the slip sheet can support the product load.
     The most popular slip sheet is manufactured from solid fiberboard, corrugated

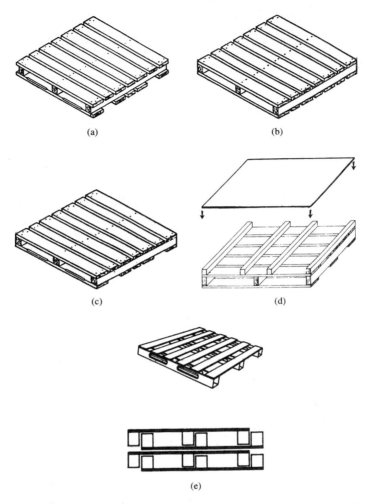

FIGURE 11.8 (*a*) Double-wing pallet board (*Courtesy of Litco Int'l, Inc.*); (*b*) reversible pallet board (*Courtesy of Litco Int'l, Inc.*); (*c*) nonreversible pallet board (*Courtesy of Litco Int'l, Inc.*); (*d*) take-it-or-leave-it pallet board (*Courtesy of Litco Int'l, Inc.*); (*e*) nestable block pallet board (*Courtesy of Armin Thermodynamics.*).

materials, or plastic shaped to the same length and width as the pallet board. But a 4-to 6-in lip extends beyond one of the sides. In most slip sheets a single lip (tab) extends forward and faces up from the slip-sheet base. Some slip sheets have double lips, three lips, or four lips. The lip (tab) permits the lift truck's push-pull device to clamp onto the lip for lifting the unit load (slip sheet and product) from the floor or pallet board.

Compared to other unit-load support devices, the slip sheet reduces the weight and space in an over-the-road transport vehicle. However, when a slip-sheet unit load is placed in a storage area, a clear space is needed for the slip-sheet tab extension beyond the base of the pallet board.

Slip-Sheet Material

The various slip-sheet materials are corrugated, fiberboard (solid kraft board), and plastic (plain-surface polypropylene or dimpled-surface polyethylene).

Corrugated Slip Sheet.   Corrugated slip sheet consists of two kraft liner board outside surfaces with a corrugated interior that is bonded together. This bond provides the required strength for clamping once or twice by the gripper bar of the lift truck's push-pull device. The corrugated slip sheet is considered a one-way slip sheet because it is easily torn.

Its disadvantages are that it is not durable, is not moisture-resistant, is not resistant to high-humidity environments, and is not used in cold storage environments. Its advantages are the low cost and one-way use.

Fiberboard (Solid Kraft Board) Slip Sheet.   The fiberboard (solid kraft board) slip sheet has several plies (layers) of solid fiberboard that are laminated together. This bonding of several flat sheets (usually three to four) increases its nontear strength. It permits the slip sheet to be used several times (at least as a two-way slip sheet) and in different temperature environments. Some fiberboard slip sheets are coated with a plastic covering that improves their use in a storage environment with high moisture or humidity. Disadvantages of fiberboard slip sheet are that it is medium-cost and not durable.

The advantages are that it has multiple uses (two-way) and can be used in humid or cold environments.

Plastic Slip Sheet.   Plastic slip sheet is made of a combination of polymerized materials that include polyethylene or polypropylene. This material gives the plastic slip sheet its greatest tear strength, so it can be used at least 12 times. This feature makes it the most durable slip sheet and the best for use in humid and cold storage environment.

The disadvantage of plastic slip sheet is the high cost. The advantages are that it lasts, the lip does not tear, and it is used in a humid and cold storage areas.

Plastic Dimpled Slip Sheet.   Plastic dimpled slip sheet (Fig. 11.9a) is a vacuum-formed polyethylene sheet that has a series of spherical dimples on very close centers. These dimples cover the entire bottom surface and provide a cushion for the product as it rides in the delivery vehicle. The dimples extend downward on the underside of the slip sheet. This dimple direction does not damage the product's exterior package and does not interfere with the push-pull device and its activity.

Compared to the plain plastic or fiberboard slip sheet, the dimpled slip sheet does not have any additional major disadvantages except that the cost is slightly higher.

The additional advantages are as follows: (1) Compared to plastic sheet, it costs less and is lightweight; (2) compared to the fiberboard sheet, it has a longer life; and (3) there is an improved cushion for the product.

Design Parameters

If you are considering purchasing and implementing a slip sheet method in your operation, these slip-sheet specifications are important: (1) size of load, (2) weight of load,

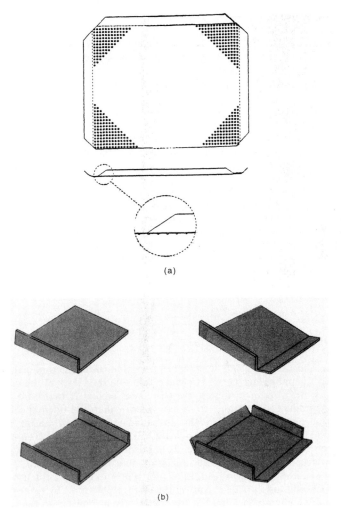

(a)

(b)

FIGURE 11.9   (*a*) Plastic dimpled slip sheet (*Courtesy of Coors Beer Distribution Technology.*); (*b*) various slip-sheet tab designs (*Courtesy of Longview Tibre Co.*).

(3) product description and characteristics (length, width, height), (4) stabilization method (including ti × hi pattern), (5) storage method with or without pallet board (if pallet board used, state type of pallet board), (6) type of push-pull device and gripper, (7) number of trips (use) or expected life, (8) storage conditions (include humidity, moisture, and temperature), (9) possible requirement for holes, (10) use on trucks, containers, or railcars (including loading pattern), (11) customer use of the slip sheet in their operation, and (12) number and location of the lips (tabs).

Slip-Sheet Designs

Slip sheets are available in most common pallet board surface sizes and dimensions (Fig. 11.9b). There are five basic designs for the tab (lip) of the slip sheet: (1) one-tab slip sheet, (2) two-tab slip sheet on opposite ends, (3) two-tab slip sheet on adjacent sides, (4) three-tab slip sheet and (5) four-tab slip sheet.

Single-Tab Slip Sheet.   In shipping slip sheets on trucks, the one-tab design is most common because in the slip-sheet truck loading pattern the tab faces the door of the truck. The delivery truck loading arrangement of the slip sheets gives the lift truck's push-pull device direct access to the tab that permits the load to be handled with efficiency by the lift truck.

Multitab Slip Sheet.   When the multitab slip sheet is used in an operation to reduce product damage and unit-load floor space requirements, the tab not being used is secured to the unit-load side or is removed from the slip sheet.

In shipping slip sheets with railcars or on vehicles that permit variations of the slip-sheet unit-load placement on the vehicle floor, the multitab (lip) sheet enables the push-pull device to handle the unit load from all four sides. This flexibility permits the unit-load placement in the delivery vehicle to maximize the cubic space and permits the unit load to be handled with efficiency.

Tips on Slip-Sheet Use

When a slip sheet is used in a warehouse operation, these factors can increase the storage capacity and reduce costs: (1) Make sure that the edges of the carton or bags on the top of the slip sheet match the slip-sheet edges. If there are open spaces between the product and cartons (ti $\times$ hi pattern), then the open spaces are in the middle of the pattern. (2) Secure the unit load with a stabilization method that reduces product movement on the slip sheet. (3) When a delivery vehicle is loaded, use dunnage material (empty pallet boards or corrugated filler) in the rounded nose of the delivery truck and between the unit loads and the vehicle walls. Allow some space between the delivery vehicle's side walls and adjacent unit loads. (4) Whenever possible, use a pallet board in the storage position. (5) After the slip sheet is in its final warehouse location, cut the tab or secure the tab to the unit load. (6) Train the lift truck operators.

Types of Lift Trucks Used to Handle a Slip Sheet

If you are considering using a slip sheet in your operation, then the lift truck's push-pull equipment is the second important component of an efficient operation. The various components of the push-pull slip-sheet handling equipment are the type of lift truck, type of push-pull device or gripper, and type of pallet board.

The lift truck or unit-load handling equipment that is adapted with a push-pull attachment includes (1) a sit-down counterbalanced lift truck (Fig. 11.10a) that handles the new unit-load weight lift requirement (lift capacity) and weight center and can enter the delivery vehicle and (2) a low-lift powered rider pallet truck (Fig. 11.10b) which enters all delivery vehicles.

(a)

(b)

FIGURE 11.10 (*a*) Counterbalanced lift truck; (*b*) pallet truck. (*Courtesy of Cascade Corp.*)

Slip-Sheet Attachments

The push-pull attachment consists of a pantographic arm that extends forward from the lift truck mast with a backrest and gripper bar. This arm is hydraulically controlled by the operator to extend over a platen, set of platens, or series of tines (a series of chisel-type forks). When the operator activates the gripper, it grips the slip-sheet lip (tab) and holds it firm (fast) between the backrest bottom and gripper bar. The operator controls the hydraulic system that orders the pantographic device to lift the slip-sheet lip upward. This lifting action raises the side of the slip sheet. In the raised position, the pantographic device and slip sheet are pulled over the platen(s) or tines to the mast of the lift truck.

With the slip-sheet under the hardened metal surface of the platen, the lift truck transports the unit load to another warehouse location, onto a pallet board or onto another vehicle.

The slip-sheet deposit activity from a lift truck is very simple. The lift truck operator positions the unit load directly in front of the pallet load. The hydraulic system raises the unit load to an elevation slightly above the pallet board. The pallet board is placed against a unit-load backstop. The slip sheet in the elevated position is moved forward until the unit-load platens are over the pallet board. In this position, the push-pull device with the unit load is extended forward until the load touches the backstop. In this action one (far) side of the unit load touches the pallet board. When this happens, the operator makes the hydraulic system gripper bar release the lip, and the backrest remains firm and extended over the near side of the pallet board. As the lift truck is moved backward, the slip-sheet unit load rests on the pallet board, and transfer is completed.

Platen Push-Pull Device.   The platen type of slip-sheet attachment consists of two 15- to 18-in-wide bottom tapered hardened metal platens (Fig. 11.10c and d). A platen resembles a flat shovel. A 4-in open space between the platens provides, at the minimum, a 34-in-wide surface to support a 40-in-wide (maximum) slip-sheet unit load. The platens are 48 in long to handle a 48-in-long slip-sheet unit load.

Tine Push-Pull Device.   The tine (chisel) slip-sheet attachment (Fig. 11.10e) consists of a series of full tapered hardened metal extensions from the base of the lift truck mast. The tine length varies from a minimum of 36 in up to a maximum of 48 in. Each tine is 4 to 5 in wide with a 4- to 5-in open space between the tines. This width and open space provide adequate support for the 40-in-wide slip-sheet load.

Bop Slip Sheet.   The bop slip sheet (Fig. 11.11a) goes on top of a slave take-it-or-leave-it pallet board. This slip sheet consists of two fiberboard full-length sleeves that are attached to the underside of the slip sheet. The sleeve on the fork-entry end has an opening with an upper guide lip. The distance between the two sleeves is on 24-in centers. These openings permit a regular 4-in-wide by 42-in-long full tapered chisel

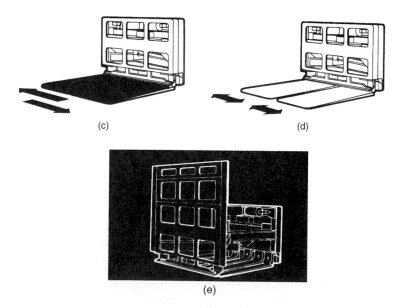

(c)   (d)

(e)

FIGURE 11.10   (*Continued*) (*c*) Single platen; (*d*) double platen. (*e*) tines. (*Courtesy of Cascade Corp.*)

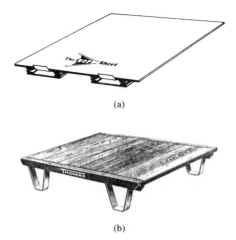

(a)

(b)

FIGURE 11.11   (*a*) Bop sheet (*Courtesy of Bop Sheet.*); (*b*) skid (*Courtesy of Thomas of Rockford, Inc.*).

fork of a lift truck to be inserted to the normal fork depth under the slip sheet. At this depth the lift truck tilts and lifts the slip-sheet unit load from the slave pallet board. For easy entry and withdrawal, the slave pallet board has four pieces of 2- by 2-in wood runners. Two runners are placed along each exterior, and two runners are placed 12 in from each exterior piece. The locations of these runners and their height permit easy lift truck fork entry and withdrawal.

The disadvantages of the bop slip sheet are the need for driver training and specially designed pallet board and the height that is added to the unit-load vertical storage stack. The advantages are that the lift truck does not require a special attachment and that it can be used in most storage systems.

Compared to pallet board, slip sheet has some disadvantages: moisture reduces the tear strength, it restricts air circulation, additional operator training is needed, the pallet board must be in the rack storage position, and it requires an additional 6- to 8-in footprint for a floor-stacked load.

The advantages of slip sheet are that it is low-cost, is handled by a lift truck with an attachment, and is easy to store empty; there are no return slip-sheet charges; and trash problems are reduced.

SKID

The next unit-load support device is the skid (Fig. 11.11*b*). The skid is designed for heavy loads; it has a load-carrying surface with two legs (runners) that run the full length of the skid on each underside. These legs support the load-carrying surface above the floor and create a two-way fork-entry opening. The skid is manufactured from wood, metal, or a combination of metal and wood. With this construction and leg support design, the skid handles heavy and compact loads. It can transport heavy loads and support unit loads in a standard rack structure.

The high-leg feature requires a lift truck, platform pallet truck, or normal pallet truck with a skid adaptor to lift and transport skids between warehouse locations or workstations.

The disadvantages of the skid are that it is difficult to stack and is more expensive and it requires an adaptor with a regular pallet truck.

The advantages are that it is durable and handles heavy, compact loads.

CONTAINER

Containers comprise the next major group of material handling devices used to transport and store product of irregular shapes and sizes and large quantities of loose, solid items and bulk granular or liquid items.

Container Types

The container has three basic forms: (1) a complete container that has a leg and stringers with a bottom and four sides, (2) a separate four-sided container that is placed on a pallet board, and (3) a bulk bag with fork-lift straps that is set on a pallet board.

The complete container has four sides made of solid fiberboard, metal, aluminum, wire mesh, welded wire, or wood slats with wire. The bottom surface is made of the same material or is a pallet board. The type of container is determined by the nature of the product (solid, granular, or liquid). The product and container weight, size, cost, and stacking height are other factors. Important design factors include (1) material handling equipment required to move the container and the fork-entry opening; (2) length, width, and height of product; (3) storage method (floor stack or rack structure); (4) method to fill or empty the container; and (5) storage environment conditions.

Containers are divided into the following classes: welded-wire containers, rigid welded-wire containers, solid-steel containers, wire-wrap-around containers, wire-bound containers, wood pallet containers, fiberboard and corrugated containers, bag with fork loops (straps), aluminum or steel containers, and plastic bins.

Welded-Wire (Wire-Mesh) Container.   The welded-wire (wire-mesh) container (Fig. 11.11c) consists of two gauge steel wire strands welded on 2-in centers on all four sides and the bottom. The corners, top edges, and bottom edges have double wire strands for strength. Legs on the four corners permit two- or four-way stacking in the storage area and clearance for the lift truck forks during product movement. Some welded-wire containers are stacked three or four high, and other models have fold-down and swing-out gates. The fold-down or swing-out gate improves employee productivity in transferring product between containers or permits the container to be used as a single-item pick position.

The disadvantage of the welded-wire container are the space needed for empty containers and the cost of the container.

The advantages are that it is used as a storage and transport device, it is durable, it holds hot parts, it is stackable and lightweight, it is see-through, access to the interior is easy, it is self-cleaning.

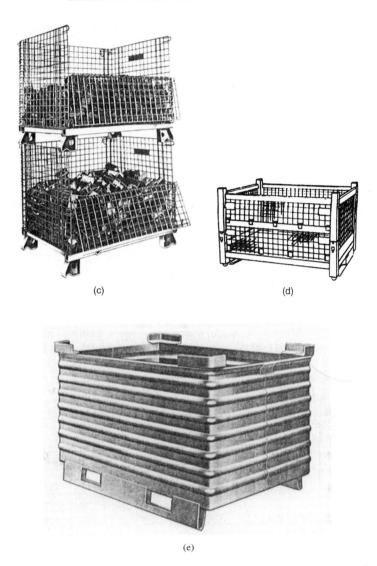

(c)

(d)

(e)

FIGURE 11.11 (*Continued*) (*c*) Welded-wire container (*Courtesy of Cargotainer.*); (*d*) rigid wire-bound container (*Courtesy of Steel King Industries.*); (*e*) solid-side container (*Courtesy of Steel King Industries*).

Rigid Wire Container. The rigid welded-wire containers (Fig. 11.11*d*) are heavy-duty containers that have four corner posts for extra strength to handle heavy loads. Some models have top and bottom horizontal members of angle iron to support long spans and to increase the stack height.

Compared to the wire-mesh container, the rigid welded-wire container has the additional disadvantage of being more costly but the additional advantage of being designed for heavy and long loads.

Solid-Steel Container.   The solid-steel container has a solid base and solid walls. The four sides are 12 to 30 in high in smooth or corrugated style. The corrugated style (Fig. 11.11*e*) provides increased rigidity and resistance to bending of the sides. The smooth style permits the maximum use of the interior for product storage. These containers are used for in-house transportation and storage functions. Some models are stacked up to six high in floor storage areas or are stored in standard racks. The full-length legs permit two-way fork entry, and four corner legs permit underclearance for lift truck fork entry.

The disadvantages are its heavier unit-load support device and that it is inverted for cleaning.

The advantages are that it is durable, it holds hot parts, and it is stackable and watertight and oiltight.

Wire-Bound Container.   Wire-bound containers, boxes, and wraparounds are lightweight boxes or crates made of thin wooden corners and slat members which are reinforced with cleats and staples. Some of these containers have their own wooden base or are placed on top of pallet boards. If the pallet board is the base, then the length and width of the container match the pallet board dimensions.

The wire-bound containers are used for in-house transportation and in the storage area. Also the wire-bound container is used for transportation between the warehouse and the customer's location. To obtain high storage density, it requires a rack structure.

The three types of wire-bound containers are the six-sided container for small parts, the base-mounted container for large parts, and the skid-mounted container for equipment.

Disadvantages of the wire-bound container are that it is not flexible, it is difficult to store outside, it may collapse if stacked beyond the limit, and one-way boxes are difficult to reuse. The advantages are that it absorbs shock, is collapsible, and stacks two high.

Wood Pallet Container.   Wood pallet containers are solid-wall containers used to transport and store granular materials or products of irregular shape and size. The pallet container consists of solid sides, solid top and bottom, and side wood board or slat members to reinforce the sides. The containers have an opening on the bottom for fork entry.

The three types of wood pallet containers are the pallet bin, pallet box, and pallet crate. The pallet bin has a pallet board base with closed sides and ends. The bin consists of a series of side boards, edge boards, and diagonal cleats. The pallet box has a pallet base, solid sides and ends, and a top. Each corner has a vertical slat that is cleated for stacking strength. The pallet crate has a pallet board base with open (slatted) side and ends and a horizontal cleat with each corner having a post for stacking strength.

The disadvantages are that they are difficult to repair and fit the pallet board. The advantages are that they are easy to move, flexible, collapsible, and reusable and they can handle many SKUs.

Fiber (Corrugated) Container.   The fiberboard (corrugated) container has solid corrugated side walls, base, and top. The container is designed to hold granular material and to sit on top of a pallet board or slip sheet for transport by a lift truck. Some containers are designed to hold furniture or household appliances and are handled by a clamp truck or lift truck with a basiloid lift device. The three most popular corrugated containers are six-sided, square bulk, and square corrugated.

The disadvantages are that moisture or high humidity reduces strength, they are difficult to repair, and there is minimum durability. The advantages are good utilization and low cost.

Bulk Bag.   The bulk bag is a cost-effective way to ship, transport, and store granular material. Bulk bags are made of plastic fabric of woven polypropylene, burlap, or kraft paper with nylon or polyester webbing added to the sides and bottom for strength. Straps or string loops are added for lift truck fork entry. Most bulk bags are the top-fill bottom-discharge type.

The disadvantages are that it serves a single purpose and that two employees are needed to place the loops on the lift truck forks. The advantages are that it costs little, it reusable, and does not require a pallet board.

Stainless-Steel Container.   When liquids, slurried materials, or granular items are transported or stored in a warehouse, the aluminum or stainless-steel container is preferred. The containers are filled at the top and discharged from the bottom. Smooth or corrugated steel bodies are available.

The disadvantages of the container are that it is not collapsible and is difficult to clean. But it has long life, is stackable, and holds liquid.

Smooth-Sidewall Container.   The smooth-sidewall container is less expensive, but does not have stacking strength.

Corrugated-Sidewall Container.   The corrugated or ribbed sidewalls of this container increase its rigidity. These containers have fork openings on the bottom and are stacked three to four high.

Plastic Container.   The plastic container (bin) provides the means to transport and store dry granular or small solid parts. These containers are filled and discharged from the top and have fork openings on the bottom. The rigid container type is made of structural molded high-density polyethylene with ribs on the exterior that add sidewall strength. This feature permits the containers to be stacked in the warehouse. The semi-rigid container is made of molded polyethylene or similar plastic. These containers have sidewalls that are smooth and flexible. This feature permits nesting of empty containers. In transportation or storage, lids are used to prevent foreign particles from getting into the container. Also lids improve security and prevent items from falling out of the container. An option during the molding process that will increase the plastic container's strength is to have metal rods inserted.

The disadvantages are that it has a lid, there is limited stackability, and it requires approval by local fire codes. The advantages are that it is reusable, nestable, and easy to handle.

Pallet Stacking Frames and Portable Racks.   Pallet stacking frames and portable racks consist of four corner metal upright posts and top horizontal members. The corners of a portable rack are attached to a wood pallet board, and the horizontal members are attached in the middle. Nails are used on each of the four corner metal baseplates to make a secured connection. The stacking frame has four corner metal posts with horizontal top members. Some stacking frames have three horizontal members that allow the frames to be nested when not in use. The portable frames have a metal base; therefore, fork sleeves (stirrups) are recommended to keep the frame from sliding on the metal forks of the lift truck. The top horizontal members permit these containers to be stacked three to four high.

These containers are used in the in-house transportation and storage warehouse functions of large items. Refer to Chap. 7 for more detail on these storage devices.

Roll Guard.   When the product to be transported or stacked in the warehouse does not lend itself to the previous containers and is round, then the roll guard can be used. The roll guard consists of wood or plastic members that have a recess on each side. These recesses lock the product, and the flat side of the members permits an additional layer of product to be stacked on top.

Container Leg Design.   One important characteristic of containers is that the base provides an opening for the lift truck fork entry and permits the container to be stacked. The fork opening provides sufficient clearance for fork entry and withdrawal and rigidity to support the container on the floor or in a storage stack.

Various Container Legs

The six various container base designs are the flat base (used on a pallet board), fork sleeve, offset channel leg, skid base leg, tapered leg, and box channel leg.

Flat Base.   This container has a flat base and is placed onto a pallet board. This arrangement requires that the container and pallet board surface dimensions match.

In using a plastic or metal container (stacking frame) that has fork openings, consideration must be given to the position of the lift truck forks. If the forks are not against the legs, then two full fork sleeves or stirrups are attached to the fork openings. During in-house transportation or movement, these sleeves prevent the container from sliding on the lift truck's metal forks.

Offset Channel Leg.   The offset channel leg is located at the four corners of the container. The offset channel leg permits four-way entry and forms stacking legs. This leg design permits easy lift truck handling, but container transport on a conveyor is limited to one direction of travel.

Skid Leg.   The skid leg provides a high opening for fork entry. This design permits two-way entry, is preferred for platform trucks, and is difficult to floor-stack. When the skid is used on a conveyor, it is conveyed in one direction.

Tapered Leg.   When containers are stacked on top of each other, the tapered leg permits the top container leg to fit into a matching receptacle on the top of the bottom container. The tapered leg is triangular and is located on the four corners. This four-corner position permits fork entry. The leg height provides at least 4 in of clearance. With this height, it is preferred to have the legs reinforced, which extends the legs' life. This design requires maximum lift height to permit the higher container leg to clear the lower container receptacle. This stacking requirement reduces lift truck operator productivity.

Box Channel Leg.   The box channel leg extends the full length of the container. It permits stacking and provides 4 to 6 in of underclearance for two-way entry. The box channel permits a lift truck to handle the container and allows the container to be transported on a roller conveyor system.

HOW TO STABILIZE UNIT LOADS

The next section reviews various methods used to stabilize unit loads which do not have fixed sidewalls and ends.

Design Factors

In unitizing a unit load onto a stacking frame (portable rack) pallet board, slip sheet, or cart, several fundamental factors must be considered to ensure effective material handling and minimal product damage: interlocking cartons, weight, shape, and height of the unit load, and good alignment of the carton stack.

Stabilization techniques used in the warehouse and distribution industry include ti × hi, tape, polypropylene (plastic) or steel bands (strapping), string, stretch wrap, shrink wrap, netting, glue and adhesive, and industrial elastic bands.

A good unit-load stabilization method as part of the unitizing or palletizing program provides the benefits of less product damage, improved space utilization, improved security, improved lift truck productivity, and improved product appearance.

Prior to the implementation of the unit-load stabilization program, review of the existing system should include the expenses associated with product damage and rework labor to ensure a good return on investment. Prior to implementation of a plastic stretch- or shrink-wrap stabilization program, the new plastic-wrapped unit loads in the storage position should be approved by the insurance underwriter and local fire authorities.

Ti × Hi

The basic stabilization method for a pallet load is the carton stacking arrangement. It ensures that the cartons of a layer (hi) and each layer (ti) are interlocked, that the alignment is straight, and that there is no overhang beyond the bottom support device. The carton unitizing (palletizing) pattern for a 48- by 40-in pallet board with an allowable overhang is 52 by 44 in. To achieve a good ti × hi pattern, the carton varies from 5 to $21\frac{1}{2}$ in in width and $6\frac{1}{2}$ to 43 in in length with $\frac{1}{2}$-in increments from the smallest to the largest dimension.

The disadvantage of the method is the shifting of the load. The advantages are that it is low-cost, there are no trash problems, it is easy to remove cartons and to see cartons.

Stretch Wrap

Stretch wrapping is a unit-load stabilizing method in which a plastic film is wrapped tightly around the four sides of a unit load. The plastic film is applied by a manual or semiautomatic method. Some of the mechanical stretch-wrap methods use prestretch film wrap that provides a more secure unit load. To wrap a unit load, the stretch-wrap film is tucked under a carton and then wrapped around the unit load and tied under another carton.

FIGURE 11.12 (*a*) Manual stretch wrap. (*Courtesy of Goodwrappers.*)

Manual. The manual portable stretch-wrap machine (Fig. 11.12*a*) consists of a hand device that has a handle grip and a spool for a roll of wrapping material. After the pallet load is in position, an employee tucks the film under a carton and walks around the unit load the required number of times to secure the product. Then the employee cuts the film and tucks it under another carton.

The second manual method employs a portable stretch-wrap machine that is constructed of welded steel members and is mounted on a 3-in-high heavy-duty caster and wheel bed. It requires a lift truck to set a pallet adjacent to the machine. The film holder is attached to the machine. After film is tucked under the product or carton, the operator walks one complete turn around the load, activates the brake, and continues to walk around the load, which stretches the film wrap. The manual portable machine handles 10 to 15 loads per hour.

Automatic. The automatic Econo-Wrap machine (Fig. 11.12*b*) is a steel machine with a film roll attached to a single post in front of the turntable with a spring-loaded top. After the lift truck or conveyor places the unit load onto the turntable, the operator tucks the film under the product or carton and turns on the machine. With completion of the activity, the operator cuts the film and tucks it under the product or carton. The automatic method handles approximately 25 unit loads per hour.

A second automatic machine is a standard model which is similar to the other automatic machine except that it has a brake and handles about 35 pallets per hour.

A third automatic method uses the automatic multiroll machine, which has the same features as the automatic standard machine except that the multiroll machine has two to three film rolls and handles 30 to 40 pallets per hour.

The disadvantages are the waste (cost) due to mistakes, a trash problem is created, for an automatic system there is increased investment and increased expense of material and labor. The advantages are that it handles odd-shaped unit loads, creates a moisture barrier, improves security, and can be used in all warehouse environments.

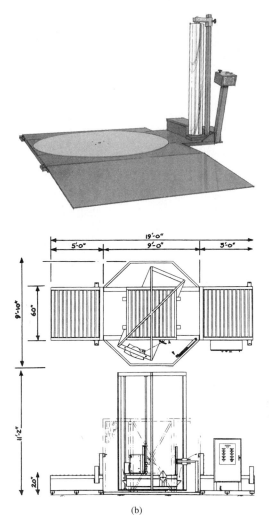

FIGURE 11.12 *(Continued)* (*b*) Automatic stretch wrap. (*Courtesy of Muller Corp.*)

Shrink-Wrap Method

The automatic shrink-wrap method (Fig. 11.13*a*) consists of a shrink-wrap tunnel that has an in-feed and out-feed conveyor system. To operate a shrink-wrap process, a pallet load is placed onto the in-feed conveyor. Prior to the shrink tunnel, a piece of plastic is draped over (covers) the unit load. The unit load is conveyed into the shrink tunnel. In the tunnel, heat is applied to the plastic material which causes the plastic to shrink around the load. In this process the plastic conforms to any shape of unit load.

Another shrink-wrap method uses the hand-held heat gun which shrink-wraps plas-

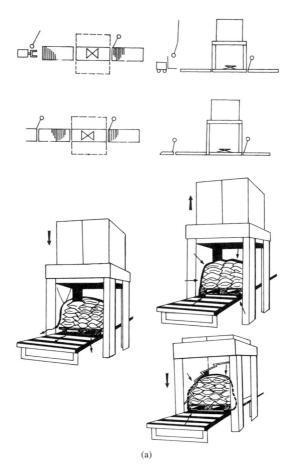

(a)

FIGURE 11.13 (*a*) Automatic shrink wrap. (*Courtesy of Shrinkfast Marketing.*)

tic to a unit load. The plastic is draped over the unit load, and then an employee uses a hand-held heat gun to direct heat against the plastic. This heat causes the plastic to conform to the unit load.

The disadvantages are that it increases the equipment investment, the floor space requirement (automatic method), and the energy (heat) costs. The advantages are the protection against moisture, dirt, and dust and less wrapping waste.

Steel or Plastic Banding

Steel or plastic (polypropylene) bands are placed horizontally around one or several layers of cartons and over the top and bottom of the unit load. The bandwidth ranges

(b)

FIGURE 11.13   *(Continued)* (*b*) Manual shrink wrap. (*Courtesy of Shrinkfast.*)

from $\frac{1}{2}$ to $1\frac{1}{4}$ in and requires a band stretcher and clamp gripper. To protect the product from damage, corrugated board material is used at the corners and top edges. Banding of the unit load is done manually or automatically with a machine.

The disadvantages are that it is labor-intensive, there may be carton damage, not all layers are banded, and cutting the bands has the potential for employee injury. The advantages are that it permits full visibility of the product, is low-cost, and is very strong.

Netting or Stretch Netting

A net is placed over the unit load and is pulled tightly around the base and top of the unit load. The netting material is made of polyethylene, and application to the unit load is similar to that for the stretch-wrap method.

The disadvantages are that it permits dirt and dust on the unit load and creates a trash problem. The advantages are that it permits circulation, allows good visibility, and stabilizes the top and bottom.

Glue and Adhesives

Glue and adhesives are used to stabilize unit loads. With modern glue (adhesives) and new application technologies and equipment, glue is applied to the top of the individual cartons or bags of the unit load. As cartons or bags with glue are placed on top of each other to form the pallet load, they stick to each other, forming an entire unit load that is stabilized with glue. When an employee requires a carton from the unit load, the bond between the cartons is broken by a simple bump or pull on the case.

The disadvantages are that there is no moisture, dirt, or dust protection and there may be damage to carton exteriors. The advantages are that it is low-cost, there is no trash problem, and visibility is good.

Elastic Bands

Elastic bands are placed around various layers of the unit load. These elastic bands are available in various widths, thicknesses, and lengths. They are manufactured in a circular form to specific lengths. These are precut and joined by a metal clip. This type of elastic band secures the unit load top and bottom as a unit load.

The disadvantages are that it is high-cost, the lower layer is difficult to apply, and it can be stretched and difficult to reuse. The advantage of the circular type is that is it reusable.

Self-Adhesive Tape

An employee applies tape to various carton layers of the unit load. In most situations the tape is on the top and bottom layers.

The tape is plastic or fiberglass-reinforced, is available in various widths from $\frac{1}{4}$ to 2 in, and can be cut with an industrial knife.

The disadvantages are that it creates a trash problem, is difficult to apply on lower layers, and creates a maintenance problem for mobile equipment wheels. The advantages are that it is low-cost, is easy to apply, and can be applied at any warehouse location.

String (Twine)

The same application method is used for string as for tape, except that string is tied by an employee around the carton layer(s) of the unit load. The string is taken from a spool or cut to predetermined lengths with a loop or metal S hook.

The disadvantages are that it is difficult to apply at lower levels, it creates a trash problem and a maintenance problem for mobile equipment wheels, and it requires additional time. The advantages are that it is low-cost and can be applied at any warehouse location.

CHAPTER 12

THE BEST INVENTORY CONTROL AND STORAGE-PICK POSITION FOR AN OPERATION

This chapter familiarizes you with inventory projection methods, stock control methods, SKU (product) identification methods, storage-pick location identification methods, SKU pick position replenishment systems, SKU cubing methods, and inventory locator systems.

This chapter examines inventory control, which is another major component of the storage-pick function. Inventory control is a tracking function that ensures through updating of inventory records that the right product is in the right place, is transferred at the right time to the right place, and is in the right quantity. It places the inventory in the location that satisfies the inventory rotation criteria and provides the lowest handling (storage-pick) costs.

In previous chapters, we examined several of the other aspects of the storage function. Each is considered a key component to the warehouse storage function. The storage function consists of four components: storage equipment, product-handling equipment, distribution facility, and inventory control.

Storage equipment ensures maximum use of space and accessibility to the product. Mobile warehouse equipment ensures maximum labor productivity. The warehouse facility ensures protection of the product and other assets.

INVENTORY CONTROL

In recent years, warehouse and distribution operations and inventory control managers have made tremendous progress in improving inventory control. This progress has been a result of five major trends in the industry: The first is a *just-in-time* (JIT) or across-the-dock program that directs your company's warehouse effect on the just-in-time raw material, goods in process or delivery of finished-goods inventory to the customer. The result has been a reduction in inventory levels and required storage area. But the new product flow system requires accurate and on-line receiving and inventory tracking functions. *Manufacturing resource planning* (MRP) has become more involved in the storage-area sizing activity due to computer technology and building and equipment investment requirements. Computer *advances and price reductions* in software and hardware have made this equipment more available to the warehouse operation. *Bar-code labels, radio-frequency devices,* and *scanning devices* provide accurate, on-line information. Last, *radio terminals* on warehouse equipment or with employees provide on-line transfer of transactions to update inventory files.

The various aspects of inventory control that are reviewed in this chapter include the inventory projection method, merchandise and document flow, storage and pick location identification methods, merchandise identification methods, warehouse activities associated with inventory control, inventory allocation method to create space, and inventory scrapping methods.

The first factor of inventory control is to design a distribution facility (building) layout, to purchase material-handling equipment, and to hire labor in quantities to handle the company's business. These resources are based on the on-hand inventory and volume projections.

INVENTORY PROJECTION METHODS

In the typical inventory projection, dollars are the basis. Dollars are an excellent denomination for your company's accounting and sale departments, but the warehouse and distribution department handles pallet loads, cartons, or single items. Therefore, the basis for the distribution facility's on-hand inventory quantity and volume is stated in pallet loads, cartons, or single items. These statistics allow you to project and track activities and inventory quantity.

If historical on-hand inventory quantities and volume are not available in units handled at your facility, then the actual sales dollars are converted to handling units. The sales dollars are converted by an average unit-dollar value that is determined by the accounting department.

The historical inventory statistics required for the necessary inventory projections include the average on-hand inventory, low on-hand inventory, peak on-hand inventory, inventory turns per year, projected sales growth per year, and projected SKU growth per year.

With these facts and the company's projected growth, you project the on-hand inventory units, storage requirements, and volume. These figures are necessary to design the storage-pick positions, to obtain the material-handling equipment, and to purchase components of the inventory control system. Two methods are used to project inventory volumes: the simple percentage increase to the total volume and the inventory stratification method which increases each SKU classification. Later in this chapter, these methods are reviewed in more detail.

If your business is in the catalog and direct-mail industry and issues several catalogs per year, then you must design SKU pick positions and inventory storage positions for the overlap of two catalogs. This overlap occurs because the new catalog SKU inventory is required on site 6 to 8 weeks prior to the discontinuation of the old catalog SKUs and inventory.

When you look at the peak catalog SKU and inventory situation, then pick and storage position utilization should be in high 80 to 90 percent.

In the retail department store distribution industry, a similar SKU increase occurs at the change of the seasons. This situation is called the *break*. In the retail department store warehouse inventory, the entire new line for the next season and a percentage of last season's SKUs remain in inventory. Compared to the middle of a season, this situation creates a higher number of SKUs in inventory. The number of pick positions should be designed for this break.

Inventory control should focus on two areas: document (purchase-order) flow and SKU (physical product) flow.

Document Flow

When the purchasing department issues the purchase order to the vendor, it sets in motion the inventory control activity. Appropriate copies of the purchase order are sent to the warehouse receiving department, or the information is entered into the receiving department's computer files. The information on the hard copy or in the files indicates the vendor, anticipated delivery date, truck delivery company, merchandise quantity, and SKU description. With today's computer network systems, this information is transferred between the purchasing department and receiving department via computer. This information permits the warehouse receiving manager to anticipate the daily receiving volume and to schedule the appropriate workforce, dock equipment, and dock positions.

After the arrival of the delivery truck at the receiving dock, the receiving employee compares the delivery truck's bill of lading, date of delivery, and product quantities to the delivery date and product quantities on the purchase order. If the information on the two documents matches, then all the product is unloaded onto the dock. If the quality and quantity verify that the product is acceptable by company standards, then the purchase order serves as the on-hand inventory update document.

If the actual unloaded quantity is verified to differ from the purchase order quantity, then the purchasing department approves the new quantity and appropriate adjustments are made to the purchase order or bill of lading. The receiving department employee has the delivery truck driver sign the receiving department's copy of the purchase order or bill of lading, to verify the actual quantity unloaded and received at the facility, and the adjusted receiving quantity becomes the adjusted inventory quantity.

When the date of delivery on the bill of lading does not match that on the purchase order, the purchasing department and receiving department managers authorize receipt of the shipment.

Inventory Update.   When the received inventory file is updated, then the receiving department uses the actual received quantities and issues the stock put-away document. The document directs the storage employee to deposit the product into a storage-pick position. After the unloading and quality and quantity checks are completed for the inbound product, the product is entered into the inventory files. This inventory entry increases the on-hand inventory quantity to reflect the inbound receipt. Two methods to update the inventory records are the delayed method and the on-line method.

Delayed Update.   Prior to the book inventory update, in the delayed-entry method the product moves from the receiving area to the storage area. With this method all inbound deposit transactions are completed, and the inventory book entry is made at a later time.

The disadvantage of the delayed-entry method is that a no-stock condition could occur with the inventory physically in the pick position because the on-hand book inventory is not updated to reflect the new receipt. A no-stock condition occurs when a customer places an order for a SKU, for the SKU the on-hand book inventory is zero, but there is an inventory quantity in the pick position. Therefore, a transaction ticket is not printed for the product demand (customer order) or pick position.

On-Line Update.   In the on-line entry method, the SKU book inventory adjustment is entered in the book inventory file as the product moves from the receiving area to the storage staging area. With this method, the SKU inventory quantity is increased to reflect the new receipt prior to the completion of all deposit transactions of the product. The disadvantage of this method is that the on-hand book inventory is updated prior to the product's being physically placed in the storage-pick position. This situation could cause a stockout. A stockout occurs when a customer places an order for a SKU and the SKU has an on-hand book inventory quantity; a transaction ticket is printed for the product requirement, but there is no inventory in the pick position. This condition is associated with low employee productivity because the employee traveled to the warehouse product storage-pick position to perform a transaction only to find no product in the position (this is nonproductive time).

Replenishment and Pick Document.   In a pallet load operation, the final document in the document flow process is the order-pick instruction document. The order pick instruction document directs the warehouse employee to withdraw the pallet from the inventory.

These inventory control documents are required in a floating (variable) pick position system: (1) The SKU put-away instruction directs the employee to place the SKU in the computer-assigned pick positions. (2) If the SKU quantity overflows the predetermined number of pick positions, then the computer prints instructions for the SKU to be put away in a fixed reserve (storage) position. (3) The SKU pick document directs the order picker to withdraw SKUs from the pick position. With the depletion of the SKU quantity in the pick position and with inventory in the reserve position, the computer prints a document to transfer the SKU from the fixed reserve (storage) position to any vacant pick position. As required, the computer prints pick instructions for the SKU in the pick position.

In a carton or single-item distribution facility with a fixed-pick-position system, the next document is a pick position replenishment document. The pick position replenishment document directs a warehouse employee to transfer a specific quantity of merchandise from the storage position to the pick position. In a distribution facility that uses a floating- or fixed-pick-position system, the final document is the order pick instruction document that directs the warehouse employee to withdraw a specific quantity of merchandise from the pick position to complete an order.

Product Flow through the Distribution Facility

Physical merchandise is received by or delivered to a distribution facility by one of the methods illustrated in Fig. 12.1. In all types of distribution operations (single-item, carton, or pallet load), the vendor receipt is an increase to inventory and the customer order quantity is a reduction to on-hand inventory.

Method of delivery	Product type		
	Pallet	Carton	Single item
Railroad car lot	X	X	—
Truck (van) lot	X	X	X
Container lot	X	X	X
Pallet load	X	X	X
Carton	—	X	X
Single item	—	—	X

FIGURE 12.1   Distribution facility.

In a pallet, carton, or single-item distribution facility, the flow of physical product is very similar to document flow. The product flow is from the receiving area to the reserve (storage) area, to withdraw (pick) the product that satisfies the customer order.

Assignment of Product Position. The first step of an inventory control program is to assign the product (SKU) to a warehouse position. The two basic methods that determine the warehouse reserve (storage) positions or deposit position are performed manually or by computer.

In the *manual* put-away position assignment method, an employee daily walks through the warehouse aisles and lists the vacant storage positions. This list is given to the receiving clerk who assigns the new received merchandise to one of these vacant storage positions. After the assignment of the inventory to the position, the clerk marks the appropriate storage position from the list. This procedure prevents the receiving clerk from assigning two SKUs (products or unit loads) to the same storage position.

The disadvantages of the manual storage position assignment method are that it is labor-intensive, errors are possible, and it handles a low volume. The advantages are that it is low-cost and handles a low volume.

To use the *computer* in assigning put-away positions, an initial survey must be made of all vacant and occupied warehouse storage and pick positions. Each of these positions is discretely numbered; this includes each storage position within a dense-storage lane. Also, this list includes the SKU description and inventory quantity in each occupied position. After the completion of this list, all items (by inventory quantity) are entered by each position and SKU into the computer. After the inventory program is on line, all deliveries and customer orders are entered into the computer and the computer adjusts the beginning SKU inventory quantity by the appropriate increase or reduction.

These transactions are made in the following sequence in a carton or single-item warehouse facility:

1. As required by each customer order, reduce the SKU on-hand inventory in the pick position.
2. When the pick position is depleted, the computer directs a replenishment employee to transfer a required pallet load, carton, or appropriate product quantity from the storage position to the pick position.
3. Update and publish all existing and new vacant storage positions that are available for inbound product. This feature allows the receiving clerk to assign these vacant positions to product received at the warehouse.

In a pallet load operation, step 2 is not required because the reserve position is the pick position.

The disadvantages of the computer put-away assignment method are the increased employee training and higher investment. The advantages of the method are that it is accurate and on-line and handles a high volume.

The manual or computer put-away assignment method is used with any of the warehouse storage philosophies that include the family group, ABC, slug, and random.

As indicated, the physical inventory control (merchandise tracking) requires accurate information flow. This information flow starts at the receiving or checking dock and ends with the order-pick verification. The receiving and checking function ensures that the actual merchandise quantity and description correspond to the company's purchase order. Also the receiving department ensures that the product meets your company's specifications (standards). After your receiving department verifies these aspects, then the product number, product description, and assigned storage identification number are placed on the product. The product identification method indicates to the warehouse employee that the product is approved for transfer from the receiving and checking area to the storage area. In a distribution facility that handles flatwear or hanging-garment operation, the product is transferred to the ticketing (price-marking) area.

Information Needed on SKU Identification.   Prior to the implementation of the product identification method, you must choose the information that appears on the product identification tag or label. Typically this includes the following: For pallet load operation, the reserve position, quantity received (ti $\times$ hi), date received, SKU (product description), and weight of the SKU are required. For carton or single-item operation, the reserve position, pick position, quantity received (ti $\times$ hi), SKU description, and weight of the SKU are needed.

PRODUCT IDENTIFICATION

The purpose of merchandise inbound identification is to identify the assigned warehouse location. In a pallet load facility, the location is a remote or reserve (pick) position. In a carton or single-item distribution facility, the warehouse location has three possibilities: remote reserve position, ready reserve position, and pick position.

Where to Put Identification Tag or Label

The means of identification is a crayon mark, tag, or label that is secured to the exterior of a pallet load, carton, or single item. In a single-item warehouse, the tag or label is applied to the side of the SKU, is slipped between two cartons, or is attached to the side of the pallet load. With hanging garments, the merchandise identification tag is placed on the lead end of the trolley or is attached to the left side of the garment. The identification tag allows an employee to recognize the destination of the merchandise.

When a pallet load is handled in the distribution facility, the label is affixed to the side of a carton on the pallet load. In this type of warehouse, the label is placed in a consistent location, such as in the lower right-hand corner of the pallet load, or in a position that allows the lift truck driver to see the label.

In a carton distribution facility that uses bar-code scanning devices, this carton is typically one of the last cartons removed from the pallet in the pick position. Also, the label is in an easy location for a lift truck operator to see and scan with a hand-held scanner because in the typical facility most employees are right-handed.

Product Identification Methods

The type of inbound merchandise identification method is determined by the identification method for the crayon mark, label, or tag used in the operation. The various methods include (1) no method, (2) no tag method, (3) manual print methods (crayon or chalk on the product or crayon or chalk on tags and labels), and (4) machine-printed tags, either readable by humans or readable by both humans and machines.

Vendor Identification.   The first merchandise identification method is no identification method at all! Your company relies upon the vendor's exterior carton markings for product identification. This method does not require that your company markings be placed on the product exterior; therefore, the majority of companies that use this method have only one type of SKU in one aisle. This storage system is usually two- or three-deep floor-stack method and there is a large quantity of one SKU.

The disadvantages of the method are that it handles a limited number of SKUs, it increases deposit and withdrawal errors, and there is low employee productivity. The advantages of the method are that there is no expense or investment and less time is spent in the receiving and staging area.

Crayon or Chalk Marks.   In this method (Fig. 12.2), the receiving clerk uses crayon or chalk to mark the product storage location on the exterior of the product. This is considered the basic method.

FIGURE 12.2   Human-made marks on tag or product. (*Courtesy of Unarco Material Handling.*)

Disadvantages of the method are that (1) hand-written numbers can be difficult to read, (2) transposition (transfer) errors can be made, (3) numbers can be smeared, (4) it increases the labor needed, and (5) with other markings on the product, the markings are difficult to identify. The advantages of this method are that it is cheap and no investment is required.

Crayon or Chalk Marks on Tag or Label.   In this method, the receiving clerk uses a crayon or chalk to write the product storage-pick location on a label or tag. The completed tag or label is secured with tape or glue or is slipped between cartons on the exterior of the load.

The advantages and disadvantages of this method are the same as those for the hand-written crayon or chalk method. An additional advantage is that the numbers are easier to identify.

Crayon or Chalk Marks on a Color-Coded Tag or Label.   In the fourth method, the reserve or pick position is hand-written on a colored label. The receiving employee writes the reserve-pick position on the face side of a color-coded label. The label is placed onto the lower right-hand corner of the carton or pallet load. The color of the label corresponds to a particular month of the year. This color code allows an employee to easily and quickly identify the age of the product.

The disadvantages and advantages of the color-coded tag method are the same as those for the label method. An additional advantage is the ability to determine the age of the physical inventory.

Machine-Printed Human-Readable Identification on a Tag or Label.   In this method (Fig. 12.3*a*), a computer-controlled printer prints the product storage-pick location on a label or tag. The location is printed in characters or digits readable by

(a)

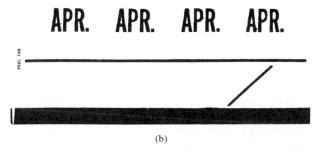

(b)

FIGURE 12.3   (*a*) Machine-printed human- and machine-readable label; (*b*) color-coded tags.

humans on a label or tag. An employee places the label or tag on the outside of the product.

The disadvantages of the method are that the label costs more, capital equipment and software investment are needed, and space in the warehouse and print time are both required.

The advantages are uniform, clear identification, medium capital investment, low labor requirement, preprinted labels, and no transposition errors.

Machine-Printed Human- and Machine-Readable Identification Marks on a Tag or Label.   The computer prints human-readable characters and digits and machine-readable symbology [bar codes or radio-frequency (RF) tags] on the label. Note that the RF tag is not a label but is a device similar to a button which is attached to the exterior of the product. Most logistic professionals refer to the bar code/human readable label on a unit of storage as the license plate for the unit of storage.

An additional disadvantage is increased print time due to the bar-code print requirement. An additional advantage is that the label is read by a scanning device.

Self-Adhesive Labels.   The self-adhesive label is widely used in the warehouse industry as the product storage-pick identification method. The self-adhesive backing provides a positive method to secure the identification label to the product. This feature ensures that the identification label remains with the product and allows the employee to recognize the label. The label can be put on a carton product either manually or by machine applicator.

Nonadhesive Labels.   There are two types of nonadhesive label: the solid type and the punched-hole type. The following procedures ensure that the product identification label or tag will remain with the product and is easily noticed by a warehouse employee who handles the product: To use the nonadhesive solid label, the label is slipped between two items or cartons. The nonadhesive punched-hole label or tag is hooked onto a plastic string or band with the loop or hook attached to the lead end of the trolley head or garment hanger.

Black-on-White Labels or Tags.   Another important feature of the inbound product location label is the colored label or the color of the printing. The color types are the plain white backing with black markings, color-coded borders, or colored printing. The label having plain white backing with black markings is widely used in a warehouse facility that is not concerned with the age of the product.

Color-Coded Label or Tag.   The color-coded product location identification label (Fig. 12.3*b*) is widely used in a warehouse facility that is concerned with the age of the product. There is a unique color for each month of the year. The color-coded label provides a quick indication of the month in which the product was received in the warehouse. In an operation that handles nonsensitive product, the color-coded tag or label verifies inventory rotation and identification of slow-moving SKUs.

PRODUCT DESTINATION IN THE WAREHOUSE

After the merchandise is unloaded and identified and the product quantity is updated in the on-hand inventory file, the physical inventory control activity in the warehouse begins. At the receiving dock, the merchandise is staged for storage location assignment and for internal transport to the storage-pick area.

After the product arrives in the storage-pick area, it is assigned to a remote position, ready reserve position, or pick position. Chapter 5 reviews the use of alphabetic characters or digits (numerics) as the means to identify storage-pick positions.

In a pallet load warehouse, then this is the reserve or pick location, and per customer order requirements, the pallet load is withdrawn from the storage position and transferred to the shipping dock for shipment to the customer.

In a carton or single-item distribution facility, the product is transferred, as required, from the reserve position to the pick position. As required on the customer order, the merchandise is withdrawn from the pick position and transferred to the packaging area or outbound staging area. Later it is loaded onto the customer delivery vehicle.

Various SKU Positions in a Warehouse Operation

As stated, during the receiving process, the merchandise receives an identification label specifying its position in the distribution operation. There are four possible warehouse operation positions: off-site remote reserve position, on-site remote reserve position, ready reserve position, and pick position.

Off-Site Remote Reserve Position.   The off-site remote reserve position is a storage position that is located in another building. This building is one of the company's own buildings or a public warehouse. As required, pallets are transported from your off-site building to your main distribution facility. Product in off-site remote reserve positions is used to replenish an on-site remote or ready reserve position. In off-site storage, the product represents an overflow from the main distribution facility. The SKU characteristics in off-site storage positions are one SKU with many pallet loads, slow-moving SKUs with a large inventory, or slow-moving SKU which has a large volume.

On-Site Remote Reserve Position.   The on-site remote reserve position is in the main distribution facility but not in the same aisle as the pick position. Typically, these remote reserve positions are dense-storage-type positions, and they contain product that is used to replenish depleted ready reserve positions. The SKU characteristics in remote reserve storage positions are one SKU with many pallet loads, slow-moving SKU with a large inventory, or slow-moving SKU which has a large volume.

Ready Reserve Position.   The ready reserve position is in the same aisle as the SKU pick position. The best ready reserve position is directly above the pick position. The next-best ready reserve position is any position adjacent to the best ready reserve position or a ready reserve position that is directly across the aisle from the SKU pick position. The typical unit quantity is the overflow from the pick position which is one or two pallet loads or cartons.

Pick Position.   The pick position is used in all types of distribution facilities. In the pick position merchandise is withdrawn by an order picker to complete the customer's order.

How to Identify Aisles and Storage-Pick Positions

In your warehouse and distribution operation, to ensure an accurate and on-line product transaction, proper signs must clearly identify the aisles and reserve or pick positions. Three methods are used to identify warehouse aisles. In all three methods, a placard with alphabetic characters or numbers is placed at the aisle entrance and exit.

One-Way Vision Placard. The first method is considered a one-way vision method (Fig. 12.4*a*). A placard is placed flat against the end of the aisle upright frame or is

FIGURE 12.4 (*a*) One-way vision placard; (*b*) two-way vision placard. (*Courtesy of Unarco Material Handling.*)

hung from the ceiling. In this arrangement the aisle identification faces outward which lets an employee entering the aisle to identify the aisle.

Two-Way-Vision Placard against the Rack.   The second aisle identification method is a rack two-way-vision method (Fig. 12.4*b*). This method is an arrangement of three placards. Two placards for each aisle are placed side by side against the face of the aisle upright frame. In this position, as the employee enters the aisle, he or she identifies the aisle. Another placard with the appropriate identification for each aisle on both sides is placed between the two other placards and extends outward toward the main traffic aisle. This placard position permits an employee from the main traffic aisle to identify the aisle.

In an option to the two-way method, one placard with the aisle number on both sides extends outward from the rack upright post into the main aisle. This arrangement permits an employee to easily identify the pick aisle from the main traffic aisle.

Two-Way Aisle Identification.   A two-way aisle identification method is used in a distribution facility that uses racks or floor-stack storage methods. One placard with four sides is ceiling-hung in front of the aisle entrance or in front of the rack. This arrangement permits an employee to identify the pick aisle from the main aisle or as the employee enters the pick aisle.

POSITION IDENTIFICATION METHODS

In a pallet load, carton, and single-item distribution facility, the method used to identify the pallet position has a direct impact on the employee's transaction productivity and accuracy. In a carton or single-item operation, using the best pick position identification method improves employee's pick position replenishment and pick transaction productivity and accuracy.

This portion of the chapter reviews the various methods used to identify the SKU reserve and pick positions. These methods are separated into the pallet load, carton, single-item, and hanging-garment groups. For each product group, the reserve and pick position identification methods are (1) no method, (2) nonhanging group (pallet load, carton, single items), and (3) hanging-garment group.

Use Characters or Digits as Identifiers

The first step is to determine what information is to appear on the storage-pick position. There are two information options.

The first option (Fig. 12.5*a*) is to use characters and digits to identify the position which is used in any type of warehouse operation that handles pallet loads, cartons, or single items. This method is considered the basic method.

The disadvantages of the method are that it is difficult to detect replenishment errors and a check system is not provided. The advantages of the method are that it requires less employee reading of the identification, the expense of the label and its attachment is incurred, it is best for a pallet load storage warehouse operation and for a floating-slot pick system.

The second option (Fig. 12.5*b*) is to use the characters, digits, and SKU description with a bar-code label to identify the position. This method is used in a fixed-position

(a) (b)

FIGURE 12.5 (*a*) Characters and digits; (*b*) characters and digits with a bar code (*Courtesy of Seton Name Plate Co.*).

warehouse operation and is considered the most sophisticated method. With the bar-code label, the application is an on-line or delayed information transfer to the computer.

The disadvantages of the method are the increased label expense, need for a large, wide label, and investment in the scanning equipment and network. The advantages of the method are that it is best for a fixed-pick-position system, it detects deposit and replenishment errors, and there is accurate, on-line information transfer.

Since this method is an on-line system, the ability of the computer system to handle the frequency and number of inputs from the scanning devices along with the other computer activities (tasks) is verified by the computer equipment suppliers. Failure to resolve this situation can mean off-line data transfer.

STORAGE-PICK LOCATION IDENTIFICATION

No Identification Method

The first warehouse reserve and pick position identification method is no identification. There are no markings on the storage-pick position rack structure or floor to identify the position. When you use this method, you rely upon your employees to match the deposit or withdrawal transaction instruction to the carton's exterior markings.

The disadvantages of the concept are that (1) it requires the SKU description of the instruction document to match the vendor carton exterior markings on the carton made by the vendor, (2) employee productivity is low, (3) the possibility of errors is increased, and (4) the employee must be able to read.

The advantage of the method is zero cost.

Nonhanging Garments

Manual Paint, Crayon, or Chalk Marks on the Position. The first nonhanging garment warehouse storage-pick position identification method uses paint, crayon, or chalk. An employee uses paint, crayon, or chalk to write the position identification numbers or alphabetic characters directly on a flat surface of the rack structure that is above, below, or to the side of the SKU position.

The disadvantages of the method are that (1) the handwriting may be unclear or confusing, (2) it is not easy to transfer, (3) it is not easy to recognize, (4) uniform characters and digits are difficult to ensure, and (5) they are not in a uniform position. The advantage is its low cost.

Manually Marked Self-Adhesive Label or Tape.   An employee places a label or tape on a flat surface of the rack structure that is above, below, or to the side of the position. Next the employee scribes with paint, crayon, or chalk the position numbers or characters onto the label or tape.

The disadvantages are the unclear or confusing handwriting and increased labor expense. The advantages are that employee productivity is improved, the label material expense is low, it is easy to transfer, the label location is uniform, and it is easy to notice.

Preprinted Self-Adhesive Label.   A computer-controlled printer or a print machine prints the number, digits, or characters for the position onto a label. The label is placed by an employee in the appropriate location at the SKU storage-pick position.

The disadvantage of the machine-printed label is the additional label expense. The advantages are the clear printing, low label labor attachment expense, uniform labels, uniform label location on the rack, and ease of transfer.

Within the self-adhesive group, there are two alternatives. The first (Fig. 12.5c) is to use an individual preprinted label for each digit or alphabetic character. These labels vary from $\frac{1}{2}$ to 3 in in height or width. As required, to complete the number or word, these labels are placed on the rack structure at the position. The disadvantages are the increased label labor expense, possible uneven label placement, and need for extra labels. The advantages are that it is easy to replace damaged labels and easy to change numbers or transfer.

The second alternative (Fig. 12.5d) is to have the entire position identification printed on one label. The disadvantages are that the printing cost is increased and damaged labels are costly to replace. The advantages include low label application labor, no label waste, and consistent label placement.

Human- and Machine-Readable Cardboard or Paper Label in a Plastic Holder.   A human- and machine-readable cardboard or paper label is placed onto the rack structure or slipped into a plastic holder which is attached to the rack. An employee secures a plastic holder to the appropriate location at the rack position. After the holder is secured, the employee inserts the preprinted label into the plastic holder.

The disadvantages are that the label material and attachment expense is increased, some holders blur the position identification, and there is potential damage to the holder. The advantages are that with a scanner it detects employee errors before they can actually affect the inventory or customer order, label presentation is uniform, and it is easy to transfer and read by bar-code scanners.

Placard Hung from the Ceiling or Embedded in the Floor.   The placard hung from the ceiling and identification that is embedded in the floor are the two alternative position identification methods most commonly used in a floor storage system. The disadvantage of the method is the increased damage to the position identification placard or to the identification numbers embedded in the floor.

Digital Display.   The digital display method (Fig. 12.6) is the most sophisticated. An indicator light is located on the metal frame of the pick position. When the customer

(c)

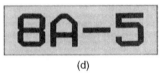

(d)

FIGURE 12.5 *(Continued)* *(c)* Preprinted individual characters and digits *(Courtesy of Brady USA.)*; *(d)* preprinted entire identification *(Courtesy of Brady USA.)*.

order requires a SKU from the position, the microcomputer (pick director) illuminates a light which directs an employee to remove the required SKUs from the position.

The disadvantages of the method are that it is used on a rack or shelf system, high capital investment is required, and it is not easy to relocate. The advantages are that the instructions are clear and understandable and there is no label material or labor attachment expense.

Hanging-Garment Identification

Printed Self-Adhesive Label. An employee places the label on the top support member of the rail system. The label is located at the beginning of each trolley storage (dynamic) lane.

Plastic Placard. A placard is hooked on the top support member of the rail.

Preformed Plastic or Cardboard Doughnut. The doughnut (Fig. 12.7*a*) is made of preformed plastic or cardboard. The plastic doughnut is a round circle with a $\frac{1}{2}$- to 1-in-wide opening in the plastic circle. The plastic circle is big enough for the position

FIGURE 12.6   Digital display. (*Courtesy of Kingston-Warren Corp.*)

identification numbers or characters. The opening is big enough for easy attachment to the storage rail. The interior of the circle is hollow to fit and slide on the storage rail.

Large Cardboard Doughnut.   The cardboard doughnut has the same characteristics as the plastic doughnut, but the circle is larger and the doughnut is made by an employee.

The problems with the preformed doughnut are that it is difficult to keep on the rail and that sometimes the identification is difficult to see. The advantage of the large doughnut is that it is tall enough to have the identification above the hangers. The advantage of both doughnut methods is that the doughnut slides along the rail to accommodate fluctuations in inventory quantity.

Rectangular Placard.   The rectangular placard (Fig. 12.7*b*) is made of plastic or cardboard. The rectangle has a slit and a hollow-circle center that allows the placard to be slipped onto the storage rail. The top part of the placard which is above the rail has sufficient space for the position identification.

The disadvantage of the placard is that it is difficult to keep on the rail. Also in a two-level rail system it is difficult to read the identification on the bottom level due to the garments hanging down from the above position (rail). The advantages of the rectangular placard are that there is more space for position identification and that it slides on the rail.

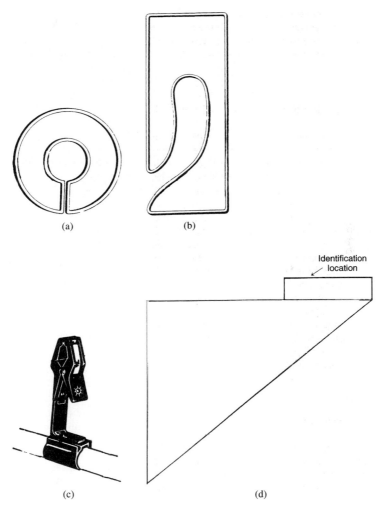

(a)   (b)

Identification
location

(c)   (d)

FIGURE 12.7   (*a*) Preformed small doughnut (*Courtesy of Railex Corp.*); (*b*) rectangular placard (*Courtesy of Railex Corp.*); (*c*) clip-on rail (*Courtesy of Railex Corp.*); (*d*) cardboard triangle.

Clip-on Method.   In the clip-on method (Fig. 12.7*c*), a plastic clamp slips onto the storage rail, and a clip device extends upward. The clip holds the identification label above the top of the rail.

The disadvantages of the clip-on method are that at the lower level it is difficult to read because garments hang down from the level above and that there is a possibility of the clip becoming damaged or of the identification falling from the clip. The advantage of this method is that for slow movers it divides the rail and prevents hangers from sliding past the pick position.

Cardboard Triangle.   When the hanging-garment storage-pick rail system has a rod pipe, cable, or rope that is parallel to the hanging-garment rail, then this handrail system permits you to use the cardboard triangle position identifier (Fig. 12.7*d*). The cardboard triangle is a precut or manually cut piece of cardboard that is used to identify the various positions on the hanging-garment rail.

On one side (bottom) of the triangle there is a slot in the middle of the side. The slot is wide enough to slip over the rail and deep enough to extend downward beyond the rail, which restricts the hanger from passing beyond the position. A second (top) side of the triangle is long enough to have its far end extend under the second rail or rope. With this side under the rope or rail, this arrangement keeps the triangle stationary. A second feature of the triangle top side has a 2- to 3-in section on the aisle side of the triangle side that extends upward above the triangle side. This 2- to 3-in-high section is used for the storage-pick position label attachment. With this feature the position identifier is above the top of the hanging garments on the rail and allows the storage-pick employee to see the location identifier.

Types of Pick Positions

In a carton or single-item distribution operation, the pick position is the warehouse location that contains product which is removed by an order picker to complete the customer's order. Inventory control professionals recognize that there are two types of pick positions: the fixed (permanent) location and the random (floating or variable) location.

When a depleted fixed SKU pick position occurs in the pick area, the SKU replenishment system in the reserve area determines how, when, and where the new SKU quantity is located to replenish (fill) the empty fixed SKU pick position.

The *fixed-pick slot* method is a SKU pick position method designed to have one SKU which is assigned to a permanent pick position. All replenishments to the pick position require the transfer of SKUs from a reserve (storage) location or from the receiving area to the SKU fixed pick position.

The disadvantages of the method are that it requires labor to transfer the product, it increases handlings and potential damage, and there are replenishment errors and possible stockouts.

The advantages of the fixed-slot method are that it increases the SKU hit (pick) concentration and hit (pick) density, permits family grouping, and requires less floor space and that there is higher replenishment and pick employee productivity.

The *floating-slot* method is a SKU pick location method designed to have a SKU pick position that is randomly located within the pick aisles of the warehouse. As the inbound merchandise or replenishment of a SKU is made to the aisle and the SKU pick position is occupied with product, the computer places the new product into a vacant pick position within the pick aisle or within another pick aisle. This new pick position and the SKU quantity are placed in computer memory. When the pickers deplete the present (occupied) pick position, the computer prints all future order pick instructions for the new pick position.

The disadvantages of the floating-slot method are that it requires larger floor space, it increases employee walk and travel time, and there is lower replenishment and order-picker productivity. The advantages are that no replenishment labor is needed and it requires fewer product handlings and reduces product damage.

PUT-AWAY AND WITHDRAWAL TRANSACTION VERIFICATION METHODS

A very important distribution facility activity is the product deposit and withdrawal transaction verification. This activity ensures that the correct SKU (product) has been deposited or withdrawn from the required storage position, at the correct time, and has been transferred to the assigned pick position, warehouse, or manufacturing workstation.

The three product transaction verification methods are the manual (visual) memory method, manual written method, and automatic method.

Human Memory

The first product deposit and replenishment transaction verification technique is that the lift truck operator mentally remembers the storage position for the SKU. This method is the basic and simplest of the transaction verification methods. The lift truck operator deposits the SKU in a storage position, and when there is a demand for the SKU, the operator remembers the storage location and transfers the SKU to the pick position, dock area, or workstation.

The disadvantages are that employee productivity is low, errors are possible, it handles a low volume and few SKUs, it is difficult to function in a large area or over two shifts, it is difficult to handle a FIFO product rotation, and it is not always the best position. The advantages of the method are its low cost and zero capital investment.

Handwritten Paper Document

The next transaction verification method is the manual handwritten method. In this method the lift truck operator uses a printed form to record the transaction. The printed form is a four-column activity sheet that has a space for the lift truck operator's name and the date. The four columns are separated into two groups of two columns each. Two columns under the deposit heading are for listing the deposit transactions. Under the withdrawal heading, the two columns are for listing the withdrawal transactions.

After the lift truck operator completes a transaction, the operator lists the storage position and the SKU identification number involved in the transaction. At the end of the shift, the activity forms are sent to the warehouse office. In the office, a clerk performs an inventory control update of the records.

The disadvantages of the method are that it handles a low to medium volume, requires additional clerk effort, adds an activity to the truck driver's work, and creates possible transposition errors. The advantages are that there is no capital investment, it is used on a two-shift operation, and it is used in a large facility.

Manual Bin File

The next transaction verification method is the bin file method. This method uses a series of reserve-position cards. Each card corresponds to a reserve position in the warehouse aisle, and the cards are placed in sequential order in the card holder. In a pallet load or carton warehouse operation, a card holder with a slot for each reserve position above the pick position is attached to the upright frame that supports the reserve position. In a single-item warehouse operation, the card holder is attached to the main aisle side of the shelf side panel.

The card has the reserve-position number printed on the top left side and has three columns. The columns are completed by the lift truck operator who performs the transaction. One column lists the product identification number involved in the transaction. The other two columns are the put-away or deposit and withdrawal or replenishment columns.

After completion of the required transaction, the lift truck driver obtains the appropriate reserve-position card from the holder. On this card, the operator lists the product identification and places a mark in the appropriate in or out column that reflects the transaction. The card is returned to the card holder for future reference.

The method has several disadvantages: handwriting may be unclear, transposition errors can be made, cards can be lost, it handles a low to medium volume, and it handles a low to medium number of SKUs. The advantages are the low cost, low capital investment, access to information in the warehouse area, and ease of implementation.

Bar-Code Scanning

The bar-code scanning device (Fig. 12.8) requires a transaction instruction format, a human- and/or machine-readable label as part of the identification, bar-code label (human- and/or machine-readable) on the warehouse position, and hand-held scanning device that transmits the transaction to the host computer or holds the transaction in memory for later entry into the computer.

In the bar-code scanning operation, prior to picking up the product in the receiving area, the lift truck operator obtains a put-away list. With the hand-held scanning device, the lift truck operator scans the product label, picks up the product, and travels to the assigned reserve position. At the reserve position, the product is placed into the reserve position, and the lift truck operator uses the hand-held scanning device to scan the position's bar-code label. The transaction information in the scanning device is transferred to the host computer for an inventory update. The transaction information transfer is on-line or delayed-entry.

When you are considering on-line transaction data entry, prior to the purchase and implementation, a review of the following facts can determine system performance: the number of scanning devices and transactions, data transmission distance and clean lines, and host computer capability to handle on-line transactions from the put-away system, and other system activities such as customer order processing and inventory receipts.

Disadvantages of the bar-code scanning method are that investment, employee training, and management control and discipline must be increased and that the computer has temporary storage of transaction or the capability to handle on-line transfer of information. Advantages of the method are that it handles a high volume, provides an accurate record, provides an accurate transfer of data, permits on-line information transfer, and results in high employee productivity.

FIGURE 12.8   Bar-code scanning. (*Courtesy of Clark Material Handling.*)

Radio-Frequency or Acoustical Wave Method

The radio-frequency wave or acoustical wave (Fig. 12.9) is used in another transaction verification method. A tag (transponder) is put on the product, and a receiver (antenna) is placed on the vehicle. The tag transmits a unique signal that identifies the SKU and is picked up by the antenna. After the antenna receives the signal, it is sent to the host computer for validation, and the small terminal on the lift truck indicates the product location.

When the product is removed from the warehouse position, the tag is thrown away except for expensive tags that are reprogrammable.

The disadvantages of the method are the increased investment, increased employee training, and high cost of the tags. The advantages are that it handles a high volume, provides an accurate record, provides an accurate transfer of data, permits on-line information flow, and can be read through dirt and grease.

VARIOUS REPLENISHMENT CONCEPTS

The next section of the chapter reviews the various methods to make replenishments from the reserve positions to a fixed pick position. This activity occurs in a carton or single-item distribution facility and ensures that the correct product is in the correct pick position for the order picker to complete a customer order.

The various product replenishment methods are random, slug, and sweep.

FIGURE 12.9   Radio-frequency wave transaction verification. (*Courtesy of I.D. Systems Magazine.*)

Random Replenishment

The first product replenishment method makes random pick position replenishments. The replenishment employee makes replenishment transactions that are based upon your employee's thoughts. To reduce travel distance and travel time, the warehouse aisles are divided into employee zones.

The disadvantages of the method are that (1) replenishment transactions do not complement the order-picker transactions; (2) management control is low; (3) employee productivity is low due to nonscheduled work activity; and (4) if the pick position is not completely depleted, then the extra product is not placed in the best location, which creates misfills or product damage. The advantage of the random replenishment method is that usually the fast-moving SKUs are handled first.

Slug Replenishment

The next product replenishment system is the slug and aisle method. Before moving to the next aisle, the replenishment employee makes all replenishment transactions within one warehouse aisle. Compared to random replenishment, the additional disadvantages are the increased stockouts and lower employee productivity due to hand stacking to make space for a new product. The advantage is the decreased employee travel time and distance between replenishment positions.

Sweep Replenishment

In the sweep method, the warehouse is divided into zones. The replenishment employee starts with one zone and predetermined location and performs in sequential order all replenishment transactions in one aisle prior to moving to the next aisle. Compared to the other replenishment methods, it has the additional advantage that the concept closely complements the order-pick activity.

Types of Storage (Reserve) Positions

Floating-Slot Method. In a pallet, carton, or single-item distribution operation, fundamental to any product storage-reserve position assignment is that all positions are floating (random or variable) assigned positions. In the floating-slot storage position method, a SKU is assigned to any vacant storage position. In this method, storage position 120 holds product A. A new delivery of product A is received at the warehouse with old product A in the storage position (120). In the storage area, the new delivery of product A is assigned to a second storage position (130). When product A is required from position 120, the computer prints a withdrawal instruction for product A in position 120. With the withdrawal of product A from position 120, the computer prints the next withdrawal instruction for product A from position 130.

Fixed-Slot Method. In an operation with the fixed-slot position, old product A is allocated to a predetermined number of storage positions. When these positions are depleted of product A, then a new inventory quantity of product A is transferred from the receiving or floating reserve area to replenish the storage positions. In a dynamic warehouse operation, with the fixed-slot method it is difficult to obtain good space utilization and good employee productivity. It is not considered for implementation in a storage area, but is considered a good method for the pick area.

Three Product Allocation Systems

This section reviews the three major product storage-pick position methods implemented in an existing or new warehouse operation. The operation is a single-item, carton, or pallet load operation. The various product storage-pick position concepts are (1) no concept, (2) ABC concept, and (3) family group concept.

The basic product storage-pick location principle is to keep the travel distance of the storage-pick employees as short as possible between two transaction locations. In a pallet load operation or in the storage area for a carton or single-item facility, these two locations are (1) between the receiving area and the reserve area and (2) between the reserve area and the shipping area.

In a single-item or carton operation, these areas are (1) between the receiving area and the reserve area, (2) between the reserve location and the pick location, (3) between the pick location and the pick location, (4) between the pick area and the pick location, and (5) from the pick location to the manifesting and shipping location.

Part of the inventory control manager's job is to ensure that the product is allocated to the warehouse location that optimizes the productivity of replenishment and pick employees. This objective is achieved by keeping the travel distance between two warehouse transaction locations as short as possible. This practice increases the dual-cycle transactions or handling in a pallet load operation. In a single-item or carton handling load operation, the shortest distance between two pick locations improves the SKU hit density (number of SKUs per aisle for a customer order) and HIT concentration (number of picks per SKU for a customer order). This feature reduces the nonproductive walk or ride time and improves the productivity of both the replenishment and order-pick employees.

No Method. No method is the first inventory location method. SKUs are randomly assigned to the storage-pick positions within an aisle. With this method, a position that has a fast-moving SKU is separated by several positions of slow-moving SKUs; therefore, the employee increases the walk or ride time and distance to complete the second transaction.

The disadvantages of no method are that employee productivity is low, it mixes SKUs from different product groups in the same aisle, and it does not maximize the use of space. The advantages of no method are that it is easy to implement and does not require management control and discipline.

ABC Method. In this method, the SKU positions in a warehouse aisle are divided into three major zones: A zone, B zone, and C zone. These zones are subdivided into microzones of (1) a zone, (2) b zone, and (3) c zone. The particular zone is restricted for the products (SKUs) that have a specific annual movement.

The first zone, the A zone, has the positions which are restricted for fast-moving SKUs. The positions within the B or second zone are allocated for the medium-moving SKUs. The final or C zone contains positions of slow-moving SKUs.

Since 80 percent of product movement (based on Pareto's law) is from zone A, the ABC method improves the aisle SKU hit density and hit concentration.

The disadvantages of the method are that it requires management control and discipline, there is the potential to mix SKUs from different family groups within an aisle, and it requires accurate estimates or actual SKU movement data. The advantages of the method are that it improves employee productivity and handles a high volume and a large quantity of SKUs.

Family Group Method. In the family group method, SKUs with similar characteristics are assigned to positions within a warehouse aisle, zone, or area. These characteristics are raw material or work-in-process, components for a finished product, physical characteristics, type of product, storage environmental conditions, material-handling characteristics, and customer's location in the retail store.

The disadvantages of this method are that it increases the need for management control and discipline, it requires additional positions for new SKUs, having the fast- and slow-moving SKUs in the same aisle means longer travel distances between two pick locations, and employee productivity is average. The advantages are that it improves employee familiarity with the product, improves customer's productivity to replenish retail shelves, and reduces errors from incorrect transactions.

What Method Is Best for Your Company?   Three product allocation methods for various industries are shown in Fig. 12.10.

Whenever possible, the preferred product allocation method is a combination of the ABC and family group methods.

Various Replenishment Quantities

The SKU replenishment activity is planned to handle the following SKU quantities: pallet (unit) load, layer of a unit load, less than a layer or several cartons, and less than a carton or one carton.

Pallet Load Quantity.   The entire pallet load is transferred from the reserve position to the pick position. To maximize space and labor, this method requires that the pick position be depleted of the SKU. With this situation, the entire pallet load is placed in the pick position. The pallet load replenishment method replenishes approximately 50 to 75 cartons per unit load and is used in a warehouse that handles cartons and sometimes single items for fast-moving SKUs. In the typical carton operation, the pallet load method is used for the fast- and medium-moving SKUs.

The disadvantages of the method are that (1) if cartons remain in the pick positions, this means low replenishment productivity and potential damage or lost SKUs (inventory) and (2) it is sometimes not on schedule to prevent stockouts. The advantages of the method are that the number of handlings is minimized, there is maximum product in the pick position, employee productivity is high, and it is used for fast- to medium-moving SKUs.

Replenishment of One Layer of a Pallet Load.   One or two layers of cartons from a unit load are removed and transferred from the reserve position to the pick position. This means that the pick position is a hand-stack rack position, case flow system, or shelf system. The method is used for slow- to medium-moving SKUs. This carton removal and transfer activity is performed in the storage area or in the pick area.

If the activity is performed in the *storage area,* then the required cartons are removed from the pallet load and placed onto a conveyor system or onto a mobile vehicle load-carrying surface for transport to the pick area. In the pick area, the cartons are transferred from the vehicle to the pick positions.

The disadvantages of the method are that the equipment cost and required floor area are increased, product is identified or separated, and product is handled twice. The advantages are that the number of unit-load movements is minimized, the exact quantity is moved, there is FIFO product rotation, vehicle movements in the warehouse are reduced, and it handles a high volume for slow- to medium-moving SKUs.

If the activity is performed in the *pick area,* then the lift operator removes a unit load from the reserve position and transports it to the pick area depalletizing station. From the pallet load, a depalletizing employee removes and transfers the required cartons from the unit load for transfer by conveyor to the pick position. After this is done,

Industry type	Product allocation method
Replacement parts and supplies	ABC
Retail grocery department store	Combination of ABC and family group
Manufacturing	Family group for unique and ABC for common parts
Catalog and direct mail	ABC and family group
Drug and novelty items	Combination of ABC and family group
Food service	Combination of ABC and family group
Spare parts	ABC
Public warehouse	Family (customer) group

FIGURE 12.10   Three product allocation methods for various industries.

the lift truck operator removes the partially depleted unit load from the depalletizing station and returns it to the storage area for placement in the assigned storage position.

The disadvantages are the double handling of the unit load, increased potential of product or equipment damage, and increased number of lift trucks required. The advantages are the ready reserve in the pick positions and the large quantity of inventory in the ready reserve position.

Replenishment of Less Than a Pallet Load.   This carton replenishment method is used for all SKUs. This method has the same operational characteristics, disadvantages, and advantages as the replenishment method for a layer of a pallet load, except that the lift truck operator picks the cartons from the unit load (stack) that remains in the storage position and travels to the depalletizing area. At this station the operator transfers the cartons from the vehicle for transport to the pick position.

Replenishment of a Carton or Less.   This replenishment method is used for very slow- to slow-moving SKUs. This method has the same operational characteristics, disadvantages, and advantages as the replenishment method of a layer of a pallet load. In some facilities, the merchandise is placed into a tote or carton for transport from the storage area to the pick position.

Single-Item Replenishment Quantities

The objective of the replenishment method is to ensure in the pick position that there is excellent space utilization, required product rotation, and available product for the order-pick activity. To satisfy these objectives in a single-item operation, the inventory control program determines the replenishment quantity that is transferred from the storage area to the pick position. The three product replenishment quantities are minimum, maximum, and capacity.

The bases for product replenishment quantities are (1) the SKU quantity per unit of product in the reserve position, (2) beginning inventory in the pick position, (3) update (deductions) for customer orders, (4) predetermined capacity of the pick position, (5) predetermined safety stock in the pick position, and (6) historical or expected demand.

Minimum Replenishment.   The minimum inventory replenishment quantity is considered the SKU safety stock quantity of pick position. When the computer pick position SKU inventory quantity reaches the minimum quantity and all the customer orders have been satisfied for the day, then for the SKU the computer creates a minimum replenishment demand. This minimum replenishment level is one reserve container or carton and is considered a low-priority replenishment because there is sufficient inventory in the pick position for the next pick wave of customer orders.

Maximum Replenishment.   The maximum inventory replenishment quantity is based on the reserve-position unit quantity and ensures excellent pick position utilization. The maximum inventory level holds the minimum inventory quantity plus one reserve unit (carton or container). Note that this maximum quantity is greater than one container due to the position capacity. The maximum replenishment requirement has a medium priority because the pick position inventory level is great enough to satisfy the majority of the next pick wave of customer orders.

Capacity Replenishment.   The capacity inventory replenishment quantity is the maximum number of a SKU inventory units that fit into the pick position. This situation is created when the pick position SKU inventory level reaches zero or is an out-of-stock condition; therefore, the product quantity required in the pick position includes the total capacity of the pick position. Capacity replenishment receives the highest priority because the pick position SKU inventory level cannot satisfy any of the customer orders.

Various Timing Methods for Replenishment

The SKU pick position replenishment activity is best performed at the moment that the pick position becomes depleted of SKUs and with the SKU quantity that maximizes the pick position space. To achieve this objective, for best results all replenishment transactions should be made by a computer.

The next major inventory control activity is to determine the appropriate time for a deposit or replenishment transaction to occur in the distribution facility. An employee moves a specific SKU quantity from a storage position to a pick position. The two methods to control or sequence a product transaction are a manually controlled method and a computer-controlled method.

Manual Control.   The manually controlled transaction method relies upon the warehouse employee to determine the time at which the product is moved from the reserve position to the pick position. This is a random method that is based upon the warehouse employee's experience and is not sequenced with the other warehouse activities of receiving, order picking, or shipping. On some occasions, the replenishment transaction is made to a partially depleted pick position. This condition requires the replenishment employee to waste time and transfer the remaining product in the pick position onto the new pallet load or into the new container.

The disadvantages of the method are that there is no control of the employee's activities, employee productivity is low, there is poor spatial utilization of the pick position, and it is not coordinated with the order-pick activity. The advantages of the method are the low cost or capital investment and absence of employee training.

Computer Control.   With the computer-controlled transaction method, the computer directs the employee to perform the product transaction between the two warehouse locations. This transaction instruction is based on two factors: the customer orders and the inventory quantity in the pick position. There are two methods to direct the employee to perform a replenishment transaction. The first alternative is a paper document that lists all the replenishment transactions which will occur on the employee's shift. On this form the SKUs are listed by SKU number, and the description in a sequential order is based on the anticipated time at which the pick position becomes depleted of product.

The first column states the SKU (product) reserve position, and the second column is for the replenishment employee's mark. The mark indicates and verifies that the withdrawal portion of the transaction was completed on schedule. The third column states the SKU (product) pick position, and the fourth column is for the employee's mark to verify completion of the replenishment transaction.

An option to the manual written (mark) verification form is the bar-code scanning of the SKU bar-code label and the reserve or pick position bar code. The bar-code scanning device transmits the information on line or delayed to the host computer for both reserve and pick position updates.

The alternative document to direct the replenishment transaction is a human- and/or machine-readable label.

Disadvantages of the computer-controlled method are that it requires management control and discipline, it increases the capital cost, and it requires employee training. The advantages are accurate transaction records, excellent employee productivity, reduction in lost inventory, improved spatial utilization, and enhanced management control.

CALCULATION OF SKU SPACE REQUIREMENTS

The next function of inventory control is to ensure that the spatial data on your SKUs and delivery packages are accurate. The SKU spatial information includes the length, width, height, and weight.

In a single-item and carton handling operation, accurate SKU spatial information and order-pick vehicle or tote carrying spatial capacity mean less nonproductive order-picker walk time and reduced product damage. This is due to the fact that the computer allocates a sufficient quantity of SKUs to maximize the vehicle load-carrying capacity and directs the employee on each trip to move the maximum amount of SKUs. In addition, an accurate delivery package and delivery vehicle capacity means the maximum load on the delivery vehicle, accurate vehicle scheduling, and an accurate shipping charge. When the customer's delivery location requires an air cargo container, then spatial information becomes very critical to obtain the lowest cost.

Vendor's Method

When you use the vendor's SKU space method, then you are relying upon the purchasing department to notify vendors that the warehouse operation requires the SKU data. After you receive these data, they are entered into the warehouse computer files and become the bases that determine work volume for all warehouse and transportation activities. These activities are related to the handling of the SKUs in the warehouse operation.

The disadvantages of the method are that (1) with loose SKUs, it does not reflect the shape of a picked SKU, (2) errors are possible, (3) the data do not change as the SKU flows to the various workstations, and (4) there is no management control. The advantages are that it does not require a large spatial expense and does not require investment in equipment.

Manual Method

The manual SKU space method requires you to assign an employee to the SKU space activity. The activity requires a space document, tape measure, and portable scale. With these items, for each SKU in the inventory, the employee physically measures and obtains the weight. On the document, the employee enters each SKU identification number, spatial measurements, and weight. The spatial information on each SKU or delivery carton is entered into the computer files. After this process, for all new SKUs that are received at the facility, a receiving employee determines the spatial information and forwards it to the computer department. This paper document or computer network is the information flow.

The disadvantages of the method are that it is time-consuming, it requires labor expense, and errors can be made. The advantages of the method are that accurate spatial data are used, it ensures that all the SKU information is in the computer, and it establishes a procedure for new SKUs.

Mechanized Method

The mechanized SKU space method (Fig. 12.11) uses two devices, a scale and a framed device, to determine the space for each SKU. If the SKU fits through the 5-ft by 5-ft frame, then the device automatically cubes the SKU to 0.04 in and determines the SKU weight to an accuracy of 0.01 lb.

When positioned in the correct location, this device determines spatial information for inbound cartons, carton or single-item pick area SKUs, and outbound delivery cartons. This spatial information is entered into the computer.

The disadvantages of the method are that a capital investment in equipment is needed, it requires an employee, and it is difficult to relocate. The advantages of the method are that it cubes all your SKUs, it is a quick and easy process, and it provides accurate data.

INVENTORY TRACKING METHODS

The most important aspect of an inventory control program is tracking the inventory and locating it as required for a warehouse transaction or customer order. Inventory tracking is a vital function for an accurate floating-slot system and for an effective, productive fixed-position pick system. There are two inventory tracking methods, manual and computer.

The manual method is considered the basic, simplest method. The three manual methods are human memory, bin file, and card slot.

FIGURE 12.11   Mechanical method for calculating SKU space. (*Courtesy of Toledo Scale.*)

Human Memory

In this inventory tracking system the warehouse employee places the product into any available reserve position. This reserve position does not necessarily agree with the warehouse location of the pick position, and the product reserve position is kept in the employee's memory. When required to make a replenishment to the pick position, the employee recalls the product's reserve position or searches all the warehouse aisle reserve positions for the product.

The disadvantages are that employee productivity is low, it is difficult to function over two shifts, and it handles a low volume and a low number of SKUs. The advantages of the human memory method are that no capital investment is needed and it is good for a very small operation.

Manual Bin File

In the manual bin file inventory tracking system (Fig. 12.12a), the warehouse employee performs the required transaction and lists the SKU identification code and type of transaction on a form. At the end of the workday or at predetermined times, the form is entered onto the bin file document (cards). The bin file is maintained at the warehouse supervisor's desk or at the end of the aisle. The bin file is arranged in SKU identification code, reserve position aisle, or storage position number sequence.

As the employee is required to perform a transaction that moves the SKU from the

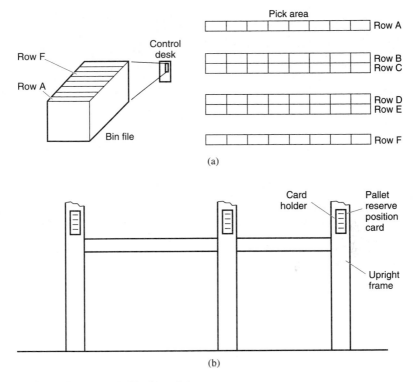

(a)

(b)

FIGURE 12.12   (*a*) Bin file; (*b*) card slot.

reserve to the pick position, the employee indicates this transaction on the form. After the shift is completed or at predetermined times, a clerk enters the information from the form onto the bin file card. These entries update the on-hand inventory location and quantity.

The disadvantages of the method are that it requires an additional clerk, it adds an activity to the employee, and transposition or writing errors are possible.

The advantages of the method are that no capital investment is needed, it handles a low volume and a low number of SKUs, and it can be used for a two-shift operation.

Card Slot

The card slot inventory tracking system (Fig. 12.12*b*) is the last manual system. With this system, as the product is placed into a reserve position, the employee marks the SKU identification code on the premarked reserve-position card. These premarked cards are in a card holder at the end of the aisle or on each warehouse storage location upright rack post. These card holders are envelope slots with sufficient slots for each reserve position with the stack.

When a depleted SKU pick position requires replenishment, the employee obtains the appropriate card from the card holder, reduces the appropriate SKU quantity from the storage column on the card, and returns the updated card to the appropriate slot in the card holder. The product is removed from the reserve position and transferred to the pick position.

The disadvantages of the method include unclear writing, transposition errors, and lost cards or damaged card holders.

The advantages are that no capital investment is needed, it is easy to implement, it functions over two shifts, it handles a low volume and a low to medium number of SKUs, and information is in the warehouse.

Computer-Controlled Tracking

In the computer-controlled inventory tracking system, each warehouse reserve and pick position is discretely identified, and each unit load has a discrete label. This unit-load label identifies the assigned reserve or pick position. As the product leaves the receiving area, this unit-load label directs the employee to deposit the unit load into a specific reserve or pick position. Upon completion of this transaction, the employee makes the appropriate mark on a paper document. The document ensures that the transaction is completed and accurate. The on-hand inventory is updated according to the paper document. For the next shift, the computer prints a replenishment transaction list for all SKU pick positions that become depleted on the shift. The paper document method ensures that the transaction is accurate and performed on schedule and that the computer has updated the on-hand inventory.

Bar-Code Scanning. With the bar-code scanning system, the paper document directs the employee to the reserve position. At the reserve position, the employee scans the unit-load reserve position, transfers the unit load to the pick position, and scans the pick position label. The scanned information is entered on line or delayed in the host computer to update the on-hand inventory.

Other labels such as radio-frequency tags are available instead of the bar-code label.

The disadvantages of the method are that investment, employee training, and management control and discipline must increase and the host computer has the on-line capability or a backup memory.

Advantages of the computer-controlled tracking method are that it handles a high volume, provides an accurate record, provides an accurate transfer of data, is on-line, and handles a large number of SKUs.

Methods of Physical Inventory Count

A very important activity of the inventory control function is the activity to verify the SKU actual on-hand inventory level. The objective of the actual on-hand inventory level verification (count) is to ensure and to establish confidence in the book on-hand inventory levels. The two types of count methods are the manual and the bar-code scanning count methods.

Manual Count. With the manual count method, an employee walks the warehouse aisles with a physical inventory form and counts the product. On the form the employee indicates the actual on-hand inventory by SKU by reserve or pick position, family group, or an entire warehouse zone. After the count is completed in the warehouse, the actual count is compared to the book inventory.

If there exists a major discrepancy between the actual count and the book count (the difference between the two exceeds the company's acceptable limit), then the book inventory is reviewed and another count is made to verify the actual count. After the second count, if a discrepancy still exists between the book count and the actual count and if the second count is the same as the first count, then the actual count is accepted as the new book on-hand inventory level. If the physical count is within the company's acceptable variance or is exact, then the physical count is the new book on-hand inventory level.

Bar-Code Scanning Count.   The bar-code scanning count is similar to the manual count method, but an employee uses a hand-held bar-code scanning device to scan bar-code labels on the product and storage-pick positions. In a unit-load warehouse, the employee scans the unit load and rack position. This represents an exact inventory count by warehouse location. In a carton or single-item operation, the employee scans the pick position label, scans the SKU bar-code label, and then enters the position count into the scanning device. All unit-load and carton reserve positions are counted with the same procedure that was used in the pallet load storage area.

When all required scans are made, the information is entered into the computer. The entry updates the on-hand inventory level by the reserve and pick position counts in the warehouse.

All major discrepancies between the book count and the on-hand inventory count require a review of the book count, and the product is recounted by another employee. After the additional count, a comparison is made to the previous count and the book inventory count. If the actual count is the same as the previous count, then the actual on-hand inventory is accepted as the new book count.

Inventory Count Types

After the company decides to implement a physical SKU on-hand inventory count program, the warehouse, data processing, and accounting managers decide on the type of count method to be used throughout the year. The alternative count procedures are random count, cycle count, and fiscal count.

Random Count.   In the random count method, an employee counts a specific SKU on-hand inventory in the warehouse. The count is made to verify the exact SKU on-hand inventory level which is a key component in a manufactured product or to a sales program.

Cycle Count.   The cycle count program is an on-hand inventory count method that separates the total SKU inventory into categories based on the SKU movement. Within a specific time period such as a quarter, the cycle count program requires a certain (predetermined) number of SKUs from each category counted by a warehouse employee. The on-hand book inventory levels are adjusted for the physical on-hand count within the company's inventory count procedures.

Fiscal Count.   The fiscal inventory count program is the major company once-a-year count for all SKUs in the warehouse. This count is made at the end of the company's accounting year and is supervised by the auditing company. This on-hand inventory count is the physical inventory count figure that is the basis for the inventory dollar amount stated on the company's balance sheet and the ending inventory figure on the company's income statement.

During the count process, each SKU warehouse location and the corresponding count are recorded by a warehouse employee. When the actual on-hand count is compared to the book count, any major discrepancies are verified by recounts prior to acceptance of the actual on-hand inventory count as the new book inventory count.

Obsolete or Scrap Inventory. In this section of inventory control we describe a distribution facility damaged and obsolete physical inventory scrapping methods. Inventory scrapping creates storage space, improves housekeeping, improves employee productivity, and reduces product damage.

Scrap inventory consists of damaged or obsolete product. Damaged product is not shipped to a customer due to poor quality or material handling damage. Obsolete product is product that is not ordered by customers for a long time.

In a distribution facility, scrap inventory is generally created from one of the following four areas: damaged product from material handling, buy-ins that see slow movement for which the supply on-hand exceeds a reasonable limit, product that does not meet the company's minimum specifications, and product that has not sold.

How to Identify Scrap Inventory. The SKUs in inventory that are candidates for scrap are determined by one of the following: (1) in the warehouse storage area, the product has accumulated dust; (2) in the warehouse storage area, an old receiving date tag is on the product; and (3) from a year-to-date product movement report, it shows slow-moving SKUs.

Four physical inventory scrapping methods are used in your warehouse operation: (1) book inventory, (2) random selection of the SKU, (3) age of the SKU, and (4) SILO method.

Book Inventory Scrap Method. The first warehouse inventory scrapping method is defined as a book inventory scrapping method (Fig. 12.13*a*). This method is based on the facts generated by the accounting and purchasing departments. From the annual projected sales, the accounting department manager allocates a specific dollar amount of sales as inventory scrap expense. From the inventory records, the purchasing manager identifies the SKUs that are candidates.

The accounting and purchasing department managers jointly ensure that the dollar value for scrap equals the budgeted inventory scrap expense. This SKU list and the quantity for each scrapped SKU are given to the warehouse manager who scraps the product from the inventory. The accounting and purchasing managers make necessary entries to remove the SKUs from the inventory and to increase the expense line item.

The disadvantages of the method are that it does not create a maximum number of openings, it does not scrap the maximum cubic feet, and there is a potential to scrap the fewest number of SKUs. The advantages are that it is easy to implement and scraps the inventory dollar amount.

Random Scrap Method. In this method (Fig. 12.13*b*), the product is randomly selected from the scrap SKU list. The disadvantages and advantages of the method are the same as those for the book scrap method.

Age-of-Product Method. In this third method (Fig. 12.13*c*), the oldest SKUs in the inventory that are on the scrap list are scrapped from the inventory. When the SKU has a limited life, the SKU age-of-product inventory method is excellent.

SKU	SKU age	Dollar value per SKU	Inventory SKU quantity	Inventory dollar value	Cubic feet per SKU	Total cubic feet of SKU inventory	SKUs to be scrapped	Dollar scrap value	Scrapped cubic feet
A	13 MO	$5	3,000	$15,000	3	9,000	—	—	—
B	16 MO	$10	500	$5,000	4	2,000	—	—	—
C	19 MO	$15	2,000	$30,000	5	10,000	—	—	—
D	22 MO	$20	1,000	$20,000	3	3,000	—	—	—
E	22 MO	$30	1,000	$30,000	3	3,000	1,000	$30,000	3,000
F	16 MO	$40	1,000	$40,000	2	2,000	1,000	$40,000	2,000
G	19 MO	$5	2,000	$1,000	6	12,000	—	—	—
H	13 MO	$10	2,000	$20,000	2	4,000	—	—	—
I	13 MO	$50	1,000	$50,000	2	2,000	1,000	$50,000	2,000
TOTAL			13,500	$220,000		47,000	3,000	$120,000	7,000
							22.2%	54.5%	14.9%
							3 SKUS		

(a)

FIGURE 12.13 (a) Typical book inventory scrap method.

SKUs to be scrapped	Dollar scrap value	Scrapped cubic feet
3,000	$15,000	9,000
500	$5,000	2,000
—	—	—
—	—	—
1,000	$30,000	3,000
—	—	—
2,000	$10,000	12,000
1,000	$10,000	4,000
1,000	$50,000	2,000
8,500	$120,000	32,000
63.0%	54.5%	68.1%
6 SKUS		

(b)

FIGURE 12.13 (*Continued*) (*b*) Random SKU selection method.

SKU	SKU age	Dollar value per SKU	Inventory SKU quantity	Inventory dollar value	Cubic feet per SKU	Total cubic feet of SKU inventory	SKUs to be scrapped	Dollar scrap value	Scrapped cubic feet
A	13 MO	$5	3,000	$15,000	3	9,000	—	—	—
B	16 MO	$10	500	$5,000	4	2,000	500	$5,000	2,000
C	19 MO	$15	2,000	$30,000	5	10,000	2,000	$30,000	10,000
D	22 MO	$20	1,000	$20,000	3	3,000	1,000	$20,000	3,000
E	22 MO	$30	1,000	$30,000	3	3,000	1,000	$30,000	3,000
F	16 MO	$40	1,000	$40,000	2	2,000	625	$25,000	1,250
G	19 MO	$5	2,000	$1,000	6	12,000	2,000	$10,000	12,000
H	13 MO	$10	2,000	$20,000	2	4,000	—	—	—
I	13 MO	$50	1,000	$50,000	2	2,000	—	—	—
TOTAL			13,500	$220,000		47,000	7,125	$120,000	31,250
							52.8%	54.5%	66.5%
							6 SKUs		

(c)

FIGURE 12.13 (Continued) (c) Age-of-SKU selection method.

SKUs to be scrapped	Dollar scrap value	Scrapped cubic feet
3,000	$15,000	9,000
500	$5,000	2,000
2,000	$30,000	10,000
500	$10,000	1,500
1,000	$30,000	3,000
—	—	—
2,000	$10,000	12,000
2,000	$20,000	4,000
—	—	—
11,000	$120,000	41,500
81.5%	54.5%	88.3%
	7 SKUs	
	(d)	

FIGURE 12.13   (*Continued*) (*d*) SILO inventory scrap method.

The disadvantages of the method are the same as those for the book scrap method, but the advantage is that it removes the oldest SKUs from the inventory.

SILO Scrap Method.   The small-in large-out (SILO) inventory scrapping method (Fig. 12.13*d*) is created from a list of the following facts: SKU velocity movement for the entire inventory, approved dollar scrap amount, cubic feet for each SKU, and on-hand inventory for each SKU (Fig. 12.14).

The facts are manually calculated or entered into a computer that lists the potential scrap SKUs with the largest space as the priority scrap candidates. The additional information on the scrap list is the total SKU cubic feet and each SKU corresponding dollar inventory scrap value. The calculation determines the SKUs that will create the maximum amount of space.

The disadvantages of the method are that it requires management discipline and control and accurate product data. The advantages of the method are that it creates the maximum number of openings, scraps the maximum number of cartons, scraps the allowable dollar value, and uses the age-of-product method.

Analysis factor	Age of SKU	Typical book	Random selection	S.I.L.O.	S.I.L.O. advantages
Number of items to be scrapped	6	3	6	7	Maximum SKUs are scrapped
Inventory dollar value	$120,000	$120,000	$120,000	$120,000	Equal total dollar value scrapped
Percentage of inventory scrapped	54.5%	54.5%	54.5%	54.5%	Equal percentage of
Number of cartons to be scrapped	7,125	3,000	8,500	11,000	Maximum number of cartons
Percentage cartons to be scrapped	52.8%	22.2%	63.0%	81.5%	Maximum percentage
Total number of cubic feet scrapped	31,250	7,000	32,000	41,500	Maximum cubic feet
Percent of total cubic feet scrapped	66.5%	14.9%	68.1%	88.3%	Maximum percentage
Storage openings based on 64 cu. ft. per pallet load	488	110	500	650	Maximum number of pallet openings
Pallet storage cost savings based on $35 per sq. ft. for each 4-pallet high stack	$68,320	$15,400	$70,000	$91,000	Maximum cost saving

FIGURE 12.14 Comparison of inventory scrap methods.

CHAPTER 13

MANUAL VERSUS AUTOMATIC IDENTIFICATION SYSTEMS

This chapter reviews the various manual and automatic identification systems available for implementation in your warehouse operation. The chapter looks at the type of operation and warehouse functions and where an identification system reduces costs and improve profits. Included is a review of other benefits, a variety of human- and machine-readable code types, code printing equipment, and data encoding devices. Also this chapter indicates the various components of an automatic identification system, including design parameters and operational characteristics.

COMPONENTS OF IDENTIFICATION METHOD

The manual identification system consists of a product with a human-readable code that contains alphabetic characters, digits, or colors. An automatic system consists of a product with a machine-readable symbology. The automatic system is a subsystem in the total warehouse operation. This subsystem has four major components: product, symbology, sensing device, and electrical network.

The sensing device, by viewing a bar-code label or other symbology, recognizes (identifies) one SKU from the other SKUs. The sensing device transfers this information via an electrical network to the decision-making component (human or machine) and record-keeping component (paper or computer). The decision-making component interprets the code and commands (directs) a device to move the appropriate SKU from one location to another, or the code is held in memory. As required, the memory component reports the activities (codes) to management.

VARIOUS IDENTIFICATION SYSTEMS

When you consider the various identification systems for your warehouse operation, the three alternatives are no method, manual method, and automatic method.

Prior to the implementation of a manual or automatic identification system in a warehouse operation, you must ensure that the proposed identification system is compatible with the material handling system, data processing system, and product type and volume, and channel of distribution segments.

An identification system is implemented in a large or small warehouse facility to improve the total facility productivity and profits. Your productivity and profit improvement opportunities are the result of improved control of SKU flow, increased labor savings, reduced operating expenses, and enhanced, accurate information flow.

No Identification Method

The first product identification method is no identification method. With this method, the company relies upon the vendor's exterior carton markings for product identification. In this method your company's markings are not found on the product exterior. The majority of the companies that use this method have one SKU in one aisle or a small number of SKUs in inventory. No identification method is *not* the preferred method for use in a dynamic warehouse operation.

The disadvantages of the method are that employee productivity is low, there is the potential of lost product, order selection and sortation errors can occur, it handles a low volume, and it handles a limited number of SKUs. The advantages of the method include no marking material or labor expense and no labels or markings on the product exterior.

Manual Identification Method

Written Label.   In this manual identification method (Fig. 13.1a), a handwritten label or markings are placed on the product's exterior. The handwritten method relies upon an employee to read the label's printing or markings that direct the product transfer between two locations. The information is noted on a paper document on a hand-held clipboard. In the office, the document information is transferred by a clerk to the files for an update. The manual SKU identification method is preferred for implementation in a manual warehouse operation that handles a low volume and a medium number of SKUs.

Disadvantages of the manual method are that some handwriting is difficult to read, transposition errors can be made, and handwriting can become smeared.

Advantages of the manual identification method are that the marking material and labor expense is low, no capital investment is required, the company's markings are on the product's exterior, it reduces selection and sortation errors, it improves inventory control, and it handles a medium volume and a medium number of SKUs.

Marking on the Product.   Manual handwritten marking on the product or on a label is considered the basic product identification method. This method has various options that enhance the ability of the identification system to satisfy your company's objectives: (1) Use self-adhesive labels for cartons or unit loads. (2) Use prepunched hole labels for hanging garments. (3) Use color-coded labels for product that requires a FIFO product rotation. (4) Use machine-printed labels. (5) Apply labels in a standard location on the product.

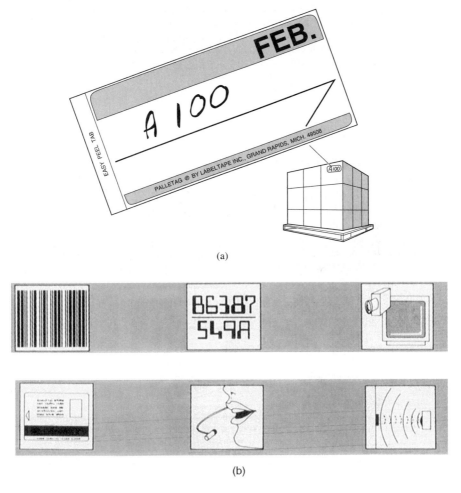

(a)

(b)

FIGURE 13.1 (*a*) Manually written label (*Courtesy of Label Tape, Inc.*); (*b*) automatic identification methods (*Courtesy of Modern Materials Handling Magazine.*).

Automatic Identification Method

The automatic identification method (Fig. 13.1*b*) consists of a human- and machine-readable code symbology (labeled) that is attached to the product. This code is read by a scanning device that transmits the bar-code data on line or delayed to a microprocessor. The microprocessor causes another machine to respond or hold this code information in memory for feedback. An automatic product identification method is the most sophisticated method and is considered for implementation in any size of warehouse operation. These facilities are manual, mechanized, or automated operations that handle a high volume.

The disadvantages of the method are that it increases capital investment and requires management control and discipline. The advantages are that it handles a high volume and a large number of SKUs, provides accurate information and on-line information flow, requires fewest employees, and reduces errors and lost inventory.

When you look at the various warehouse functions, the following functions are considered as potential candidates for an automatic identification system: receiving, transportation, storage, picking, sortation, security, inventory control and flow, manifest and shipping, and maintenance.

VARIOUS AUTOMATIC SYSTEMS

The automatic identification systems available for implementation in a warehouse operation are listed below. Some of the identification methods are used on transport (sortation) systems, and others are used to identify SKUs (product). These identification concepts are (1) wire prong, (2) photoreflective, (3) bar code, (4) magnetic strip, (5) optical character recognition, (6) radio frequency, (7) machine vision, (8) voice recognition, and (9) surface acoustic wave.

Wire-Prong Identification

Wire-prong identification (Fig. 13.2a) is used in a product transportation system. The product is conveyed on a powered conveyor surface, in a container or on a trolley that is on a powered chain system. Wire-prong identification consists of these components: powered conveyor system, wire prong on the container or trolley, wire-prong reading device, and divert mechanism.

The wire-prong identification system is most commonly used on a powered trolley system used to transport hanging garments or single items in a hanging basket. On a powered trolley system, the trolleys are automatically diverted from the main traffic line onto a second main traffic or onto a branch gravity spur line. These divert angles or trolley paths are at 30° or 45° angles from the main traffic line. After the divert location on the branch line, the trolley system requires a trolley accumulation line with full-line sensors and conveyor system shutdown controls.

When a wire-prong system is used on the conveying system, the trolley's lead spool has a code or sliding pin index, and a reading device has a set prong. In the wire-prong method, one or two sliding pins provide up to 99 settings, and the wire prongs of the reading device are matched to the number of sliding pins. Each setting represents a divert location.

In wire-prong identification, the reading device is located prior to the divert location. As the trolley lead spool passes the wire-prong sensing device, the wire prong detects the corresponding pin (divert) setting or does not detect the trolley pins. If the sensing device detects the pins, then the sensing device activates the divert mechanism to divert the trolley from the main line to the spur. If the pins are not detected, then the trolley continues travel on the conveyor system until it is read by the appropriate sensing device.

The disadvantages of the system are that it increases exposure to missettings, is not used to identify SKUs, limits the application (usually to use on a trolley or container

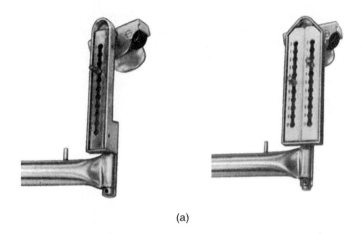

(a)

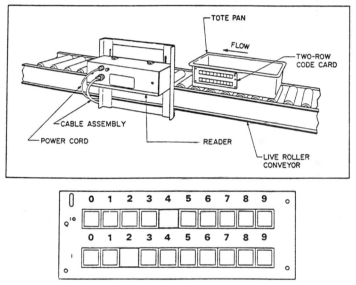

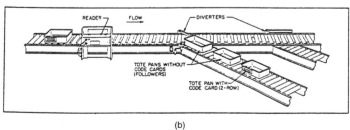

(b)

FIGURE 13.2   (a) Wire-prong (*Courtesy of W & H Systems, Inc.*); (b) photoreflective identification (*Courtesy of Markem Corp.*).

system), and creates potential maintenance problems. The advantages of wire-prong identification are that it is durable and reusable, it does not require print time, the operation is simple, capital investment is low, and there is no paper trash problem.

Photoreflective Identification

The photoreflective tab (Fig. 13.2b) is used on a conveyor system to transport containers or trolleys. Photoreflective identification consists of these components: powered conveyor system, photoreflective tabs or sliding tab on a non-light-reflective surface, light source, beam reader, and divert mechanism.

The photoreflective tab identification system is used on a trolley conveyor system or a package conveyor system that has a container (package) captive to the warehouse operation. The photoreflective system requires that the reflective tab(s) be placed in a standard position on the trolley or container side. Prior to the divert location, a light source is focused to concentrate its beam onto the trolley or container front or side and to hit the tab location; the beam reader receives the photoreflective tab's reflected light. The photoreflective tab system is designed with one, two, or three sliding tabs that provide up to 999 settings (divert locations).

In photoreflective identification, as the trolley or container travels past the light beam source station, the light beam reflects off the photoreflective tab onto the light beam reader. If the reflective tab setting corresponds to the light beam's reader setting, then the divert mechanism is activated to divert the item from the main travel path. If there is not a match, then the container passes the divert location.

The disadvantages of the system are that there is a potential for missettings, maintenance problems may arise, it is difficult to identify the SKU, it requires a light source or line of sight, and the container or trolley is made of non-light-reflective material and is captive to the system. The advantages of the method are the reusable code, durable code, and low investment cost.

Bar-Code Identification

Bar-code identification (Fig. 13.3a) is the most popular automatic identification system used in the warehouse and distribution industry. Its popularity is a result of many things: It is a simple operation, has a high degree of flexibility to handle a large number of SKUs, costs little, is very accurate, is high-speed, and has the ability to be implemented in any key warehouse functions. Also, the bar-code symbology characteristics allow implementation in all segments of your product channel of distribution. This channel starts with your product supplier, at your warehouse receiving dock or in your customer's location (workstation). Numerous bar-code types are available for implementation in your warehouse facility. These bar-code symbologies and scanning devices are reviewed later in this chapter.

The bar-code automated identification system is a simple design. A scanner focuses an intense (laser) light beam over a bar-code symbology (a series of black bars and white spaces that vary in width) which is placed onto an object. The light of the scanning device that is reflected back to the scanner from the white spaces which exist between the black bars is sensed by the scanning device. This reflection is a signal that is sent to a microprocessor. The microprocessor activates another mechanical

(a)

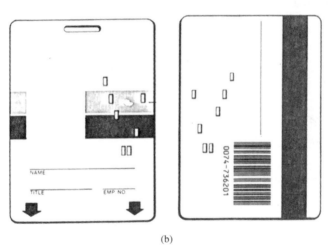

(b)

FIGURE 13.3 (*a*) Bar-code (*Courtesy of I.D. Systems Magazine.*); (*b*) magnetic stripe identification (*Courtesy of Identicard Systems, Inc.*).

device to divert the object from the conveyor system or stores these bar-code data in its memory.

Disadvantages of the bar-code are that it requires a line of sight and a clear label, the label is not durable, it is difficult to recode the label, the code is not human-readable but the label can contain human-readable characters, and nonconveyable items require an operator.

The advantages of the bar code are that it is on-line, it is used in many warehouse functions, it has a high degree of accuracy, read distance is up to 10 ft, the cost per bar-code symbology (label) is low, it reads label printed by a wide variety of methods, cost per scanner is low to medium, it is used in a high-speed sortation system with high read rates, its fixed position does not require an operator, and it handles a high volume.

Magnetic-Stripe Identification

The magnetic stripe (Fig. 13.3b) is most commonly used in security and personnel warehouse functions. The magnetic stripe system consists of a small magnetic stripe that is on the reverse or front side of a card. The card is the size of a credit card or personnel badge. A large quantity of information on the card is encoded onto the magnetic stripe. The data transfer or encoding is much like a tape recorder that records onto a blank tape.

For a dynamic warehouse function, the disadvantage of the magnetic stripe is that it has limited application in the various warehouse functions that handle product. This limitation is due to the fact that the magnetic stripe is read by a contact scanning device which has a slow reading speed. Compared to a bar-code symbology, the advantage of the magnetic stripe is that it reads through a film of dirt or grease. But magnetic stripe identification is very useful for personnel identification badges that control access to security areas and regulate employee attendance.

The disadvantages of the system are that it requires a contact reader, the read speed of 1 to 2 seconds is slow, it is limited to magnetic stripes, it requires a line of sight, and it is difficult to handle a high volume. The advantages are that the readers are low-cost, the label cost is low, the label is recodable, the data density is high, and it reads through dirt or grease.

Radio-Frequency Identification

Radio-frequency identification (Fig. 13.4a) is the next system used in the storage function of a warehouse. A tag (symbology) is attached to the product and there is a receiver (antenna) on the storage vehicle. The tag is a transponder that is placed onto the product and transmits a radio signal. This radio signal is produced from an integrated circuit located inside the tag. The tag with a film of dirt or grease transmits the radio signal. The radio signal that is transmitted uniquely identifies the product. The receiver (antenna) or reader on the storage vehicle picks up the radio signal from the tag. The transmission distance is up to 30 ft. After the antenna picks up the radio signal, the reader decodes and validates the signal for transmission to the host computer, which identifies the storage position.

The tags are either passive or programmable. The passive tag is permanently coded, stores up to 20 alphanumeric characters, and is relatively inexpensive. The programmable tag is recodable, stores up to 2000 characters, contains a battery, and is more expensive.

In manual or automated dense-storage systems, use of radio-frequency identification improves inventory control.

The disadvantages are that the tag is not human-readable, tags are produced by a special machine, and the cost per tag is high. The advantages are that it does not require a line of sight, it reads up to 30 ft, the read rate of 0.1 second per tag is high,

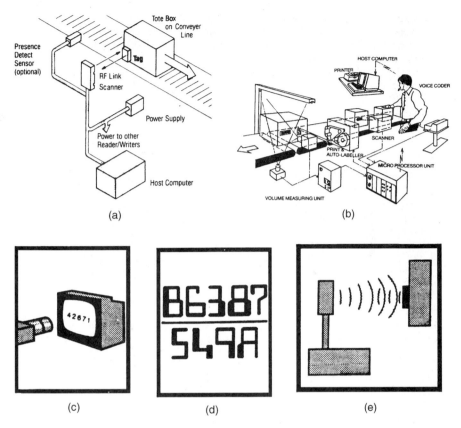

FIGURE 13.4   (*a*) Radio-frequency; (*b*) machine vision; (*c*) voice recognition (*Courtesy of I.D. Systems Magazine.*); (*d*) optical character recognition (*Courtesy of I.D. Systems Magazine.*); (*e*) surface acoustic wave identification systems (*Courtesy of I.D. Systems Magazine.*).

some tags are recodable while others are permanent, the scanner costs little, and it reads through dirt or grease.

Machine Vision Identification

Machine vision (Fig. 13.4*c*) is the next automatic product identification system used in a warehouse sortation function. The machine vision system consists of a camera with a scanner, a secondary light source (warehouse light could be sufficient), and a symbology. The symbology is a bar code or alphanumeric characters that are on a SKU (product) which travels on a conveyor system. The machine vision scanners (camera) are solid-state image sensors of two types. The first type, the linear type, is arranged in a vertical or horizontal line to the product's direction of travel. The second type, the area scanner, uses two cameras.

The machine vision system takes up to 4000 snap pictures per second. With this speed, machine vision can be used on high-speed conveyor sortation systems. Also, hand-held solid-state cameras are used to identify the nonconveyable SKUs or in other warehouse functions.

The disadvantages of the system are that it requires a line of sight, the capital cost is high, and it is difficult to read a label with dirt or grease. The advantages are that the nominal read distance is 10 ft, it does not require an operator, it reads all printed produced labels, it reads alphanumeric and bar-code labels, it reads at high speeds of less than 0.5 second per label, and it is available in portable hand-held devices.

Voice Recognition Concept

Voice recognition identification (Fig. 13.4d) is commonly used in the warehouse receiving and shipping and sortation functions. The voice recognition system consists of a headset microphone on an employee and alphanumeric characters on the product which is on a conveyor system. The voice recognition design ensures that the employee's hands are free to handle the product and has a vocabulary of 200 to 300 words. As your employee speaks into the headset microphone, the system makes a bar-code label for the SKU or activates a divert mechanism.

There are three types of speakers: speaker-dependent, speaker-independent, and continuous speech.

The speaker-dependent system recognizes one word at a time from an employee who uses up to a 250-word vocabulary. The unique feature of this system is that it is programmed for the speaker's accent and dialect. The speaker-independent system recognizes one word at a time but is limited to one speaker and is more costly than the speaker-dependent system. The continuous-speech system recognizes words without intervening pauses but is programmed for each employee.

Compared to the other automatic identification systems, voice recognition requires the employee to be trained but the employee's accent and dialect are recognized. The disadvantages are that it is expensive, labels are human-readable, the read rate of 8 seconds per label is low, and it requires an employee. The advantages are that it frees up the employee's hands and does not require a line of sight.

Optical Character Identification

The optical character recognition (OCR) system (Fig. 13.4d) is used in most warehouse functions. Optical character recognition consists of stylized digits or alphabetic characters that are read by a scanning device. The hand-held scanning device requires a light source and relies on a sensor to distinguish the printed number or characters. These characters and numbers have nonreflective shapes which appear from the light-reflecting background. To read the data, a hand-held scanner is swept across the data.

Compared to bar-code identification, OCR has human- and machine-readable symbology (characters) which allows the bar code to appear on the same label. This feature permits the product to be identified at any warehouse function. The system requires a clean, clear label.

The disadvantages of the system are that it requires contact to read, it requires a clean and clear label, the read rate of 4 to 6 second per label is moderate, a line of sight is needed, the label is not durable, the label cannot be recoded, and page readers have a high cost.

The advantages are that hand-held scanners cost little, it reads a human-readable symbology, the cost per label is low, and it reads labels printed by all printers.

Surface Acoustic Wave Identification

The *surface acoustic wave* (SAW) is used in the warehouse storage function to identify unit loads. The surface acoustic wave (Fig. 13.4*e*) consists of a low-powered radar transmitter and reader and tags. These tags contain a lithium niobate crystal which, when exposed to a radar signal, emits an electromagnetic signal. This signal is read by a signal antenna on the storage vehicle.

Compared to radio-frequency tags, the surface acoustic wave tags are permanently encoded, smaller, and limited to a binary code of 32 bits of data.

The disadvantages are that it reads at 5 ft, the tag price is high, it is not recodable and not human-readable.

The advantages are that it is low-cost, smaller tags are needed, it does not require a line of sight, it reads through dirt and grease, it is long-lasting and fast-reading.

COMPONENTS OF AUTOMATIC IDENTIFICATION SYSTEM

An automatic identification concept has several additional components: SKU or SKU carrier, symbology (code) printer, symbology or label, encoding device, and mechanical and structural equipment.

SPECIFYING THE SKU, SKU CARRIER, AND STRUCTURE

The first component of a SKU identification system is the structure, product, or product carrier. The product is a single item, package, carton, or pallet load. The structure is the rack upright post, vehicle, or badge. The product carrier is a tote or trolley. The characteristics of these items are that they move or are tracked as they flow through the warehouse and distribution operation or are in a fixed location for a transaction.

SYMBOLOGY (CODE)

The second element of a SKU identification method is the symbology (code), or the information that is to appear on the label. The code permits the SKU to carry information and is identified in terms of the following: what it is, where it came from, where (when) it is going to, where its location is, what is in it, and other meaningful company information.

Your first major consideration concerning a symbol that appears on a label is to make sure that the symbols used in the SKU identification system are understood in all warehouse functional areas, by customers and by vendors.

The three symbology options are human-readable, machine-readable, and human- and machine-readable.

Human-Readable Code.   The human-readable code consists of alphabetic characters or numeric digits. This language permits employees or machine vision scanners to read the code, or the code can be a colored tag or colored location on a human-readable label which is recognized by an employee.

Machine-Readable Code.   The second code (symbology) is the machine-readable code (Fig. 13.5*a*). What is the bar-code, tag, prong, or magnetic-stripe code that is read by a reading device? The unique feature is that without the scanning device the employee cannot interrupt the code.

Human- and Machine-Readable Code.   The third code type is the human- and machine-readable code (Fig. 13.5*b*). This code has both human-readable and machine-readable language.

When you review all your warehouse functions (activities), you realize several things. First, employees are involved in all functions. Second, there exists a good possibility that warehouse and distribution suppliers and customers have human-readable requirements. Third, if the automatic system goes down, you must rely on your employees to perform the activity. These factors make the combination of the human- and machine-readable code preferred for your label face.

Purpose of the Code

Your next major consideration is the application or purpose of the code (symbology). In your warehouse facility, there are two uses of symbologies, permanent and expendable.

Permanent Code.   The permanent code is used to track or to inventory capital assets, mobile equipment, and employees or to identify a storage-pick position. These codes or tags are expected to last a long time. With this criterion, the paper codes have a special surface (treatment) to protect the black bars and white spaces. This treatment is a lamination of clear mylar.

Expendable Code.   The expendable code (symbology) is printed directly onto the product or onto a pressure-sensitive self-adhesive paper that is placed by an employee or label applicator machine onto the product. The product with the bar-code label is transported to any warehouse location or is sent to the customer.

CODE GENERATION ALTERNATIVES

Your third question prior to printing the label involves code generation, or how to print the label. The three alternatives are batch, sequential, and random. The batch

(a)

RESERVE LOCATION:

SKU NO.:

DESCRIPTION:

QUANTITY:

ACTIVE LOCATION:

123456789043

(b)

FIGURE 13.5 (*a*) Machine-readable code (*Courtesy of Modern Materials Handling Magazine.*); (*b*) human- and machine-readable code.

generation method is used for codes that have the same bar code or bar formats. The sequential generation method is for bar codes printed in a consecutive numerical order such as 001, 002, 003. Most commonly, both these bar-code generation methods have bar codes printed by off-site printers and are for permanent labels.

In the third bar-code generation method, random bar-code generation, the printer prints the input that was supplied by an on-line computer. This input data are a warehouse transaction request or a customer order label for a SKU. The random generation is determined by several factors that vary such as several customer orders for many SKUs in your inventory; therefore, each code or series of codes printed varies by its content. Another feature of random generation is the ability to print human-readable alphanumeric characters. These factors make random bar-code generation a popular method in the warehouse and distribution industry.

Where to Print Bar-Code Labels

After these factors have been determined, your next important choice is whether your symbology (bar-code labels) is to be printed off-site or on-site. When we mention off-site printing or on-site printing, we are referring to the alternative locations to print pressure-sensitive self-adhesive human- and machine-readable labels. Most radio-frequency or surface acoustic wave tags are produced off-site. Some optical character recognition and magnetic-stripe symbologies are produced on-site, but these automatic identification methods do not satisfy the needs of a dynamic material handling system in a warehouse facility. The machine vision and bar-code identification methods are printed off-site and on-site.

Off-Site.   By definition, off-site label printing occurs when the warehouse facility purchases preprinted bar-code labels from another company. These companies have the hardware, software, and lamination equipment to specialize in symbology production. The two factors that affect whether you print off-site or on-site are the cost (economics) and when you require the bar-code labels in the warehouse operation.

The most typical situations that call for off-site printing occur (1) when the information on the bar code is fixed and a few hundred labels are required per day and (2) when a large quantity of bar-code labels (fixed or sequential order) are required by the operation.

Examples are labels for pick position identification or asset identification projects. The various off-site preprinted label methods are photographic, laser, electrostatic, dot matrix, letterpress, offset, flexographic, rotogravure, and screen.

On-Site.   On-site printing of bar-code labels is very common in the warehouse and distribution industry. In on-site printing, your company purchases the bar-code label printing hardware, software, self-adhesive paper stock, and ink applicator (printer). With these items, your company prints the required bar-code labels in its own facility.

This label printing approach gives your company complete quantity and quality control of the label printing process. This flexibility includes the ability to change the type of bar-code symbology or alphanumeric information that appears on the face of the label.

Bar-Code Printing Design Parameters

Prior to selection of the on-site label print method, consideration is given to several factors. These factors are (1) the type of symbology (code), (2) volume (how many labels you require per minute, shift, day and whether you can preprint labels). These volumes are based on the average and peak activity days, (3) flexibility (ability of the system to provide the format, the number of characters or lines you require), (4) resolution (whether the density of the code and the print quality meet your requirements), (5) economics (the investment cost, paper, and ink expenses), (6) space (whether the printing equipment requires an air-conditioned space), and (7) spares (whether you require a spare printer).

Various On-Site Systems

The various on-site bar-code printing methods include (1) dot-matrix impact, (2) formed-character impact, (3) ion deposition, (4) laser, (5) ink jet, (6) thermal, and (7) thermal transfer.

Dot-Matrix Printer.   The dot-matrix impact printing technique (Fig. 13.6*a*) is one of the most common systems in the warehouse industry that is used to print labels. The machine prints labels onto continuous-form pages or onto pressure-sensitive label paper.

The dot-matrix impact printer uses a line of wire pins that are driven by electromagnets. These pins strike an inked ribbon which produces a series of black dots on a paper page or pressure-sensitive label. The black bars of the bar code or alphanumeric characters and digits are built up from these lines of dots. The dot-matrix impact printer is controlled by a programmable microcomputer. This gives the flexibility to easily change bar codes or human-readable language. Some new models have the ability to print various bar codes because the necessary bar-code fonts are built into the printer.

When dot-matrix printed bar-code labels are used with the built-up dots from inked or reinked ribbons that have reached their useful life, some bar edges become ragged and create no-reads for a scanning device. Another reading problem is created from using poor-quality paper or a different-quality paper. These labels may look black and of good quality to the employee's eye, but to the scanning device these black bars look gray and create a no-read.

The disadvantages of the dot-matrix system are that the print bar width is 10 to 21 mils or greater, it prints 6 to 7 characters per inch and 30 to 50 labels per minute (low to medium volume), and it has fair to low bar edge definition.

The advantages of this method are that it prints all alphanumeric characters, bar codes, and OCR labels which gives high flexibility, it prints on a demand basis, printer cost is average, label cost is average, there is medium resolution, and it can use reinked ribbons and print documents.

Formed-Character Impact Printing.   In the formed-character impact printer (Fig. 13.6*b*), bars are etched in reverse on the surface of a drum (print wheel) that is continuously rotated at high speed. Together a carbon ribbon and a strip of label paper (polyester or other label stock) are fed over this drum. During this action, an electromagnetically driven hammer presses the label paper and the carbon ribbon against the drum. Timing marks on the rotating drum are sensed and determine the character that

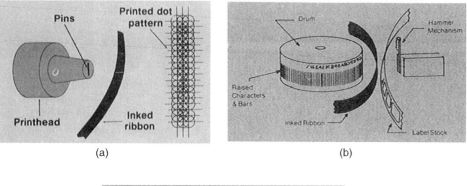

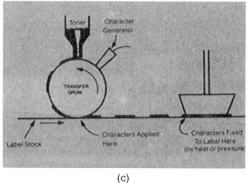

FIGURE 13.6   (*a*) Dot-matrix printer; (*b*) formed-character printer; (*c*) ion deposition (electrostatic) printer. (*Courtesy of I.D. Systems Magazine.*)

is under the drum. When the desired character passes under the hammer, the hammer is quickly activated to strike the ribbon. Each hammer strikes does print an entire bar code (black bar) onto the label paper.

The formed-character printed labels have very few printing errors, but it prints the bar widths (characters) that are engraved on the drum.

The disadvantages of the method are that there is low flexibility, it is noisier than the dot matrix, it cannot print documents, and it is difficult to print large human-readable characters. The advantages of the method are that it prints a medium volume (40 to 100 labels per minute) and uniform codes, there are 9.4 characters per inch, the bar widths are as narrow as 0.0075 in, the cost per label is low, there is high resolution, the cost per printer is average, and there is excellent bar edge definition.

Ion Deposition (Electrostatic) Printer.   The next label print machine is the ion deposition (electrostatic) printer (Fig. 13.6*c*) which is similar to the office copier (copy machine). The ion deposition print method is a two-step process. First, a pattern of electrostatic charges is produced on the surface of a drum. Particles of ink (toner) are attracted to the charged drum area and develop an image of the bar code and alphanumeric characters on the drum. Second, the ink image is transferred from the drum face

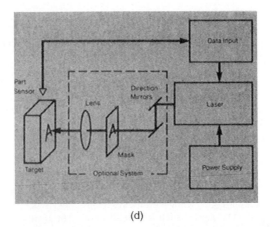

(d)

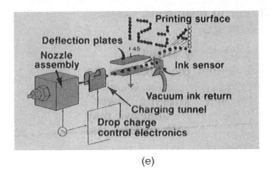

(e)

FIGURE 13.6 (*Continued*) (*d*) Laser printer; (*e*) ink jet printer. (*Courtesy of I.D. Systems Magazine.*)

onto the paper stock. The bar-code image is fixed to the paper by the pressure of the ink drum against the paper.

The disadvantages of the method are that the label cost is average, the printer cost is high, it requires an original copy of the label, and label change is slow. The advantages are that the print speed is high (200 to 300 labels per minute), flexibility is high, there is medium resolution, and the bar width has any required width.

Laser Printing. Laser printing (Fig. 13.6*d*) is similar to an office xerographic copier. In the print machine a laser beam scans over an electronically charged photosensitive drum. The electronic charges are neutralized at the points or at the location where the laser beam strikes the drum. When ink or toner is spread across the drum surface, the ink sticks to the charged area and develops an image. The ink image is transferred from the drum to the paper and is fixed onto the paper stock surface by heat and pressure.

The laser printer produces a high volume of high-quality bar-code labels.

The disadvantages of the method are that printing is limited to batches and the cost of the equipment and of the label is medium. The advantages are that it imprints through coatings, the bar edge definition is excellent, the minimum bar width is 0.0003 in, it is very flexible, there is high resolution, and the print speed of 2000 per minute is high.

Ink Jet Printing.   The ink jet printing machine sprays drops of thin ink from its nozzle onto label stock, paper, or cartons. This ink stream is moved electrostatically. The bar-code or alphanumeric characters are built up from the dots formed when the moving ink stream hits the label's stock surface or the exterior of a carton.

The disadvantages of the method are the low bar edge definition, low resolution, use on carton sides, and increased maintenance. The advantages are that it prints a high volume of 2000 per minute, there is good flexibility, and label cost is low.

Thermal Printing.   In thermal printing (Fig. 13.6*f*), the machine prints onto special heat-sensitive label stock. Special chemicals in the label stock change color as heat is applied to the area. The paper stock is moved over a set of electrically heated pins. As the pins are selectively heated, dots are formed on the label stock. Thermal printing produces a high-quality bar-code label.

The disadvantages of the method are the low volume at 10 to 18 labels per minute, low flexibility, low cost per printer, and high cost for paper. The advantages are the low cost per label, production of high-quality label, medium or high resolution, and excellent bar edge definition.

Thermal Transfer Printing.   In thermal transfer printing (Fig. 13.6*g*), a thin film or paper-based ribbon has a waxlike coating. A heating element is located behind the ribbon. The thermal elements melt the coating, and the coating is then rolled onto the paper. Thermal transfer printing produces a bar-code label that has clear dot resolution.

The disadvantages are the single ribbon pass, low flexibility, and medium cost. The advantages are the medium volume of 25 to 100 per minute, wide variety of stock to print on, and quiet, high resolution.

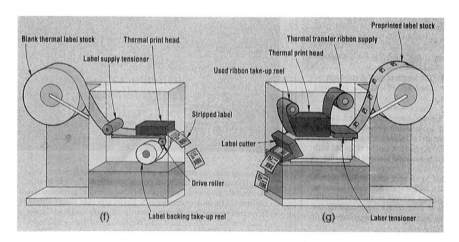

FIGURE 13.6   (*Continued*) (*f*) Thermal printer; (*g*) thermal transfer printer. (*Courtesy of Modern Materials Handling Magazine.*)

BAR-CODE TYPES

The next major component is the type of code used on the label. The bar-code label is placed onto the product or rack position. These bar-code labels are read by the employee or by an optical scanning device.

Various symbologies (codes) are available for implementation in a warehouse facility:

- Manual readable symbologies: handwritten, machine-written, and colors
- Machine-readable symbologies: UPC, EAN, 2 of 5, Telepan, code 93, code 3 of 9, code 128, codabar, I 2 of 5, and code 11.

MANUAL READABLE CODES

If you have a manual warehouse operation, three labels are used in your facility. The first label has handwritten alphabetic characters or numeric digits on a preprinted or on-demand manually printed label. The second label contains the information that the machine (computer) printed on the label face. These labels have ink and paper options: black ink on white paper, black ink on colored paper, or colored ink on white paper. The third label is solid-colored paper used to identify the product or warehouse location. In a color-coded system, the number of different SKUs or warehouse locations is restricted by the number of colors in the spectrum.

MACHINE-READABLE CODES

If you have a mechanized or automated facility, then the machine-readable bar-code label which is widely used in the warehouse industry is best for your operation. The bar-code label consists of black ink, in the form of bars, that is spread in a predetermined pattern onto a white background (paper). The pattern varies as to the bar width and the white space between the black bars, which is determined by the bar code (symbology). In most warehouse applications for optimum reads, the bar codes have quiet zones between both label edges and each exterior black bar. Also it is good to have a square bar code in which has the length of the bars is equal to or longer than the width of the total black bars and white spaces.

Universal Product Bar Code

The *universal product code* (UPC) is a numeric code that appears on almost every product sold to a customer in a retail outlet. The UPC usually consists of 12 digits, but one version contains 5 digits. In this code, each character has two black bars and two white spaces. The bar code is separated into two zones. If there is a left zone and a right zone which permits bidirectional scanning, then the code is a minimum of 6 digits long.

The UPC is not recommended for use in a warehouse operation. The code is not printed by a formed-character impact or dot-matrix impact printer, contains a self-checking digit, and has a fixed length and a variable width (density of 13.7 characters per inch).

European Article Number

The *European article number* (EAN) code is similar to and compatible with the universal product code. The unique feature of EAN code is that there is a country flag that identifies the code's (product) origin. The extended EAN code version is EAN 13 NS; the compressed version is EAN 8.

A feature of the European article number is that it has similar applications and operational characteristics as the universal product code. It is used in the retail industry on the retail floor.

I (Interleaved) 2 of 5 Bar Code

The I 2 of 5 code is a complete numeric code used in the warehouse facility on the sortation, shipping, and SKU identification functions. With this label, sometimes the white space between two black bars carries information. With this feature, the label requires a large number of bars per character. It has a wide/narrow ratio. If the width ratio is 3 to 1, then the bar code requires 14 units of space per character which means a long label. The narrow width ratio of 2-to-1 bar-code requires 12 units which means a short label. The bar code is comprised of five black bars and four white spaces. Other features of the I 2 of 5 code are that it is self-checking and bidirectional.

The I 2 of 5 bar-code label gets its name from the fact that two black bars are packed together to represent the first character and the white spaces represent the second character. Each number in the bar code is represented by five bars and four spaces.

The code characteristics are that it is highly compressed, numeric only with different start and stop characters, the density is 17.8 characters per inch, the length is fixed with check digits, and it is self-checking and bidirectional. A unique characteristic is that the code has an even number of characters. If your code has an odd number of characters, then to make an even number of characters, you add zeros to the leading part of the character. A character is equal to five bars and four white spaces.

Telepan Bar Code

The Telepan code is used for the SKU identification function in the textile, insurance, and milk processing industries. This bar code is not widely used in the single-item, carton, or pallet load warehouse industry, because the code is owned by a private company. But the owner company permits licenses for its use. The Telepan code has the capability of encoding the full ASC 11 set because it has 96 keyboard characters and 32 computer control commands. A character is defined by a 16-element symbol or in a double-density numeric mode that has 8 elements per character. Other features are that it has variable length, there are 12.5 characters per inch in an alphanumeric mode and

25 characters per inch in double-density numeric mode. It is not self-checking but is bidirectional.

Code 93 Bar Code

Code 93 is an alphanumeric bar-code symbology. The code has the capacity of 43 data characters composed of numbers 0 to 9, letters A to Z, six symbols, space, plus four control characters and a start-stop character. With two check (start-stop) characters, Code 93 is almost error-free. It has the ability to have a high density of 13.9 characters per inch. This results from a character of 5 bars and 4 white spaces. The code is used in the warehouse functions of sorting, receiving, and shipping. In the manufacturing industry, Code 93 is used for component and work-in-process verification. Code 93 is bidirectional.

Code 3 of 9 Bar Code

The code 39 (3 of 9) has three wide black bars and six white spaces. This feature gives the code its name and provides 39 data digits or 43 characters. These characters are start-stop, 10 digits (0 to 9), 26 letters (A to Z), and six symbols. Using symbols and characters, the code encodes the entire ASC 11 set.

In the United States the 3 of 9 code is self-checking, bidirectional alphanumeric code. The code is used in the warehouse industry and in the Department of Defense as part of the LOGMARS program.

Compared to Code 93, code 3 of 9 (39) has a wider (longer) label.

Code 128 Bar Code

Code 128 contains the full 128 characters in the ASC 11 set. All characters consist of three bars and three spaces. Each bar or space can contain one to four elements. The bar-code density for alphanumeric is 9.1 characters per inch, and the numeric is 18.1 characters per inch. The special features of this bar code are that it has a black bar and white space character parity for character integrity, a function character for symbol linking, and a spare function character that permits suspension or unique applications.

Bar code 128 has 107 characters, 3 separate and different start characters, and 1 stop character. The code is not self-checking but is bidirectional. Utilization of the three start characters provides the flexibility of three different subsets in the code: subset A for alphanumeric characters with control and specialized characters, subset B for alphanumeric characters with lowercase alphabetic and special characters, and subset C for 100 digit pairs, 00 to 99.

Codabar Bar Code

Codabar is a numeric bar code that has the flexibility to have additional characters at the start and the end. The bar code contains up to 32 characters at 10 characters per inch and has self-checking and bidirectional features. To achieve this flexibility and versatility, it uses 18 different black-bar and white-space widths. This feature and the fact that each character has four bars and three spaces with a maximum character set of 20 make the codabar a complex code.

In the United States the codabar has been used by several transportation companies, American Association of Blood Banks, libraries (Library of Congress), photofinishing, inventory control, SKU pricing, and distribution industries.

Code 11 Bar Code

Code 11 is a numeric code. The code has a very high density of 15 characters per inch with the black bars and white spaces that are very narrow. The bar code has 11 different characters: 10 digits and a dash (-). When one check digit is used in the code, then there are nine digits. If two check digits are used, then there are eight digits. Other features are that it is bidirectional and is not self-checking.

Code 11 is used to identify communications equipment and electric and electronic components.

VARIOUS PRODUCT IDENTIFICATION ENCODING DEVICES

The next key consideration in a product identification system is the product identification encoding device. The encoding device (induction) obtains the product identification from the pin, voice, or bar code and transmits the product identification to a microprocessor or divert device.

The various encoding devices include

- Manual types (key pad, voice, and slot reader)
- Mechanical types [sliding tabs (pins) or retroreflective]
- Special product identification devices [optical character recognition or vision machine (camera)]
- Automatic types: hand-held (contact or noncontact), fixed-position fixed-beam, and fixed-position moving-beam

Encoding Requires a Gap

To have an effective and efficient operation, the encoding (induction) system requires a gap (open space) between two product identifications. The open space permits the encoder to handle the volume, adjust for the next product identification, and obtain a clear start-stop of the product identification and permits the sortation device to divert the object from the main travel path.

Hand-Held Scanning Device. With hand-held scanning devices, the gap is created by the employee's moving the scanning device between two product identification codes. The codes are on a stationary structure (rack post or object) or on a dynamic object that is being moved by a vehicle, human, or conveyor.

Overhead Trolley System.   On an overhead trolley system, the distance between trolleys creates the gap. The trolleys are on nominal 10-ft centers. With a 42- to 48-in-long trolley, there is sufficient space between two product identifications.

Brake Belt and Meter Belt on a Package Conveyor System.   On a container or package conveyor system, the open space is created between two objects with product identification codes traveling over a series of belt conveyors. These belts are the brake belt (momentary stop), metering belt (creates the space), and scanning belt (presents the bar code to the scanner).

Tilt Tray.   Objects on a tilt tray system automatically are provided with a gap between two product identifications because the trays are on 17- to 27-in centers. The centers and the open space between the two trays provide the necessary gap between bar codes.

Manual Encoding Devices

The manual product identification devices require an employee to read the code and encode the identification into the control system.

Keypad.   The keypad (Fig. 13.7a) is the basic manual encoding device. It requires a human-readable label with the minimum number of digits or alphabetic characters on the label face. The alphabetic characters and numeric digits are clearly visible and face the employee's induction station.

The keypad is electronically operated and is on a pedestal (console device) located adjacent to the induction section of the conveyor system. The keypad consists of a series of keys with digits (0 to 9) or alphabetic characters and cancel, repeat, clear, enter, and send buttons. To maximize the induction (operator's productivity) the digits of the product identification used at the induction station are as few as possible and in large characters or bold type. The preference is one or two digits. Also the induction belt network is controlled by the operator's foot pedal or by a send button on the keypad.

The disadvantages of the keypad are that employee productivity is low (20 to 30 per minute), it handles a low volume, it requires employee training, and it requires the maximum number of employees and a human-paced system. The advantages are that the capital cost is low, it does not require a bar code, and it is an excellent backup system.

Slot Reader.   The slot reader is designed to read magnetic stripes on personnel cards. The slot reader is located adjacent to the door frame and has a communications link to a microprocessor that controls access to the room. The slot reader is an environmentally sealed container; a card passes through the slot, and the reader reads the card with the magnetic stripe. At the bottom of the slot is a read head which reads the stripe.

The disadvantages are that it is limited to magnetic stripes and sees limited application in a warehouse. The advantages are that it reads magnetic stripes and there are no encoding errors.

Voice Recognition.   The voice recognition encoding device consists of a headset microphone. The employee is trained to speak into the microphone which encodes the product identification. To encode an employee's speech, the system is programmed to accept the employee's accent and dialect.

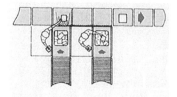

FIGURE 13.7 (*a*) Manual keypad (*Courtesy of Acco Systems.*); (*b*) contact pad (*Courtesy of I.D. Systems Magazine.*).

The disadvantage are that it has a limited vocabulary, it is a human-paced system, and there is high capital investment. The advantages are that it does not require a bar code and frees up the employee's hands.

Mechanical Encoding Devices

The second group of product identification encoding devices is the mechanical group. An employee mechanically sets and resets the product identification. The sliding pin and retroreflective sliding tab are read by the encoding device which encodes the identification to the control system.

Sliding Pin.   The sliding-pin device is used on a carrier or tote that travels over a fixed path. Fixed-pin readers are along the fixed path. The pin or pins extend outward from the lead end of the carrier or trolley. As the appropriate set pin passes through the corresponding set reader, it triggers the divert device to operate. The sliding pins are available as single or double sliding pins.

Retroreflective Sliding Tab.   The retroreflective system consists of three components. The first is the sliding retroflective tabs or tags on the lead end of a trolley or tote that has dark nonreflective material. The second component is a light source. The third is a decoder (the receiver) that reflects light from the photoreflective material. The position of the tag or tab and light that is returned to the decoder determines the product identification.

Automatic Code Reading

The next major group is the automatic bar-code scanning devices, of which there are two types. These scanning devices automatically encode the product identification information into the microprocessor.

Optical Recognition.   The optical character recognition (OCR) scanning device is a fixed-position fixed-beam device. This fixed-position beam device reads alphabetic characters or stylized digits. The sensor distinguishes the printed characters which have reflective shapes.

Vision Machine (Camera).   The second is the vision machine (camera) device with a scanning device that requires a secondary light source. The device is a fixed-position fixed-beam scanning device that reads alphanumeric characters.

Various Automatic Bar-Code Devices

The next group of encoding devices is the automatic bar-code label reader which requires a bar code to be placed on an object or carrier. These bar-code scanning devices automatically encode the bar-code (product identification) information into the microprocessor. The three basic scanning groups are the hand-held scanner, fixed-position fixed-beam scanner, and fixed-position moving-beam scanner.

Hand-Held Scanning Device. The hand-held scanning device is considered the manual encoding device of the automatic bar-code label readers. Hand-held scanners are functionally similar to writing with a pencil or using a hair dryer because the employee holds and moves (points) the scanning device across the bar-code label. These scanning devices are used in a warehouse and distribution facility that handles a low volume, at a workstation (activity) that has the employee sit or stand to perform the work, as a backup scanning device for a fixed-position scanning device, or at any of the key warehouse functions that handle nonconveyable SKUs.

There are two basic types of hand-held scanning devices, contact (pen) and noncontact (gun).

Contact (Pen) Scanning Device. The contact light pen (Fig. 13.7*b*) consists of a light source, a lens or fiber-optic bundle, a photosensor, and a housing that is similar to a pencil with a metal or jeweled tip. The light pen receives its power from a battery or direct current.

There are three types of light sources for a light pin. (1) The incandescent light source has a filament type of lamp which has a light output that is similar to a household bulb. These lights have a bulb that shatters easily, requires a great power supply, and have a low resistance to shock or vibration. (2) The infrared light source is a small solid-state light system of *light-emitting diodes* (LEDs). These infrared sources have a low light output and read bar codes that have carbon inks. (3) A combination of the incandescent and infrared light sources permits the pen to read bar codes with a low carbon content.

To operate, the employee moves (strokes) the light pen tip across the bar-code label that picks up the bar code. The data are stored in memory or are transmitted to a microcomputer. The various light pen systems are pocket light pen, light pen with hand-held terminal, and light pen with fixed terminal.

Noncontact (Gun) Scanning Devices. The noncontact hand-held laser wand or gun is more costly and complex than the light pens (see Fig. 13.3*a*). The two types of hand-held scanners are the fixed-beam model and the moving-beam model. The fixed-beam device requires the operator to move the light beam across the bar code. The moving-beam scanning device requires the operator to point the gun at the bar-code label, and the device automatically directs the light across the bar code. When not in use, to ensure best operator performance and reduced damage to the gun, the employee has a holster that holds the gun.

Both models have battery power and an internal decoder. The scanner emits a light beam that is visible to the operator. These scanners convert the output to a computer-acceptable message.

The gun's disadvantages are that it is more expensive and heavier. The advantages are the read distance in noncontact, its high speed, it reads borderline-quality labels, etched labels, and through overlay material.

Fixed-Position Fixed-Beam Scanning Device. The fixed-beam fixed-position scanning device (Fig. 13.8*a*) directs a light beam at a specific location, and the bar code is in position for reading by the scanning device. The scanning device is a reflective reader or reflex reader.

The reflective reader has a separate light source for the optical source to detect the black codes on a white label. These codes come in two types. The parallel code has the entire code read at one time. This requires several scanner heads banked together, and the code has a trigger or start code (mark) that indicates to read. The sequential codes (series of codes) are read one at a time on one or more labels or code tracks that

MAXIMUM PERMISSIBLE CODE MISORIENTATION
RELATIVE TO SCANNING BEAM

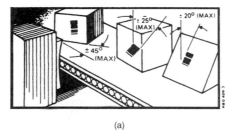

(a)

FIGURE 13.8   (*a*) Fixed-position fixed-beam scanner. (*Courtesy of Accu-Sort Systems, Inc., and Modern Materials Handling Magazine.*)

each have start and stop read code marks. In this situation each level requires one scanner and transmit circuit.

The reflex reader (coaxial reader) has an illuminating source and optic receiver that share the same optic system and housing shell.

These scanners focus the light beam on a single area or point on a conveyor system and look for a bar code on the object as it passes the predetermined location. This means that the bar-code label passes the same location each time. The label position is in one of three modes: (1) the bull's-eye (the black bars and white spaces are in a circular pattern), (2) the ladder (the black bars and white spaces are in a parallel pattern to the direction of travel), and (3) the picket fence (the black bars and white spaces are in a vertical pattern to the direction of travel).

The fixed-beam reader is made up of individual scanners with a stationary light source and stationary light sensor with an incandescent or light-emitting diodes.

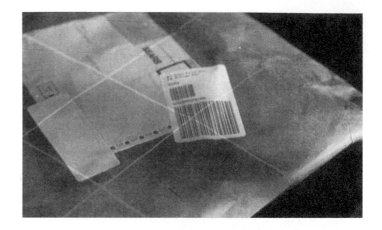

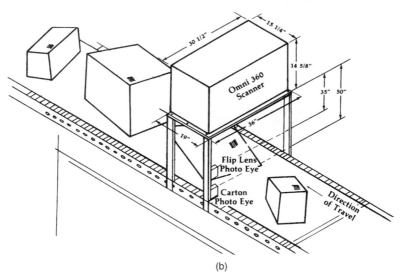

(b)

FIGURE 13.8 (*Continued*) (*b*) Omnidirectional scanner. (*Courtesy of BRT Corp. and Accu-Sort Systems, Inc.*)

Of the automatic scanning devices, the fixed-beam fixed-position scanning device is the simplest type and the least expensive. In a pallet and carton handling warehouse, the fixed-beam fixed-position device is used to scan labels on pallets and packages on a conveyor system on the side, front end, top, or bottom.

The disadvantages of the method are that it increases the cost; the label is scanned from the side, top, front, or bottom; there is a low depth of field; and it reads the ladder, picket fence, or bull's-eye. The advantages are that it is used to scan moving objects, requires no employees, reads a small bar code, increases accuracy, gives immediate information flow, and has a longer life.

Fixed-Position Moving-Beam Scanners.   This device consists of a low-power helium and neon laser beam that is reflected off a rotating multifaceted head to project a light beam onto a conveyor surface. The two types of scanners are the raster scanner and the omnidirectional scanner.

Raster Scanner.   The raster scanner is the first fixed-position moving-beam scanner that directs one light beam onto the conveying surface. This permits the scanning device to read a bar code that faces any direction on the conveying surface.

Omnidirectional Scanner.   The omnidirectional scanner (Fig. 13.8*b*) is a fixed-position moving-beam scanner that directs one light beam perpendicular to the product's direction of travel and two oscillating (waves) of light beams parallel to the direction of travel. The three-light-beam pattern allows the bar code to be oriented in any direction of travel (ladder or picket fence) or in any direction of the compass. The omnidirectional scanner directs these light beams to cover the entire area of a conveyor surface.

Star Burst and Laser Track Scanners.   This new technology in fixed-position moving-beam scanning equipment overcomes some of the shortcomings of the raster and omnidirectional scanning devices. The star burst and laser track scanning devices (Fig. 13.9*a*) have multiple beams that are thrown in an elliptical pattern over the conveyor surface. This light pattern detects a smaller bar code that is oriented in any direction of travel with a greater label skew or tilt.

Omnix-DRX Scanner.   This new-technology fixed-position moving-beam scanner (Fig. 13.9*b*) is similar to the star burst scanner except for two things. First, the Omnix-DRX reads at faster speeds. Second, if it reads a label with a partial code, then the scanner system recreates a partial label which reduces the number of no reads.

The fixed-position moving-beam scanning device's disadvantage is its high cost. The advantages are that there is maximum flexibility of label orientation and pitch, there is maximum depth of field, and it reads a small label.

When to Use the Ladder Bar-Code Orientation.   If your operation has a fixed-beam fixed-position scanner with no label window on the SKU (label procedure) and your bar-code label travels past a specific location, then for the maximum number of reads your bar-code label orientation should be in the ladder presentation to the scanner (Fig. 13.10*a*). Compared to the picket fence, with the ladder bar-code label orientation there is an increased possibility of reading the label because the scanner reads the entire SKU surface. This feature allows flexibility of bar-code placement (in any position) on the SKU. With the picket fence orientation to a line scanner, if the bar code is not in the path of the scanner line, which is the width of the bar code (small window), then it has a greater potential to cause a no-read condition.

Use the Postage Stamp Label Placement Method with the Picket Fence Bar-Code Orientation.   If your operation has a fixed-beam fixed-position scanner with no label window on the SKU and no procedure and your bar-code label travels past a specific location, then for the maximum number of reads with the picket fence bar-code orientation (Fig.13.10*b*) (presentation) to the scanner, use the "postage stamp" label placement technique. The postage stamp technique is a manual label procedure in which employees place the bar-code label in the carton's upper right-hand corner. This label location is similar to the postage stamp location on an envelope. It ensures proper bar-

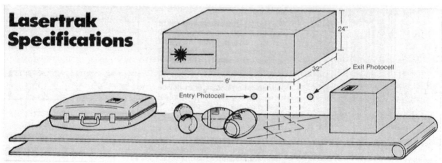

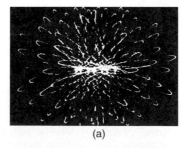

(a)

FIGURE 13.9   (*a*) Star burst or laser track scanner. (*Courtesy of BRT Corp.*)

DRX reconstructs
and decodes partial
scans providing valid
bar code data.

(b)

FIGURE 13.9   *(Continued)* (b) Omnix-DRX scanner. *(Courtesy of Accu-Sort Systems, Inc.)*

code presentation to the scanner because the carton's two edges provide a guide for the two label edges. Also when the carton is placed onto a conveyor system, the label is in one position.

Whenever possible, use a label window on the SKU. If your operation uses a fixed-position fixed-beam scanner and the bar orientation on the carton is the picket fence or ladder, then a label window on the carton increases the number of reads. This increase is due to that fact that the label window ensures proper bar-code label presentation to the scanner.

Ensuring the Greatest Number of Reads.
To ensure the maximum number of reads from a fixed-beam or moving-beam scanning device, follow these guidelines: (1) When you scan without a check digit, then a partial scan of a label by a variable-length programmed scanner might produce an error that creates a missort. The bar-code scanning device encodes all numbers. (2) The bar code permits several scans, at least four to eight, which is redundant before the code is validated. (3) The moving-beam scanner operates by detecting variations in contrast between code black marks and the label background white color or carton surface. (4) The difference in light reflected from the bars and white spaces of the label is used to distinguish the code elements. This means that the contrast between the black bars and white spaces is very critical to obtain a high scan rate. (5) When the bar length equals the width of all the black bars, then the bar code approaches, an over squared label improves the number of reads of the bar code.

MECHANICAL AND ELECTRICAL COMPONENTS

The final components of the product identification system consist of the mechanical (moving) equipment and the electrical equipment.

The mechanical equipment is the object on a conveying surface or rack surface. The electrical equipment of a system consists of relays, solenoids, programmable controllers, microprocessor, minicomputer, magnetic tapes, video displays, horns, whistles, start-stop controls, emergency stops, and mimic panel displays.

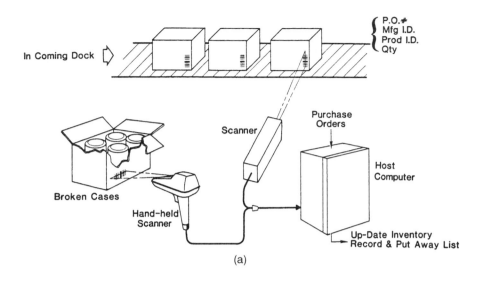

(a)

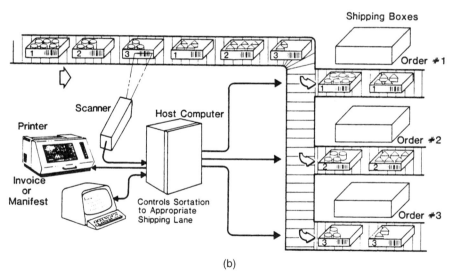

(b)

FIGURE 13.10 (*a*) Ladder bar-code orientation and (*b*) picket fence bar-code orientation. (*Courtesy of Accu-Scan.*)

Bar-Code Design Parameters

When you consider a product identification system for implementation in your warehouse operation, then prior to equipment selection you must clarity a number of design factors. (1) Choose the operational procedures and economics of each identification system. (2) Decide on label location and placement on the object. The alternatives are top, front, side, and bottom. (3) Choose a code size (the number of characters influences the code size and bar-code selection). (4) Select a conveying speed and the bar-code travel speed past the scanner. (5) Select the depth of field (space between the scanner and bar code on smallest object). (6) Choose the field of view (scanning the space between the two bar codes on a conveying surface). (7) With fixed-beam scanning devices, select pitch of bar code to the beam or Y axis, skew of the bar code to the beam or Y axis, and tilt of the bar code to the X axis and to the direction of travel. (8) Test the light in the area and your label that is printed on your selected stock and by your selected printer.

Verifying that Bar Code Meets Standards

In most warehouse bar-code applications, the bar-code reading requirements are very high, and your bar code is printed from a standard that was established many months in the past. If your bar code is poor-quality, then your scanners have a low read rate, and this reduces your material handling system's ability to handle the business.

A poor-quality bar code has a problem with one or several of these characteristics: (1) dimensional tolerances (width for each black bar and white space, width between two black bars, and overall width for all the black bars and white spaces); (2) spots or accidental black spots in the white spaces; (3) voids that are white spaces, no ink, within a black bar; (4) rough edges of a black bar which has the black edge not straight; and (5) reflectiveness and contrast, which is the difference in the shades of color in the black bar to the white space or the darkness of the black and brightness of the white space.

Bar-Code Verification Methods

After you established the minimum bar-code standard, how do you verify that the printed bar-code dimension and characteristics meet your standards? Use a scanning device or a verifier machine to ensure that the printed bar-code widths and lengths meet your standards.

If you use a scanning device on the material handling system and do not adhere to your bar-code printing and reading standards, then you can increase the number of no-read or misread conditions. This situation increases the cost of doing business and decreases the volume handled by the operation.

The two types of verifier devices are the circular aperture and rectangular aperture. These are manually or automatically controlled devices.

Other practices to ensure good-quality printed bar codes require maintaining the paper and ink that established your standard and the ribbon change standards set by the printer manufacturer.

CHAPTER 14

DECREASING PRODUCT FLOW TIME AND HANDLING COSTS WITH AN ACROSS-THE-DOCK OPERATION

This chapter reviews the warehouse function of handling flow-through merchandise. Included in the chapter is a review of various single-item, carton, and unit-load (pallet) methods and layouts.

In the retail store distribution industry, across-the-dock operations are considered one of a company's steps to establish a *just-in-time* (JIT) replenishment or *quick-response* (QR) program.

In the past, to handle flow-through product, retailers have used a private trucking company's break-bulk terminals or their own break-bulk facility. These facilities were designed to receive a large quantity of one product (SKU) that was separated by each customer location and transferred to the appropriate customer shipping dock location. Later, the product was loaded onto the customer's delivery truck for transportation to the customer's location.

OBJECTIVE OF AN ACROSS-THE-DOCK OPERATION

Using an across-the-dock operation for a retail distribution company is just one of its methods to reduce the number of days required for product to travel from the vendor to the retail store shelf (channel of distribution). When a retail distribution company uses a conventional retail distribution facility and stores delivery trucks to handle across-the-dock product, then an across-the-dock operation costs the least.

The future for retail store distribution lies in organizing an across-the-dock operation to handle single items, cartons, or pallet loads of product. In an across-the-dock operation, the vast majority of the product is handled by several warehouse functions, shown in Fig. 14.1.

Warehouse function	Product handling characteristics		
	Single item or hanging garment	Carton	Pallet
Receiving	X	X	X
Sort (distribution)	X	X	X
Ticketing and packing	X	X	
Transfer	X	X	X
Loading and shipping	X	X	X

FIGURE 14.1   Several warehouse functions.

WHAT IS REQUIRED

To maximize return on investment and provide the best service to retail store customers, with an across-the-dock operation you develop (1) quality and quantity standards, (2) a paper flow system, (3) a product flow system, and (4) the physical layout of material handling equipment.

When the retail store distribution company develops an across-the-dock operation, it combines its across-the-dock operation with its conventional retail store distribution company operation. This distribution concept increases the return on investment and service to customers and requires the vendor (manufacturer) to increase his or her involvement in the retail company distribution system. These areas of vendor involvement concern (1) high-quality product and exact quantities, (2) smaller product quantities per individual store requirement and assurance that each package has the appropriate retail store identification, (3) improved product identification (automatic identification) that is used by the company and price tickets on the SKUs, and (4) conveyable packages.

FROM YOUR COMPANY, WHO IS INVOLVED?

As your retail distribution facility becomes involved in an across-the-dock program, the warehouse operation establishes good communications and becomes the center of communications between the following groups: purchasing department, vendors, warehouse and transportation departments, and retail stores or customers.

After your retail store distribution company has decided to undertake an across-the-dock operation, there are two alternative strategies. The first strategy is to have a pass-through operation that handles a pallet load or carton products as pallet loads or cartons. The second strategy is to handle large SKU quantities that are broken down into single items or cartons. These items are ticketed and distributed in smaller individual quantities to the retail stores.

In an across-the-dock (pass-through) carton and pallet program, the purchasing department is required to send to the vendor the total product quantity required for the sales program. If the company has multiple distribution facilities, then the purchase order is for the entire company's sales program, but is separated into required product subtotals for each distribution facility. Each distribution facility is responsible for separating the product by the quantity that accommodates each retail store's projected sales requirements.

THE VENDOR'S RESPONSIBILITY

The vendor's responsibility is divided into two parts. The first part is to provide quality product in the exact quantities and to clearly identify the product on its exterior. This requires either the vendor to mark the product description on the carton exterior or, at the receiving dock, an employee on demand to print product and store identification labels which are placed on the carton's exterior.

The second part of the vendor's responsibility is for each package of the total purchase order to identify each individual retail store number (address) on the package exterior. This alternative does not require the employee at the receiving dock to mark or label the carton, activity which reduces product flow and increases expenses.

Three methods are used to identify the package exterior: human-printed human-readable label, machine-printed human- and machine-readable label, and machine-printed human- and machine-readable label with price tags.

Human-Readable Labels

The minimum exterior carton markings by the vendor include the description of the enclosed product: size, color, and SKU description. With this product marking method, the receiving department verifies the SKU destination and quantity and then physically marks the appropriate retail store address on the exterior of each carton. This store address is a crayon marking or printed on demand onto a human- and machine-readable label that is applied to the exterior of each carton. After this identification process, the receiving employee transfers the carton to the manual or mechanized sortation concept for transport or transfer to the appropriate customer loading dock area.

The disadvantages of this method are that it increases errors; employee productivity is low; it requires 1 to 2 days to handle the product, a larger dock area and number of dock doors, and the largest number of employees; it reduces the payback of a mechanized (conveyor or sortation) system; some markings are not clear; it handles a low volume; and it increases employee injuries.

The advantages are that there is a very low product cost savings from the vendor not marking the cartons and that it handles all product types.

Human- and Machine-Readable Identification

If your product is identified with a machine-printed human- and machine-readable label (Fig. 14.2) that includes the retail store number, then your operation uses a manual or mechanized receiving and manual or mechanized sortation concept. These methods verify that the product was received and transferred to the appropriate customer loading area.

In the manual system, employees physically mark and move product as described

THRU:

To:

0-1008-055927-0

1008　DATE　LABEL#　000000　VOID
　　　 04/26　　3L
CANNON　　　M/F 04/13 11145

M 7006

FIGURE 14.2　Human-readable identification.

in previous systems. In this alternative system an employee uses a hand-held scanning device to verify receipt of the product by scanning the machine-readable bar-code label. After this process, the carton is transferred by the employee onto a conveyor system for transport or transfer to the retail store's appropriate staging area.

The disadvantages of the system are that employee productivity is low, 1 to 2 days are needed to handle product, a larger dock area and a number of dock doors are needed, it requires the largest number of employees, it increases injuries, and there is the increased investment in bar-code scanners.

The advantages are that it handles a medium volume, handles all SKUs, provides accurate information, has readable markings, and is used on a mechanized (conveyor) system.

With a machine-printed human-readable label that includes the appropriate retail store identification (customer) number on a conveyable carton's exterior, the operation uses a conveyorized sortation system.

The carton with the bar-code label in the proper orientation of travel is placed onto the conveyor in-feed system that transports the carton past an automatic scanning device. The automatic scanning device reads the bar-code label and sends the information to the host computer for receiving department recordkeeping and receipt preparation and to the conveyor system microprocessor and tracking device. As the carton travels on the conveyor system, the microprocessor or tracking device tracks the carton and, when required, activates the appropriate divert device to transfer the carton from the sortation conveyor onto the appropriate customer's loading conveyor.

It is clearly understood that the nonconveyable product is handled by the manual hand-held label scanning system and manual transfer.

The disadvantages are that it requires a nonconveyable system, a higher capital investment, and a human- and machine-readable label.

The advantages are that it handles a high volume, takes 1 day or less to move product through the facility, reduces the dock area and number of docks, requires the fewest number of employees, and handles surges and peaks.

When the vendor supplies your warehouse with conveyable cartons that have the appropriate individual retail store (customer) number as a machine-printed human- and machine-readable identification label that has price tags for the individual SKUs inside the carton, then your retail distribution company maximizes its return on investment. This return is realized from the distribution facility's across-the-dock labor and space savings and the retail store's price-marking labor and material expense savings.

Compared to the previous system, the additional disadvantages are the slight increase in vendor cost and possibility of damaged price tickets.

The additional advantages include less expense (material, equipment, labor) at the retail store level, companywide uniform printed price tags, improved store productivity, and a uniform price tag location on all merchandise in the entire market area.

MATERIAL HANDLING SYSTEMS AND PRODUCT TYPES

Carton or Pallet Load SKUs

A carton or pallet load (pass-through) across-the-dock warehouse facility's receiving, unloading, and sortation systems are based upon the handling unit (SKU). The material handling systems used at a facility are shown in Fig. 14.3.

The description of operations, systems layouts, and advantages and disadvantages of these systems were reviewed in the following chapters: Receiving, Chap. 4; Sortation and Distribution, Chaps. 6, 7, 8; Staging and Loading, Chap. 4.

RECEIVING AND SHIPPING AREAS ARE MOST IMPORTANT

When the distribution facility becomes involved in a pass-through (across-the-dock) operation, the receiving and shipping functions become more important than the stor-

| | Handling unit, SKU | |
Warehouse function and system	Carton	Pallet
Receiving		
Nonpowered conveyor	X	—
Powered extendable conveyor		
Carton	X	—
Pallet and unit load	—	X
Manual powered equipment with stacking frames	X	X
Sortation and distribution		
Manual (walk)	X	—
Conveyor (package, tilt tray, tilt slat, SBIR, gull wing)	X	—
Mobile unit-load handling equipment		
AGV	X	X
Towline	X	X
Cartrac	X	X
Staging and loading, temporary hold and direct load		
Manual	X	—
Gravity skatewheel and roller	X	X
Powered extendable conveyor	X	X
Powered mobile pallet handling equipment	—	X

FIGURE 14.3   Material handling systems used at a facility.

age and picking warehouse functions. With new emphasis on the flow of product, there is an increase in the number of docks and the dock staging areas which have a larger area. The increased number of docks permits the facility to handle a large number of trucks. The larger dock areas permit the facility to unload, sort, stage, and load a greater volume of product through the warehouse.

With warehouse emphasis on product flow, the new across-the-dock facility is designed with separate receiving and shipping docks. The other building design features include a rectangular building (Fig. 14.4) that has a short distance between the receiving and shipping areas. These areas require a low ceiling height.

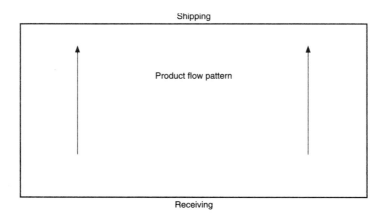

FIGURE 14.4   Rectangular building.

In a carton across-the-dock operation, a material handling system transports the product (pallets and cartons) from the receiving area, sorts the product, and moves it to the appropriate customer shipping area. This system can be manual or mechanized.

If the receiving department handles cartons and pallets and the receiving function is an across-the-dock operation, then the receiving department's objective is to unload the delivery truck on schedule and to verify that the total delivered-product quantity matches the total purchase-order quantity. The detail objective of a carton operation verifies that the actual individual store carton quantity matches the purchase order quantity. If there is not a match, then the receiving and purchasing departments make the necessary individual store carton quantity adjustments to ensure that each individual store receives the appropriate carton quantity.

VARIOUS CARTON DISTRIBUTION SYSTEMS

The various distribution concepts are manual, package conveyor, SBIR, tilt slat, tilt tray, and gull wing.

With dual induction on a SBIR, tilt tray, tilt slat, or gull wing sortation system in a loop design, it handles the sortation of loose, single items on one side (straight-run) and shipping carton sortation on the other (dock) side (straight-run).

Single-Item Across-the-Dock System

If your across-the-dock operation handles single items (flatwear, hanging garments, or other items) that are separated by individual store location or are price-ticketed and packed by employees, then you require processing (ticketing and packing) areas. These areas are added to the facility. These warehouse functions (activities) are located between the unloading/receiving area and the shipping/loading area.

Single-Item Sequence of Operations. In a single-item ticketing, order-pick (distribution), and packing across-the-dock warehouse facility, the receiving department separates each delivered quantity into SKU (style), color, and size. This product separation by style, color, and size ensures that the ticketing and order-pick (distribution) activities are efficiently performed by employees.

After the product in cartons is unloaded from the delivery truck, the cartons are placed onto the assigned opening lanes. If there is no space available, then the product is placed into temporary storage positions. These storage positions are floor-stacked pallet boards, stacking frames, storage racks, or carton conveyor lanes.

When the required opening lanes become available (and as required by the operation), this product is transferred from the temporary storage position to the assigned vacant opening lane. The two types of opening systems are the tabletop and the gravity flow lane.

SKU Ticketing. All preticketed product flows directly to the order pick (distribution) area. Prior to arrival at the pick (distribution) area, all nonticketed product receives a ticket.

Tabletop Ticketing. The tabletop system consists of a 6-ft by 30-in flat-surface table that has a full pallet of cartons placed on the right end. As required, the employee places one carton on the table and removes the SKUs from the carton. During this process, the employee verifies the SKU quantity and attaches a price ticket to each SKU. The SKUs are returned to the carton which is placed on an empty pallet board on the left side of the worktable. The full pallet board is transferred to the order pick (distribution) area. Other transport devices are carts, roller and skatewheel conveyor, and overhead trolleys.

The disadvantages of this system are that it handles a low volume, employee productivity is low, and there is increased product handling. The advantages are the low investment and need for little employee training.

Gravity Flow Conveyor Ticketing. The second processing method consists of a series of gravity flow conveyors, powered zero-pressure accumulation and powered belt conveyors, conveyable containers, and work shelves and platforms.

The cartons are placed onto the opening lanes in an arrangement that has one SKU (size, color, and style) occupy one lane. A lane is a gravity flow conveyor that permits the carton to flow from the receiving and temporary-hold area to the opening positions. On the opening side of the lane is a fixed end stop. If there is more than one item per pallet load (which means there are only one or two cartons per SKU on the purchase order), then the SKU arrangement is by color and style with the largest size on the flow lane's right-hand side. The right side is the right side of the receiving person who faces the conveyor lane(s).

The opening employee removes the product from the carton, counts the merchandise, and places the merchandise in a tote (container). The merchandise is placed in the tote in an arrangement that has the merchandise ticketing location in the tote's direction of travel. The opening employee enters all merchandise counts onto a tally sheet. The empty carton and other trash material are thrown into the trash take-away system. The merchandise tickets for the entire SKUs are placed in the first tote. A merchandise identification tag is attached on the lead end of the lead (first) tote. The totes, large size first, are transferred to a second set of gravity conveyor lanes that have adjustable end stops. These lanes are the opening take-away conveyor lanes. When the entire purchase order is opened, counted, and placed in totes and the totes are placed on the lanes, then the lane release employee releases the totes from the gravity lane to the powered conveyor system for transport to the ticketing lane.

As the totes arrive in the ticketing area, the lane control employee directs the lead tote with the identification tag and price tickets into one of the pricing conveyor lanes. The totes are allowed to flow into the appropriate flow lane until the next tote with a lead identification tag and price tickets arrives in the ticketing in-feed area.

The totes arrive at the pricing station end of the price ticketing gravity conveyor lane, and the merchandise in the totes has the pricing location facing the pricing employee. The pricing (ticketing) employee removes the price tickets from the lead tote and leaves the merchandise in the tote, tickets each individual item, and leaves the merchandise identification tag on the lead tote. After all the merchandise is ticketed, the tote is released to the distribution take-away conveyor lane. At the end of the lane is a tote end stop.

When all the styles, sizes, and colors of merchandise are ticketed, then the lane release employee releases the full totes from the ticket lanes to the merchandise distribution lanes. As the totes arrive at the distribution gravity lanes, the lane control employee directs the totes into one of the vacant gravity lanes. As required, additional adjacent lanes are used until the lane is full or the next lead tote with identification tag arrives at the transfer station.

The disadvantages of the gravity flow method are that the capital investment is high, it requires employee training, and it requires management control and discipline. The advantages are that it handles a high volume, there is work position flexibility, and it permits queues of the product.

MERCHANDISE ORDER-PICK METHODS

When the tote arrives at the distribution area, the processing/ticketing department has completed its activities. The next activity is the individual store order pick (sortation or distribution).

The heart of an across-the-dock operation is the merchandise order pick (sortation and distribution) method. The merchandise order pick (distribution or across-the-dock operation) activity ensures that the customer receives the ordered merchandise quantity. The various methods include the manual system, waterfall and store dump, carousel, rapid pack hotel or continuous, indexing power and free conveyor, S. I. Cartrac, Bomb bay drop, SBIR, flap sorter, and tilt tray.

Manual System

The first distribution method is a manual system that has a shipping carton for each store located on the floor. As required, the employee physically carries the merchandise from the distribution area and places the customer's required merchandise quantity into the appropriate shipping container.

The disadvantages of the concept are that it handles a low volume and a low number of customers, employee productivity is low, and it requires the greatest number of employees and a large dock area. The advantages are that it requires low capital investment and little employee training.

Store Dump

The second across-the-dock operation distribution method is the store dump (Fig. 14.5a). The employee's pick instructions are in the reverse arrangement of the packing area. As required by the customer order, the distribution employee removes the required quantity from the tote in the distribution lane and places the merchandise into a multishelf trolley basket, cart, or tote. Each shelf holds one customer order quantity by size for a specific style and color that appear on a distribution document. After completion of the store picks, the pick document is placed on top of the merchandise, and the next order is filled by the employee.

The store dump method entails a full shipping container take-away conveyor system, an employee aisle, an empty shipping container replenishment system, and a double-shelf arrangement. This shelf arrangement holds one shipping container per shelf. Each shipping container is identified with the store number. A staging area is provided prior to each store-dump aisle. The transport device is moved from the staging area into the distribution area in an employee aisle located between the shelf lanes and the conveyor path. Walking in the aisle, the employee looks at the pick instructions, removes the appropriate merchandise, and transfers the merchandise from the transport device to the store's shipping container. For good employee productivity, the high-volume stores are located at the end of the store packing line. With several crossover aisles, this layout permits empty carts, trolleys, and containers to be removed from the distribution area and merchandise to flow through the system with minimal queues.

Empty shipping containers are supplied to the packing area by an overhead conveyor system or an employee with a two-wheel hand truck. All full shipping cartons are pushed forward from the packing shelf onto a take-away conveyor for transport to the sealing and manifest station.

The disadvantages of the system are that it increases employee walking, it is difficult to maintain productivity, and it requires management control and discipline. The advantages are that it handles a medium volume and a medium number of stores and it reduces the need for employees to lift cartons.

Carousel and Indexing Power and Free Conveyor

The carousel or indexing inverted power and free conveyor systems (Fig. 14.5b) are similar. The carousel system consists of a merchandise container in-feed conveyor, an empty-carton conveyor, and three to four horizontal carousel units. Each group of carousel units has a visual display or paper distribution document pick instruction from in the merchandise container and a shipping take-away conveyor. The inverted power and free conveyor system consists of container-carrying platforms that revolve in a closed-loop path past the distribution stations.

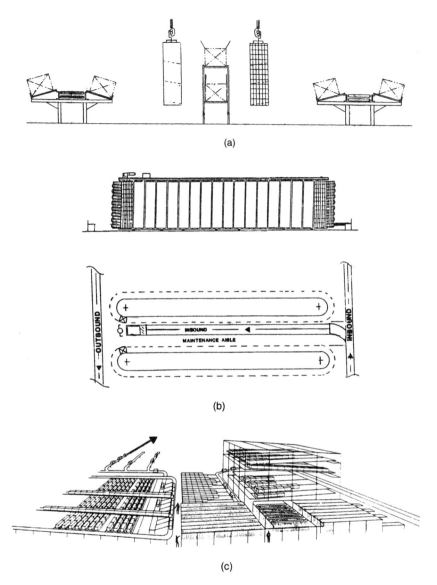

FIGURE 14.5   (*a*) Store-dump method (*Courtesy of Sedlak Management Consultants.*); (*b*) carousel; and (*c*) hotel (rapid pack) method (*Courtesy of SDI.*).

Full totes and cartons of ticketed merchandise travel on an inbound conveyor to a distribution station. At the distribution station the totes are stopped, and the tote label that identifies the merchandise is read by a scanning device. The scanning device sends the information to the host computer. At the packing station, an appropriate scanning device reads the store identification label on the carousel or power or free conveyor carton. This information is sent to the host computer. The host computer with these two pieces of information indicates to the packing employees on a visual display terminal the merchandise requirement for each store. The paper pick instruction form is an alternative in these systems.

After the employee removes the merchandise from the inbound tote, the employee transfers the merchandise into the store carton in the carousel or in the carton on the power or free platform. When the employee completes the required transfer of merchandise for each store at this location, the employee moves the carousel or power or free conveyor index to the next store position. The distribution activity is repeated for the new group of stores. With several carousels or power or free units facing one workstation, the machine-paced work activity is very dynamic with high employee productivity in a small area.

When all the merchandise is depleted from the inbound container, the employee advances a new full container with additional merchandise to complete the other retail store orders or repeats the activity for new merchandise.

All full totes are released onto a take-away conveyor system. Empty shipping containers are supplied by a separate conveyor system.

The disadvantages of the system are that the capital investment is high and it requires two to four carousel and power and free units.

The advantages are that it improves employee productivity, handles a high volume, is paperless or paper picking, handles a large number of stores, reduces errors, reduces employee walking, requires the fewest number of employees, and permits bulk order picking.

Hotel (Rapid Pack) Method

The next distribution method is the hotel (rapid pack) method with two alternative designs (Fig. 14.5c). The designs are different because in one all the customer's containers travel continuously past the pick stations and in the other a slug of customer containers travels past the pick stations.

In the first design, all customer shipping containers continuously travel past pick stations on a closed-loop conveyor. In the second design, a specific group of customers (slug of shipping containers) from a temporary holding area travels past the pick station and returns to the temporary holding area.

In these systems, the pick conveyor transports the shipping container past a series of gravity flow lanes that contain merchandise. As the shipping container arrives at the pick area (specific zone) and prior to the beginning of the pick area, a scanning device reads the customer label on the container and sends the label information to the host computer. The host computer sends the customer SKU quantity for each lane in the zone to a controller that lights each lane indicator (pick) light with the required customer quantity. The distribution employee is directed by each lane's digital display terminal (which was activated by the scanning device and host computer). An alternative pick instruction form is a distribution document. Both methods provide the distribution employee with the merchandise quantity that is picked from the gravity flow lane and transferred to the tote. All full shipping boxes are properly identified and pushed onto a shipping take-away conveyor. From the empty shipping box (K D carton) all empty shipping boxes are transferred to the pick position from the shipping box (K D carton) in-feed system.

The disadvantages of the method include the high capital investment and increased management control and discipline.

The advantages are that it handles a high volume and a large number of customers, it provides high employee productivity, it reduces errors, it reduces lifting injuries, it requires the fewest employees, and only a small area is needed.

Various Methods That Require Induction

Mechanized distribution systems were reviewed in Chap. 6 on small-item order pick and sortation. These systems are the tilt tray, flap sorter, S. I. Cartrac, SBIR, gull wing, and Bombay drop.

What Happens to Residual Across-the-Dock Merchandise

With an across-the-dock distribution system, all extra (undistributed) merchandise is transferred to the storage area or active pick area. In the future, as the customers require restocking of merchandise, the merchandise is picked en masse and sent to the across-the-dock distribution system. This merchandise is handled as a typical across-the dock SKU.

PACKING VERIFICATION

In an across-the-dock operation, the packing operation verifies that all the required merchandise has arrived at the packing station or loading dock. The systems used to determine product quantity are reviewed in Chap. 6. These systems are random check, 100 percent check, manual scanning, and bar-code scanning.

Sealing of Across-the-Dock Cartons. When small items are packed into a shipping container, the cardboard or plastic container is sealed to protect the merchandise from damage or being lost. These methods are reviewed in Chap. 6: gummed tape, self-adhesive tape, strapping, and plastic seals.

Addressing Across-the-Dock Cartons. During the packing process, the final step is the placing of the ship-to address on the carton or plastic container. The various methods reviewed in Chap. 6 are the handwritten, the ink stamp, and the machine-printed self-adhesive or non-self-adhesive label applied with a glue pot or clear tape.

Loading Across-the-Dock Cartons onto a Delivery Vehicle. After the packing employee seals and weighs the shipping container (box), the customer's package is sent to a temporary holding area or is placed on a delivery truck. As the shipping box leaves the packing area, the box passes a manifest station that verifies that the box has left the facility, records the customer identification number, and verifies that the box was placed on a delivery vehicle. Methods of loading the shipping vehicle are reviewed in Chap. 6: manual, gravity skatewheel or roller conveyor, powered extendable conveyor, and powered pallet truck.

HANDLING HANGING GARMENTS IN AN ACROSS-THE-DOCK OPERATION

The next merchandise that is handled at a retail across-the-dock facility is hanging garments and garments in boxes. In an operation involving hanging garments and garments in boxes, these warehouse activities are performed: Hanging garments are received on rolling racks or a trolley on a cart or with an extendable boom, and garments in boxes are received on skatewheel and/or roller conveyors, powered extendable or manual or powered pallet handling equipment. Then garments are sorted and ticketed. Next is store sortation (distribution) on a rail loop, rapid pack, or Promech system. Finally shipping is done by rope, cart, or box.

There are two ways to receive hanging garments at the across-the-dock facility: the garments are on hangers from a rope delivery truck or the garments in boxes (flat) are removed from the box and placed on hangers.

Unloading of Hanging Garments

There are three ways to unload hanging garments at a distribution facility. First, manually pushed rolling racks can be used. These racks are pushed from the delivery truck through the warehouse to the sorting area. This method handles a low volume, requires a low capital investment, and is used in a facility without a rail system. Second, a trolley is hung onto a trolley cart. In the warehouse, the trolley is discharged from the cart onto the trolley rail system. This method is used for an operation that has a rail system, handles a low to medium volume, and represents a low capital investment. Third, the trolley is hung on an extendable trolley boom so that the trolleys merge onto the rail system. This method is used for a facility that has a rail system, handles a high volume, and requires a medium investment.

Handling Garments in Boxes

When garments in boxes are received at the distribution facility, there are three ways to unload the cartons. First, the gravity skatewheel and roller conveyor is used for dock unitizing onto pallets or carts or into a stacking frame. This method is used for a low-volume operation, requires a low investment cost, and handles a floor-stacked delivery truck. Second, the powered conveyor is similar to the gravity skatewheel and roller conveyor except that it handles a high volume and has a higher investment cost. The cartons are transported to the warehouse dock area for unitizing on a pallet or cart. Third, manual or powered pallet handling equipment is used to unload palletized loads. This method handles a high volume and requires a medium cost.

Unloading Objective

When hanging garments or garments in boxes are unloaded, the objective is to unload the delivery vehicle and to obtain a total piece carton count. For efficient product handling at the other hanging-garment processing stations, the hanging garments are separated onto trolleys or unit loads by style, color, and size. This requires that the garments be transferred to a garment open, count sort, ticket, and distribution area.

Hanging-Garment Sortation

The hanging-garment sortation (distribution) methods of (1) rail loop, (2) rapid pack, and (3) Promech were reviewed in the hanging section of Chap. 6.

Hanging-Garment Shipping

Hanging garments are shipped to your customer by one of the following:

Shipping method	Delivery method
Rope	Company fleet
Cart	Company fleet
Box	Public or company fleet

CHAPTER 15
FACILITY SITE SELECTION

This chapter reviews important geographic and demographic site selection factors, center-of-gravity site selection method, systematic site selection method, site and facility size determination methods, centralized (single) distribution or decentralized (multiple) distribution location strategy, and existing facility factors.

The underlying premise of this chapter is that the entire process undertaken to locate a new site or facility for a warehouse and distribution operation must be designed with a single company objective: to find a location for the day-to-day operations of a warehouse and distribution facility that operates at the lowest possible cost and provides the best service to customers. The site selection process is a complex, time-consuming project that is a small expense compared to the millions of dollars spent on capital expenditures to build and equip the new facility and the associated annual facility operating costs. The project team evaluates prospective sites in relationship to transportation, labor and land costs, the availability and costs of energy and utilities, taxes, and many noneconomic factors.

CENTRALIZED OR DECENTRALIZED OPERATIONS

To achieve a balance of the economic investment, lowest operating costs, and best customer service, the first step of the site selection strategy is to determine whether the company requires one facility (centralized operation) or at least an additional facility (decentralized operation). This company distribution strategy is finalized prior to, or is the first step in, the warehouse site selection process. This is evident for any size company in any industry group.

The decision to have a centralized operation or a decentralized operation is based on two factors: the impact of the choice on the company's net profits and the effect on the company's ability to provide the best service to its customers.

Centralized

In the centralized system (Fig. 15.1a), one facility provides service to the customer group. This theory holds that one distribution facility has a more effective and efficient allocation of warehouse space, labor, material handling equipment, transportation fleet and transportation services, purchase of inventory, and control of inventories and costs to earn a profit and provide customers with the best service.

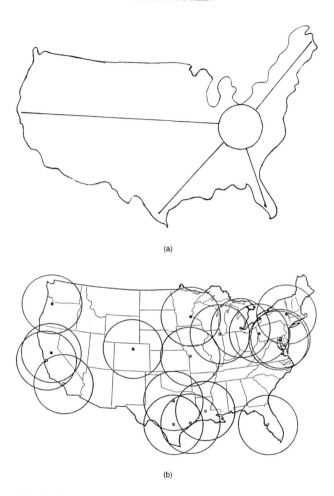

FIGURE 15.1   (*a*) Centralized theory; (*b*) decentralized theory.

Decentralized

The decentralized theory (Fig. 15.1*b*) maintains that multiple distribution facilities provide the best service to a customer group. This theory holds that your company's large volume is not effectively and efficiently handled at one large facility or building or you cannot expand the existing building to handle the volume. This latter situation is referred to as being *landlocked* and requires your company to become decentralized or to build a larger facility on a new site. When a warehouse operation become decentralized, there are several satellite facilities with material handling equipment, labor, transportation fleet, and inventory to service the customer group at a profit in a specific geographic area.

FACTORS IN THE DECISION

When company business has grown to such a volume that a decision is required about a centralized or decentralized operation, several operational cost factors influence the decision: labor availability; material handling equipment (systems) capabilities; building capacity; inventory volume; transportation distances; taxes, utilities, and other operating costs; and delivery time (life of the SKU).

Assign a Dollar Value to Each Factor and Rank the Five Factors.   The second step in the site selection process is to assign a dollar value and determine the estimated costs for each site selection factor. The major site selection factors are transportation, labor, taxes, land, and energy or utility. These five factors are reviewed in more detail later in this chapter.

NONECONOMIC (SECONDARY) SITE SELECTION FACTORS

These noneconomic (secondary) site selection factors do not have a tangible dollar value but can have an effect on how the new operation functions. To assist in the evaluation of various sites, these factors are ranked with a quantitative value which permits comparison of the potential sites:

1. Transportation factors, vehicle restrictions, rail facilities (companies, spur, piggyback operation location), access to good roads, access to waterway (rivers or harbors), access to the main highway, adjacency to the main highway, location relative to a major airport, location relative to a free-trade zone, location of truck terminal (FX, UPS, USPS, and common carrier), and location of existing and future customers and vendors

2. Land factors; current and proposed zoning restrictions and requirements; insurance and local fire requirements and building codes (UBC or SBC); annexation potential; pollution laws (noise and odor); shape and size of land; shadow laws; setback, berm, and green-area requirements; topography of the land to show grade and slope, including highest and lowest points, easements, and rights-of-way; legal description; total acreage; past use of existing buildings and condition; high-water marks (100-year flood level); zoning regulations for parking requirements (office and warehouse) and occupancy allowed in the area; average number of rainy days (maximum and minimum, including the month), average degree-days (coldest and hottest, including the month); expansion capability; and type of business neighbors (product and union)

3. Tax and incentive factors, insurance rates, inventory taxes, property taxes, relocation incentives (local and state), employee training programs, employee workers' compensation, and free-trade zone status

4. Labor factors, availability of labor, educational level, medium income, size of family, number of workdays lost to work stoppages, number of holidays per year, right-to-work state, percentages of union and nonunion employees, percentages of males and females, willingness to travel to work, required amenities, attitude of community toward the business, availability of public transportation, availability of good restaurants and hotels, availability of gasoline and diesel fuel, availability

and cost of housing, community resources (architects, engineers, construction companies), material handling equipment vendors, and availability of community services (fire, EMS, hospital, trauma, and police)

5. Utilities and energy factors, name of gas company and size of pipe, name of electric company and size of service, name of water and sewer company and size of pipe, size of water reservoir, number of thunderstorms, loss of electric power, and telephone company

6. Quality of life, type of crime and crime rates, number of schools and colleges, number of hospitals and doctors, recreational activities, churches, hotels and motels, cultural activities, major shopping centers, and news coverage

YOUR BUSINESS AND ITS RELATIONSHIP TO THE SITE SELECTION FACTORS

The nature of your company's business determines the ranking (importance) assigned to each site selection factor. The importance assigned to a factor is determined by its effect on the business operation, projected annual operating expenses, and customer service. The following examples show the importance of a site selection factor to a company's business (industry). (1) A catalog and direct-mail distribution operation would not assign high importance to a location close (150 to 200 miles) to the customer's location or to a short delivery truck travel time to the customer's location. (2) A labor-intensive operation assigns high importance to the availability and quality of labor. (3) A food or retail department store distribution facility considers the ability to provide one-day delivery service (150 to 200 miles) to the retail store's customers as a very important factor. (4) A nonunion operation considers the right-to-work state (union-free) as a high-value factor.

SITE SELECTION PROJECT TYPES

There are three basic types of site selection project: (1) international (foreign country), (2) macro (national, state, region, or major city), and (3) micro (within a state, region, or major city).

International Site

In an international site selection project, you determine the best foreign country for your company's new operation. The international project includes the five major economic site selection factors (transportation, labor, land, energy and utilities, and taxes and incentives) and many unique site selection factors. The unique international site selection factors are (1) value of the U.S. dollar and the host country's currency, (2) stability of the host (foreign) government, (3) stability of the host (foreign) currency, (4) ability to take profits out of the host country and availability of barter agreements, (5) population attitude toward a foreign company, (6) government attitude toward a foreign company, (7) import and export regulations, (8) availability of required material handling equipment, (9) free trade or most-favored-nation status with the United States, and (10) culture and customs of the host country.

Macro Site

The macro (national, state, regional, or major city) site selection project reviews the quality of the factors for a region of the country or a particular state or city.

In a site selection study, divide the country into well-defined regions (areas). An example of these regions in the United States is the following eight sections:

Northeast	Southwest
Midatlantic	Southeast
South atlantic	West
Midcentral	Midwest

Transportation Cost.   In a particular state or major city site selection project, identify the various metropolitan cities of interest to your company. To determine the transportation costs for each major city, make a complete list of cities that you consider as potential site locations. For each potential city, list the miles or delivery days between each potential site location (city) and each major city that has a company customer. This information is obtained from atlases specifying the miles between two cities or between two postal Zip codes. Or measure the mileage from a scale on a map. Then project the company's delivery volume to each metropolitan city. Now calculate a delivery cost per average customer order from each potential metropolitan city. Last, for each potential city, calculate the delivery cost based on the projected volume from each potential city to each major city with a customer.

Census Information.   Next, obtain and review the latest census track information for each proposed major city. U.S. census information is available at your local public library. This census information is demographic (population characteristics) and is in a consistent format for each major area in the country. This feature permits one major city to be compared with another.

Local Government and Business Development Agencies.   The next step is to contact the state and local economic development agencies and utility companies for population, geographic, and infrastructure information. The information supplied by these agencies includes highway networks, educational institutions, industrial neighbors, high and low temperatures (degree-days) and frequency of rain and snow, location and names of transportation companies, and locations of airports, railroads, and ports.

Professional Associations and Businesses in the Area.   The next step is to contact professional associations and companies that are doing business in the potential locations. This information provides you with the most recent business experiences of the businesses moving into the proposed area and the current business conditions there.

Labor Availability and Cost.   After you determine the quantity of labor that you require for the proposed warehouse operation, project the labor economics for each potential region and city. For each city, the availability of labor and an average hourly wage rate for each warehouse job classification can be obtained from *U.S. Census Track,* state and local development agencies and businesses and professional organizations in the area.

Taxes and Incentives.   During the preliminary macro project phase, from the various state and city development agencies you obtain for each potential state and city the following information: income tax, property tax, inventory tax, workers' compensation laws, franchise tax, and relocation incentives.

Based on your operational and economic projections, you estimate your company's tax burden. The state and local incentives are a reduction of the state and local tax burden for locating the business in that state or city.

Availability and Costs of Energy and Utilities.   As a portion of your macro regional study, contact the utility companies in the potential site area. These companies can supply you with sufficient information to determine along with your projected usage the following costs: electrical, water, sewer, telephone, and natural gas.

Building and Land Requirements and Costs.   During the macro site selection process, the site selection team finalizes the building and land criteria for the proposed operation. The relationship between the land and building components has two parts: (1) the size, shape, and quality of the land influence the building design and (2) the building design affects the land size.

In estimating the building size for a new facility, several major factors affect the building's size and shape: (1) inventory quantity (average and peak), (2) selected material handling equipment, (3) type of operation (storage and hold or across-the-dock), (4) type of business, (5) type of product handled and average and peak daily volumes, (6) inbound and outbound delivery methods, (7) number of warehouse and office employees per shift, (8) desired amenities, and (9) size of office and administrative areas.

The major factors in the land requirement are (1) building size, (2) number of inbound and outbound vehicles staged on site, (3) sprinkler and water-holding requirements, (4) desired expansion, (5) number of employee parking spaces, (6) local codes for green areas and setbacks, (7) shadow laws, (8) product flow through the facility and site, and (9) delivery methods.

From the local utility companies and commercial real estate companies in the potential site area, obtain the average cost per acre. This figure and your land requirement determine the estimated land cost.

The building cost is determined from a local architect or building construction company or from R. S. Mean's *Dodge Construction* estimate manuals. These provide you with an average cost per square foot for warehouse space, office space, mezzanine space, and landscape or hardscape.

If your business is in the catalog, direct-mail, or processing industries, then these site selection factors provide you with sufficient information to make a decision of the preferred major city. The preferred major city is the city that has the lowest total for the five site selection cost factors and ranks highest in the majority of the secondary site selection factor categories.

Micro Site

If your business is in the retail store distribution industry that services customers who are scattered or are in cluster group(s), then your project team continues with the site selection process to include a micro site selection project.

In the micro site selection project, the project team identifies the site that best ser-vices all the retail customers within a small geographic area. To select a preferred site, you project team is required to make exact projections for the variable site selection fac-tors. These factors include store delivery transportation cost and time and land costs.

The other site selection factors are considered fixed because the potential sites are within the same geographic area. These are labor, taxes, and utility and energy.

Your micro site selection project covers a county or is within a 150- to 200-mile radius of a major city. One prospective site is compared with alternative sites. During this site selection process, your project team determines the exact land size, building size, and the geographic location for the new operation.

To complete a micro site selection project, you require a more systematic approach to the project because the process provides exact information for each site and for each of the major variable economic factors (transportation and land costs).

After the macro project team has chosen the best region, state, or major city, then the team proceeds with a micro site selection project.

Four basic micro site selection methods can be used to determine the exact retail store delivery (transportation) costs and delivery times to all your customers (store locations); (1) random method, (2) serve-a-cluster-of-customers method, (3) center-of-gravity (demand pull or weighted-average) method, and (4) systematic site selection method.

Random Method.   The random site selection method is not based on specific cus-tomer order or delivery criteria or methodology. Basically, a random choice for a new facility is similar to a person throwing darts at a dart board. The dart board is a map of the micro site selection study geographic area. This result is totally by chance, and there could be no potential site in the selected area. The random method does not fine-tune the transportation costs for the alternative sites, but it is an easy, quick method.

Serve-a-Cluster-of-Customers Method.   In this site selection method, the project team chooses a site that is reasonably close to most of your customer (delivery) loca-tions. Ideally, the site chosen is in the center of the store (customer) cluster. This loca-tion reduces the delivery truck's traveling time and miles from the distribution center to various customer locations within the cluster.

Since there may be no available real estate for a site on this location, the method gives a theoretical site location. An actual review of available real estate (site loca-tions) may indicate that the available site does not serve your cluster of customers.

The method's disadvantages are that the transportation costs are not fine-tuned between the proposed sites and present and future customers and that the actual site location is subject to available real estate.

The advantages of the serve-a-cluster-of-customers method are that it is easy to use, it is quick, and it provides the geographic center of the cluster of customers (stores).

Center-of-Gravity (Weighted-Average or Demand Pull) Method.   The center-of-graity (weighted-average or demand pull) method is a systematic and detailed site selection method. This site selection method locates a grid square on a sheet of mylar that overlays a map. This grid square location optimizes the company transportation costs and service to a cluster of its customers and to the scattered group of customers. The center-of-gravity method indicates the best grid square or map location according to each retail store's frequency of delivery trucks that travel from a theoretical dis-tribution facility to each customer location. This site location is at the center of gravity and is ideal for the distribution center. The delivery truck transportation cost per trip is determined from the proposed site location to each of the customer's existing and future locations. Therefore, the project team concentrates the site location search in the area identified within the center-of-gravity grid.

The center-of-gravity site selection process requires the project team to (1) determine the number of delivery trucks needed to service each existing and future customer location, (2) develop a customer location map, (3) develop a grid sheet for an overlay of the map, (4) indicate for each grid square the frequency of delivery trucks and total number of deliveries, (5) total each column and line of the grid sheet, and (6) identify the grid square where both the highest line and column totals intersect.

The intersection of the grid line and column in that grid square on the map location is the center of gravity for the deliveries of product to all existing and future customer locations.

To determine the annual transportation costs for the proposed location, the project team estimates the number of miles between the grid square with the proposed site and each grid square midpoint that has a customer location. Each of these mileage figures is multiplied by the number of deliveries and the delivery cost per mile. The resulting figure is the annual transportation cost to deliver product to all present and future customer locations.

Suppose your company has a backhaul program. To assure top management that this center-of-gravity square does optimize the company's total operation, review the impact of this new site on the backhaul program. Identify vendor locations on a map. This overlay determines what grid square is in a cluster of vendor locations. To determine the projected backhaul income, do the mileage and backhaul income calculations (which is the same as the store delivery calculation method. Figure 15.2 illustrates the center-of-gravity site selection method.

If the center-of-gravity site selection method produces two or more grid squares, then reduce the size of the grid square. This step refines the center-of-gravity process to identify the optimum grid square. Figures 15.3 and 15.4 show the refinement process.

Disadvantages of the center-of-gravity site selection are that (1) it identifies a theoretical or ideal map location that might not have real estate or a site location for the required facility size, (2) it does not consider the truck delivery times between the new facility and store locations, and (3) some scattered stores could be beyond the truck's delivery mileage or driving parameters.

Advantages of the center-of-gravity site selection method are that (1) it identifies the theoretical optimum map location for a new warehouse and distribution facility,

	Grid map overlay					
						Total
	1	0	1	1	2	3	8
	0	2	1	1	5	4	13
	5	3	1	2	2	1	14
	1	1	0	4	2	1	9
	0	1	0	2	2	2	7
	0	0	0	0	2	2	4
Total	7	7	3	10	15	13	

FIGURE 15.2 Center of gravity site selection grid.

Grid map overlay						Total
7	8	4	6	9	5	39
5	3	5	7	10	9	39
4	3	3	3	1	6	20
1	1	2	8	6	5	23
5	12	1	2	3	7	30
6	7	4	3	10	3	33
10	8	7	1	3	4	33
Total						
38	42	26	30	42	39	

FIGURE 15.3   Refined center of gravity site selection grid.

Refined grid overlay map												Total
1	3	2	2	1	1	2	1	2	2	1	1	19
2	1	2	2	0	2	3	0	3	2	2	1	22
3	2	1	1	5	0	2	1	3	3	1	7	29
0	0	1	0	0	0	3	1	3	1	1	0	10
1	1	3	0	2	1	0	1	1	0	2	0	12
1	1	0	0	0	0	1	1	0	0	2	2	8
0	0	1	0	1	0	2	2	5	1	2	1	15
0	1	0	0	0	1	2	2	0	0	1	1	8
1	1	3	3	0	1	1	6	0	2	3	0	21
1	2	3	3	0	0	0	1	1	0	0	4	15
1	2	1	3	2	0	1	1	5	1	1	1	19
1	2	1	1	0	2	1	0	1	3	1	0	13
3	2	2	2	6	0	1	0	1	1	2	2	22
3	2	2	2	0	1	0	0	0	1	0	0	11
Total												
18	20	22	19	17	9	19	17	25	17	19	20	

FIGURE 15.4   Additional refined center of gravity site selection grid.

(2) it is easy to calculate, (3) it gives quick results, (4) location does provide service to existing and future stores in a cluster and to scattered stores, (5) it does consider vendor locations, and (6) it does consider the frequency of truck deliveries.

Systematic Site Selection Method.   This site selection method locates an actual real estate site that satisfies your building requirements and optimizes your company's transportation and land costs. The selected site provides the best service to your cluster of stores (customers) and scattered stores. The systematic site selection method identifies the preferred site from a proposed group of sites. This selection is a result of each store's frequency of deliveries from each proposed site.

The systematic site selection method is a four-part process. (1) List all proposed real estate locations (sites) within the geographic area. (2) Determine the number of truck deliveries to present and future stores. (3) Calculate the annual transportation cost from each site to all existing and future store cluster groups or scattered stores. (4) Estimate a standard delivery truck cost per mile.

The systematic site selection method entails the following steps: (1) Determine the number of required delivery trucks for the existing and future stores. (2) Develop a store (customer) location map. (3) Develop a grid sheet to overlay the map. (4) Identify the location for each proposed site on the map. (5) Determine the truck delivery mileage on existing highways from a proposed site and to each of present and future customer locations. Alternative methods use the miles between two zip codes or to the map scale. (6) Multiply the estimated store truck delivery miles by the frequency of deliveries and cost per mile.

To identify the preferred site, total the transportation costs, energy and utility costs, taxes, labor costs, and land costs. The results of this calculation indicate the proposed (preferred) site that has the lowest total cost to your company. To assure top management that the site is the best and optimizes the total company operations, show them a vendor location map with each proposed location relative to all the vendor locations.

Advantages of this method are that (1) it identifies the optimum existing site location (real estate) from a proposed group of existing sites, (2) it calculates the annual transportation costs, (3) it totals the five site selection economic factors, (4) it ensures that the preferred site services all the company's existing and future store locations, (5) it does consider vendor locations, (6) it does consider required truck delivery frequencies, and (7) it does consider delivery truck travel times.

LOOK AT EXISTING BUILDINGS TO REMODEL (RETROFIT)

During the macro and micro site selection project, the project team considers the possibility of making an existing vacant industrial facility a new facility location. Compared to a green-field site or undeveloped land, the existing industrial building offers two advantages: lower start-up expenses and ability to be operational in less time. These advantages mean improved company profits and improved customer service.

Critical Factors for an Existing Building

The critical factors in the final decision about an existing facility are the conditions of the existing land and building. If the existing land and building fit the company's operational (material handling system) requirements, then it is considered a potential alternative. If not, then the existing building and land are useless to the company. A word of caution: Have the site reviewed for any covenants, easements, subterranean problems (tanks or old pond area), or past use of building or product handled in it. Also, if top management chooses the location, then there is no alternative available site. The list of unique factors for the land and building include (1) the size of the property; (2) number and condition of buildings on the site; (3) number of floors; (4) acreage for expansion; (5) load-bearing capacity of the floors and roof; (6) property condition and past use; (7) ceiling height; (8) the space of the columns, size of the columns, and direction of the bays; (9) sprinkler and fire protection system; (10) levelness, condition, and re-bar depth of the floor; (11) office and other administrative facilities within the facility; (12) truck and rail dock facilities and ramp locations and number; (13) air conditioning, heating, and ventilation; and (14) available utilities.

HOW TO SERVICE REMOTE CUSTOMERS

When your warehouse and distribution company considers expanding its market area, the basic choice is to maintain one centralized facility or to decentralize the operation.

If you decide to maintain one facility, then your logistics strategy is less complex. This strategy requires your company to service your remote customer locations by using contract carriers or by using a series of company-owned truck terminals with a company-owned delivery truck fleet.

Contract Carrier

With the contract carrier method, your company rents trucks (tractors and trailers) and drivers from a rental company to deliver product from your main distribution facility to remote customer locations. This is a one-way trip from the main distribution facility to remote customer locations.

The disadvantages of the method are that there is minimal control, it increases costs, and it limits backhaul opportunities. The advantages of the contract carrier are that the trip costs are one-way, it requires the fewest employees, and your company does not purchase and maintain trucks.

Remote Company Truck Terminal

With the remote truck terminal system, your company uses a company-owned truck terminal in the remote market area. The company acquires or rents land at a truck terminal in the remote market area. This space is used by the main facility drivers to drop off (leave) a remote market area store delivery truck. A driver who lives in the remote market area takes the truck from the remote truck terminal to the retail store and returns the empty truck to the terminal. The main facility truck driver returns with the previous day's delivery (empty) truck from the terminal to the main distribution facility.

If your remote market area contains product supplier (vendor) locations that sell product to your company, then after the customer delivery with the empty truck the remote driver picks up a backhaul (your company's purchase of product from the vendor) at the supplier and returns to the remote terminal. The main facility truck driver brings the backhaul trailer from the remote facility to the main distribution facility. The backhaul represents a vendor company making payment to your company for the delivery of product from the supplier's location to your company on your company truck. This payment reduces the product costs to your company.

Disadvantages of the remote terminal method are that (1) it requires that a terminal be maintained; (2) trucks must be purchased and maintained; (3) if there is no backhaul, then a dead-head expense is incurred; (4) the security of the trucks must be guarded; and (5) additional employees are needed.

The advantages of the method are the increased management control and backhaul opportunities.

Remote Distribution Facility

An alternative strategy is to invest in a distribution facility in your remote market area. When the company makes this decision, then there are two strategies that you can pursue to maintain a remote warehouse; a full-line stock warehouse or a fast-moving, large space, or heavy SKUs (stock-spot) warehouse.

Full-Line Stock Warehouse.   The full-line stock remote warehouse has a complete inventory of SKUs and a representative on-hand inventory level to support market area sales. Your supplier or company trucks make product deliveries to the remote warehouse, and company employees perform all key warehouse functions.

Disadvantages of the full-line warehouse are that it increases the number of company employees, increases maintenance, and requires purchase of assets. The advantages are increased management control, lower costs, and improved security.

Large-Volume SKUs in the Remote Warehouse.   An alternative is to maintain in the remote area a warehouse for the large-volume, heavy SKUs in inventory. Vendor trucks deliver fast-moving, heavy, or large product to the remote warehouse. At the remote warehouse employees perform the required warehouse functions. To complete a remote store order, the main facility picks and loads the slow- and medium-moving SKUs onto a truck. A main facility truck driver delivers the load to the remote customer location or to the remote warehouse location. If the main facility truck has available space, then the delivery truck is dropped off at or stops at the remote facility and the remote distribution facility tops off (adds product to) the main facility delivery truck. If the main distribution facility truck is full, it goes directly to the store or is dropped at the remote facility; then the remote warehouse operation picks and loads SKUs onto a separate truck for the store delivery.

Whether the main facility driver delivers the load to the customer location or drops the trailer at the remote location is a function of allowable driver time and backhaul opportunity.

First, if the main facility driver does not have sufficient (allowable) trip time to complete a two-way trip including unloading, then the load is dropped at the remote warehouse.

Second,if a backhaul exists in the remote market area and the main facility driver does not have sufficient time to complete a two-way trip and pick up the backhaul, then the main facility driver drops the delivery truck at the remote facility.

If the main facility driver has enough time to complete a two-way trip and handle an additional activity (unload at the customer location or backhaul pickup) then the additional activity is part of the trip.

If the backhaul exists in the remote market area, then the benefits to the company are increased. This delivery method maximizes the use of space on delivery trucks and creates a lower delivery cost.

The disadvantages are the increased number of employees and the need to purchase and maintain assets. The advantages are moderate control, improved security, backhaul opportunity, and low cost.

INDEX

ABOUT THE AUTHOR

David E. Mulcahy is managing director of Logistics Planning/Systems International, a company established to provide warehouse and distribution seminars and publications. He is an award-winning designer of warehouse facilities and the author of numerous articles on warehouse facilities and management techniques. Mr. Mulcahy previously worked as a project engineer for Joseph A. Sedlak Consultants, Inc., and as a materials handling engineer for Montgomery Ward Co.